AF357327

# Numerical Study of Concrete

# Numerical Study of Concrete

Editor

**Vipul Patel**

MDPI • Basel • Beijing • Wuhan • Barcelona • Belgrade • Manchester • Tokyo • Cluj • Tianjin

*Editor*
Vipul Patel
La Trobe University
Australia

*Editorial Office*
MDPI
St. Alban-Anlage 66
4052 Basel, Switzerland

This is a reprint of articles from the Special Issue published online in the open access journal *Crystals* (ISSN 2073-4352) (available at: http://www.mdpi.com).

For citation purposes, cite each article independently as indicated on the article page online and as indicated below:

LastName, A.A.; LastName, B.B.; LastName, C.C. Article Title. *Journal Name* **Year**, *Volume Number*, Page Range.

**ISBN 978-3-0365-1976-0 (Hbk)**
**ISBN 978-3-0365-1977-7 (PDF)**

# Contents

# About the Editor

**Dr Vipul Patel** is a Senior Lecturer of Structural Engineering at La Trobe University, Australia. Dr Patel obtained his PhD degree from Victoria University, Australia in 2013. Dr Patel was a research associate in the University of New South Wales. Dr Patel has published two research books and published over 60 research articles in the prestigious journals.

# Preface to "Numerical Study of Concrete"

Concrete construction material is widely used in model composite structure due to their distinguished featured. Concrete has a high compression strength and low tensile strength. This special issue presents the recent development in the research area of concrete structures. The special issue provide the research development for practicing structural and civil engineers, academic researchers and undergraduate and postgraduate students in civil engineering who are interested in the latest computational technologies for concrete structures.

**Vipul Patel**
*Editor*

*Article*

# Durability Assessment of PVA Fiber-Reinforced Cementitious Composite Containing Nano-SiO$_2$ Using Adaptive Neuro-Fuzzy Inference System

Ting-Yu Liu [1], Peng Zhang [1,*], Qing-Fu Li [1], Shao-Wei Hu [1,2] and Yi-Feng Ling [3]

[1]  School of Water Conservancy Engineering, Zhengzhou University, Zhengzhou 450001, China; tingyliu@163.com (T.-Y.L.); qingfuli66@163.com (Q.-F.L.); hushaowei@cqu.edu.cn (S.-W.H.)

[2]  College of Civil Engineering, Chongqing University, Chongqing 400045, China

[3]  Department of Civil, Construction and Environmental Engineering, Iowa State University, Ames, IA 50011, USA; yling@iastate.edu

*  Correspondence: zhangpeng@zzu.edu.cn

Received: 15 April 2020; Accepted: 27 April 2020; Published: 28 April 2020

**Abstract:** In this study, the durability of polyvinyl alcohol fiber-reinforced cementitious composite containing nano-SiO$_2$ was evaluated using the adaptive neuro-fuzzy inference system (ANFIS). According to the structural characteristics of the cementitious composite material and some related standards, the classification criteria for the evaluation indices of cementitious composite materials were clarified, and a corresponding structural framework of durability assessment was constructed. Based on the hypothesis testing principle, the required test data capacity was determined under a certain degree of accuracy, and durability experimental data and expert evaluation results were simulated according to statistical principles to ensure that there were sufficient datasets for ANFIS training. Using an environmental factor submodule as an example, 14 sets of actual test data were used to verify that the ANFIS can quickly and effectively mimic the expert evaluation reasoning process to evaluate the durability of cementitious composites. Compared with other studies related to the durability of cementitious composites, a systematic evaluation system for the durability of concrete was established. We used a polyvinyl alcohol fiber-reinforced cementitious composite containing nano-SiO$_2$ to conduct a comprehensive evaluation of cementitious composites. Compared with the traditional expert evaluation method, the durability evaluation system based on the ANFIS learned expert experience, stored the expert experience in fuzzy rules, and eliminated the subjectivity of expert evaluation, thereby making the evaluation more objective and scientific.

**Keywords:** cementitious composite; nano-SiO$_2$; PVA fiber; durability evaluation; adaptive neuro-fuzzy inference system

## 1. Introduction

Reinforced concrete has served in the construction industry for 140 years [1] since the French engineer Abeck first manufactured reinforced concrete floors in 1879. However, many instances of prematurely failed reinforced concrete have occurred for various reasons [2], resulting in structures failing and failing to reach their specified service life and causing many casualties and property losses. Concrete cannot be used normally and requires much manpower and material resources for maintenance when it loses durability [3]. In addition, structures cannot operate normally and result in great economic losses [4]. Therefore, it is extremely important and necessary to improve the durability of reinforced concrete structures. The most common factor affecting the durability of concrete structures is the crack propagation [5]. The reason for concrete cracking is mainly the structure and components of a cementitious composite material being subjected to compressive stress. However, in most cases, components develop internal cracks resulting from tensile stress owing to temperature deformation, shrinkage deformation, composite creep, chemical erosion, and mechanical load [6]. Environmental factors also directly affect the durability of cementitious composite materials, such as freeze–thaw cycles [7], chloride penetration [8], and changes in temperature and humidity [9]. Furthermore, cementitious composites have heavy weight, high brittleness, and low tensile strength, which can lead to the brittle fracture and sudden failure of structures and components. These disadvantages have largely limited the wide application of gelled composites and affected the durability of gelled materials [10]. To solve the shortcomings of cementitious composites in the tensile state and to improve the durability of structures and components, researchers have incorporated various fibers into cementitious composites to enhance the toughness of their matrices and to improve their properties. The effects of fibers such as polyvinyl alcohol (PVA) fiber [11], discontinuous microfibers [12], polypropylene fiber [13], steel fiber [14], carbon fiber [15], polyester fiber [16], and nano silica [17] have been investigated. Among the various fibers, PVA fiber is a commonly used gelled composite fiber [18]. Scholars have conducted research on nanometer-doped PVA fiber-reinforced cementitious composites, including work performance, crack resistance, basic mechanical properties, bending resistance [19], durability [20], and microscopic mechanism [21]. A large number of research results show that the incorporation of nanoparticles improves the frost resistance and impermeability of cementitious composites materials [22].

However, at present, there have been relatively few studies on the durability evaluation of PVA fiber-reinforced cementitious composites containing nano-$SiO_2$, and the durability evaluation of cementitious composites is not perfect [23]. The traditional expert evaluation method relies too much on the experience of experts. Inevitably, the durability assessment results of a cementitious composite material will be deviated from the actual situation according to the subjective opinions of experts. It is necessary to establish a more objective, scientific, and effective evaluation method for gelled composite materials [24]. Although expert evaluation has been the longest and most widely used durability assessment method [25], two other methods of durability assessment of cementitious composites exist. One is the comprehensive evaluation by means of neural networks [26], and the other is reliability theory based on reliability mathematics [27]. Zhou et al. (2017) utilized gray system theory to evaluate the durability of concrete. Their calculation process was simple and suitable for practical engineering applications, but the value of its resolution coefficient needs to be further optimized and verified [28]. Yu et al. (2017) proposed a probabilistic framework for the durability assessment of concrete structures using reliability and sensitivity analysis based on the uncertainties of the environment and materials in a marine environment [29]. The adaptive neuro-fuzzy inference system (ANFIS) proposed by Jang in the 1990s is a fuzzy inference system that combines the organic combination of fuzzy logic and neural networks [30]. A fuzzy inference system is suitable for expressing fuzzy experience and knowledge but lacks an effective learning mechanism [31]. Neural networks have a self-learning function but cannot express the reasoning of human brain [32]. The ANFIS uses the back-propagation algorithm and least squares method to learn to adjust the premise parameters and conclusion parameters, which can automatically generate If-Then rules. The expert experience contained in the fuzzy rules provides a certain physical meaning to the neural network and allows it to eliminate the black box disadvantages while avoiding the poor learning ability of the fuzzy inference system [33]. At present, many scholars are applying the ANFIS for condition assessment and performance prediction [34–38]. Xu et al. (2016) developed an underwater structural condition assessment system for a bridge based on the ANFIS to provide a good application effect [39]. Wang et al. (2015) used the ANFIS inference system to establish a prediction model for the compressive strength of hollow concrete block masonry. The accuracy of the prediction was significantly better than that of the current standard calculation model [40].

This study applied the ANFIS inference system to assess the durability of PVA fiber-reinforced cementitious composites containing nano-$SiO_2$. According to the structural characteristics of cementitious composites and some related standards [41–44], the classification criteria for the evaluation indices of cementitious composite materials were clarified and a structural framework for durability evaluation was constructed. According to the hypothesis testing principle, the required test data capacity was determined under a certain degree of accuracy. The durability experimental data and expert evaluation results were simulated according to statistical principles to ensure that there were sufficient datasets for the ANFIS training. Using the environmental factor submodule as an example, 14 sets of actual test data were used to verify that the ANFIS can quickly and effectively mimic the expert evaluation reasoning process to evaluate the durability of cementitious composites. Japanese scholars improved the traditional expert evaluation method and proposed a comprehensive evaluation method for buildings. The comprehensive evaluation through three surveys reduced the subjective influence of experts, but the overall evaluation cost was high, and the workload was large [45]. Xu et al. (2019) used the durability evaluation of concrete aqueducts in Gansu Province and the fuzzy analytic hierarchy process (FAHP) to establish a multi-level and multi-indicator evaluation model for the durability of concrete buildings. Their model provided an improved characterization of the durability grade of hydraulic concrete structures and had practical application value [23]. Compared with the traditional expert evaluation method, the durability evaluation system based on the ANFIS was more objective and scientific and lowered the evaluation cost and workload. Most studies have been based on the prediction and evaluation of a certain durability index [46]. In this study, a systematic durability evaluation system of PVA fiber-reinforced cementitious composite containing nano-$SiO_2$ was established. The ANFIS compensates for the shortcomings of the black box of neural networks and the lack of learning ability of fuzzy systems. It describes the fuzzy relationship between durability and the many uncertain factors that affect the durability of cementitious composites, has better applicability, and provides a new method for durability evaluation of cementitious composites.

## 2. Principle and Structure of ANFIS

### 2.1. ANFIS Principle

#### 2.1.1. ANFIS Structure

The typical structure of the ANFIS can be illustrated by two input vectors and one output vector. The structure diagram is shown in Figure 1, where $x_1, x_2$ is the input vector of the two input nodes; $R_1$, $R_2$, $Z_1$, and $Z_2$ are membership functions, which fuzzify the input vector and then obtain membership degrees corresponding to different levels; $\Pi$ is the fixed node mark, and the membership degrees $\mu$ in the second layer are multiplied to obtain the trigger strength weight $\omega_1, \omega_2$ of each rule; $N$ denotes the calculation of the normalized credibility, the normalization of the strength of each rule, and then the obtainment of the trigger weight $\overline{\omega}_1, \overline{\omega}_2$ of each rule in the overall rule.

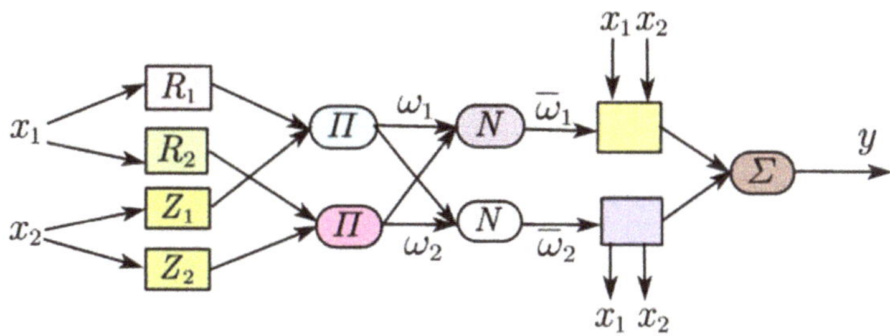


**Figure 1.** Model structure of adaptive neuro-fuzzy inference system (ANFIS).

The ANFIS algorithm is described as follows [47]:
Layer 1: Fuzzy processing

The membership node function is used to obfuscate the vector $x_1, x_2$ of the input nodes to obtain different membership degrees $\mu$. The shape of the membership function depends on the ranking and value of the previous parameter. Layer 1 is an adaptive unit with the function

$$O_{1,j} = \begin{cases} \mu_{Aj}(x), j = 1, 2 \\ \mu_{B(j-2)}(x_2), j = 3, 4 \end{cases}. \tag{1}$$

Layer 2: Rule-based reasoning

The memberships $\mu$ of the first layer are multiplied to get the trigger strength of each rule as

$$\omega_j = \mu_{Aj}(x_1) \times \mu_{Bj}(x_2), \ j = 1, 2. \tag{2}$$

Layer 3: Normalization

The trigger strength of each rule obtained in the second layer is normalized to obtain the trigger proportion of the rule in the entire rule base, that is, the probability of applying the rule in the entire reasoning process, which is calculated

$$\overline{\omega}_j = \frac{\omega_j}{\omega_1 + \omega_2}, \ j = 1, 2. \tag{3}$$

Layer 4: Defuzzification

The output of the fuzzy rules is calculated and the output characteristic parameters of the antecedent are linearly combined to obtain the output as

$$O_{4j} = \overline{\omega}_j f_j = \overline{\omega}_j (f_{1j} x_1 + f_{2j} x_2 + f_{3j}), \tag{4}$$

where, $\overline{\omega}_j$ is the proportion of the rule relative to the overall rule, and $\{f_{1j}, f_{2j}, f_{3j}\}$ is the set of linear parameters at the nodes.

Level 5: Output

The calculation result of each rule in the fourth step is deblurred to obtain the exact output. The normalized triggering degrees of each rule are presented as a weighted average as

$$O_{5j} = \sum_j \overline{\omega}_j f_j = \frac{\sum_j \omega_j f_j}{\sum_j \omega_j}. \tag{5}$$

2.1.2. ANFIS Learning Algorithm

The ANFIS uses a hybrid learning algorithm in which parameter learning and adjustment is performed simultaneously in the forward transfer and reverse transfers. In forward propagation, the forward parameters are fixed, and when passed to the fourth layer, the backward parameters are updated by least squares estimation. In back propagation, the backward parameters (parameters in the rules) are fixed, the partial derivatives of the forward parameters are calculated according to the loss function (using the chain rule), and the parameters are updated from the reverse direction of the gradient direction.

In the ANFIS learning algorithm [48], the measurement error is the sum of the mean square errors determined as

$$E_{rror} = \sum_{i=1}^{N} E_{rrori} = \sum_{i=1}^{N} (T_i - O_{rji})^2, \tag{6}$$

where $E_{rrori}$ is the mean square error of the i-th data output, $T_i$ is the expected output of the i-th data group, and $O_{rji}$ is the r-th component of the i-th actual output group.

According to the chain rule, we obtained the partial derivative of $E_{rror}$ for each parameter as

$$\frac{\partial E_{rror}}{\partial O_{rji}} = -2(T_i - O_{rji}), \tag{7}$$

$$\frac{\partial E_{rror}}{\partial \gamma} = \sum_{O' \in S} \frac{\partial E_{rror}}{\partial O'} \frac{\partial O'}{\partial \gamma}, \tag{8}$$

where {S} is the forward element set, $O'$ is any of the elements, and $\gamma$ is the forward parameter.

### 2.2. System Topology

Based on the idea of rounding to zero, the system consisted of five subsystems to greatly reduce the complexity of the ANFIS algorithm [49]. The entire system had a tree structure [50]. The parent node of each subsystem and its child nodes formed a subnet. The parent node was the master node, and child nodes were slave nodes. Figure 2 is a tree-like network topology diagram of the durability evaluation of cementitious composites. The concentric circles represent the soundness of the output durability evaluation indices of the cementitious composites, the large circles represent the evaluation result of the single index, and the small circles represent the evaluation indicators of the test items.

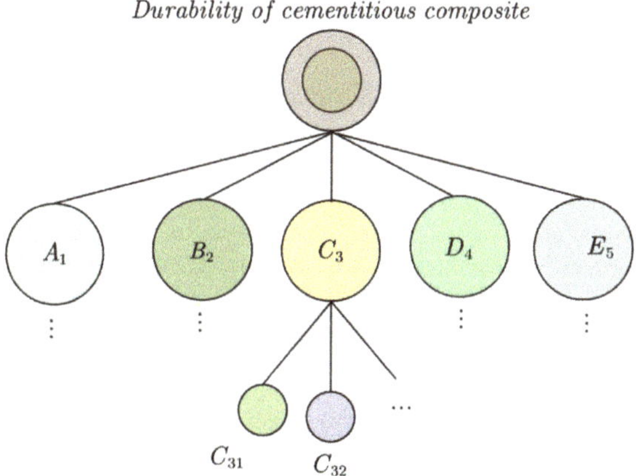


**Figure 2.** Tree network topology of composite durability evaluation.

## 3. Durability Evaluation System Based on ANFIS

The durability evaluation system of cementitious composites based on the ANFIS took full advantage of the complementary advantages of fuzzy systems and neural networks [51]. The ANFIS used a hybrid algorithm of a back-propagation algorithm and the least squares method to learn to adjust the premise and conclusion parameters and automatically generate If-Then rules that contain expert experience in the fuzzy rules. According to the durability evaluation indices and evaluation system, the ANFIS sequentially evaluated from the bottom test indices to the high-level index. This allowed the neural network to eliminate the black box disadvantages, gain certain physical significance, and avoid the poor learning ability of the fuzzy inference system.

Figure 3 presents a framework diagram of the ANFIS for the evaluation of a single indicator. The inference system is a tree-like network topology structure in which the ANFIS submodules of a single evaluation indicator are interconnected. In the figure, the wavy line in the input represents the membership function, and $\Sigma$ represents fuzzy synthesis. In the calculation process, the input data were processed by the membership function to obtain memberships of different grades. Memberships were calculated by fuzzy rules with a certain trigger strength and trigger weight, the results of each rule were obtained, and then the evaluation result of the single index was obtained.

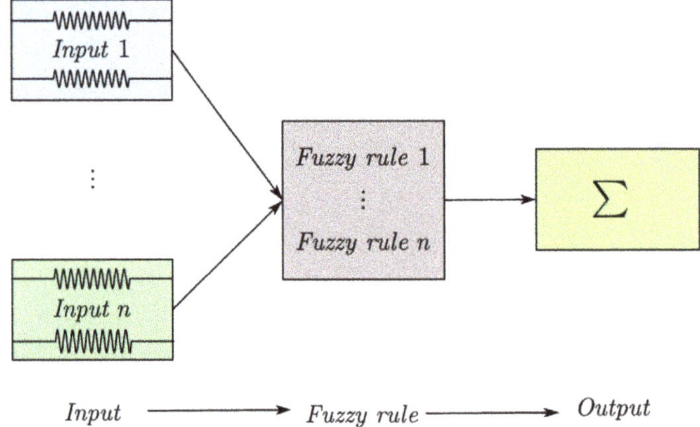


**Figure 3.** ANFIS framework of a single evaluation index.

*3.1. Evaluation Index System and Index Grading Standards*

### 3.1.1. Evaluation Index System

The selection of the durability indices of a cementitious composite material followed the principles of hierarchy, a combination of qualitative and quantitative indicators, and compatibility combined with actual engineering problems and references [24,52,53]. Each box in Figure 4 represents an indicator, and indicators are determined by one or more input data. The analytic hierarchy process was used to establish a durable evaluation system. Five intermediate indicators—raw materials $A_1$, construction and maintenance $B_2$, environmental factors $C_3$, mix ratio $D_4$, and work performance and strength $E_5$—were selected and decomposed one by one to obtain 26 leaf indicators. The durability evaluation index system of cementitious composite materials is presented in Figure 4. The evaluation was performed from the leaf indices to the intermediate index, and finally the durability evaluation of the cementitious composite material was obtained.

### 3.1.2. Classification Standards

A complete evaluation standard for the durability of gelled composites was established using a percentage system. Considering that some uncertain factors cannot be accurately measured, the smaller the number of classifications, the higher the reliability of the results of the durability evaluation. Therefore, the evaluation standard was divided into four or five classification levels. The five levels were excellent (80–100), good (60–80), qualified (40–60), poor (20–40), and dangerous (0–20). The four levels were excellent (75–100), good (50–75), qualified (25–50), and dangerous (0–25). The index classification was performed in accordance with some related standards [41–43] for detailed quantitative evaluation. The results are presented in Table 1.

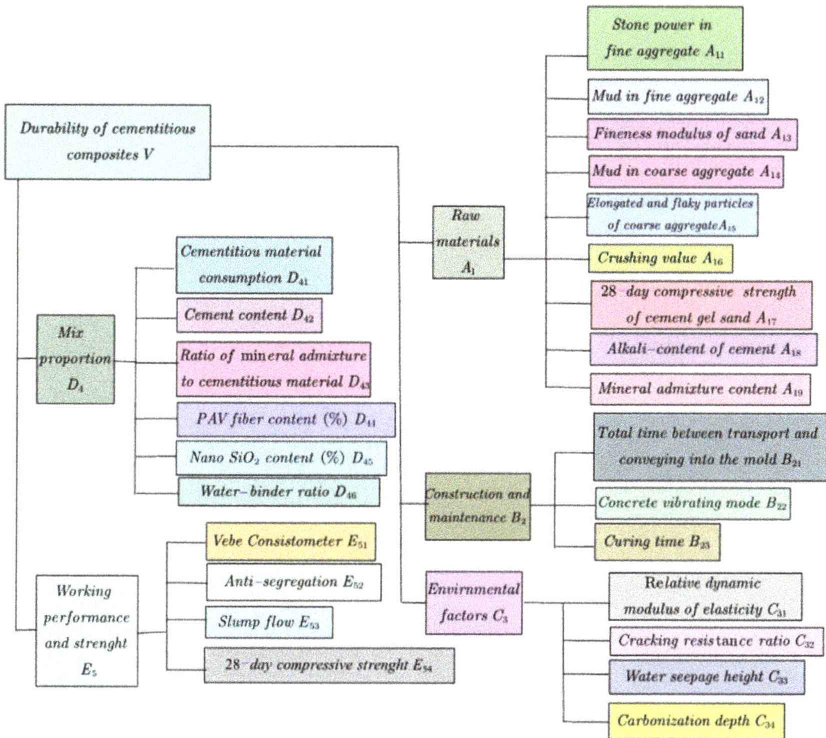

**Figure 4.** Evaluation index system for the durability of cementitious composite.

**Table 1.** Evaluation indices and grading standards.

| No | Index | Classification | | | | |
|----|-------|------------|------------|-----------|------------|-----------|
| | | First Grade | Second Grade | Third Grade | Fourth Grade | Fifth Grade |
| 1 | $A_{11}$ | $(7.0, 10.0]$ | $(5.0, 7.0]$ | $(3.0, 5.0]$ | $(2.0, 3.0]$ | $(0, 2.0]$ |
| 2 | $A_{12}$ | $(7.0, 10.0]$ | $(5.0, 7.0]$ | $(3.0, 5.0]$ | $(2.0, 3.0]$ | $(0, 2.0]$ |
| 3 | $A_{13}$ | $[3.1, 3.7]$ | $[2.3, 3.0]$ | $[1.6, 2.2]$ | $[0.7, 1.5]$ | $(0, 0.6]$ |
| 4 | $A_{14}$ | $(2.0, \infty)$ | $(1.0, 2.0]$ | $(0.5, 1.0]$ | $(0, 0.5]$ | |
| 5 | $A_{15}$ | $(25, \infty)$ | $(15, 25]$ | $(8, 15]$ | $(0, 8]$ | |
| 6 | $A_{16}$ | $(16, \infty)$ | $(12, 16]$ | $(10, 12]$ | $(0, 10]$ | |
| 7 | $A_{17}$ | $(\infty, 32.5]$ | $(32.5, 42.5]$ | $(42.5, 52.5]$ | $(52.5, 62.5]$ | $(62.5, \infty)$ |
| 8 | $A_{18}$ | $[1.2, \infty)$ | $(0.8, 1.2]$ | $(0.4, 0.8]$ | $(0, 0.4]$ | |
| 9 | $A_{19}$ | $(0.55, 0.65]$ | $(0.45, 0.55]$ | $(0.35, 0.45]$ | $(0.30, 0.35]$ | $(0, 0.30]$ |
| 10 | $B_{21}$ | $(240, \infty)$ | $(210, 240]$ | $(180, 210]$ | $(90, 180]$ | $(0, 90]$ |
| 11 | $B_{23}$ | $(0, 3]$ | $(3, 7]$ | $(7, 14]$ | $(14, 28]$ | $(28, \infty)$ |
| 12 | $C_{31}$ | $(50, 60]$ | $(60, 70]$ | $(70, 80]$ | $(80, 90]$ | $(90, 100]$ |
| 13 | $C_{32}$ | $(75, 80]$ | $(80, 85]$ | $(85, 90]$ | $(90, 95]$ | $(95, 100]$ |
| 14 | $C_{33}$ | $(35, 45]$ | $(30, 35]$ | $(25, 30]$ | $(20, 25]$ | $(5, 20]$ |
| 15 | $C_{34}$ | $(30, 100]$ | $(20, 30]$ | $(10, 20]$ | $(0.1, 10]$ | $(0, 0.1]$ |
| 16 | $D_{44}$ | $[0, 0.3]$ | $(0.3, 0.6]$ | $(0.6, 0.9]$ | $(0.9, 1.2]$ | |
| 17 | $D_{45}$ | $(0, 0.5]$ | $(0.5, 1.0]$ | $(1.0, 1.5]$ | $(1.5, 1.0]$ | $(2.0, 2.5]$ |
| 18 | $D_{46}$ | $(0.55, 0.6]$ | $(0.50, 0.55]$ | $(0.45, 0.50]$ | $(0.40, 0.45]$ | $(0, 0.40]$ |
| 19 | $E_{51}$ | $[31, \infty)$ | $[21, 30]$ | $[11, 20]$ | $[6, 10]$ | $[3, 5]$ |
| 20 | $E_{53}$ | $(0, 200] \cup (700, \infty)$ | $(200, 350) \cup (500, 700)$ | $[350, 450)$ | $[450, 500]$ | |
| 21 | $E_{54}$ | $(0, 40)$ | $(40, 50]$ | $(50, 60]$ | $(60, 70]$ | $(70, 80]$ |

### 3.1.3. Input Items for Durability Evaluation

The indicators used for evaluating the durability of cementitious composites included raw materials, construction and maintenance conditions, environmental factors, mix ratios, and other data that can be obtained through experimentation or observations. Two types of input data were used for the evaluation of the underlying indicators. The first type provided a description of the state of the cementitious composite material and simply divided it into grades. Such indicators were qualitative indicators, and the values in 0.1 increments within the range of 0.1–1.0 given by the gelatinous composite testers or observers were used as input data. The larger the dataset, the better the durability of the cementitious composite. The second type of data were a single numerical index whose test index was an interval value, such as, to name two, carbonization depth and water seepage height. This type of data was imported directly into the system. Depending on the structural characteristics and durability influencing factors of the cementitious composite material, the system selected the 26 input items shown in Table 2 to evaluate five underlying indicators.

**Table 2.** Input items for bottom indicators.

| Assessment Indicators | Input Item |
| --- | --- |
| Concrete working behavior | Collapsibility, Vebe consistometer, segregation resistance, 28-day compressive strength |
| Environment function | Relative dynamic modulus of elasticity, cracking resistance ratio, water seepage height, carbonization depth |
| Mix proportion | Cementitious material consumption, cement content, ratio of mineral admixture to cementitious material, PVA fiber content, nano $SiO_2$ content, water–binder ratio |
| Construction and maintenance | Total time between transportation and feeding, concrete vibrating mode, maintenance time |
| Raw material | Stone power in fine aggregate, mud in fine aggregate, fineness modulus of sand, mud in coarse aggregate, elongated and flaky particles of coarse aggregate, crushing value, 28-day compressive strength of cement mortar, alkali content of cement, type of mineral admixtures |

### 3.2. Fuzzy Rules

The fuzzy rule represented the mapping relationship between the input data and the evaluation indices explained here with the construction and maintenance $B_2$ module [54]. The indicator value (construction and maintenance sound value) was obtained from three input data (total time interval of transport and conveyance into the mold, concrete vibrating mode, and maintenance time). The form of the fuzzy rule is:

$$\text{if } x \text{ is } a, \ y \text{ is } b \text{ and } z \text{ is } c, \text{ then } r = px + qy + mz + n \tag{9}$$

where x, y, and z are the input sets, which are the total time interval of transport and conveyance into the mold, concrete vibrating mode, and maintenance time, respectively; and r is an output variable (construction and maintenance status) automatically generated by the training process in actual use. Taking environmental factor $C_3$ as an example, this submodule had 625 fuzzy rules, all of which belonged to one fuzzy rule group representing the evaluation index of environmental factors. Other fuzzy indicators were similar to environmental factor $C_3$. The entire system had five fuzzy rule groups of ANFIS submodules and one fuzzy rule group for the overall durability evaluation.

### 3.3. Membership Function

Membership function is the basis of fuzzy control. Common types are Z-type, trapezoidal, Gaussian, bell-shaped, triangular, and S-type. In practical applications, the appropriate type of membership function is generally selected based on expert experience and actual conditions. We performed a normality test on the existing data. Due to the small number of samples, the normality test was based on the Kolmogorov–Smirnov (K-S) results. The test results show that the significances were between 0.7 and 2 greater than 0.05, so the durability of cementitious composites basically followed the normal distribution. The Gaussian function has the characteristics of smooth symmetry and non-linear continuous differentiability. After a series of experiments, the system used a Gaussian membership function to describe the input index as

$$f(x, \theta, \mathbf{a}) = e^{-\frac{(x-a)^2}{2\theta^2}} \tag{10}$$

where $x$ is the input index learned during training. Figure 5 provides a schematic diagram of a Gaussian membership function.

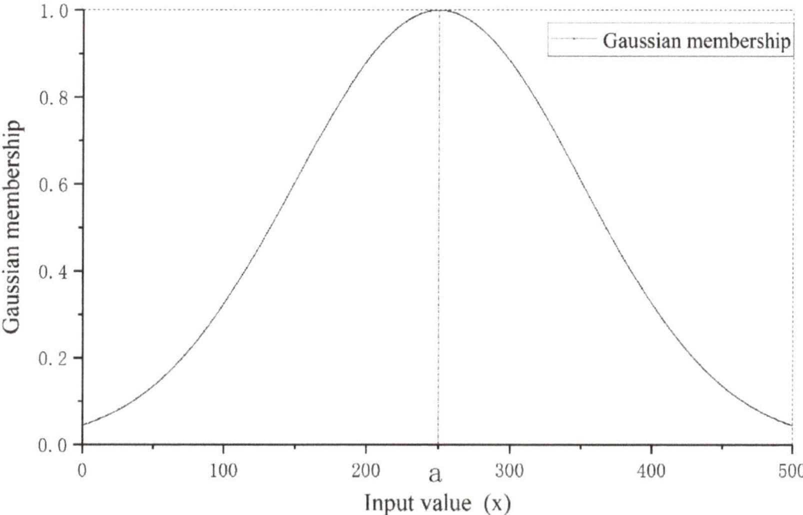


**Figure 5.** Curve of the Gaussian membership function.

## 4. Description of ANFIS Submodule

### 4.1. Simulation of Test Data

Generation of Simulation Test Data

The more data obtained from the test, the more accurately the trained ANFIS system can reflect the results of expert evaluation. However, the experimental dataset cannot be too large because obtaining high-quality large-scale experimental datasets requires much manpower and many material resources. Using environmental factor $C_3$ as an example, according to the determination method of the sample size in the hypothesis test, at a 95% confidence level, when the environmental factor score was 4.99 and the acceptable error range was 1, the required sample capacity was $n = (1.96 \times 4.99/1)^2 = 95.65$ [55], of which 1.96 was the critical value at a 95% confidence level. Therefore, 100 sets of simulation data were capable of meeting the ANFIS training requirements.

The simulation data were the original 14 experimental data, the initial data were all derived from reference [56] as shown in Table 3, and compositions of mixtures were shown in Table 4. The mixtures 1–5 and 10–13 were prepared to study the influence of PAV fiber on the durability of cementitious composites. The mixtures 6–9 were prepared to study the influence of nano-$SiO_2$ on the durability of cementitious composites. Mixture 14 was taken as the control mixture. The data generated by the simulation included endurance test results and expert evaluation results. Using the environmental factor $C_3$ subsystem as an example, the test dataset included five data elements, each group having four test data, and an expert evaluated the results. The method of generating random numbers by normal distribution and generating expert evaluation results by certain derivation rules to generate simulation test datasets can solve the problem of a lack of high-quality test data [34]. Therefore, to obtain sufficient training samples and ignore the interaction between the cementitious composites, the normal distribution method was used to randomly simulate the experimental data of the cementitious composites. The mean and standard deviation refer to the normal distribution pattern of the original data as shown in Table 5.

**Table 3.** Initial data.

| No. | Relative Dynamic Modulus of Elasticity/% | Cracking Resistance Ratio/% | Water Seepage Height/mm | Carbonization Depth/mm | Score |
|---|---|---|---|---|---|
| S1 | 78.2 | 85.0 | 40.2 | 13.9 | 13.17767 |
| S2 | 86.1 | 87.2 | 23.1 | 13.0 | 21.28441 |
| S3 | 90.1 | 92.9 | 21.2 | 12.3 | 24.03373 |
| S4 | 91.1 | 97.4 | 17.5 | 11.3 | 25.97210 |
| S5 | 88.9 | 99.8 | 15.4 | 11.8 | 26.05322 |
| S6 | 91.6 | 97.5 | 13.7 | 10.8 | 26.68299 |
| S7 | 92.0 | 97.8 | 12.1 | 10.5 | 27.08332 |
| S8 | 93.0 | 98.4 | 9.9 | 10.0 | 27.78438 |
| S9 | 94.2 | 99.2 | 9.0 | 9.0 | 28.47065 |
| S10 | 87.2 | 94.9 | 12.3 | 12.3 | 25.17145 |
| S11 | 90.3 | 97.0 | 10.4 | 11.5 | 26.62340 |
| S12 | 91.3 | 99.7 | 9.9 | 10.1 | 27.58360 |
| S13 | 83.0 | 52.6 | 8.5 | 8.9 | 22.46571 |
| S14 | 78.2 | 85.0 | 40.2 | 13.9 | 13.17767 |

**Table 4.** Mix proportions of polyvinyl alcohol (PVA) fiber cementitious composites.

| Mix No. | Water | Cement | Fly Ash | Quartz Sand | Water Reducing Agent | PVA Fiber | Nano-SiO$_2$ |
|---|---|---|---|---|---|---|---|
| | kg/m$^3$ | kg/m$^3$ | kg/m$^3$ | kg/m$^3$ | kg/m$^3$ | % | % |
| S1 | 380 | 650 | 350 | 500 | 3 | 0 | 0 |
| S2 | 380 | 650 | 350 | 500 | 3 | 0.3 | 0 |
| S3 | 380 | 650 | 350 | 500 | 3 | 0.6 | 0 |
| S4 | 380 | 650 | 350 | 500 | 3 | 0.9 | 0 |
| S5 | 380 | 650 | 350 | 500 | 3 | 1.2 | 0 |
| S6 | 380 | 644 | 350 | 500 | 3 | 0.9 | 1 |
| S7 | 380 | 640 | 350 | 500 | 3 | 0.9 | 1.5 |
| S8 | 380 | 637 | 350 | 500 | 3 | 0.9 | 2 |
| S9 | 380 | 634 | 350 | 500 | 3 | 0.9 | 2.5 |
| S10 | 380 | 637 | 350 | 500 | 3 | 0.3 | 2 |
| S11 | 380 | 637 | 350 | 500 | 3 | 0.6 | 2 |
| S12 | 380 | 637 | 350 | 500 | 3 | 1.2 | 2 |
| S13 | 380 | 637 | 350 | 500 | 3 | 0 | 2 |
| S14 | 380 | 637 | 350 | 500 | 3 | 0 | 0 |

**Table 5.** Mean and standard deviation of initial data.

| Relative Dynamic Modulus of Elasticity/% | Cracking Resistance Ratio/% | Water Seepage Height/mm | Carbonization Depth/mm | Environmental Factor Score |
|---|---|---|---|---|
| 88.23 | 91.74 | 17.39 | 11.38 | 23.97 |
| 5.14 | 12.44 | 10.63 | 1.61 | 4.99 |

Using the relative dynamic modulus of elasticity as an example, we selected 100 random numbers with an average value of 88.23 and a standard deviation of 5.14. The distribution of random numbers is shown in Figure 6, where $C_{31}$ is the random number set of the relative dynamic modulus of elasticity and $f$ is the number of random occurrences. The simulation process was implemented by SPSS (Statistical Product and Service Solutions, Version 24.0, IBM Corp., Armonk, NY, USA).

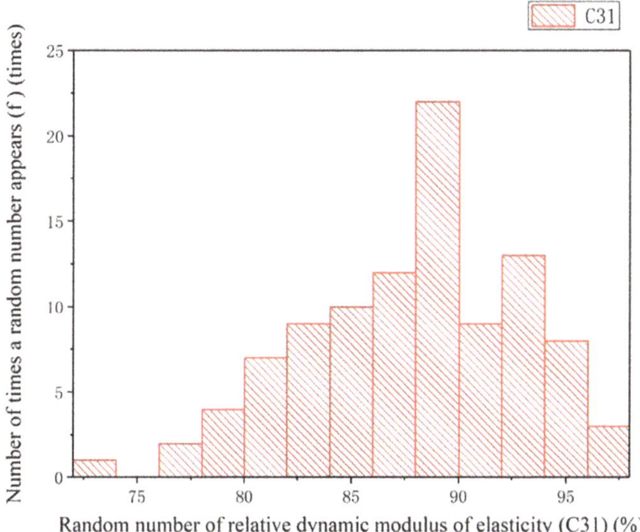


**Figure 6.** Probability distribution histogram of a random number.

The expert evaluation results to verify system performance were derived from the system's original output in accordance with certain rules. There are two generation rules for simulating expert evaluation results. The first rule is that the normal distribution of expert evaluation data applies directly to the original data, which is the system output before training. The second rule is that the expert evaluation data of the original data are independently shifted by different small increments, which conform to a normal distribution. We first used the second rule to generate twice the number of expert evaluation results required for the dataset and then divided it into two groups. The small offset increments were consistent with a mean of zero and a standard deviation of 0.5. The actual data were used to train the two sets of expert evaluation results to obtain the two trained groups of expert evaluation results. Then, the two groups of data trained each other to obtain the final simulated expert evaluation data.

### 4.2. Implementation of ANFIS Submodule

Using the environmental factor $C_3$ as an example to illustrate the structure of the ANFIS submodule, Figure 7 shows the ANFIS structural model of $C_3$. There were five layers of network structure, including four inputs and one output [57]. According to the classification of the durability evaluation index of cementitious composite material, the membership function structure of the four input units was 5, 5, 5, 5, corresponding to the 20 neural nodes of the second layer; that is, the hierarchical status of each input unit was 5, 5, 5, 5. The membership function curve is shown in Figure 8. The third and fourth layers each had 625 neuron nodes corresponding to the structure's 625 fuzzy rules and the output of each fuzzy rule. The partial membership function rules are shown in Figure 9. The fifth layer had a sound value corresponding to $C_3$ of the neuron node. After the ANFIS system was set up, the simulation experimental data were input into the system for training, and the training process was implemented by MATLAB software. After sufficient training, the training error converged to 0.03. The algorithm training completed at epoch 2. The results of experiments show that the algorithm converges quickly and has satisfactory training accuracy.

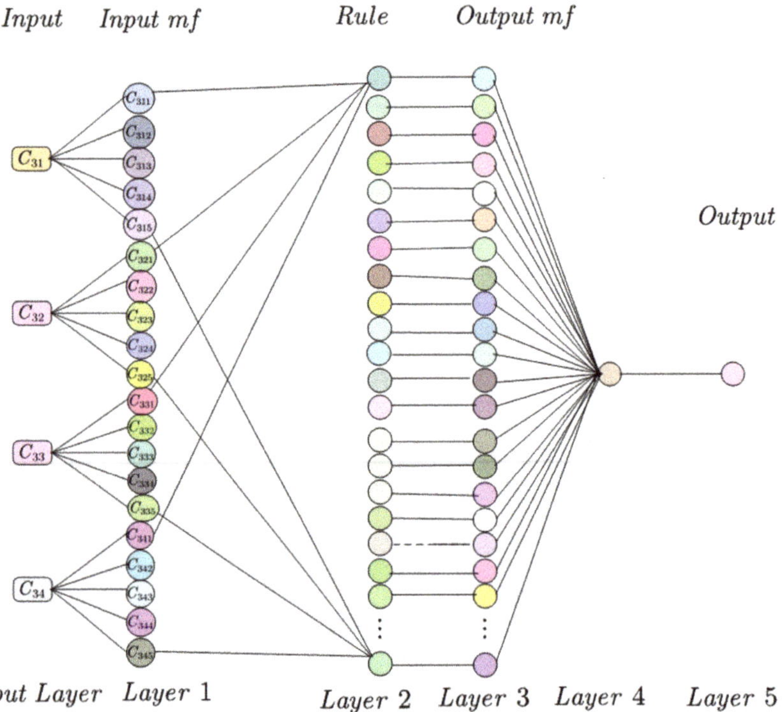


**Figure 7.** Proposed ANFIS model structure ($C_{31}$: relative dynamic modulus of elasticity, $C_{32}$: cracking resistance ratio, $C_{33}$: water seepage height, and $C_{34}$: carbonization depth).

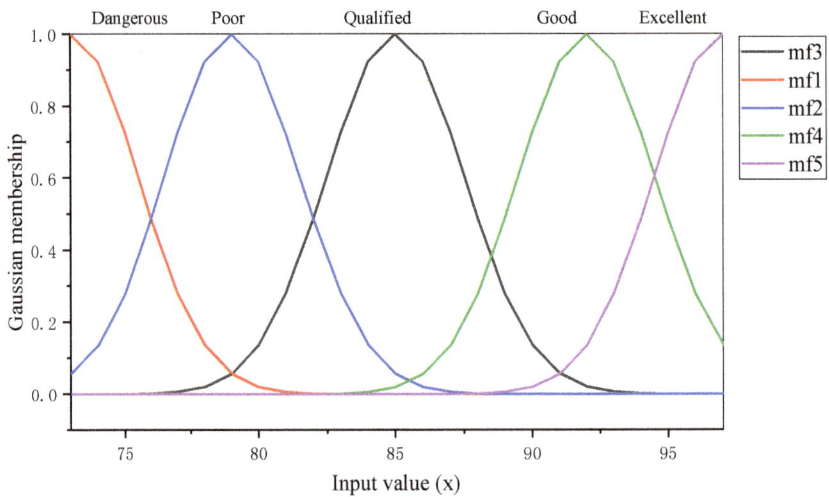


**Figure 8.** Degree function curve.

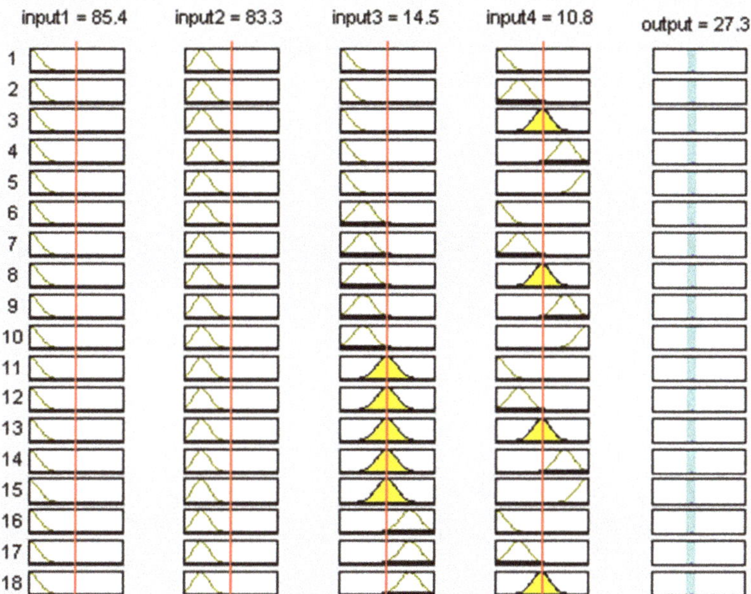


**Figure 9.** Partial fuzzy rules.

*4.3. Verification of System Learning Ability*

The reasoning ability of the ANFIS was measured by the similarity between the initial expert evaluation result of the test dataset and the evaluation result of the reasoning system. This measurement reflects the ability of the ANFIS to learn and adjust the parameter value of the membership function and other parameters according to the given dataset. To verify the reasoning ability of the proposed system for practical problems, using environmental factors as an example, the above-mentioned ANFIS system was verified with 14 groups of data obtained from the experiment.

Table 6 shows the comparison of original evaluation results, system output before training and system output after training. Taking five scores as one grade difference, the number of output status grades of the system after training that were consistent with the expert evaluation opinions increased from 5 before training to 12, the number of levels with a difference of 2 or more decreased from 8 before training to 0. In the end, only two data differed by one level. The coincidence rate of the state grade of the indices evaluated by the system before training was 35.7%, and the grade after training is increased to 85.7%. The average test deviation of the direct application of the original expert evaluation data was 10.39, and the average test deviation using the trained expert evaluation data was reduced to 2.93. This indicates that the learning and reasoning ability of the ANFIS after training was improved to some extent. Pearson correlation analysis [58] was used to analyze the expert evaluation and the system output after training to judge the degree of interdependence between the two vectors. The Pearson correlation analysis results are presented in Table 7. It can be perceived from Table 7 that the conspicuousness was 0.947. Therefore, according to Pearson correlation analysis, the correlation was significant [59,60]. The deviation between the reasoning results after ANFIS training and the expert evaluation results in the test data is shown in Figure 10, where the x-axis is the composite material test number. After the system was trained, the output results were closer to the expert evaluation results.

**Table 6.** Comparison of original evaluation results: system output before training and system output after training.

| Evaluation Indicators | Expert Evaluation Data | System Output after Training | System Output before Training |
|---|---|---|---|
| S1 | 13.18 | 12.8 | 8.25 |
| S2 | 21.28 | 19.34 | 24.31 |
| S3 | 24.03 | 22.87 | 10 |
| S4 | 25.97 | 23.64 | 21.1 |
| S5 | 26.05 | 24.75 | 17.02 |
| S6 | 26.68 | 24.11 | 19.2 |
| S7 | 27.08 | 24.83 | 16.07 |
| S8 | 27.78 | 25.05 | 16.51 |
| S9 | 28.47 | 24.81 | 50.89 |
| S10 | 25.17 | 20.17 | 19.99 |
| S11 | 26.62 | 20.43 | 24.84 |
| S12 | 27.58 | 26.25 | 16.09 |
| S13 | 22.47 | 19.36 | 8.97 |
| S14 | 13.18 | 12.8 | 8.25 |

**Table 7.** Results of the correlation analysis.

| Project | Correlation Coefficient | Saliency | Number of Cases |
|---|---|---|---|
| Expert evaluation data | 1.0 | 0.947 | 14 |
| System output after training | 1.0 | 0.947 | 14 |

At 0.01 level (double tail), the correlation was significant.

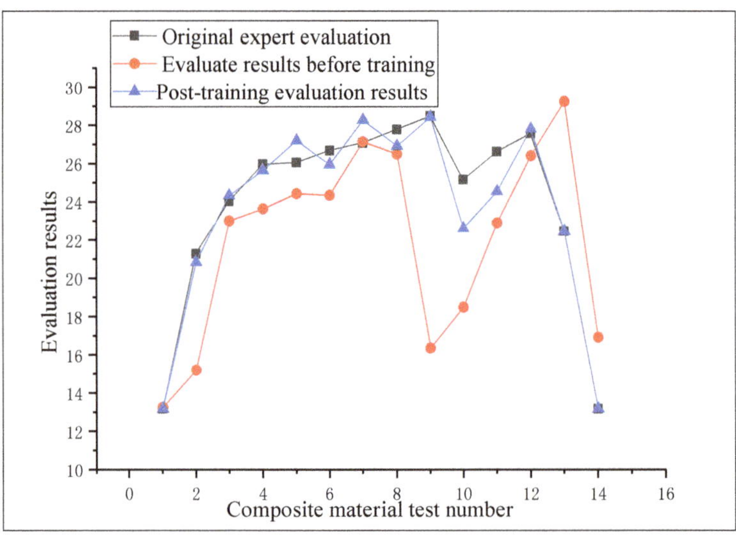


**Figure 10.** Comparison between system output before and after training and expert opinion.

The results show that the evaluation results of the system after training agreed well with the expert evaluation results, and the system after training can quickly and accurately imitate the nonlinear expert reasoning process. The durability evaluation results of cementitious composite materials obtained by the ANFIS reflected the durability of concrete to a certain extent. It served as a guide to identify the weak links, determined the reliability of the existing concrete in time, and acted against the weak links to strengthen the construction quality and optimize the working performance of the cementitious composite materials. For instance, the score of the S1 evaluation index was 12.8, which belongs to the third level and is relatively low. If the score is lower than the level required by the corresponding specifications, measures need to be taken to improve it. For example, a secondary mixing method, sand enveloped with cement paste or a sand and stone binding method, could be adopted to improve the workability, water retention, and concrete strength.

## 5. Conclusions

(1) The durability of PVA fiber-reinforced cementitious composites containing nano-$SiO_2$ was evaluated by using a method based on the ANFIS. Moreover, the classification criteria for the evaluation indices of cementitious composite materials were clarified, and a corresponding structural framework of durability assessment was constructed.

(2) The results show that 85.7% of the ANFIS evaluation results were consistent with the expert evaluation results, and only two data differed by one level, which demonstrates that the system simulated the nonlinear expert reasoning process quickly and accurately and had good parameter mathematics. The evaluation results of the ANFIS can serve as a reliable guide for actual concrete construction and maintenance applications.

(3) Using the ANFIS to evaluate the durability of cementitious composite materials is feasible in engineering applications. By comparing the results of ANFIS evaluation with a corresponding evaluation system, the evaluation results can be obtained intuitively. Through the establishment of a corresponding evaluation index system, the system cannot only be used to evaluate the durability of a PVA fiber-reinforced cementitious composite with $SiO_2$, but also can be used to evaluate the durability of ordinary concrete and other cementitious composite materials.

**Author Contributions:** Conceptualization, T.-Y.L. and P.Z.; methodology, T.-Y.L., Q.-F.L. and S.-W.H.; formal analysis, T.-Y.L., P.Z., Q.-F.L. and S.-W.H.; investigation, T.-Y.L., P.Z. and Q.-F.L.; resources, P.Z. and S.-W.H.; writing—original draft preparation, T.-Y.L. and P.Z.; writing—review and editing, Q.-F.L., S.-W.H. and Y.-F.L.; visualization, Y.-F.L.; supervision, P.Z.; project administration, Q.-F.L.; funding acquisition, P.Z. All authors have read and agreed to the published version of the manuscript.

**Funding:** This research was funded by the CRSRI Open Research Program (Grant No. CKWV2018477/KY), National Natural Science Foundation of China (Grant No. 51678534), Open Projects Funds of Dike Safety and Disaster Prevention Engineering Technology Research Center of Chinese Ministry of Water Resources (Grant no. 2018006) and Program for Innovative Research Team (in Science and Technology) in University of Henan Province of China (Grant No. 20IRTSTHN009).

**Conflicts of Interest:** The authors declare no conflict of interest.

## References

1. Kong, X.L. Application of Distributed Optical Fiber Sensing Technology in Monitoring of Steel Corrosion. Master's Thesis, Dalian University of Technology, Dalian, China, 2011. (In Chinese).
2. Hooton, R.D. Future directions for design, specification, testing, and construction of durable concrete structures. *Cem. Concr. Res.* **2019**, *124*, 105827. [CrossRef]
3. Zhao, J. Time Variation and Similarity of Chloride Diffusion Properties of Concrete under Dry Wet Cyclic Environment. Master's Thesis, Zhejiang University of Technology, Hangzhou, China, 2019. (In Chinese).
4. Zhang, L.L. Study on durability of reinforced concrete structure. *Sichuan Build. Mater.* **2012**, *38*, 40–42.
5. Liu, H.Z.; Zhang, Q.; Li, V.; Su, H.Z.; Gu, C.S. Durability study on engineered cementitious composites (ECC) under sulfate and chloride environment. *Constr. Build. Mater.* **2017**, *133*, 171–181. [CrossRef]
6. Ling, Y.F.; Zhang, P.; Wang, J.; Chen, Y.Z. Effect of PVA fiber on mechanical properties of cementitious composite with and without nano-$SiO_2$. *Constr. Build. Mater.* **2019**, *229*, 117068. [CrossRef]
7. Zhang, P.; Wittmann, F.H.; Vogel, M.; Müller, H.S.; Zhao, T. Influence of freeze-thaw cycles on capillary absorption and chloride penetration into concrete. *Cem. Concr. Res.* **2017**, *100*, 60–67. [CrossRef]
8. Fu, C.Q.; Ye, H.L.; Jin, X.Y.; Yan, D.M.; Jin, N.G.; Peng, Z.X. Chloride penetration into concrete damaged by uniaxial tensile fatigue loading. *Constr. Build. Mater.* **2016**, *125*, 714–723. [CrossRef]

9.  Zhao, H.T.; Jiang, K.D.; Yang, R.; Tang, Y.M.; Liu, J.P. Experimental and theoretical analysis on coupled effect of hydration, temperature and humidity in early-age cement-based materials. *Int. J. Heat. Mass. Tran.* **2020**, *146*, 118784. [CrossRef]

10. Dong, W.K.; Li, W.G.; Shen, L.M.; Sun, Z.H.; Sheng, D.C. Piezoresistivity of smart carbon nanotubes (CNTs) reinforced cementitious composite under integrated cyclic compression and impact. *Compos. Struct.* **2020**, *241*, 112106. [CrossRef]

11. Li, L.; Cao, M.L. Influence of calcium carbonate whisker and polyvinyl alcohol- steel hybrid fiber on ultrasonic velocity and resonant frequency of cementitious composites. *Constr. Build. Mater.* **2018**, *188*, 737–746. [CrossRef]

12. Zhang, D.; Yu, J.; Wu, H.L.; Jaworska, B.; Li, V.C. Discontinuous micro-fibers as intrinsic reinforcement for ductile Engineered Cementitious Composites (ECC). *Compos. Part B* **2020**, *184*, 107741. [CrossRef]

13. Deb, S.; Mitra, N.; Majumder, S.B.; Maitra, S. Improvement in tensile and flexural ductility with the addition of different types of polypropylene fibers in cementitious composites. *Constr. Build. Mater.* **2018**, *180*, 405–411. [CrossRef]

14. Li, H.; Mu, R.; Qing, L.B.; Chen, H.S.; Ma, Y.F. The influence of fiber orientation on bleeding of steel fiber reinforced cementitious composites. *Cem. Concr. Compos.* **2018**, *92*, 125–134. [CrossRef]

15. Badanoiu, A.; Holmgren, J. Cementitious composites reinforced with continuous carbon fibres for strengthening of concrete structures. *Cem. Concr. Compos.* **2003**, *25*, 387–394. [CrossRef]

16. Rostami, R.; Zarrebini, M.; Mandegari, M.; Mostofinejad, D.; Abtahi, S.M. A review on performance of polyester fibers in alkaline and cementitious composites environments. *Constr. Build. Mater.* **2020**, *241*, 117998. [CrossRef]

17. Murthy, A.R.; Ganesh, P. Effect of steel fibres and nano silica on fracture properties of medium strength concrete. *Adv. Concr. Constr.* **2019**, *7*, 143–150.

18. Pakravan, H.R.; Ozbakkaloglu, T. Synthetic fibers for cementitious composites: A critical and in-depth review of recent advances. *Constr. Build. Mater.* **2019**, *207*, 491–518. [CrossRef]

19. Zhang, P.; Ling, Y.F.; Wang, J.; Shi, Y. Bending resistance of PVA fiber reinforced cementitious composites containing nano-SiO$_2$. *Nanotechnol. Rev.* **2019**, *8*, 690–698. [CrossRef]

20. Zhang, P.; Li, Q.F.; Wang, J.; Shi, Y.; Ling, Y.F. Effect of PVA fiber on durability of cementitious composite containing nano-SiO$_2$. *Nanotechnol. Rev.* **2019**, *8*, 116–127. [CrossRef]

21. Wang, J.Q.; Dai, Q.L.; Si, R.Z.; Guo, S.C. Investigation of properties and performances of Polyvinyl Alcohol (PVA) fiber-reinforced rubber concrete. *Constr. Build. Mater.* **2018**, *193*, 631–642. [CrossRef]

22. Khan, M.I.; Abbas, Y.M.; Fares, G. Review of high and ultrahigh performance cementitious composites incorporating various combinations of fibers and ultrafines. *J. King Saud Univ. Eng. Sci.* **2017**, *29*, 339–347. [CrossRef]

23. Xu, C.D.; Yao, Z.P.; Li, Z.; Zhu, X.L.; Wang, M.Y.; Wang, Y. Durability evaluation model and application of concrete structure based on fuzzy analytic hierarchy process. *Hydropower Energ. Sci.* **2019**, *37*, 95–99.

24. Lai, Y.C.; Liu, D.W.; Huang, L.J. Durability evaluation of machine-made sand concrete structure based on cloud model. *Silicate Bull.* **2019**, *38*, 3305–3308.

25. Jin, W.L.; Zhao, Y.X. *Durability of Concrete Structure*; Science Press: Beijing, China, 2014; pp. 299–320.

26. Parichatprecha, R.; Nimityongskul, P. Analysis of durability of high performance concrete using artificial neural networks. *Constr. Build. Mater.* **2009**, *23*, 910–917. [CrossRef]

27. Zhang, H. Durability reliability analysis for corroding concrete structures under uncertainty. *Mech. Syst. Signal. Process.* **2018**, *101*, 26–37. [CrossRef]

28. Zhou, X.L.; Yuan, Y.C.; Chen, L. Durability evaluation of in-service concrete structure based on grey correlation degree. *J. Wuhan Univ. Eng.* **2017**, *39*, 169–174.

29. Yu, B.; Ning, C.L.; Li, B. Probabilistic durability assessment of concrete structures in marine environments: Reliability and sensitivity analysis. *China Ocean Eng.* **2017**, *31*, 63–73. [CrossRef]

30. Jang, J.S.R.; Sun, C.T.; Mizutani, E. Neuro-fuzzy and soft computing, a computational approach to learning and machine intelligence. *Automat. Contr.* **1997**, *42*, 1482–1484. [CrossRef]

31. Yazdanbakhsh, O.; Dick, S. A systematic review of complex fuzzy sets and logic. *Fuzzy. Set. Syst.* **2018**, *338*, 1–22. [CrossRef]

32. Xu, X.Z.; Cao, D.; Zhou, Y.; Gao, J. Application of neural network algorithm in fault diagnosis of mechanical intelligence. *Mech. Syst. Signal. Pr.* **2020**, *141*, 106625. [CrossRef]

33. Boulkaibet, I.; Belarbi, K.; Bououden, S.; Chadli, M.; Marwala, T. An adaptive fuzzy predictive control of nonlinear processes based on Multi-Kernel least squares support vector regression. *Appl. Soft. Comput.* **2018**, *73*, 572–590. [CrossRef]

34. Di, W.; Xu, X.; Zhang, Z.C. Application of adaptive neural fuzzy inference system in bridge state assessment. *J. Zhejiang Univ. (Eng. Edit.)* **2008**, *42*, 2016–2022.

35. Jie, M.; Niu, H.Y.; Qi, D.Y. Application of adaptive neuro fuzzy inference system in traffic pollutant concentration prediction. *Fuzzy Syst. Math.* **2019**, *33*, 144–152.

36. Wang, Y.L.; Ma, G.; You, H.H. Carbon content modeling of pulverized coal boiler fly ash based on adaptive neural fuzzy inference system. *Therm. Power Gener.* **2018**, *47*, 27–31.

37. Guo, H.W.; Dong, H.Y. Wind speed prediction based on fuzzy neural network. *Electr. Drive Autom.* **2012**, *34*, 1–13.

38. Hou, T. Application of T-S fuzzy neural network in surface water quality assessment of Sanchuan River. *Energ. Energ. Sav.* **2011**, *6*, 54–56.

39. Xu, J.Y.; Pan, X.Y.; Li, S.; Chen, Y. Research and development of bridge underwater structure state assessment system based on ANFIS. *Technol. Highw. Transp.* **2016**, *5*, 73–78. (In Chinese) [CrossRef]

40. Wang, F.L.; Zhou, Q.; Zhu, F. Prediction of compressive strength of concrete hollow block masonry based on ANFIS. *J. Wuhan Univ. Eng. Edit.* **2015**, *3*, 324–326.

41. *Concrete Durability Inspection and Evaluation Standard*; JG/J 193-2009; National Standard of the People's Republic of China: Shenzhen, China, 2009.

42. GB/T 50746-2008. Code for durability design of concrete. National Standard of the People's Republic of China. In Proceedings of the International Conference on Durability of Concrete Structures, Hangzhou, China, 26 November 2008.

43. *Test Method for Chloride Diffusion Coefficient of Cement*; JC/T 1086-2008; National Standard of the People's Republic of China: Shenzhen, China, 2008.

44. Poston, R.W.; Rabbat, B.G. *ACI 318-14. Building Code Requirements for Structural Concrete and Commentary*; ACI Committee 318; National Standard of the United States of America: Farmington Hills, MI, USA, 2014.

45. Jin, W.L.; Zhao, Y.X. Review and Prospect of research on durability of concrete structures. *J. Zhejiang Univ. Eng. Edit.* **2002**, *36*, 371–379.

46. Zuo, J.B. Prediction of Chloride Diffusion Coefficient of Fly Ash Concrete Based on Artificial Neural Network. Master's Thesis, Nanhua University, Chiayi County, Taiwan, 2012. (In Chinese).

47. Çakıt, E.; Olak, A.J.; Karwowski, W.; Marek, T.; Hejduk, I.; Taiar, R. Assessing safety at work using an adaptive neuro-fuzzy inference system (ANFIS) approach aided by partial least squares structural equation modeling (PLS-SEM). *Int. J. Ind. Ergon.* **2020**, *76*, 102925. [CrossRef]

48. Wang, J.S. Parameter optimization of ANFIS model based on particle swarm optimization. *J. Pet. Univ.* **2007**, *20*, 41–44.

49. Wei, Y.L. Topology optimization of wireless sensor network based on adaptive artificial immune network algorithm. *Elect. Meas. Techno.* **2020**, *43*, 85–88.

50. Bao, Z.S.; Wang, K.X.; Zhang, W.B. Practical Byzantine fault tolerant consensus algorithm based on tree topology network. *J. Appl. Sci.* **2020**, *38*, 35–49.

51. Zhang, Z.H.; Han, Z.; Xiao, S.X.; Chang, G.Z.; Gan, S.W. Real time prediction of traffic flow based on ANFIS and its realization in MATLAB. *J. Chongqing Jiaotong Univ.* **2007**, *26*, 112–115.

52. Seco, A.; Urmeneta, P.; Prieto, E.; Marcelino, S.; Miqueleiz, L. Estimated and real durability of unfired clay bricks: Determining factors and representativeness of the laboratory tests. *Constr. Build. Mater.* **2017**, *131*, 600–605. [CrossRef]

53. Pan, H.K.; Yang, Z.S.; Xu, F.W. Study on concrete structure's durability considering the interaction of multi-factors. *Constr. Build. Mater.* **2016**, *118*, 256–261. [CrossRef]

54. Srokosz, P.E.; Bagińska, M. Application of adaptive neuro-fuzzy inference system for numerical interpretation of soil torsional shear test results. *Adv. Eng. Softw.* **2020**, *143*, 102793. [CrossRef]

55. Rice, J.A. *Mathematical Statistics and Data Analysis*; Academic Press: Cambridge, MA, USA; Mechanical Industry Press: Beijing, China, 2014; pp. 228–248.

56. Dai, X.B. Study on Durability of Nano Particle and PVA Fiber Reinforced Cement-Based Composite. Master's Thesis, Zhengzhou University, Zhengzhou, China, 2017. (In Chinese).

57. Wu, X. Improvement of Fuzzy System and ANFIS and Its Application in Intelligent Selection of Machining Parameters. Master's Thesis, Jilin University, Changchun, China, 2007. (In Chinese).
58. Erdeljić, V.; Francetić, I.; Bošnjak, Z.; Budimir, A.; Likić, R. Distributed lags time series analysis versus linear correlation analysis (Pearson's r) in identifying the relationship between antipseudomonal antibiotic consumption and the susceptibility of Pseudomonas aeruginosa isolates in a single Intensive Care Unit of a tertiary hospital. *Int. J. Antimicrob. Agents.* **2011**, *37*, 467–471.
59. Williams, L.L.; Quave, K. *Quantitative Anthropology*; Academic Press: Pittsburgh, PA, USA, 2019; pp. 105–114.
60. Ross, S.M. *Introductory Statistics*; Academic Press: Pittsburgh, PA, USA, 2017; pp. 489–518.

*Article*

# Effect of Aggregate Type and Specimen Configuration on Concrete Compressive Strength

**Sherif Yehia *, Akmal Abdelfatah and Doaa Mansour**

Civil Engineering Department, American University of Sharjah (AUS), Sharjah PO Box 26666, UAE;
akmal@aus.edu (A.A.); g00045494@alumni.aus.edu (D.M.)
* Correspondence: syehia@aus.edu

Received: 25 June 2020; Accepted: 17 July 2020; Published: 19 July 2020

**Abstract:** In this paper, concrete mixes utilizing two sizes of natural aggregate and two sources of lightweight and recycled aggregates were used to investigate the effect of aggregate type and specimen size and shape on the compressive strength of concrete. In addition, samples from ready-mix concrete producers with different strengths were evaluated using standard size cylinders and cubes. Results were obtained on the 7th, 28th, and 90th day. In addition, flexural strength, split tension, and modulus of elasticity were evaluated on the 28th and 90th day. Statistical analyses were conducted to examine the significance of the difference between the compressive strength values for each two mixes using tests of hypotheses. Moreover, other mechanical properties as a function of compressive strength were discussed and compared to those predicated by the American Concrete Institute (ACI) specifications. Results indicate specimen shape has a noticeable effect on the compressive strength as the Cylinder/Cube ratio on the $90^{th}$ day was ranging between 0.781 and 0.929. The concrete compressive strength and modulus of elasticity were significantly affected by the aggregate type. The flexural strength and split tensile strength were less affected by the aggregate type, which was also confirmed by the values predicted with the ACI equations.

**Keywords:** aggregate type; specimen shape; specimen size; compressive strength; concrete mechanical properties

---

## 1. Introduction

Concrete compressive strength ($f'_c$) provides an indication about the ability of a specific mix to resist axial compression loads. Traditionally, axial compression test utilizes standard cubes and cylinders to determine $f'_c$ and became the most commonly used test in the construction industry. In addition, other mechanical properties could be predicted as a function of $f'_c$ and design codes and standards for concrete structures consider the concrete compressive strength as the main indicator of the concrete resistance to loading. Concrete compressive strength is greatly affected by cement content, water-to-cement ratio (w/c), and aggregate size and type. Other factors, summarized in Table 1, influence the test results or indirectly affect the concrete compressive strength [1–16].

**Table 1.** Factors affecting compressive strength of concrete [1–16].

| Factors Affecting Compressive Strength | Comments |
| --- | --- |
| Type of cement | Different concrete applications according to exposure conditions might require cement with specific properties. Five basic types of Portland cement are commonly used in the construction industry. A blend of cement and cementitious materials is also available in the market. The main bonding material and responsible for strength. |
| Supplementary Cementitious materials | Used to reduce heat of hydration, to improve workability and durability depending on the materials. They also ensure quality of concrete during the stages of mixing, transporting, placing, and curing in adverse weather conditions. |
| Water/cement ratio (w/c) | A low water/cement ratio increases resistance to weathering, provides a good bond between successive concrete layers. Excess water increases the porosity and permeability of the concrete and reduces strength. |
| Aggregate | Concrete strength is affected by the strength, surface texture, grading, and maximum size of the aggregate. |
| Mixing water | Impurities in the mixing water might affect concrete set time, strength, and long-term durability. It is generally thought that the pH of water should be between 6.0 and 8.0. |
| Curing | Moisture during curing: prolonged moist curing leads to the highest concrete strength. 3- and 7-day moist curing period will lead to 60% and 80% of the strength of the continuously cured concrete. |
| | Temperature during curing: increasing curing temperature while preventing moisture loss leads to an increase in the rate of hydration and, consequently, the rate of strength development. In cold weather, if concrete freezes soon after it has been placed will have a severe strength loss. |
| Age of concrete | Concrete gains strength with age. |
| Maturity of concrete | It indicates the progress of hydration and it is the relationship between strength gain, time, and concrete temperature. |
| Rate of loading | The axial compressive strength is reduced by about 75% of the standard test strength under slow rates of loading. Whereas, at high rates of loading, the strength increases, reaching 115% of the standard strength tests. |
| Specimen shape and size | Cube specimen generally gives more strength than that of cylinder specimen. Moreover, increasing the specimen size decreases the compressive strength. |
| Admixtures | Air-entraining admixtures increase resistance to freeze and thaw cycles by adding a stable bubble of air which might lead to a reduction of compressive strength. However, target compressive strength could be achieved by careful mix proportioning and reduced w/c ratio. |
| | Using superplasticizers can improve the strength, marginally, when using recycled aggregates or GGBFS. For self-compacting concrete, the use of superplasticizers has a slight effect on the concrete strength and that effect is more apparent in the early stages. |

As shown in Table 1, several factors might influence the strength of concrete, especially aggregate since it represents about 40–60% per volume of any concrete mix. In general, for the same cementitious materials, w/c ratio and curing conditions, aggregate strength, texture, absorption, size, and gradation affect the failure mechanism of the concrete, which in turn affects the compressive strength. Failure of concrete depends mainly on three factors: bond strength between the aggregate and the cement paste, strength of the cement paste, and strength of the aggregate.

## 1.1. Effect of Aggregate Type on Compressive Strength

In general, compressive strength of normal weight aggregates depends on the strength of the parent rock that ranges from 35 MPa to 350 MPa [1]. Aggregate strength influences the production of high-strength concrete (HSC). Such mixes have high strength of the paste and the bond between aggregate and cement is improved. Accordingly, cracks may extend through the aggregate under loading, which makes use of the full strength of the aggregate and hence affects the concrete strength. However, for normal strength concrete, bond strength between the aggregate and the cement paste is affected by the physical properties of the aggregate, which in turn influences the concrete compressive strength. Therefore, failure tends to occur in the cement paste and in the interfacial zone between aggregate particles and the paste, before happening in the aggregate [7]. Other aggregate types, lightweight and recycled aggregates, and their effect on compressive strength of concrete are summarized in Table 2 [17–28]. Lightweight and recycled aggregates usually have lower strength and higher absorption than that of natural aggregate. High porosity of lightweight aggregate leads to less strength and high absorption, while recycled aggregate properties are affected by processing, previous loading, and exposure conditions, as illustrated in Figure 1. Compressive strength of concrete prepared with lightweight aggregate is affected by aggregate strength, w/c ratio, surface texture, mechanical interlock, and bond with the cement paste. In addition, variation in absorption capacities leads to formation of interfacial transition zones (ITZ), which have different microstructure. However, mixing procedure and the use of supplementary materials improve the ITZ and the concrete compressive strength. Similarly, concrete strength produced utilizing recycled aggregate is affected by the previous factors, in addition to the percentage of recycled aggregate used in the mix.

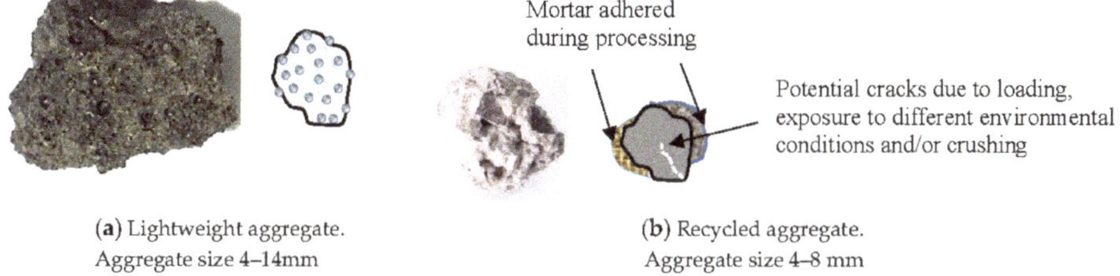


(**a**) Lightweight aggregate.
Aggregate size 4–14mm

(**b**) Recycled aggregate.
Aggregate size 4–8 mm

**Figure 1.** Schematic of lightweight and recycled aggregate particles.

The effect of petrographic characteristics of aggregates on the quality of concrete has been considered by References [29,30]. It was evident that concrete strength is significantly affected by the mineralogy and microstructure of the coarse aggregates. In addition, the aggregate strength and mechanical properties are influenced by the secondary products of serpentinites and andesites.

## 1.2. Effect of Aggregate Size and Shape

Aggregate gradation (size distribution), shape (flat, elongated aggregate, round), and texture (rough and smooth) have different effects on the mechanical properties according to concrete type being prepared Normal Strength Concrete (NSC), High Strength Concrete (HSC), Lightweight Concrete (LWC), and Recycled Aggregate Concrete (RAC). Several research efforts evaluated the effect of aggregate size and shape on the mechanical properties of concrete are summarized in Table 3 [6,7,31–34]. Aggregate size and shape can affect the cement-aggregate bond strength; for example, large aggregate size creates high stress concentration in the cement paste, leading to increased cracking; therefore, small aggregate size is recommended for HSC. Similar recommendation is suggested for LWC to avoid crushing of big size aggregates while loading.

**Table 2.** The effect of aggregate type on concrete compressive strength [13,17–28].

| References | Aggregate Used in the Study | Effect on Strength (Compressive and Modulus of Elasticity) | Comments |
|---|---|---|---|
| Chi et al. [17] | Three types of LWA | Both compressive strength and elastic modulus were affected by the aggregate type and decreased with an increase in w/c ratio. | Different concrete strength, Portland cement, w/c ratio 0.3, 0.4, and 0.5 |
| Wasserman and Bentur [18] | Fly ash LWA | Concrete strength does not depend only on the aggregate strength but affected by other characteristics such as its rough surface, porosity, and pozzolanic nature. | Portland cement, w/c ratio 0.4 |
| Lo and Cui [19] | LWA | Porous surface of LWA improved the interfacial bond between aggregate and cement paste, resulting in better interlocking. Development of initial strength was much higher in LWC than in NWC. | Portland cement, w/c ratio 0.36 |
| Topçu, and Uygunoğlu [20] | Normal crushed limestone and three types of LWA | Compressive strength increased with w/c decrease. Replacement of natural aggregate with LWA cause decrease in modulus of elasticity. Pumice is a useful aggregate for production of SCLC with high interlocking in ITZ and high compressive strength. | Different concrete strength. Portland cement, w/c ratio 0.36, 0.4, 0.43, 0.46, and 0.48 |
| Tabsh and Abdelfatah [13] | RA | Concrete strength is affected by the source of the recycled aggregate. In addition, bond strength is affected by the quantity and the shape of the old cement mortar adhered on the aggregate surface. | |
| Poon and Lam [21] | Natural crushed aggregate and three types of RCA | Compressive strength increased when aggregate/cement ratio decreased. Using RCA instead of NCA reduced density and strength. The compressive strength was directly proportional to strength of the blended aggregate | |
| Seo and Choi [22] | RCA | The amount of the used recycled aggregate has a remarkable effect on the compressive strength. If amount of old cement mortar is about 10%, the mortar can improve the bond quality. | Portland cement |
| Xiao et al. [23] | Natural and RCA | Both compressive strength and modulus of elasticity reduced as the RCA percentage increased. | |

**Table 2.** *Cont.*

| References | Aggregate Used in the Study | Effect on Strength (Compressive and Modulus of Elasticity) | Comments |
|---|---|---|---|
| Tsounami et al. [24] | Natural and RCA | For RCA replacement percentage higher than 25%, compressive strength reduces as replacement percentage increases. | Portland cement |
| Katz [25] | RCA | Concrete with 100% recycled aggregate was weaker than that of the natural aggregate, and about 25% reduction of the compressive strength was observed. | Portland cement |
| McNeil et al. [26] | RCA | RCA is less dense, more porous, and has higher water absorption than NCA due to the residual adhered mortar on RCA, which weakens the interfacial zone and reduces both compressive strength and modulus of elasticity. | |
| Wardeh et al. [27] | Recycled gravel with 30%, 65%, and 100% replacement ratio. | The use of recycled aggregate up to 30% caused a 14% reduction in the compressive strength, while increasing the replacement ratio had increased the cement content, causing an increase in the compressive strength that counterbalances the negative effect of the recycled aggregate. | C35, CEM I CALCIA 52.5 N CE CP2 NF cement, w/c ratio 0.5–0.52 |
| Choi et al. [28] | RA | Enhancing the ITZ help suppress the occurrence of microcracks and improve the mechanical performance of the aggregate, resulting in improved strength, permeability, and durability of concrete. | Normal strength concrete, Portland cement, w/c ratio 0.55 |

**Table 3.** The effect of aggregate size and shape on concrete compressive strength [6,7,31–34].

| References | Aggregate Used in the Study | Cementitious Materials (Mineral Admixtures) | Effect on Strength (Compressive and Modulus of Elasticity) | Comments |
|---|---|---|---|---|
| Loannides and Mills [7] | Natural aggregate with different sizes | Air entraining admixtures | Smaller aggregates tended to shear under loading while larger aggregate was more prone to pull out. | Normal strength concrete Portland cement w/c ratio 0.5 |
| Rocco and Elices [31,32] | Mullite aggregate with different sizes | | Fracture energy increased with increasing the aggregate size and not affected significantly by the aggregate shape. Modulus of elasticity decreased as the aggregate size increased. Using crushed aggregates (with strong matrix-aggregate interfaces) showed higher modulus of elasticity than that while using spherical ones of the same size. | Normal strength concrete Portland cement w/c ratio 0.42, 0.7 |
| Ajamu and Ige [33] | Aggregate with different sizes | | Compressive strength increased as the aggregate size increase from 13mm to 19mm. | Normal strength concrete Portland cement w/c ratio 0.65 |
| Meddah et al. [34] | Aggregate with different size and distribution | HRWRA | Compressive strength increased as the aggregate size increased in normal strength concrete. While in HSC, Compressive strength increased with decreasing the aggregate size. Increasing coarse aggregate content in NSC enhanced the compressive strength, while in HSC there is an optimum content. Effect of aggregate content and particle size is more significant in HSC than NSC. | Normal and High-strength concrete w/c ratio 0.58, 0.4 |
| Yaqub and Bukhari [6] | Aggregate with different sizes | | Smaller aggregate sizes exhibited higher compressive strength in the case of HSC. | High-strength concrete Portland cement w/c ratio 0.23–0.35 |

### 1.3. Effect of Specimen Configuration

Cylinders (100 × 200 mm, 150 × 300 mm) and cubes (100 × 100 × 100 mm, 150 × 150 × 150 mm) are commonly used to monitor compressive strength of concrete. The variation in concrete compressive strength due to shape and size might be attributed to variability in aggregate physical properties, different friction between concrete surfaces and loading platen, and variation of crack propagation and localized failure zone. Localization of the damage in a certain zone is affected by the slenderness of the sample and the boundary restraint between the loading platens and the specimen [12,35,36]. For cylinder specimens, it is expected that compressive strength should not be affected by the sample size as long as the ratio of height to maximum lateral dimension (h/d) is maintained 2:1 [37]. However, several studies [9,10,38,39] showed that compressive strength of small-size cylinders is slightly higher than that obtained using the 150/300 mm cylinders. This is due to the smaller contact area between the specimen surface and steel platen of the testing machine, which results in lower friction-based shear forces. Smaller specimens are also denser as they have less number of micro-cracks and defects, which strengthen their compressive strength. The effect of specimen shape on concrete compressive strength is, however, recognized in the difference between cylinder and cube strength. The concrete nominal strength ($f'_c$) has a great effect on the specimen shape factor. It positively correlates with the cylinder/cube strength ratio as it is found that increasing the concrete strength decreases the specimen shape effect [40]. European standard (ENV 206:1990) and (BS 1881: Part 120) specifications [41,42] recommend a 0.8 ratio for the cylinder/cube strength and this ratio reaches 1.0 as concrete strength increases. Several studies showed that it is considerably hard to adopt a simple ratio for the cylinder/cube strength [40,43–51]. Compressive strength of concrete utilizing lightweight (LWA) and recycled aggregate (RA) are significantly affected by the size and aspect ratio of specimens due to the decrease in the concrete unit weight. This also could be attributed to poor crack distribution and localized failure zone due to the deteriorated aggregate interlock, which causes further decrease in the compressive strength with the increase of the specimen size [1,35].

In this study, effect of aggregate type and specimens' configuration on compressive strength of concrete is investigated. The aggregate type includes natural aggregate, recycled aggregate, and lightweight aggregate. The specimens' configuration includes two sizes for cubes and cylinders. The lab tests were conducted at 7, 28, and 90 days. Tests were conducted according to the American Society for Testing and Materials (ASTM) specifications and British Standards (BS). In addition, statistical analyses were conducted to examine the significance of the difference between the compressive strength values for each two mixes using tests of hypotheses.

### 1.4. Research Significance

Due to the depletion of and increased demand for natural resources, several natural aggregates were introduced to the construction industry. Therefore, one of the authors' goals is to evaluate the quality of the natural aggregates and compare the results to the previous well-established facts. In addition, another goal was to evaluate the suitability of the recycled and lightweight aggregates available in the market to produce concrete comparable to that produced by natural aggregate, hence, establish a base for comparison with concrete produced with natural aggregate.

## 2. Experimental Program

The main objective of the experimental program is to investigate the effect of aggregate type and specimen size and shape on the concrete compressive strength. In addition, correlation between compressive strength and aggregate strength is examined. The experimental program focuses on the evaluation of samples prepared in the laboratory utilizing six aggregate types, in addition to samples of five concrete strength (C45_1 (Ready mix 1-1), C75 (Ready mix 1-2), C45_2 (Ready mix 2-1), C60 (Ready mix 2-2), and C80 (Ready mix 2-3)) collected from ready-mix producers in Dubai and Sharjah, UAE. The use of the ready-mix concrete is considered to validate the results obtained based on the mixes prepared in the laboratory while conducting this research. Laboratory samples were prepared twice to ensure consistency and to account for variability of aggregate properties, especially recycled aggregate. The evaluation criteria included compressive strength of standard size cylinders and cubes, modulus of elasticity, flexural strength, and split tensile strength. Testing was conducted according to ASTM specifications [52–58].

### 2.1. Materials

All mixes prepared in the laboratory have Ordinary Portland cement type I (SG = 3.15), silica fume (SG = 2.2), Ground Granulated Blast-Furnace Slag "GGBS" (SG = 2.85), and tap water. In addition, normal weight dune sand (particle size 100% passing 0.6 mm, SG 2.60) and coarse sand (maximum particle size 4.75 mm, SG 2.60) were used as fine aggregates. Six types of coarse aggregate (Figure 2) were used to produce the concrete mixes in the lab: 10 mm and 20 mm sizes limestone (10 mm Nat., 20 mm Nat.), sintered pulverized–fuel ash ($LWA_1$) and pumice ($LWA_2$) lightweight aggregates with aggregate size 4 to 8 mm, and recycled aggregate ($RCA_1$, $RCA_2$) from two sources with size 4–14 mm. In addition, several mechanical and physical properties of the coarse aggregates were evaluated and summarized in Table 4. Detailed investigation of the aggregate properties has been reported in previous research [59,60]. In addition, all ready-mix producers' mixes used a mix of 10 mm and 20 mm natural aggregates, cement, GGBS, silica fume, fine sand, and w/cm = 0.39.

**Figure 2.** Coarse aggregate used in the investigation.

**Table 4.** Summary of the coarse aggregate properties.

| Physical Property / Aggregate Type | Absorption % | Bulk Dry Sp. Gr.* | Moisture Content% | LA Abrasion | | |
|---|---|---|---|---|---|---|
| | | | | Grade B % | Grade C % | Grade D % |
| Natural Aggregate | 0.65 | 2.685 | 0.558 | 22.38 | | |
| Sintered pulverized–fuel ash | 24.32 | 1.336 | 0.564 | | | 25.97 |
| Pumice | 12.6 | 1.605 | 0.604 | | | 27.27 |
| Recycled Aggregate - 1 | 3.94 | 2.36 | 0.719 | 35 | 31 | |
| Recycled Aggregate - 2 | 3.35 | 2.47 | 0.84 | 24.97 | | |

* Bulk Dry Specific Gravity.

Los Angeles abrasion: results from the Los Angeles (LA) abrasion test could be used as an indicator of aggregate strength and a relative measure of resistance to crushing under a gradually applied compressive load. A high LA abrasion number reflects a larger portion of crushed aggregates (fine) compared to the original mass. However, in case of recycled aggregate, high LA values could be attributed to the mortar adhered to the aggregate during crushing of recycled concrete. In general, aggregate's composition, texture, and structure affect its strength [1]; therefore, it is recommended to evaluate the aggregate performance in concrete samples to have a better indication of its strength.

Absorption: lightweight and recycled aggregates from both sources have high absorption capacity than that of the NWA. Therefore, the LWA and RCA were pre-wet 15 min prior to mixing with part of the mixing water (about 15% of the LWA aggregate weight and 5% of the RCA aggregate weight) to compensate for the high absorption capacity and to avoid impact on short-term workability. In addition, about 5% of the cement and cementitious materials weight were added during the pre-wet time to enhance the bond strength between the aggregate and the cement paste [59,60].

## 2.2. Mix Proportioning

All mixes prepared in the lab were proportioned using the absolute volume method and were based on a normal weight self-consolidated concrete mix [61] with a target strength of 70 MPa. Volume fractions, Table 5, for all materials and w/c ratio were the same for all mixes prepared in the laboratory; however, weights of course aggregates were adjusted to account for the difference in specific gravities. The total volumetric fraction of the cement and supplementary cementitious materials was 16%. A low cement content (8%) per volume was selected as an approach to achieve eco-friendly concrete mixes. The volume fraction of the GGBS (6%) and silica fume (2%) are the commonly used ratios by the ready-mix producers. In addition, the w/c ratio was selected to achieve the target strength and durability requirements [61].

**Table 5.** Mix proportioning.

| Material | Absolute Volume |
|---|---|
| Cement | 0.08 |
| GGBS | 0.06 |
| Silica fume | 0.02 |
| water | 0.18 |
| Coarse aggregate | 0.33 |
| FA (1/2 dune and 1/2 crushed) | 0.33 |
| Total Volume | 1 |

## 2.3. Testing Program

Cylinders 100 × 200 mm (4 × 8 in.) ($Cy_1$), 150 × 300 mm (6 × 12 in.) ($Cy_2$) and cubes 100 × 100 × 100 mm (4 × 4 × 4 in.) ($Cu_1$) and 150 × 150 × 150 mm (6 × 6 × 6 in.) ($Cu_2$) were prepared from each mix to evaluate the compressive strength development. In addition, flexural, splitting tensile strength, and modulus of elasticity were evaluated at 28- and 90-day for all mixes except mix $LWA_1$. There was not enough material from $LWA_1$ to produce all samples, and the source was not available in the market. Therefore, only samples for compressive strength evaluation, for $LWA_1$, were prepared and included in the related discussions. All samples were cured using wet burlaps for three days and were left to dry in ambient room temperature. Table 6 summarizes the tests, number of samples, sample size, age at testing, and specifications followed during testing.

**Table 6.** Summary of the experimental investigation.

| Test | Test Specifications | Specimen Size (mm) | No. of Specimens Per Test | Testing Events (day) |
|---|---|---|---|---|
| Compressive strength | ASTM C39/C39M-17 [37] | 150 × 300 cylinder<br>100 × 200 cylinder<br>150 × 150 × 150 cube<br>100 × 100 × 100 cube | 2 cylinders*<br>2 cylinders*<br>2 cubes*<br>2 cubes* | 7th, 28th and 90th |
| Modulus of Elasticity | ASTM C469/C469M-14 [58] | 150 × 300 cylinder | 1 cylinder | 28th and 90th |
| Splitting tensile strength | ASTM C496/C496M-11 [56] | 150 × 300 cylinder | 2 cylinders | 28th and 90th |
| Flexural strength | ASTM C78/C78M-16 [57] | 100 × 100 × 500 beam | 2 beams | 28th and 90th |

* Two sets of samples were prepared for a total of 4 samples of each cube/cylinder.

## 3. Results

### 3.1. Compressive Strength

Table 7 provides a summary of the compressive strength results for different concrete mixes at 7-, 28- and 90-day. In addition, Figures 3 and 4 show the average compressive strength for cube and cylinder specimens, respectively. The results show that generally the compressive strength for all the concrete mixes increased with age, as expected. The cube specimen of the natural concrete mixes gave 75% of its target strength (70 MPa) on the 7-day, 97% on the 28-day, and exceeded it on the 90-day. Both the lightweight aggregate and recycled aggregate concrete mixes resulted in 15–20% lower compressive strength than that of the natural mixes in the 7-, 28-, and 90-day due to the lower aggregate strength. Cylinder specimen exhibited lower strength than that of the cube as a result of the specimen shape effect. In addition, sample failure modes of cubes and cylinders from all mixes are shown in Figure 5. In addition, failure modes of other samples are illustrated in Annex A. These failure modes indicate the variability of the aggregate-cement bond strength, which is affected by the aggregate type. Similar results were reported by References [13,16,19,62,63]. Nonetheless, the lightweight and recycled aggregates showed improved bond strength due to the addition of cementitious materials during the pre-wet process before mixing, which was discussed elsewhere by References [59,60].

**Table 7.** Summary of compressive strength for different concrete mixes at 7th, 28th, and 90th day.

| Test Date<br>Mixes | 7-day | | | | 28-day | | | | 90-day | | | |
|---|---|---|---|---|---|---|---|---|---|---|---|---|
| | $Cy_1$ | $Cy_2$ | $Cu_1$ | $Cu_2$ | $Cy_1$ | $Cy_2$ | $Cu_1$ | $Cu_2$ | $Cy_1$ | $Cy_2$ | $Cu_1$ | $Cu_2$ |
| 10mm Natural | 46.06 | 40.9 | 52.1 | 52.83 | 63.31 | 56.05 | 59.47 | 67.93 | 63.01 | 58.35 | 64.07 | 79.13 |
| 20mm Natural | 59.67 | 48.8 | 61.97 | 58.31 | 65.19 | 65.65 | 67.6 | 72.63 | 72.68 | 54.16 | 76.55 | 77.01 |
| $LWA_1$ | 44.9 | 43.2 | 43.49 | 49.2 | 53.54 | 42.4 | 52.89 | 60.92 | 54.67 | 50.25 | 57.37 | 61.65 |
| $LWA_2$ | 38.47 | 41.15 | 37.78 | 45.1 | 47.65 | 46.4 | 52.28 | 60.57 | 51.69 | 43.76 | 52.94 | 59.57 |
| $RCA_1$ | 44.89 | 41.15 | 51.4 | 48.44 | 52.34 | 51.4 | 52.72 | 60.93 | 56.9 | 60.46 | 57.49 | 66.4 |
| $RCA_2$ | 45.56 | 38.35 | 45.98 | 54.97 | 54.56 | 53.1 | 60.52 | 66.63 | 59.82 | 48.8 | 66.56 | 69.6 |
| Ready mix 1-1 | 35 | 31.3 | 43.7 | 45 | 36.57 | 35.75 | 44.43 | 52.2 | 37.66 | 37.05 | 55.6 | 56.59 |
| Ready mix 1-2 | 48.17 | 53.1 | 60.66 | 59.32 | 58.98 | 57.85 | 62.51 | 68.13 | 58.09 | 57.7 | 67.61 | 76.83 |
| Ready mix 2-1 | 43.61 | 41.87 | 46.39 | 47.91 | 50.49 | 49 | 54.35 | 59.11 | 54.28 | 49.3 | 59 | 61 |
| Ready mix 2-2 | 53.08 | 48.55 | 49.49 | 54.07 | 58.23 | 57.85 | 58.7 | 66.4 | 60.47 | 55.5 | 55.76 | 62.35 |
| Ready mix 2-3 | 64.4 | 65.14 | 63.19 | 64.64 | 67.91 | 82.8 | 75.74 | 83.19 | 78.49 | 80.89 | 81.11 | 86.47 |

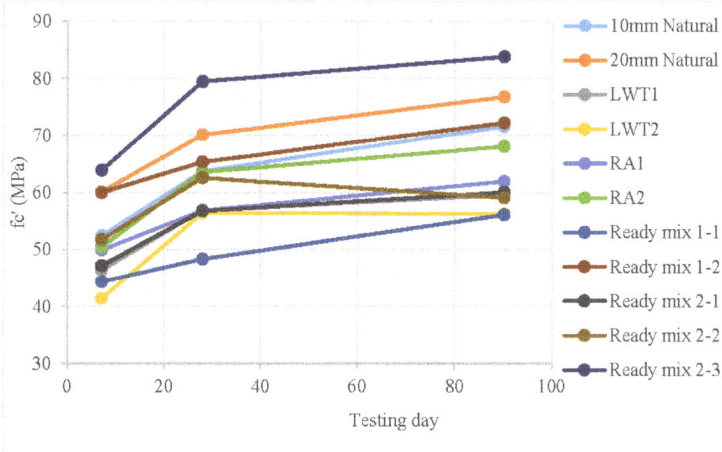


**Figure 3.** Average cube compressive strength.

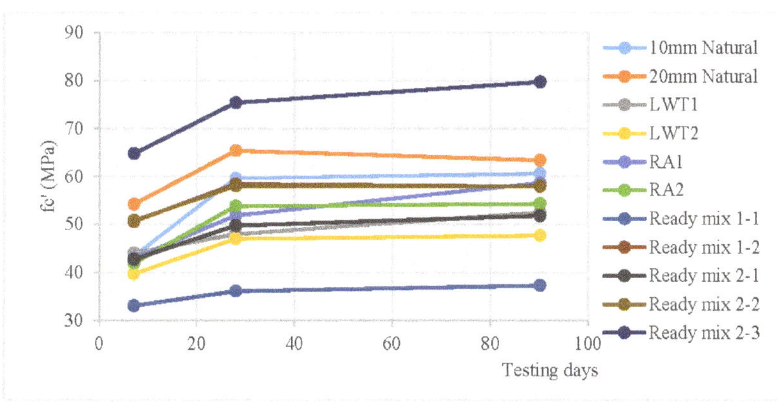


**Figure 4.** Average cylinder compressive strength.

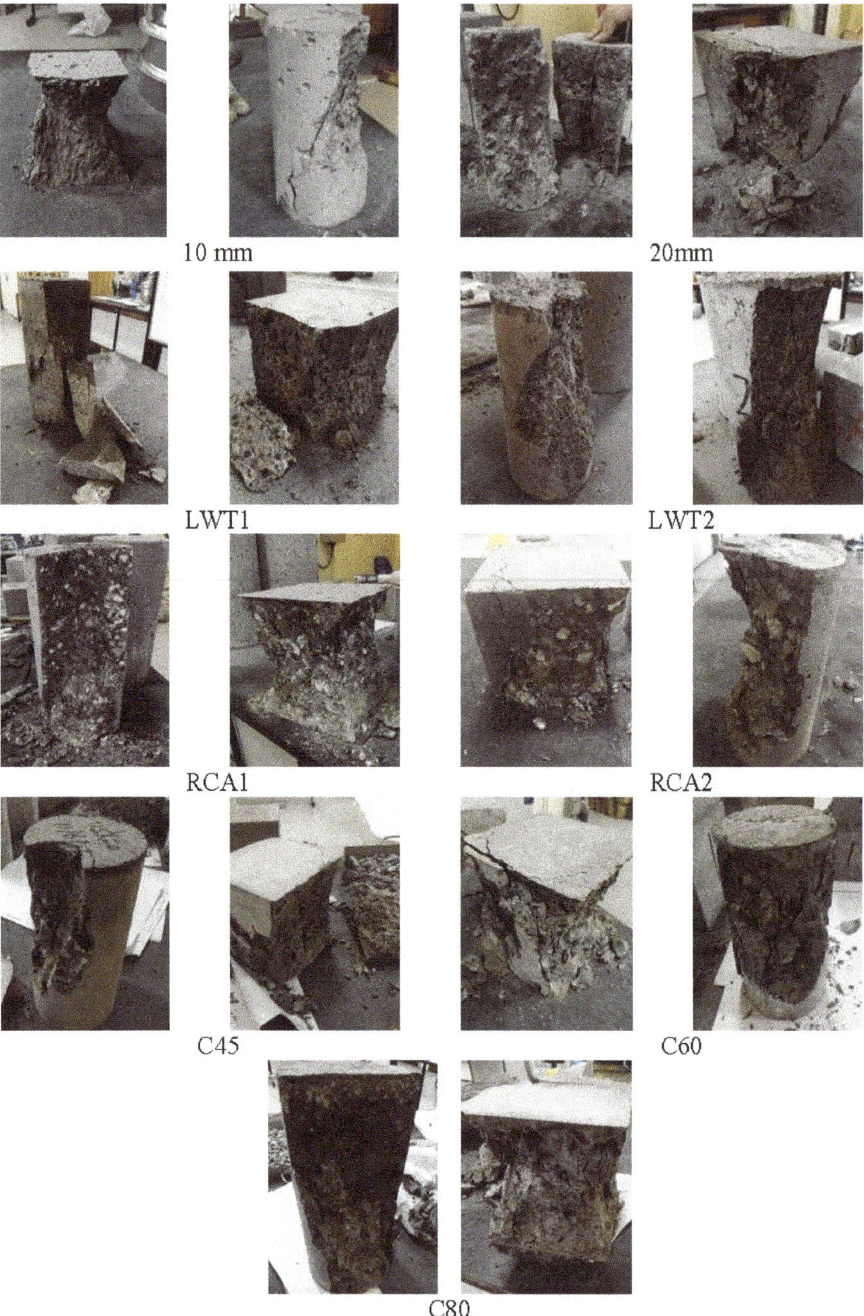


**Figure 5.** Failure modes of cube and cylinder specimens from all mixes.

### 3.2. Modulus of Elasticity, Split Strength, and Flexure Strength Results

Figures 6–8 summarize the test results of modulus of elasticity, split tensile strength, and flexure strength for all mixes except mix $LWA_1$. The results were significantly affected by the aggregate type. Moreover, the modulus of elasticity and flexural strength test results for all the concrete mixes had increased from 28-day to 90-day, as shown in Figures 6 and 8. The modulus of elasticity was less than that of the natural aggregate and ready-mix concrete by about 25% (for RCA) and 40% (for the LWA). However, the results of split tensile strength test indicate a significant variation among different concrete mixes as illustrated in Figure 7. This could be attributed to variation of the aggregate-cement bond properties, which are greatly influenced by the aggregate type [19,39,64,65]. It should be noted that the results in Figures 6–8 are based on two (for modulus of elasticity) or four (for split test) specimens only. Therefore, the results are considered preliminary results.

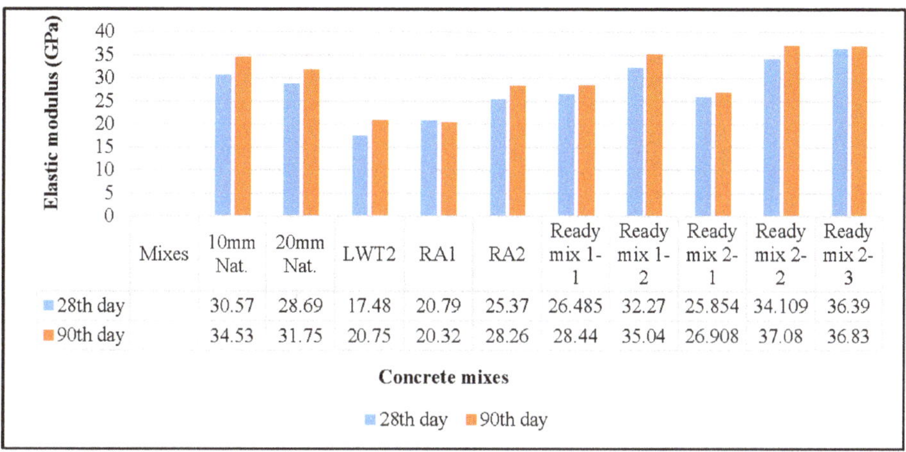

**Figure 6.** Elastic modulus for different concrete mixes at 28- and 90-day.

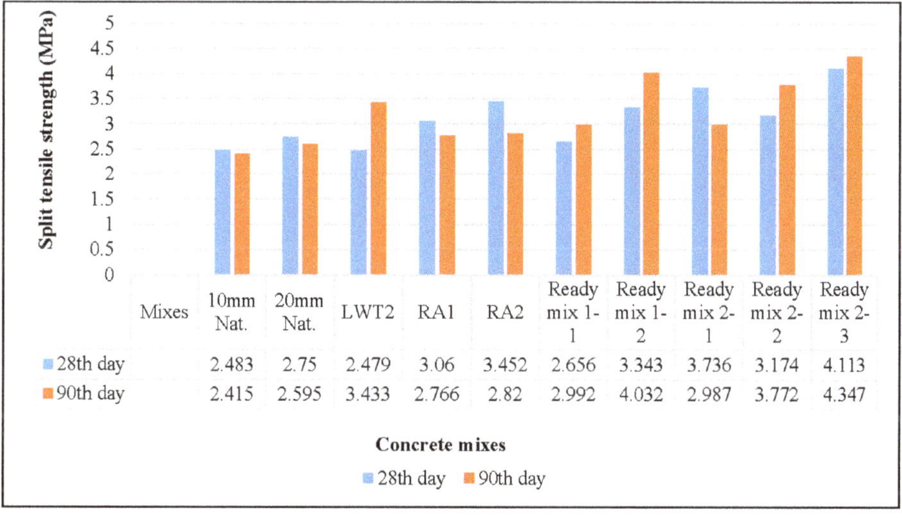

**Figure 7.** Split strength for different concrete mixes at 28- and 90-day.

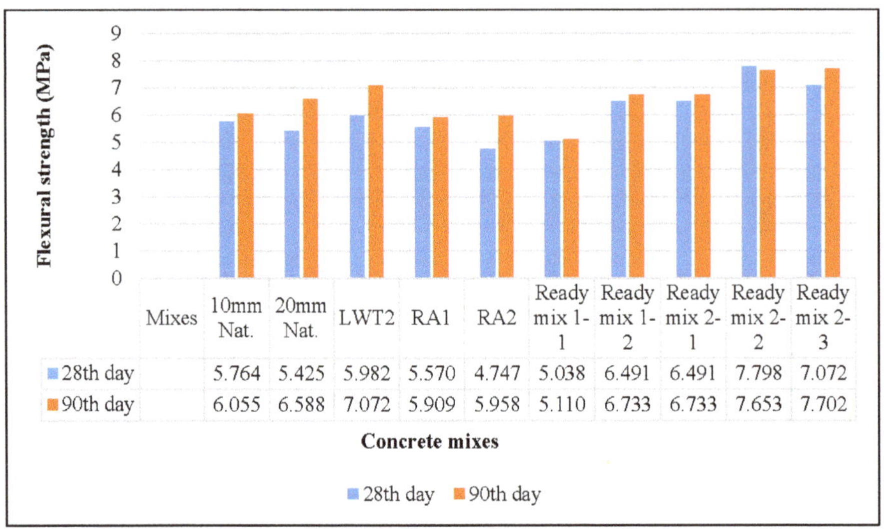



| Mixes | 10mm Nat. | 20mm Nat. | LWT2 | RA1 | RA2 | Ready mix 1-1 | Ready mix 1-2 | Ready mix 2-1 | Ready mix 2-2 | Ready mix 2-3 |
|---|---|---|---|---|---|---|---|---|---|---|
| 28th day | 5.764 | 5.425 | 5.982 | 5.570 | 4.747 | 5.038 | 6.491 | 6.491 | 7.798 | 7.072 |
| 90th day | 6.055 | 6.588 | 7.072 | 5.909 | 5.958 | 5.110 | 6.733 | 6.733 | 7.653 | 7.702 |

**Concrete mixes**

■ 28th day ■ 90th day

**Figure 8.** Flexure strength for different concrete mixes at 28- and 90-day.

## 4. Discussion

To evaluate the effect of sample configuration (size and shape) and aggregate strength on the concrete compressive strength, t-pooled statistical analysis was conducted on the test results. For a specific sample configuration and testing date, most of the results reported in Table 7 are average of four samples. The results for the same sample size, ratios between $f'_c$ at 7-day and 28-day, $f'_c$ at 90-day and 28-day, $f'_c$ at 7-, 28-, and 90-day and the target compressive strength, and average shape factor (ratio between the cylinder and cube compressive strength values) at 7-, 28-, and 90-day are summarized in Table 8. The results indicated that there was a continuous hydration and formation of C-S-H, which is reflected by the strength gain, and most of the ratios for $Cy_1$, $Cy_2$, $Cu_1$, and $Cu_2$ specimens for different mixes were less than 1 for 7- to 28-day ratio. Previous studies [16,59,60] conclude that the natural concrete mixes typically achieve 0.7-0.8 of the 28-day strength in the first 7 days. On the other hand, most of the ratios are more than 1 for 90- to 28-day ratio. This highlights the importance of long-term evaluation especially for lightweight and recycled aggregate concrete, which is also in agreement with the recommendations for concrete with GGBS, lightweight, and recycled aggregate concrete [15,19,60]. Some scattered results in Table 8 show lower compressive strength for the 90-day than the 28-day, especially for $Cy_2$ specimen; nonetheless, the majority of the results indicate higher strength, which is similar to results reported in the literature [31,32,44]. Table 9 illustrates a summary of the ratio between the cylinder and cube strength from the literature and from the current study.

### 4.1. Shape Factor Analysis

Aggregate type, target strength, and sample configuration (cylinder or cube) and sample size might affect the compressive strength values of the concrete mixes and the shape factor (the ratio of cylinder to cube compressive strength). Therefore, statistical analysis was conducted to examine the effect of these parameters on the shape factor.

Table 8. Summary of compressive strength ratios.

| Test Date / Mixes | Ratio between 7th /28th- Day Compressive Strength | | | | Ratio between 90th /28th-Day Compressive Strength | | | | $Cu_2$/TS Ratios | | | Average Cy/Cu Ratio | | |
|---|---|---|---|---|---|---|---|---|---|---|---|---|---|---|
| | | | | | | | | | 7th | 28th | 90th | 7th | 28th | 90th |
| | $Cy_1$ | $Cy_2$ | $Cu_1$ | $Cu_2$ | $Cy_1$ | $Cy_2$ | $Cu_1$ | $Cu_2$ | $Cu_2/T_S$ | $Cu_2/T_S$ | $Cu_2/T_S$ | Cy/Cu | Cy/Cu | Cy/Cu |
| 10 mm Natural | 0.73 | 0.73 | 0.88 | 0.78 | 1.00 | 1.04 | 1.08 | 1.16 | 0.75 | 0.97 | 1.13 | 0.839 | 0.922 | 0.839 |
| 20 mm Natural | 0.92 | 0.74 | 0.92 | 0.80 | 1.11 | 0.82 | 1.13 | 1.06 | 0.83 | 1.04 | 1.10 | 0.925 | 0.925 | 0.866 |
| $LWA_1$ | 0.84 | 1.02 | 0.82 | 0.81 | 1.02 | 1.19 | 1.08 | 1.01 | 0.70 | 0.87 | 0.88 | 0.946 | 0.806 | 0.878 |
| $LWA_2$ | 0.81 | 0.89 | 0.72 | 0.74 | 1.08 | 0.94 | 1.01 | 0.98 | 0.64 | 0.87 | 0.85 | 0.957 | 0.829 | 0.818 |
| $RCA_1$ | 0.86 | 0.80 | 0.97 | 0.80 | 1.09 | 1.18 | 1.09 | 1.09 | 0.69 | 0.87 | 0.95 | 0.872 | 0.934 | 0.929 |
| $RCA_2$ | 0.84 | 0.72 | 0.76 | 0.83 | 1.10 | 0.92 | 1.10 | 1.04 | 0.79 | 0.95 | 0.99 | 0.855 | 0.836 | 0.781 |
| Ready mix C45_1 | 0.96 | 0.88 | 0.98 | 0.86 | 1.03 | 1.04 | 1.25 | 1.08 | 1.00 | 1.16 | 1.25 | 0.747 | 0.748 | 0.666 |
| Ready mix C75 | 0.82 | 0.92 | 0.97 | 0.87 | 0.98 | 1.00 | 1.08 | 1.13 | 0.79 | 0.91 | 1.02 | 0.844 | 0.894 | 0.802 |
| Ready mix C45_2 | 0.86 | 0.85 | 0.85 | 0.81 | 1.08 | 1.01 | 1.09 | 1.03 | 1.06 | 1.31 | 1.35 | 0.906 | 0.877 | 0.863 |
| Ready mix C60 | 0.91 | 0.84 | 0.84 | 0.81 | 1.04 | 0.96 | 0.95 | 0.94 | 0.90 | 1.11 | 1.04 | 0.981 | 0.928 | 0.982 |
| Ready mix C80 | 0.95 | 0.79 | 0.83 | 0.78 | 1.16 | 0.98 | 1.07 | 1.04 | 0.81 | 1.04 | 1.08 | 1.013 | 0.948 | 0.951 |

$Cy_1$: cylinder 4x8 in, $Cy_2$: cylinder: 6x12 in, $Cu_1$: cube 4x4x4 in, $Cu_2$: cube 6x6x6 in, TS – Target Strength.

**Table 9.** Summary of cylinder vs. cube strength ratios from the literature compared to the current study. [43,45–51].

| Reference | Average cylinder/cube Ratio | Remarks |
|---|---|---|
| Cormack [43] | 0.87 | Study focused on high-strength concrete. Few data were generated for $f_c'$<41 MPa |
| Evans [45] | 0.77–0.96 | Lower-strength concrete had generally lower cylinder/cube strength ratios |
| Sigvaldason [46] | 0.71–0.77 <br> 0.76–0.84 | Segregating concrete <br> Non-segregating concrete |
| Gyengo [47] | 0.65–0.84 | Variation due to changing coarseness of aggregate grading |
| Gonnerman [48] | 0.85–0.88 | Tests performed using standard cylinders and 6″ and 8″ cubes |
| Plowman, Smith, and Sheriff [49] | 0.74 <br> 0.64 | Water-cured specimens <br> Air-cured specimens <br> In both cases, portions of steel bars were embedded in cylinder specimens |
| Raju and Basavarajaiah [50] | 0.61 <br> 0.51 | Using 150 mm cubes <br> Using 100 mm cubes |
| Lasisi et al. [51] | 0.67–0.76 <br> 0.55–0.86 | Landcrete specimens (small agg. From lateric soil) <br> Concrete specimens |
| Current study | 0.66–0.98 <br> 0.82–0.88 <br> 0.78–0.93 | Normal and high-strength concrete. <br> Lightweight aggregate concrete. <br> Recycled aggregate concrete. |

### 4.1.1. Effect of Aggregate Type on Shape Factor

The statistical testing was conducted to compare the compressive strength of the 6-inch and the 4-inch standard specimen sizes for the same specimen shape (cube and cylinder) and concrete mix. Testing was conducted for available "n1 = n2 = 4" samples of results and was carried out for the compressive strength at 7-, 28- and 90-day. The result of this test showed that the majority of the compared standard specimen sizes gave equal compressive strength, which indicates a negligible effect of using different standard sizes on the concrete compressive strength. Accordingly, the data for the cube and cylinder have been combined (having a size of 8 samples). The average concrete compressive strength is calculated as the trimmed average of 6 samples (after excluding the highest and the lowest value). The specimen shape factors "$C_y/C_u$" for six concrete mixes were compared using the t-pooled hypothesis testing, for a 95% confidence level, to investigate the effect of aggregate type on the shape factor. Table 10 shows the summary of the hypothesis testing results for the shape factors at 90-day, which is considered to represent the long-term performance of the concrete. It should be noted that the limit for "t" is ±2.228. The results shown in Table 9 indicate that the aggregate type has a significant effect on the shape factor and the following observations were concluded: (i) there is no significant difference in the shape factor values when comparing natural and lightweight aggregates. However, there was a significant difference between the two types of lightweight aggregates, which was confirmed by Reference [59]; (ii) recycled aggregate concrete mixes show a statistically significant difference in the shape factors when compared to other types of aggregate. This can be attributed to the fact that the recycled aggregate properties cannot be controlled due to the high variability of their sources even in the same batch, which may have different quality and strength [13,21,59,66].

**Table 10.** Summary of hypothesis testing results for Cy/Cu at 90-day.

| Mixes | t-Value | | | | | |
|---|---|---|---|---|---|---|
| | 10 mm Nat. | 20 mm Nat. | LWA$_1$ | LWA$_2$ | RCA$_1$ | RCA$_2$ |
| 10 mm Nat. | | 0.860 | 1.301 | −0.652 | 3.104 | −1.808 |
| 20 mm Nat. | | | 0.534 | −2.087 | 3.130 | −3.531 |
| LWA$_1$ | | | | −2.971 | 3.108 | −4.525 |
| LWA$_2$ | | | | | 5.989 | −1.640 |
| RCA$_1$ | | | | | | −7.393 |
| RCA$_2$ | | | | | | |

### 4.1.2. Effect of Concrete Target Strength on Shape Factor

A summary of calculated specimen shape factors $Cy_1/Cu_1$ and $Cy_2/Cu_2$ of the five ready-mix concrete with different target strengths at 7-, 28-, and 90-day are shown in Table 11. The results for samples collected from the same ready-mix producer show that, in general, when increasing the target strength level of the concrete mix, the specimen shape factor increased. Similar results were reported by other researchers [12,40].

**Table 11.** Summary of $Cy_1/Cu_1$ and $Cy_2/Cu_2$ ratios at 7-, 28-, and 90-day.

| Test Date / Mixes | 7-day | | 28-day | | 90-day | |
|---|---|---|---|---|---|---|
| | $Cy_1/Cu_1$ | $Cy_2/Cu_2$ | $Cy_1/Cu_1$ | $Cy_2/Cu_2$ | $Cy_1/Cu_1$ | $Cy_2/Cu_2$ |
| Ready mix 1-1 (45 MPa) | 0.80 | 0.70 | 0.82 | 0.68 | 0.68 | 0.65 |
| Ready mix 1-2 (75 MPa) | 0.79 | 0.90 | 0.94 | 0.85 | 0.86 | 0.75 |
| Ready mix 2-1 (45 MPa) | 0.94 | 0.87 | 0.93 | 0.83 | 0.92 | 0.81 |
| Ready mix 2-2 (60 MPa) | 1.07 | 0.90 | 0.99 | 0.87 | 1.08 | 0.89 |
| Ready mix 2-3 (80 MPa) | 1.02 | 1.01 | 0.90 | 1.00 | 0.97 | 0.94 |

### 4.2. Effect of Aggregate Type on Concrete Compressive Strength Analysis

Compressive strength of six concrete mixes was compared using the pooled t-test method to investigate the effect of the aggregate type on the concrete strength. Table 12 summarizes the t-pooled analysis for each specimen shape "Cy and Cu" at 90-day. For the cylinder specimen, Table 12, the results showed that compressive strength was affected by both the aggregate type and specimen shape. To investigate the aggregate type effect separately, the 90-day specimen shape factors were applied to the compressive strength of the cylinders and then re-compared using the t-pooled testing method. Results in Table 12 show that generally the concrete compressive strength was highly affected by the aggregate type. The following observations were concluded:

**Table 12.** Summary of **t** pooled analysis results at 90-day.

**Cy of Six Concrete Mixes**

| t-Value / Mixes | 10 mm Nat. | 20 mm Nat. | LWA$_1$ | LWA$_2$ | RCA$_1$ | RCA$_2$ |
|---|---|---|---|---|---|---|
| 10 mm Nat. | | 0.880 | -4.079 | -4.263 | -3.837 | -0.616 |
| 20 mm Nat. | | | -3.247 | -3.510 | -2.899 | -1.188 |
| LWA$_1$ | | | | -0.581 | 0.865 | 2.534 |
| LWA$_2$ | | | | | 1.412 | 2.866 |
| RCA$_1$ | | | | | | 2.072 |
| RCA$_2$ | | | | | | |

**Cu of Six Concrete Mixes**

| Mixes | 10 mm Nat. | 20 mm Nat. | LWA$_1$ | LWA$_2$ | RCA$_1$ | RCA$_2$ |
|---|---|---|---|---|---|---|
| 10 mm Nat. | | 1.346 | -3.065 | -4.773 | -1.138 | -2.040 |
| 20 mm Nat. | | | -3.075 | -4.175 | -1.941 | -2.511 |
| LWA$_1$ | | | | -1.717 | 2.056 | 0.623 |
| LWA$_2$ | | | | | 3.814 | 2.175 |
| RCA$_1$ | | | | | | -1.153 |
| RCA$_2$ | | | | | | |

**Adjusted Cy of Six Concrete Mixes**

| Mixes | 10 mm Nat. | 20 mm Nat. | LWA$_1$ | LWA$_2$ | RCA$_1$ | RCA$_2$ |
|---|---|---|---|---|---|---|
| 10 mm Nat. | | 0.736 | -3.390 | -4.060 | -2.677 | -0.705 |
| 20 mm Nat. | | | -3.663 | -4.217 | -3.067 | -1.404 |
| LWA$_1$ | | | | -0.800 | 0.894 | 3.199 |
| LWA$_2$ | | | | | 1.711 | 4.018 |
| RCA$_1$ | | | | | | 2.340 |
| RCA$_2$ | | | | | | |

(1) For normal strength concrete, changing the aggregate size did not affect the concrete compressive strength when using natural aggregate, similar findings were reported by References [7,34].

(2) Both lightweight and $RCA_1$ mixes show less compressive strength than the two natural concrete mixes. This is attributed to their lower aggregate strength. This conclusion is compatible with the results reported by previous research [13,27], which conclude that compressive strength has decreased with replacing the natural aggregate with recycled aggregate.

(3) The $RCA_2$ mix resulted in an equivalent compressive strength to those of the natural concrete mixes for both cylinder and cube specimens, which matches their corresponding aggregate strength and similarly gives higher compressive strength than those of the lightweight aggregate mixes.

(4) In accordance with the higher strength shown by the $RCA_2$ than $RCA_1$, its concrete mix exhibited a slightly higher compressive strength than that of $RCA_1$ concrete mix in the cube specimen and an equivalent strength in the cylinder case. This difference could be attributed to the variation of the recycled aggregate sources even within the same batch, which may have different properties and strength.

### 4.3. Correlation between Aggregate Strength and Concrete Compressive Strength

Regression analysis was conducted to investigate the relationship between the aggregate strength of the six types considered in this study represented with their weight loss percentage that resulted from the LA abrasion test, and their concrete mixes' trimmed average of the equivalent compressive strength. Analysis was done for both cylinder and cube specimens' compressive strength at 28-day. The results indicate a good correlation between the aggregate strength and their concrete mix compressive strength. The $R^2$ values, for a second-order equation, was around 0.60 for the cylinder and cube compressive strength, as shown in Figure 9. These results refer to the great role played by the aggregate type in affecting the concrete compressive strength. This contradicted with several previous studies [67–71] that estimate concrete compressive strength based on the water-cement ratio and ignoring the effect of aggregate type/strength in their models. Findings from the current study are in agreement with other studies [6,9,17,44,62–64] that aggregate type should be considered while predicting the concrete compressive strength.

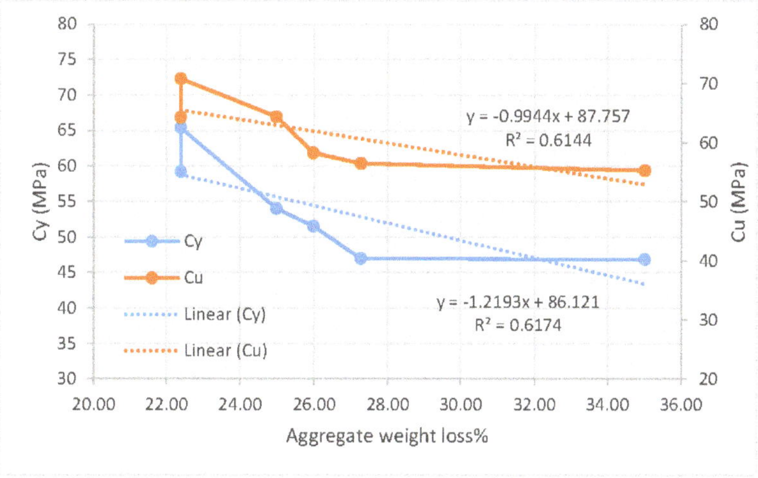


**Figure 9.** Correlation between compressive strength and aggregate strength.

*4.4. Comparing Test Results to ACI Standards*

ACI363R-10 [72] and ACI318-19 [73], Equations (1) to (6) [72,73], utilize cylinder compressive strength to predict other mechanical properties.

$$\text{Modulus of elasticity } E = 3320 \sqrt{f_c'} + 6900 \text{ for } f_c' \text{ 21 MPa } < f_c' < 83 \text{ MPa} \tag{1}$$

$$E = 4700 \sqrt{f_c'} \tag{2}$$

$$\text{Split tensile strength } f_{spt} = 0.59 \sqrt{f_c'} \text{ for } f_c' \text{ 21 MPa } < f_c' < 83 \text{ MPa} \tag{3}$$

$$f_{spt} = 0.56 \sqrt{f_c'} \tag{4}$$

$$\text{flexural strength } f_r = 0.94 \sqrt{f_c'} \text{ for } f_c' \text{ 21 MPa } < f_c' < 83 \text{ MPa} \tag{5}$$

$$f_r = 0.62 \lambda \sqrt{f_c'} \tag{6}$$

where,

$f_c'$ is the cylinder Cy$_2$ compressive strength at 28-day in MPa.
$\lambda$ is a reduction factor to account for aggregate type. $\lambda = 0.85$ for sand lightweight concrete.

In the current study, modulus of elasticity, split tensile strength, and flexural strength were calculated using equations from Reference [72] for natural aggregates and equations from Reference [73] for lightweight and recycled aggregate concrete mixes. The main goal of the comparison is to check the applicability of using ACI equations to predict the mechanical properties of concrete with different aggregate types. Table 13 summarizes the comparison of the calculated and tested values for the modulus of elasticity, the flexure strength, and the split strength. It is observed that the tested modulus of elasticity for the ready mixes and the mixes with natural aggregate ranged between 0.85 and 1.06 of the calculated values according to ACI equations. However, in the case of lightweight and recycled aggregates, the ACI equations overestimated the modulus of elasticity, which indicates unsuitability of these equations with RCA and LWA samples. Values from the tested samples were much lower than their standard predicted values, which were in the range of 0.55 to 0.74 of the calculated values. Lightweight aggregate and recycled aggregate samples require more curing time, which might affect the test results at 28 days, as discussed earlier in the paper.

The tested split tensile strength for the majority of the concrete mixes is lower than the standard predicted value, especially for the 10mm and 20mm concrete mixes. This indicates weak bond strength, resulting in an early failure in the lateral direction. In addition, ready-mix concrete mixes have resulted in tested flexural strength that is comparable to the calculated values (ratio of 0.83 to 1.09). The natural aggregates resulted in a lower flexure strength than the calculated values. The pumice lightweight aggregate and recycled aggregate concrete mixes exhibited higher flexural strength than those predicted, especially for the pumice lightweight aggregate concrete mix. This could be a result of its small size and rough surface, which increased the interlocking between aggregate and the cement paste and the special presoaking procedure followed for both lightweight and recycled aggregate concrete mixes, which enhanced the bond in the ITZ.

Table 14 summarizes some of the results of the modulus of elasticity, split strength, and flexural strength from the literature and the current study results, which came compatible with each other. Variation in values is attributed to the different types of aggregate used and the strength level targeted in those studies.

**Table 13.** Summary of calculated and tested values.

| Mix | Modulus of Elasticity (GPa) | | | Split Strength (MPa) | | | Flexure Strength (MPa) | | |
|---|---|---|---|---|---|---|---|---|---|
| | Calculated | Tested | Ratio+ | Calculated | Tested | Ratio+ | Calculated | Tested | Ratio+ |
| 10 mm Nat. | 31.76 | 30.57 | 0.96 | 4.42 | 2.48 | 0.56 | 7.04 | 5.76 | 0.82 |
| 20 mm Nat. | 33.80 | 28.69 | 0.85 | 4.78 | 2.75 | 0.58 | 7.62 | 5.43 | 0.71 |
| LWA2 | 32.02 | 17.48 | 0.55 | 3.81 | 2.48 | 0.65 | 3.59* | 5.98 | 1.67 |
| RCA1 | 33.70 | 20.79 | 0.62 | 4.01 | 3.06 | 0.76 | 4.45 | 5.57 | 1.25 |
| RCA2 | 34.25 | 25.37 | 0.74 | 4.08 | 3.45 | 0.85 | 4.52 | 4.75 | 1.05 |
| Ready mix 1-1 | 26.75 | 26.49 | 0.99 | 3.53 | 2.66 | 0.75 | 5.62 | 5.04 | 0.90 |
| Ready mix 1-2 | 32.15 | 32.27 | 1.00 | 4.49 | 3.34 | 0.74 | 7.15 | 6.49 | 0.91 |
| Ready mix 2-1 | 30.14 | 25.85 | 0.86 | 4.13 | 3.74 | 0.90 | 6.58 | 6.49 | 0.99 |
| Ready mix 2-2 | 32.15 | 34.11 | 1.06 | 4.49 | 3.17 | 0.71 | 7.15 | 7.80 | 1.09 |
| Ready mix 2-3 | 37.11 | 36.39 | 0.98 | 5.37 | 4.11 | 0.77 | 8.55 | 7.07 | 0.83 |

*$\lambda$ factor of 0.85 was adopted to account for the aggregate type; + Ratio = tested/calculated.

**Table 14.** Results of mechanical properties from literature and the current study results. [13,20,21,25–28,33,34,62,63,74–80].

| Reference | Flexural Strength (MPa) | Split Tensile Strength (MPa) | Modulus of Elasticity (GPa) | Remarks |
|---|---|---|---|---|
| Topcu and Uygunoglu [20] | | 3.7–3.9<br>1.6–1.7 | 33–39<br>17–18 | Natural aggregate concrete.<br>Lightweight aggregate concrete. |
| Tabsh and Abdelfatah [13] | | 4<br>2.9–4 | | Natural aggregate concrete.<br>Recycled aggregate concrete. |
| Poon and Lam [21] | | 3–4.2 | | Normal and high-strength concrete. |
| Katz [25] | 5.4 | 3.1 | 11.3 | Recycled aggregate concrete. |
| McNeil et al. [26] | 10.2<br>8.9–9.7 | 3.3<br>2.7–3 | | Natural aggregate concrete.<br>15%, 30%, and 50% RCA. |
| Beshr et al. [62] | | 2.4–4 | 21–28 | High-strength concrete. |
| Beushausen and Dittmer [63] | 2.66–2.93 | 3.74–4.35 | 27.03–44.81 | High-strength concrete. |
| Kilic et al. [74] | 5.2–17.3 | | | High-strength concrete. |
| Wu et al. [75] | | 5–5.3 | 31–39.5 | w/c = 0.44. |
| Zhou et al. [76] | | | 18.6–51.3 | High-performance concrete. |
| Ozturan and Cecen [77] | 4.7–5.3 | 3.9–5.2 | | Normal and high-strength concrete. |
| Aitcin and Mehta [78] | | | 31.7–37.9 | High-strength concrete. |
| Ezeldin and Aitcin [79] | 7.4–9.2 | | | High-strength concrete. |
| Sengul et al. [80] | | 2.59–3.88<br>4.44–8.14 | 25.3—38<br>36.3–51.1 | Normal strength concrete.<br>High-strength concrete. |
| Ajamu and Ige [33] | 4.4-4.93 | | | Normal concrete with different aggregate sizes. |
| Meddah et al. [34] | | | 28.5–37 | Normal and high-strength concrete. |
| Wardeh et al. [27] | 4.9<br>3.95–4.75 | 3.6<br>3–3.3 | 39.5<br>30–36 | Natural aggregate concrete.<br>300%, 65%, and 100% RCA. |
| Choi et al. [28] | | 1.8–2.5 | | Recycled aggregate concrete. |
| Current study | 5.5–7.8<br>6<br>4.75–5.57 | 2.5–4.11<br>2.5<br>3–3.5 | 28–36<br>17.5<br>21–25 | Normal and high-strength concrete.<br>Lightweight aggregate concrete.<br>Recycled aggregate concrete. |

## 5. Conclusions

This paper presents a preliminary experimental investigation of the effect of specimen configuration and aggregate type on the concrete compressive strength. Six concrete mixes with different aggregate types were cast and tested for compressive strength, elastic modulus, splitting tensile strength, and flexural strength. In addition, samples from five ready-mix concrete were collected from two producers and were evaluated for the same mechanical properties. Two specimen shapes, cylinder and cube, were used with $100 \times 200$ mm ($4 \times 8$ in.) and $150 \times 300$ mm ($6 \times 12$ in.) sizes. Compressive strength was evaluated on the 7-, 28- and 90-day, while the elastic modulus, split tensile strength, and flexural strength were evaluated on the 28- and 90-day. The results from the current study are based on a small number of samples, however, the following could be concluded from the findings:

For the same cementitious materials, w/c ratio and curing conditions, compressive strength of the recycled and lightweight aggregate concrete was about 15–20% lower than that of natural aggregate and ready-mix concrete mixes. The modulus of elasticity was less than that of the natural aggregate and ready-mix concrete by about 25% (for RCA) and 40% for the LWA. These lower values can be attributed to the fact that these aggregates have lower strength than natural aggregates.

For all the concrete mixes in the study, the concrete compressive strength shows significant correlation with the aggregate strength (with an R2 of 0.61). Both flexural and split tensile strengths were less affected by the aggregate type than the compressive strength and elastic modulus.

The specimen size and aggregate unit weight effect on compressive strength is negligible regardless of the specimen shape and the aggregate type. In addition, for normal strength concrete and for the aggregate gradation used in the study (10 mm and 20 mm), aggregate size has no effect on the concrete compressive strength.

The cylinder/cube ratio ranged between 0.781 and 0.929 for the 90-day, which is in agreement with previous research. Using different types of aggregate in concrete mixes resulted in a cylinder/cube ratio influenced by the aggregate strength and, in turn, was affected by the specimen shape.

Based on the preliminary results obtained in this research, it seems that ACI equations [72,73] overestimated modulus of elasticity and split tensile strength values for LWC and RAC, which indicates inapplicability of such equations for lightweight and recycled concrete.

The presoaking procedure followed during mixing of lightweight and recycled concrete enhanced the ITZ, which was reflected by the improved bond, interlocking with the cement paste and better flexural performance. In this study, pumice lightweight aggregate and recycled aggregate concrete mixes exhibited higher flexural strength than those predicted by the ACI standard [69,70].

**Author Contributions:** Conceptualization, S.Y. and A.A.; methodology, S.Y. and D.M.; software, A.A. and D.M.; validation, S.Y., A.A. and D.M.; formal analysis, D.M., S.Y. and A.A.; investigation, D.M.; resources, S.Y. and A.A.; data curation, D.M.; writing—original draft preparation, S.Y., A.A. and D.M; writing—review and editing, S.Y. and A.A.; visualization, S.Y. and A.A.; supervision, S.Y.; project administration, S.Y. and A.A.; funding acquisition, S.Y. All authors have read and agreed to the published version of the manuscript.

**Funding:** The APC was funded, in part, by Open Access Program from the American University of Sharjah.

**Acknowledgments:** The authors acknowledge the support provided by the research office at the American University of Sharjah (AUS). The work in this paper was supported, in part, by the Open Access Program from the American University of Sharjah. In addition, the authors would like to acknowledge Naved Sayed from Gulf Ready Mix for providing timely assistance during the samples' preparation. In addition, the authors would like to acknowledge Mr. Mohamed Ansari, AUS Laboratory Technician for helping during samples' preparation and testing.

**Conflicts of Interest:** The authors declare no conflict of interest.

**Disclaimer:** This paper represents the opinions of the author(s) and does not mean to represent the position or opinions of the American University of Sharjah.

## References

1. Neville, A.M. *Properties of Concrete*; Addison Wesley Longman: London, UK, 1995.
2. MacGregor, J.G.; Wight, J.K. *Reinforced Concrete: Mechanics and Design*; Prentice-Hall: Singapore, 2016.

3. Mamlouk, M.S.; Zaniewski, J. *Materials for Civil and Construction Engineers*, 3rd ed.; Pearson Education, Inc.: Upper Saddle River, NJ, USA, 2016.

4. Nikbin, I.M.; Beygi, M.H.; Kazemi, M.T.; Vaseghi, J.A.; Rabbanifar, S.; Rahmani, E.; Rahimi, S. A comprehensive investigation into the effect of water to cement ratio and powder content on mechanical properties of self-compacting concrete. *Constr. Build. Mater.* **2014**, *57*, 69–80. [CrossRef]

5. Toutanji, H.; Delatte, N.; Aggoun, S.; Duval, R.; Danson, A. Effect of supplementary cementitious materials on the compressive strength and durability of short-term cured concrete. *Cem. Concr. Res.* **2004**, *34*, 311–319. [CrossRef]

6. Yaqub, M.; Bukhari, I. Effect of size of coarse aggregate on compressive strength of high strength concrete. In Proceedings of the 31st Conference Our World in Concrete & Structures, Singapore, 16–17 August 2006.

7. Loannides, A.M.; Mills, J.C. *Effect of Larger Sized Coarse Aggregates on Mechanical Properties of Portland Cement Concrete Pavements and Structures*; FHWA/OH-2006/10A; Ohio Department of Transportation Office of Research and Development: Columbus, OH, USA, 2006.

8. Prem, P.R.; Bharatkumar, B.H.; Iyer, N.R. Influence of curing regimes on compressive strength of ultra high performance concrete. *Sadhana* **2013**, *38*, 1421–1431. [CrossRef]

9. Del Viso, J.; Carmona, J.R.; Ruiz, G. Shape and size effects on the compressive strength of high-strength concrete. *Cem. Concr. Res.* **2008**, *38*, 386–395. [CrossRef]

10. Yi, S.-T.; Yang, E.I.; Choi, J.-C. Effect of specimen sizes, specimen shapes, and placement directions on compressive strength of concrete. *Nucl. Eng. Des.* **2006**, *236*, 115–127. [CrossRef]

11. Tokyay, M.; Ozdemir, M. Specimen shape and size effect on the compressive strength of higher strength concrete. *Cem. Concr. Res.* **1997**, *27*, 1281–1289. [CrossRef]

12. Abd, M.K.; Habeeb, Z.D. Effect of specimen size and shape on compressive strength of self-compacting concrete. *Diyala J. Eng. Sci.* **2014**, *7*, 17–29.

13. Tabsh, S.W.; Abdelfatah, A.S. Influence of recycled concrete aggregates on strength properties of concrete. *Constr. Build. Mater.* **2009**, *23*, 1163–1167. [CrossRef]

14. Sharma, R. Effect of wastes and admixtures on compressive strength of concrete. *J. Eng. Des. Technol.* **2020**. [CrossRef]

15. Zhang, S.Y.; Fan, Y.F.; Li, N.N. The Effect of Superplasticizer on Strength and Chloride Permeability of Concrete Containing GGBFS. *Adv. Mater. Res.* **2013**, *804*, 12–16. [CrossRef]

16. Mardani-Aghabaglou, A.; Tuyan, M.; Yılmaz, G.; Arıöz, Ö.; Ramyar, K.; Yilmaz, G. Effect of different types of superplasticizer on fresh, rheological and strength properties of self-consolidating concrete. *Constr. Build. Mater.* **2013**, *47*, 1020–1025. [CrossRef]

17. Chi, J.; Huang, R.; Yang, C.; Chang, J. Effect of aggregate properties on the strength and stiffness of lightweight concrete. *Cem. Concr. Compos.* **2003**, *25*, 197–205. [CrossRef]

18. Wasserman, R.; Bentur, A. Effect of lightweight fly ash aggregate microstructure on the strength of concretes. *Cem. Concr. Res.* **1997**, *27*, 525–537. [CrossRef]

19. Lo, Y.T.; Cui, H. Effect of porous lightweight aggregate on strength of concrete. *Mater. Lett.* **2004**, *58*, 916–919. [CrossRef]

20. Topçu, I.B.; Uygunoğlu, T. Effect of aggregate type on properties of hardened self-consolidating lightweight concrete (SCLC). *Constr. Build. Mater.* **2010**, *24*, 1286–1295. [CrossRef]

21. Poon, C.S.; Lam, C.S. The effect of aggregate-to-cement ratio and types of aggregates on the properties of pre-cast concrete blocks. *Cem. Concr. Compos.* **2008**, *30*, 283–289. [CrossRef]

22. Seo, D.; Choi, H. Effects of the old cement mortar attached to the recycled aggregate surface on the bond characteristics between aggregate and cement mortar. *Constr. Build. Mater.* **2014**, *59*, 72–77. [CrossRef]

23. Xiao, J.; Li, W.; Fan, Y.; Huang, X. An overview of study on recycled aggregate concrete in China (1996–2011). *Constr. Build. Mater.* **2012**, *31*, 364–383. [CrossRef]

24. Tsoumani, A.A.; Matikas, T.E.; Barkoula, N.M. Influence of Recycled Aggregates on Compressive Strength of Concrete. In Proceedings of the 2nd International Conference on Sustainable Solid Waste Management, Athens, Greece, 12–14 June 2014.

25. Katz, A. Properties of concrete made with recycled aggregate from partially hydrated old concrete. *Cem. Concr. Res.* **2003**, *33*, 703–711. [CrossRef]

26. McNeil, K.; Thomas, H.; Kang, K. Recycled Concrete Aggregates: A Review. *Int. J. Concr. Struct. Mater.* **2013**, *7*, 61–69. [CrossRef]

27. Wardeh, G.; Ghorbel, E.; Gomart, H. Mix Design and Properties of Recycled Aggregate Concretes: Applicability of Eurocode 2. *Int. J. Concr. Struct. Mater.* **2014**, *9*, 1–20. [CrossRef]
28. Choi, H.; Choi, H.; Lim, M.; Inoue, M.; Kitagaki, R.; Noguchi, T. Evaluation on the Mechanical Performance of Low-Quality Recycled Aggregate Through Interface Enhancement Between Cement Matrix and Coarse Aggregate by Surface Modification Technology. *Int. J. Concr. Struct. Mater.* **2016**, *10*, 87–97. [CrossRef]
29. Petrounias, P.; Giannakopoulou, P.P.; Rogkala, A.; Stamatis, P.M.; Tsikouras, B.; Papoulis, D.; Lampropoulou, P.; Hatzipanagiotou, K. The Influence of Alteration of Aggregates on the Quality of the Concrete: A Case Study from Serpentinites and Andesites from Central Macedonia (North Greece). *Geosciences* **2018**, *8*, 115. [CrossRef]
30. Petrounias, P.; Giannakopoulou, P.P.; Rogkala, A.; Stamatis, P.M.; Lampropoulou, P.; Tsikouras, B.; Hatzipanagiotou, K. The Effect of Petrographic Characteristics and Physico-Mechanical Properties of Aggregates on the Quality of Concrete. *Minerals* **2018**, *8*, 577. [CrossRef]
31. Elices, M.; Rocco, C. Effect of aggregate size on the fracture and mechanical properties of a simple concrete. *Eng. Fract. Mech.* **2008**, *75*, 3839–3851. [CrossRef]
32. Rocco, C.; Elices, M. Effect of aggregate shape on the mechanical properties of a simple concrete. *Eng. Fract. Mech.* **2009**, *76*, 286–298. [CrossRef]
33. Ajamu, S.O.; Ige, J.A. Effect of Coarse Aggregate Size on the Compressive Strength and the Flexural Strength of Concrete Beam. *J. Eng. Res. Appl.* **2015**, *5*, 67–75.
34. Meddah, M.S.; Zitouni, S.; Belâabes, S. Effect of content and particle size distribution of coarse aggregate on the compressive strength of concrete. *Constr. Build. Mater.* **2010**, *24*, 505–512. [CrossRef]
35. Sim, J.-I.; Yang, K.-H.; Kim, H.-Y.; Choi, B.-J. Size and shape effects on compressive strength of lightweight concrete. *Constr. Build. Mater.* **2013**, *38*, 854–864. [CrossRef]
36. Van Mier, J.G.M. Strain-Softening of Concrete under Multiaxial Loading Conditions. Ph.D. Thesis, Eindhoven University of Technology, Eindhoven, The Netherlands, 1986.
37. ASTM Standards. *Standard Test Method for Compressive Strength of Cylindrical Concrete Specimens*; Annual Book of ASTM Standards (ASTM C39–01); American Society for Testing and Materials: Philadelphia, PA, USA, 2001.
38. Yazıcı, Ş.; Sezer, G.I. The effect of cylindrical specimen size on the compressive strength of concrete. *Build. Environ.* **2007**, *42*, 2417–2420. [CrossRef]
39. Kampmann, R.; Roddenberry, M.; Ping, W.V. Contribution of Specimen Surface Friction to Size Effect and Rupture Behavior of Concrete. *ACI Mate. J.* **2013**, *110*, 169–176.
40. Elwell, D.J.; Fu, G. *Compression Testing of Concrete: Cylinders vs. Cubes*; New York State Department of Transportation, State Campus: Albany, NY, USA, 1995; Report FHWA/NY/SR-95/119.
41. BS 1881: Part 120 Method of Determination of the Compressive Strength of Concrete Core. *Mag. Concr. Res.* **1983**, *27*, 161–170.
42. ENV 206. *Concrete: Performance, Production, Placing, and Compliance Criteria*; European Standard: London, UK, 1990; ISBN 0-580-20943-1.
43. Cormack, H.W. *Notes on Cubes Versus Cylinders*; New Zealand Engineering: Wellington, New Zealand, 1956; Volume 11, pp. 98–99.
44. De Brito, J.; Kurda, R.; Da Silva, P.R. Can We Truly Predict the Compressive Strength of Concrete without Knowing the Properties of Aggregates? *Appl. Sci.* **2018**, *8*, 1095. [CrossRef]
45. Evans, R.H.; Abeles, P.W.; E Reynolds, C.; Konyi, K.H.; Squire, R.H. Correspondence. the Plastic Theories for the Ultimate Strength of Reinforced Concrete Beams. *J. Inst. Civ. Eng.* **1944**, *22*, 383–398. [CrossRef]
46. Sigvaldason, O.T. The influence of testing machine characteristics upon the cube and cylinder strength of concrete. *Mag. Concr. Res.* **1966**, *18*, 197–206. [CrossRef]
47. Gyengo, T. Effect of Type of Test Specimen and Gradation of Aggregate on Compressive Strength of Concrete. *J. Am. Concr. Inst.* **1938**, *34*, 269–282.
48. Gonnerman, H.F. Effect of Size and Shape of Test Specimen on Compressive Strength of Concrete. In *Proceedings of ASTM*; ASTM International: West Conshohocken, PA, USA, 1925; Volume 25, pp. 237–250.
49. Plowman, J.M.; Smith, W.F.; Sheriff, T. *Cores, Cubes, and the Specified Strength of Concrete*; The Institution of Structural Engineer: London, UK, 1974; Volume 52, pp. 421–426.
50. Raju, N.K.; Basavarajaiah, B.S. Experimental Investigations on Prismatic Control Specimens for Compressive, Flexural, and Tensile Strength of Concrete. *J. Inst. Eng. (India) Civil Eng. Div.* **1976**, *56*, 254–257.

51. Lasisi, F.; Osunade, J.A.; Olorunniwo, A. Strength Characteristics of Cube and Cylinder Specimens of Laterized Concrete. *West Indian J. Eng.* **1987**, *12*, 50–59.

52. ASTM C33–01. *Standard Specification for Concrete Aggregates*; ASTM International: West Conshohocken, PA, USA, 2002.

53. ASTM C136/C136M-14. *Standard Test Method for Sieve Analysis of Fine and Coarse Aggregates*; ASTM International: West Conshohocken, PA, USA, 2014.

54. ASTM C127–15. *Standard Test Method for Relative Density (Specific Gravity) and Absorption of Coarse Aggregate*; ASTM International: West Conshohocken, PA, USA, 2015.

55. ASTM C131/C131M-14. *Standard Test Method for Resistance to Degradation of Small-Size Coarse Aggregate by Abrasion and Impact in the Los Angeles Machine*; ASTM International: West Conshohocken, PA, USA, 2006.

56. ASTM C496/C496M-11. *Standard Test Method for Splitting Tensile Strength of Cylindrical Concrete Specimens*; ASTM International: West Conshohocken, PA, USA, 2004.

57. ASTM C78/C78M-16. *Standard Test Method for Flexural Strength of Concrete (Using Simple Beam with Third-Point Loading)*; ASTM International: West Conshohocken, PA, USA, 2016.

58. ASTM C469/C469M-14. *Standard Test Method for Static Modulus of Elasticity and Poisson's Ratio of Concrete in Compression*; ASTM International: West Conshohocken, PA, USA, 2014.

59. Yehia, S.; Helal, K.; Abusharkh, A.; Zaher, A.; Istaitiyeh, H. Strength and Durability Evaluation of Recycled Aggregate Concrete. *Int. J. Concr. Struct. Mater.* **2015**, *9*, 219–239. [CrossRef]

60. Yehia, S.; Alhamaydeh, M.; Farrag, S. High-Strength Lightweight SCC Matrix with Partial Normal-Weight Coarse-Aggregate Replacement: Strength and Durability Evaluations. *J. Mater. Civ. Eng.* **2014**, *26*, 04014086. [CrossRef]

61. Yehia, S.; Abudayyeh, O.; Bhusan, B.; Maurovich, M.; Zalt, A. Self-Consolidating Concrete Mixture with Local Materials: Proportioning and Evaluation. *Mater. Sci. Res. J.* **2009**, *3*, 41–64.

62. Beshr, H.; Almusallam, A.; Maslehuddin, M. Effect of coarse aggregate quality on the mechanical properties of high strength concrete. *Constr. Build. Mater.* **2003**, *17*, 97–103. [CrossRef]

63. Beushausen, H.; Dittmer, T. The influence of aggregate type on the strength and elastic modulus of high strength concrete. *Constr. Build. Mater.* **2015**, *74*, 132–139. [CrossRef]

64. Jones, R.; Kaplan, M.F. The effect of coarse aggregate on the mode of failure of concrete in compression and flexure. *Mag. Concr. Res.* **1957**, *9*, 89–94. [CrossRef]

65. Zimbelmann, R. A contribution to the problem of cement-aggregate bond. *Cem. Concr. Res.* **1985**, *15*, 801–808. [CrossRef]

66. Yehia, S.; Abdelfatah, A. Examining the Variability of Recycled Concrete Aggregate Properties. In Proceedings of the International Conference on Civil, Architecture and Sustainable Development, London, UK, 1–2 December 2016.

67. Feret, R. *On the Compactness of Hydraulic Mortars (in French—Sur la compacité des mortiers hydrauliques)*; C. Dunod: Paris, France, 1892.

68. Abrams, L.D. *Properties of Concrete*, 3rd ed.; Pitman Publishing Ltd.: London, UK, 1919.

69. Powers, T.; Brownyard, T. *Studies of the Physical Properties of Hardened Portland Cement Paste*; American Concrete Institute (ACI): Farmington Hills, MI, USA, 1946; Volume 18, pp. 669–712.

70. Popovics, S.; Ujhelyi, J. Contribution to the concrete strength versus water-cement ratio relationship. *J. Mater. Civ. Eng.* **2008**, *20*, 459–463. [CrossRef]

71. Behnood, A.; Behnood, V.; Gharehveran, M.M.; Alyamaç, K. Prediction of the compressive strength of normal and high-performance concretes using M5P model tree algorithm. *Constr. Build. Mater.* **2017**, *142*, 199–207. [CrossRef]

72. ACI363R-10. *Report on High-Strength Concrete. Farmington Hills*; American Concrete Institute: Farmington Hills, MI, USA, 2010.

73. ACI318–19. *Building Code Requirements for Structural Concrete. Farmington Hills*; American Concrete Institute: Farmington Hills, MI, USA, 2019.

74. Kılıç, A.; Atiş, C.; Teymen, A.; Karahan, O.; Özcan, F.; Bilim, C.; Ozdemir, M.; Kılıç, A. The influence of aggregate type on the strength and abrasion resistance of high strength concrete. *Cem. Concr. Compos.* **2008**, *30*, 290–296. [CrossRef]

75. Wu, K.-R.; Chen, B.; Yao, W.; Zhang, N. Effect of coarse aggregate type on mechanical properties of high-performance concrete. *Cem. Concr. Res.* **2001**, *31*, 1421–1425. [CrossRef]

76. Zhou, F.; Lydon, F.; Barr, B. Effect of coarse aggregate on elastic modulus and compressive strength of high performance concrete. *Cem. Concr. Res.* **1995**, *25*, 177–186. [CrossRef]
77. Özturan, T.; Çeçen, C. Effect of coarse aggregate type on mechanical properties of concretes with different strengths. *Cem. Concr. Res.* **1997**, *27*, 165–170. [CrossRef]
78. Aitcin, P.C.; Mehta, P.K. Effect of Coarse Aggregate Characteristics on Mechanical Properties of High-Strength Concrete. *ACI Mater. J.* **1998**, *95*, 252–261.
79. Mehta, P.; Ezeldin, A.; Aitcin, P.-C. Effect of Coarse Aggregate on the Behavior of Normal and High-Strength Concretes. *Cem. Concr. Aggreg.* **1991**, *13*, 121–124. [CrossRef]
80. Sengul, O.; Tasdemir, C.; Mehmet Ali Tasdemir, M.A. Influence of Aggregate Type on Mechanical Behavior of Normal and High Strength Concretes. *ACI Mater. J.* **2002**, *99*, 528–533.

Article

# Enhanced Performance of Concrete Composites Comprising Waste Metalised Polypropylene Fibres Exposed to Aggressive Environments

Rayed Alyousef [1,*], Hossein Mohammadhosseini [2,*], Fahed Alrshoudi [3], Mahmood Md. Tahir [2], Hisham Alabduljabbar [1] and Abdeliazim Mustafa Mohamed [1]

1   Department of Civil Engineering, College of Engineering, Prince Sattam bin Abdulaziz University, Alkharj 11942, Saudi Arabia; h.alabduljabbar@psau.edu.sa (H.A.); a.bilal@psau.edu.sa (A.M.M.)
2   Institute for Smart Infrastructure and Innovative Construction (ISIIC), School of Civil Engineering, Faculty of Engineering, Universiti Teknologi Malaysia (UTM), Skudai 81310, Johor, Malaysia; mahmoodtahir@utm.my
3   Department of Civil Engineering, College of Engineering, King Saud University, Riyadh 12372, Saudi Arabia; Falrshoudi@ksu.edu.sa
*   Correspondence: r.alyousef@psau.edu.sa (R.A.); mhossein@utm.my (H.M.)

Received: 28 July 2020; Accepted: 11 August 2020; Published: 12 August 2020

**Abstract:** The utilisation of waste plastic and polymeric-based materials remains a significant option for clean production, waste minimisation, preserving the depletion of natural resources and decreasing the emission of greenhouse gases, thereby contributing to a green environment. This study aims to investigate the resistance of concrete composites reinforced with waste metalised plastic (WMP) fibres to sulphate and acid attacks. The main test variables include visual inspection, mass loss, and residual strength, as well as the microstructural analysis of specimens exposed to aggressive environments. Two sets of concrete mixes with 100% ordinary Portland cement (OPC) and those with 20% palm oil fuel ash (POFA) were made and reinforced with WMP fibres at volume fractions of 0–1.25%. The results revealed that the addition of WMP fibres decreased the workability and water-cured compressive strength of concrete mixes. The outcomes of the study suggest that the rate of sulphate and acid attacks, in terms of mass losses, was controlled significantly by adding WMP fibres and POFA. The mutual effect of WMP fibre and POFA was detected in the improvement in the concrete's resistance to sulphate and acid attacks by the reduction in crack formation, spalling, and strength losses. Microstructural analysis conducted on the test specimens elucidates the potential use of POFA in improving the performance of concrete in aggressive environments.

**Keywords:** concrete composites; waste metalised polypropylene fibres; durability; sulphate and acid attacks; palm oil fuel ash

---

## 1. Introduction

The management of several types of waste has become a challenge owing to the vast quantities of waste materials produced worldwide. Cleaner productions can be achieved by minimising the generation of industrial and domestic solid wastes and the reduction in the consumption of raw resources [1]. According to Prem et al. [2], several kinds of non-biodegradable solid wastes such as plastic wastes will remain for hundreds or thousands of years in the environment. The aspiration to attain a green society is commonplace, regardless of the lack of appropriate waste management systems in both developed and developing countries. Although there is room for enhancement in existing practices, the transition to green production needs the identification of context-specific leverage points that can effect change, particularly for the informal divisions. Abdel-Gawwad et al. [3] stated that the utilisation and recycling of the waste materials generated are the key issues of waste management considerations. Moreover, as the idea of cradle-to-cradle is of growing significance nowadays for a green society, sectors such as concrete construction must plan to reuse solid waste instead of raw materials. Consequently, green construction with lesser cost will be introduced, as pointed out by Evi Aprianti [4] and Jittin et al. [5].

The manufacture of several types of plastics has grown-up enormously globally. According to Gu and Ozbakkaloglu [6], the global manufacture of plastics in various forms rose to approximately 300 million tons in 2014. Among the total plastic waste generated, approximately half of the plastics made were used only once, which began the generation and disposal of a massive amount of waste plastics, as pointed out by Aryan et al. [7]. Accordingly, inadequate controlling and mismanagement of waste materials result in harmful effects on human life as well as the environment. Nonetheless, the mainstream of plastic wastes can be reused and reprocessed thermally or chemically, although not all sorts of waste plastics, as reported by Longo et al. [8]. Metalised plastic film wastes are generated and sent to landfills worldwide. These are mostly polymer-based and covered with a thin layer of aluminium; they are used primarily in foodstuff packing. Among all plastic wastes, WMP is unsuitable for reprocessing and recycling [9]. As an appropriate method for the recycling of such a widespread amount of waste plastics does not exist, they are consequently sent to a landfill to be burnt [10]. Consequently, reliable approaches to disposal that can replace the existing methods have become essential. Besides, POFA is an industrial waste containing pozzolanic materials. POFA is attained by burning palm fruits in a palm oil mill as fuel. As specified by Khankhaje et al. [11], in Malaysia, which is the second producer of the palm oil products in the world, in 2010, approximately 4 million tons of POFA was generated and sent to landfill. The waste POFA ashes generated from mills are now categorised as a good pozzolanic material with suitable features, which have potential to be used as cementing materials in concrete to improve the performance of concrete, as pointed out by Juenger and Siddique [12].

Generally, due to the availability of raw materials, ease of production, adequate durability and strength properties, concrete is a well-known construction material globally [13]. Nevertheless, owing to the wide applications of concrete, particularly in the industrial sectors, it is frequently exposed to chemical attacks. These attacks can be natural, e.g., by chemicals such as sulphates in soils, groundwater and seawaters, or chemical attacks made by a humans, for example, drainage wastewater and industrial sewage, which destructively impact the performance of concrete structures and shorten their service life. In this regard, Brown and Badger [14] stated that the deterioration of concrete might occur by chemical reactions owing to the exchange of ions under the impact of such aggressive environments. Consequently, the results of these attacks are in the form of degradation in the hydration products as we as the formation of harmful products such as gypsum and ettringite in the matrix. It, therefore, develops the cracks, spalling, and disparities in the microstructure of concrete and, finally, reduce the strength and durability of concrete structures.

According to Santhanam et al. [15] and Bulatovic et al. [16], sulphate particles in the solid form do not attack the concrete severely; nevertheless, with the existence of water, these sulphate particles find a way to enter into the cavities in concrete components and chemically react with the hydration products of cement. Sotiriadis et al. [17] also reported that the attack of sulphate particles on calcium aluminate hydrate results in the creation of expansive gypsum and ettringite in the concrete matrix. Besides, Monteny et al. [18] described that acidic rain in the urban area is another serious harm that affects the concrete structures. Acid rain is mostly due to the burning of fossil fuels and addition of $CO_2$ and sulphuric acid to the atmosphere in the form of gases. Therefore, these pollutant gases have harmful influences on concrete components when mixed with rain.

According to Palankar et al. [19], most of the acids can either slowly or rapidly decompose the concrete components, depending on the type and concentration of acid. In the concrete components exposed to acid, $Ca(OH)_2$ is the most vulnerable constituent. However, other hydration products, such as Calcium-silicate-hydrate (C-S-H) crystals, are also affected by acid. As pointed out by Monteny et al. [18], the deterioration of concrete components exposed to sulphuric acid can be described through the chemical reactions, as shown in Equations (1)–(3). Consequently, the outcomes of the given reaction between sulphuric acid and concrete are gypsum and ettringite, which are made on the external face of concrete specimens. These harmful products are leads in the creation of cracks and spalling, and therefore, loss in strength of concrete specimens. Prepacked aggregate concrete has a unique construction process, and it has been investigated for its mechanical properties. Nevertheless, there has been a lack of research on the deformation and strength behaviour of prepacked aggregate concrete reinforced with waste carpet fibres and also containing POFA. Given the quarrel as mentioned above, the purpose of this work was to inspect the mutual effects of waste polypropylene fibre and POFA on the strength and deformation properties of new prepacked aggregate concrete, as well as the microstructural analysis. As the construction of PAC requires unique skills and experience that most workers do not have, the current paper offers beneficial information on the manufacturing of PAFRC that can help both engineers and workers. The consumption of waste PP fibre and POFA in the manufacture of PAFRC could be beneficial from both an environmental and economic point of view.

$$Ca(OH)_2 + H_2SO_4 \quad \rightarrow \quad CaSO_4.2H_2O \tag{1}$$

$$3CaO.Al_2O_3.12H_2O + 3(CaSO_4.2H_2O) + 14H_2O \quad \rightarrow \quad 3CaO.Al_2O_3.3CaSO_4.32H_2O \tag{2}$$

$$CaSiO_2.2H_2O + H_2SO_4 \quad \rightarrow \quad CaSO_4 + Si(OH)_4 + H_2O \tag{3}$$

As the development of cracks is one of the major problems of plain concrete exposed to chemical attacks, a substantial technique that declines the brittleness of concrete components is desired. For this purpose, several researchers proposed the reinforcement of concrete with short fibres to enhance the performance of concrete under aggressive environments. Afroughsabet et al. [20] and Mastali and Dalvand [21] specified that by including short fibres into the concrete mix, the brittle nature of concrete enhanced significantly. Besides, Ranjbar et al. [22] and Hossain et al. [23] pointed out that using POFA as cementing materials up to a certain level, moderately improved the acid resistance of concrete by reducing the rate of mass loss and strength loss. Therefore, based on the existing literature, the utilisation of pozzolanic materials and short fibres have been recommended to enhance the performance of concrete exposed to chemical attacks. Amongst all recommended fibres, polymeric based fibres such as polypropylene (PP) are more beneficial in contact with acid and sulphate solutions [24]. Thus, the current paper aimed to investigate the potential use of WMP fibre and POFA in the manufacture of concrete exposed to sulphate and acid attacks. In addition to the compressive strength, the residual properties of concrete exposed to acid and sulphate after one-year exposure were evaluated in terms of visual inspection, mass loss and strength loss. Besides, to assess the effects of chemical attacks on the microstructure of concrete, scanning electron microscopy (SEM) was carried out on the cement passed particles. Moreover, the consumption of waste materials such as plastic wastes in concrete would help to decrease the waste generated, and preserve the environment from pollution.

## 2. Experimental Program

### 2.1. Materials

In this study, a type I cement (TASEK CEMENT, Ipoh, Malaysia) that complies with the specifications of ASTM C 150-2007 was used. In addition, OPC was substituted by POFA (PPNJ Kahang Palm Oil Mill, Johor, Malaysia) at a substitution level of 20%. The raw palm oil fuel ash particles were collected as waste from the local mill industry. The ash particles were dried at the temperature of $100 \pm 5\ °C$ and then sieved to remove the larger particles of over 150 μm. Subsequently, the small size of ash particles was kept in a crushing machine, and the grinding process was continued for about two hours per each 4 kg of ash. Then, the grounded POFA particles were tested following the specifications of ASTM C618-15 and BS 3892: Part 1-1992 to achieve the desired properties. The ashes which passed the standard requirements with the desired chemical compositions and physical properties, as given in Table 1, were then used as cementing materials.

**Table 1.** Physical properties and chemical composition of ordinary Portland cement (OPC) and palm oil fuel ash (POFA).

| Material | Physical Properties | | Chemical Composition (%) | | | | | | | |
|---|---|---|---|---|---|---|---|---|---|---|
| | Specific Gravity | Blaine Fineness (cm$^2$/g) | SiO$_2$ | Al$_2$O$_3$ | Fe$_2$O$_3$ | CaO | MgO | K$_2$O | SO$_3$ | LOI * |
| OPC | 3.15 | 3990 | 20.3 | 5.3 | 4.21 | 62.5 | 1.53 | 0.005 | 2.13 | 2.35 |
| POFA | 2.42 | 4930 | 62.7 | 4.8 | 8.14 | 5.8 | 3.54 | 9.07 | 1.17 | 6.28 |

* Loss on ignition.

Natural river sand having a specific gravity of 2.6 g/cm$^3$, 2.3 fineness modulus, 0.7% water absorption and a maximum size of 4.75 mm was used. The coarse aggregates of crushed granite with 0.5% water absorption, a specific gravity of 2.7 g/cm$^3$ and a maximum size of 10 mm was used. The polypropylene (PP)-type metalised plastics with aluminium metallisation treatment used in food packing were collected as waste and washed to clean from any contaminations. Subsequently, as shown in Figure 1a–c, the waste films were torn into fibres of 2 mm width and 20 mm length. As the aspect ratio (length/diameter) is an important parameter in the effectiveness of the fibres in the concrete matrix, and due to the rectangular cross-section of WMP fibres, to calculate the aspect ratio, the diameter was determined following the specifications of EN 14889-1:2006 given in Equation (4). Therefore, the fibres with the aspect ratio (l/d) of 47 were used in this study. The typical engineering properties of WMP fibre used in this study are shown in Table 2.

$$d = \sqrt{(4.w.t)/\pi}\tag{4}$$

where $w$ (mm) and $t$ (mm) are the width and thickness of fibre.

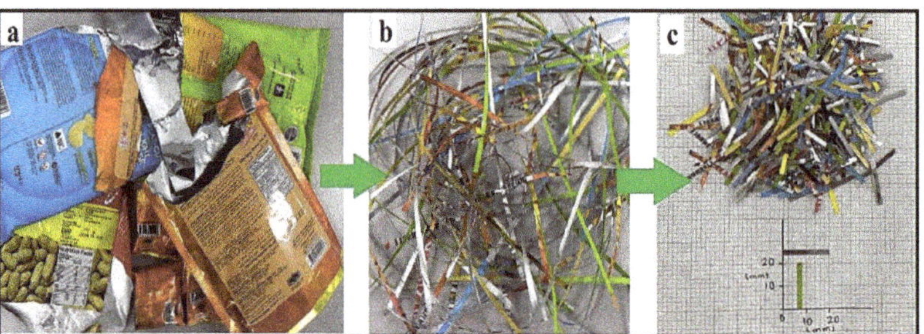


**Figure 1.** (a) Metalised plastic wastes; (b) Preparation of fibres; (c) 20 mm length waste metalised plastic (WMP) fibres.

**Table 2.** Engineering properties of WMP fibres.

| Resin Type | Plastic Type | Size (W * L) (mm) | Density Range (kg/m$^3$) | Thickness (mm) | Tensile Strength (MPa) | Elongation (%) |
|---|---|---|---|---|---|---|
| Polypropylene | LDPE | 2 * 20 | 0.915–0.945 | 0.07 | 600 | 8–10 |

## 2.2. Mix Proportions

The mix proportions of different materials used in the production of concrete composites are illustrated in Table 3. In this study, in total, twelve mixtures with different WMP fibre dosages of 0, 0.25%, 0.50%, 0.75%, 1.0% and 1.25% were prepared. Six mixes, namely B1–B6, were made with 100%

OPC, while another six mixes of B7–B12 were cast with 20% POFA as cement replacement. In addition, a water/binder ratio of 0.48 was maintained in all concrete mixes.

**Table 3.** Mix details of concrete composites.

| Mix | Cement (kg/m$^3$) | POFA (kg/m$^3$) | Water (kg/m$^3$) | Fine Aggregate (kg/m$^3$) | Coarse Aggregate (kg/m$^3$) | WMPF (%) |
|---|---|---|---|---|---|---|
| B1 | 445 | - | 215 | 830 | 860 | 0.0 |
| B2 | 445 | - | 215 | 830 | 860 | 0.25 |
| B3 | 445 | - | 215 | 830 | 860 | 0.50 |
| B4 | 445 | - | 215 | 830 | 860 | 0.75 |
| B5 | 445 | - | 215 | 830 | 860 | 1.0 |
| B6 | 445 | - | 215 | 830 | 860 | 1.25 |
| B7 | 356 | 89 | 215 | 830 | 860 | 0.0 |
| B8 | 356 | 89 | 215 | 830 | 860 | 0.25 |
| B9 | 356 | 89 | 215 | 830 | 860 | 0.50 |
| B10 | 356 | 89 | 215 | 830 | 860 | 0.75 |
| B11 | 356 | 89 | 215 | 830 | 860 | 1.0 |
| B12 | 356 | 89 | 215 | 830 | 860 | 1.25 |

### 2.3. Production of Specimens

The workability test of the fresh concrete composite was conducted according to BS EN 12350-2:2009 for slump test and BS EN 12390-3:2009 for VeBe time. Besides, cubic concrete specimens measuring 100 mm were fabricated following BS EN 12390-2:2009 and BS EN 12390-3:2009. Following the moulding procedure, the specimens were de-moulded after 24 h and were subsequently placed into a water tank for curing until they were needed for testing purposes. The average ambient temperature recorded in the laboratory was 27 ± 2 °C, where the relative humidity (RH) was 85 ± 5%. Based on the existing literature, there are few techniques to simulate and evaluate the acid and sulphate attacks for plain and fibre reinforced concrete and mortar. Consequently, the immersion method for sulphuric acid and magnesium sulphate in 5% and 10% solutions were adopted in this study, respectively. The immersion procedures were also followed by the methods proposed by Monteny et al. [18] and Sideris et al. [25], who used 10% $MgSO_4$ solution and 5% $H_2SO_4$ solution in their studies for these tests.

Similar to the compressive strength test procedure, overall 36 concrete cubes of 100 mm were made for each sulphate and acid tests. After 24 h from casting, the concrete cubes were removed from moulds and kept in water tank for 28 days. Then, the specimens were removed from the water tank and prepared for immersion. The specimens were dried, and the mass in the saturated surface dry (SSD) condition was measured as the initial mass of specimens. The concrete specimens were then submerged in $MgSO_4$ solution with 10% concentration for the sulphate resistance test, and $H_2SO_4$ solution of 5% concentration for the sulphuric acid resistance test, for 365 days. The pH values of the solutions were controlled and fixed as 2.5 and 8.5 for sulphuric acid and magnesium sulphate solutions, respectively. Subsequently, the immersed test specimens were removed after 365 days and cleaned by using a dry towel and left in a testing room for about one hour. The visual assessment was done through the critical observation on the shape, size and colour of the specimens immersed in the acid and sulphate solutions and were compared with those cured in water. To evaluate the variation in the mass of specimens, the mass of all concrete specimens were measured and noted as the residual mass after one-year immersion. The percentage of loss in the mass of samples was calculated by using Equation (5):

$$ML_t = \frac{M_t - M_i}{M_i} \times (100) \tag{5}$$

where $ML_t$ is the mass change after 365 days exposure, $M_t$ is the residual mass after the exposure period (gr) and $M_i$ is the original mass before immersion (gr).

In this study, the losses in the strength of concrete specimens after exposure to acid and sulphate solutions were evaluated in terms of strength loss factor (*SLF*) by using Equation (6) as follow:

$$SLF = \frac{F_{cw} - F_{cs}}{F_{cw}} \times (100) \qquad (6)$$

where $F_{cw}$ is the cube compressive strength of water-cured companion specimen, and $F_{cs}$ is the cube compressive strength after 365 days exposure in acid and sulphate solutions.

## 3. Results and Discussion

### 3.1. Workability

The workability of concrete composites in terms of slump test and VeBe time test was piloted. The outcomes of the workability tests are illustrated in Figure 2a,b. It was discovered that the inclusion of WMP fibres reduced the slump values such that the highest slump value recorded was 190 mm for the control mix (B1), as shown in Figure 3a,b. For the concrete mix reinforced with 0.5% fibre (B3), the slump values recorded were 80 mm. It was clear that the VeBe time increased linearly by adding WMP fibres. Furthermore, the results indicated that the sump of concrete mixes with POFA was less than that of OPC mixes. Likewise, the addition of WMP fibres demonstrated a similar effect on the slump values of POFA mixes. For instance, values of 170 and 50 mm were noted for a slump of B7 and B9 mixtures, respectively. Moreover, the VeBe times increased for the POFA-based concrete, and values of 3.4 and 7.3 s were noted for mixtures of B7 and B9, respectively.

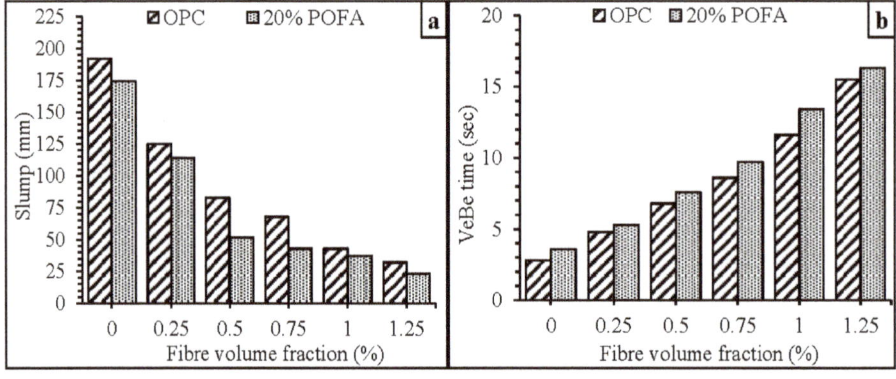


**Figure 2.** Effects of WMP fibres on (**a**) slump and (**b**) VeBe time values.

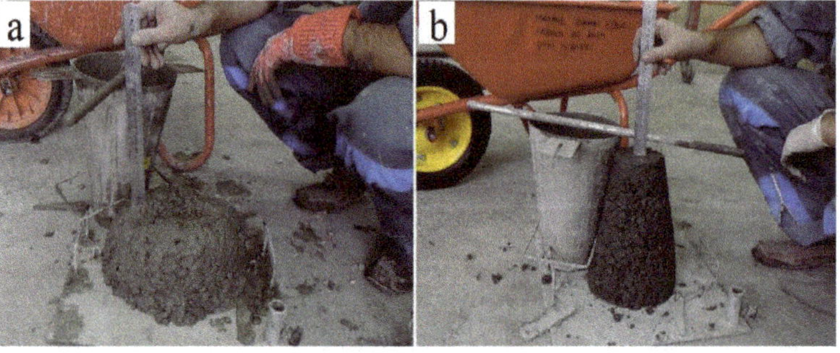


**Figure 3.** The slump test of (**a**) plain OPC mix, and (**b**) POFA-based mix reinforced with WMP fibres.

### 3.2. Compressive Strength of Water-Cured Specimens

Figure 4 reveals the values attained from the compressive strength test for concrete mixes at various curing ages. The results showed that the compressive strength dropped by including WMP fibre into the concrete. The outcomes exposed that the obtained values of strength were higher in POFA mixes than those of OPC mixes at the ages beyond 180 days. The reinforcement of OPC concrete mixes by 0.25%, 0.5%, 0.75%, 1% and 1.25% WMP fibres, resulted in the reduction of compressive strength values by about 10.4%, 14.5%, 17.7%, 20.4% and 23.9%, respectively, as compared to the plain control mix (B1) at the age of 28 days. Likewise, at 28 days, in the POFA-based samples with the same fibre content, the compressive strength dropped by 14.6%, 18.8%, 24.7%, 26.4% and 27.9%, respectively. The higher reduction in the strength of POFA mix specimens the early age of 28 days could be due to the slow hydration rate of POFA, which results in lower strength values. Besides, the existence of voids in the concrete specimens, due to the lack of hydration products in the POFA-based specimens as well as the balling effect owing to the higher dosage of fibres, results in comparatively lower strength values [26].

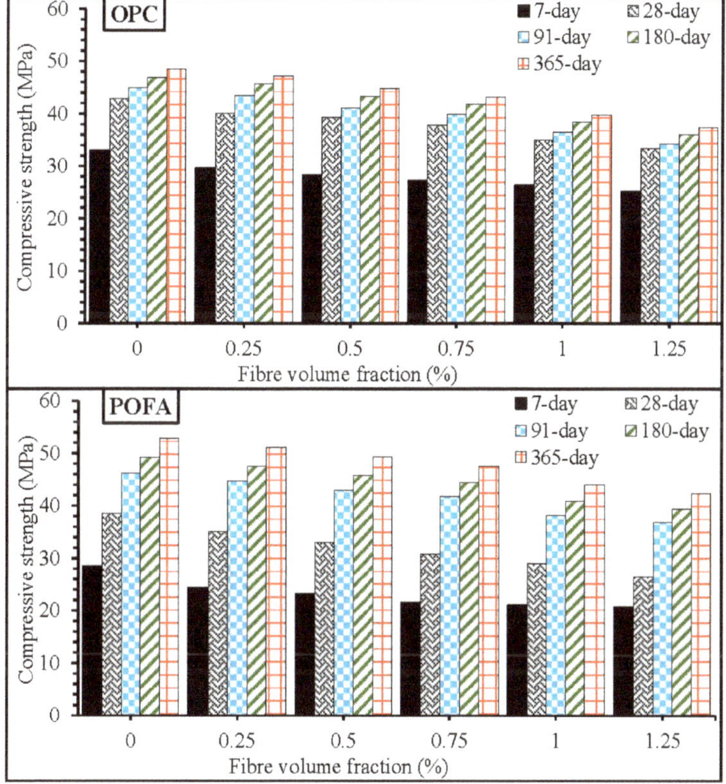


**Figure 4.** Compressive strength of concrete specimens cured in water.

However, beyond 180 days curing of specimens in water, owing to the consumption of POFA and its high pozzolanic nature, the attained strengths of concrete samples were found to be remarkably greater than those of the early age values. For example, at 365 days, the compressive strength values of 52.9, 51.2, 49.3, 47.6, 44, and 42.5 MPa were recorded for the POFA-based mixes with WMP fibre contents of 0%, 0.25%, 0.5%, 0.75%, 1% and 1.25%, respectively. The values obtained are comparatively higher than those values of 48.6, 47.2, 44.8, 43.2, 39.8, and 37.4 MPa, which were recorded for the OPC-based mixes with same fibre contents. It could be observed that the POFA mixes show higher strength values, in association with the OPC specimens for the similar fibre contents. The findings of this study are similar to those reported by Ranjbar et al. [22], who described the significant improvement in the compressive strength of concrete by using POFA as a partial cement replacement with longer curing periods.

The microstructural analysis was done using scanning electron microscopy (SEM) to investigate the variations in the hydration process of concrete with the existence of POFA. Figure 5 displays the distribution of C-S-H gel in the pastes at the curing period of 365 days for OPC and POFA mixes. The C-S-H gel was homogeneously dispersed in the paste of POFA mix as related to the OPC paste. The formation of additional C-S-H gels in the POFA mixes due to pozzolnic reactions, as shown in Equations (7) and (8), resulted in a dense microstructure, particularly, and therefore, enhanced the strength of concrete [27]. The reactive silica ($SiO_2$) present in POFA particles chemically reacted with the liberation of calcium hydroxide as a result of OPC hydration in the presence of moisture, and consequently formed an extra C-S-H gel in the matrix [28]. This reaction was relatively slow; however, the enhancement in the strength of POFA mixes was more evident at the ultimate ages. Besides, Figure 6 displays the interfacial transition zone between fibre and cement paste in the concrete matrix after one year of water curing. The SEM image reveals a robust bond among WMP fibre and blended cement paste. This strong bonding between fibres and paste provides a reliable solid microstructure in the matrix, which results in the reduction in the crack formation at the interface zone. Consequently, a lower volume of cracks were caused in the strength development and, therefore, high ductility and durability performance in aggressive environments was achieved [29].

Hydration reaction of OPC: $C_2$ or $C_3S + H_2O \quad \rightarrow \quad$ main C-S-H gel + $Ca(OH)_2$     (7)

Pozzolanic reaction of POFA: $Ca(OH)_2 + SiO_2$ (POFA) + $H_2O \quad \rightarrow \quad$ Additional C-S-H gel     (8)

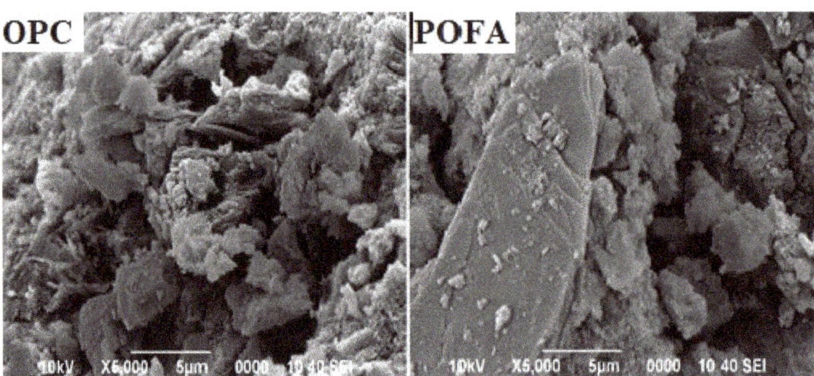


**Figure 5.** Scanning electron microscopy (SEM) images of OPC- and POFA-based pastes cured in water for 365 days.

*3.3. Sulphate Resistance*

3.3.1. Visual Inspection

The visual assessment was done on the concrete cubes exposed to 10% $MgSO_4$ solution for 365 days to assess the effects of sulphate on concrete, as illustrated in Figure 7a,b. In this regard, the changes in the shape, size and the surface colour of the cubical samples were inspected and analysed. The colour of concrete specimens can provide comprehensive information on how the sulphate influenced the concrete. Table 4 displays the overall visual inspection of the concrete cubes with and without fibres. At the end of the exposure period, a whitish precipitate was detected on the surface of all concrete specimens, which was the result of sulphate attack. However, these precipices were found to be lower on the surface of POFA-based specimens as compared to those of OPC specimens. Due to the higher amount of lime in the OPC specimens, as a result of sulphate attack, the lime particles leached and deposition, and, therefore, calcium sulphate ($CaSO_4$) was formed in higher quality, and the whitish precipitate was formed on the outer surface of specimens [30]. It was also observed that cracks were formed on the surface of plain concrete specimens.

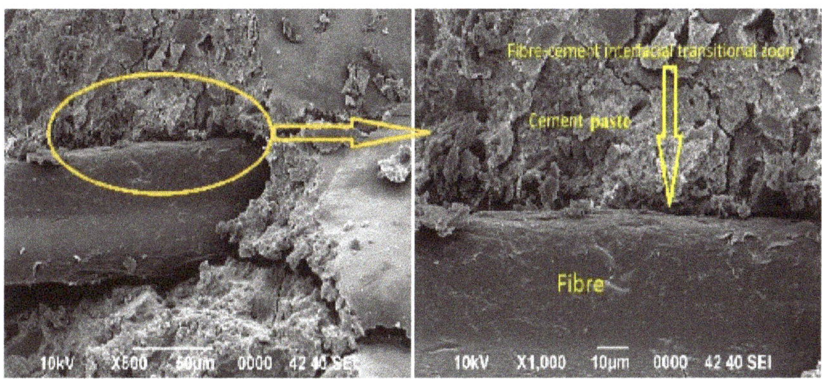


**Figure 6.** Interfacial transition zone with strong bonding between paste and fibre cured in water for 365 days.

**Figure 7.** Concrete specimens of (**a**) OPC-based and (**b**) POFA-based mixes at 365-day of exposure in $MgSO_4$ solution.

**Table 4.** The outcomes of visual inspection of typical cubical concrete samples exposed to $MgSO_4$ solution for 365 days.

| Mix | Surface Texture | Size | Colour | Edge | Shape |
|---|---|---|---|---|---|
| **OPC** | | | | | |
| B1 | Slightly deteriorated | Marginally increased | Whitish deposit | Fine cracks | Perfect cube |
| B2 | Smooth | Marginally increased | Whitish deposit | Fine cracks | Perfect cube |
| B3 | Smooth | No change | Whitish deposit | Perfect | Perfect cube |
| **POFA** | | | | | |
| B7 | Slightly deteriorated | Marginally increased | Light grey | Fine cracks | Perfect cube |
| B8 | Slightly deteriorated | Marginally increased | Light grey | Fine cracks | Perfect cube |
| B9 | Smooth | No change | Light grey | Perfect | Perfect cube |

Although the cracks detected on the POFA specimens were found to be finer than those cracks developed on the OPC specimens, in the reinforced specimens with 0.25% and 0.5% fibres very fine cracks were formed in both OPC and POFA mixes. From the observations made, it can be concluded that the inclusion of WMP fibres caused in the superior prevention of crack development on the surface of the specimens. This could be owing to the interlocking of concrete constituents with the help of fibres, which arrest the microcracks and prevent the further formation of cracks [31]. It was also noted that no physical distortion was found on the specimens reinforced with WMP fibres, and the specimens with POFA performed better due to the more extended period of test and excellent pozzolanic activity of POFA after one year.

### 3.3.2. Mass Gain

In this study, the influence of sulphate attack on the variation in the mass of concrete specimens after one year of immersion was assessed. Figure 8 displays the changes in the mass of cubical samples at one year of immersion in a sulphate solution. A similar tendency was observed for all specimens reinforced with WMP fibres, in which the masses of all specimens were gained by absorbing the sulphate particles after one year of exposure. It can be seen that the masses of plain OPC and POFA concrete mixes gained by about 1.56% and 0.9%, respectively. Nevertheless, the addition of WMP fibres up to 1% fibre content into the concrete mix considerably prevented the gain in the mass of the cube samples. For instance, the masses of OPC-based concrete specimens reinforced with 0.25%, 0.5%, 0.75% and 1% fibres gained by 0.53%, 0.6%, 1.1% and 1.43%, respectively, whereas for the POFA-based specimens the masses gained by 0.38%, 0.545, 0.88% and 1.2% for the same fibre content, respectively. The outcomes exposed that the substitution of OPC by 20% POFA resulted in a significant reduction in the mass gain of concrete specimens with and without fibres. However, a further increase in the fibre content beyond 1% increased the mass gain of the samples immersed in a sulphate solution. The lower percentage of mass gain in the specimens reinforced with WMP fibres could be due to the development of a grid structure in the matrix by fibres, which prevented the diffusion and disruption of harmful constituents into the concrete specimens. Therefore, using POFA as a partial cement replacement in concrete under a sulphate environment is beneficial to provide a dense microstructure and filled up the pores in the matrix by superior pozzolanic action [32,33].

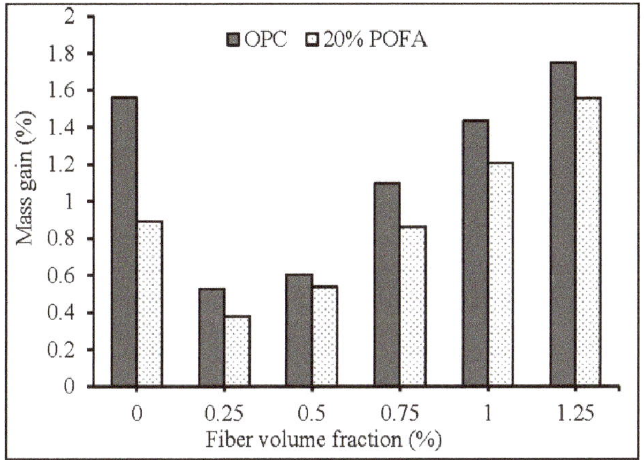


**Figure 8.** Effect of WMP fibre on the mass gain of concrete specimens exposed to MgSO$_4$ solution.

### 3.3.3. Residual Compressive Strength

In this study, to assess the effects of the sulphate attack on the strength performance of concrete, the residual compressive strength of specimens after one-year immersion in MgSO$_4$ solution was carried out. The results of the test and comparison amongst the water-cured specimens and those specimens exposed to sulphate solutions are demonstrated in Figure 9. Besides, the variation in the strength values of specimens cured in water and those exposed to sulphate attack was termed strength loss. It was noted that the compressive strength of all concrete mixes was affected by sulphate attack. When the companion samples gained their strength, and the hydration process was continued in the water at room temperature, the test samples were suffered in sulphate solution and lost their strength. The results obtained also indicated that the rate of strength losses was higher in OPC-based concrete specimens than those of POFA-based specimens for all fibre volume fraction mixes. For instance, the strength loss values of 10.6, 7.1, 5.6, 10.8, 13.8 and 16.3 MPa were recorded for OPC-based specimens reinforced with 0%, 0.25%, 0.5%, 0.75%, 1% and 1.25% fibres, respectively, while in POFA-based specimens, comparatively lower strength loss values of 8.6, 6.3, 4.1, 9.4, 11.8, and 13.5 MPa were noted for the same fibre dosages, respectively. This reduction in the strength of all mixes was due to the long-term immersion of specimens in the MgSO$_4$ solution. Moreover, the formation of cracks in the OPC-based specimen as a result of sulphate attack leads to accelerate the loss in strength of concrete. However, the combination of POFA and WMP fibres minimised the development of cracks and prevented the spalling of concrete specimens. The finding of this study is comparable to those results reported by Al-Rousan et al. [34] on the potential use of polypropylene fibres against the sulphate attack in concrete.

The strength loss factor (SLF) of concrete mixes owing to the deterioration by sulphate attack was also evaluated. The outcomes of SLF after one year of exposure to the sulphate solution are demonstrated in Figure 10. It was observed that SLF developed for all concrete mixes, which indicates the effect of sulphate attack on the strength of concrete. The results show that the SLF values are comparatively higher for the OPC-based specimens as compared to those of POFA mixes. It is interesting to note that a lower SLF value was found for the POFA mix containing 0.5% WMP fibre, at 8.2%. However, the SLF values increased with the rise in fibre content.

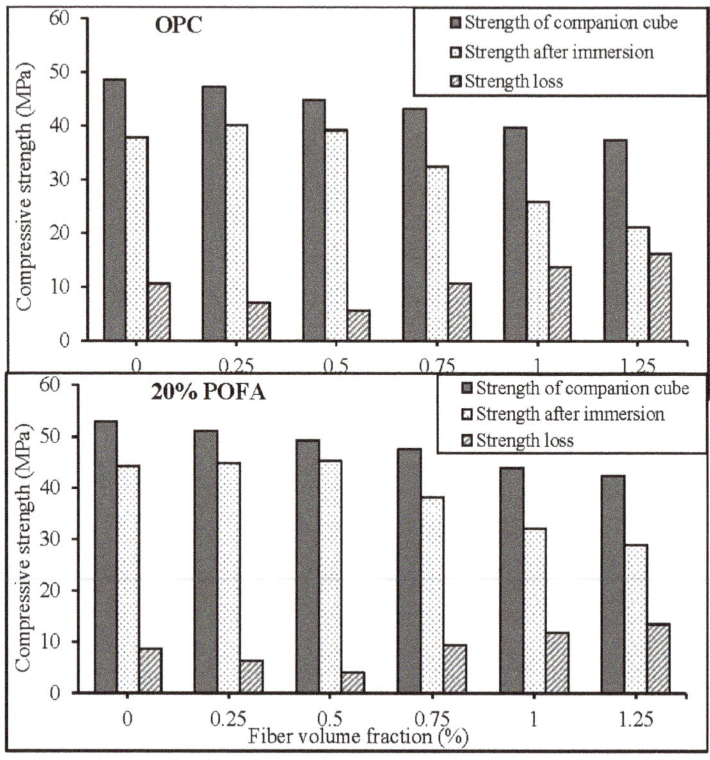

**Figure 9.** Effect of WMP fibres on the compressive strength of water-cured specimens and those exposed to $MgSO_4$ solution after 365 days.

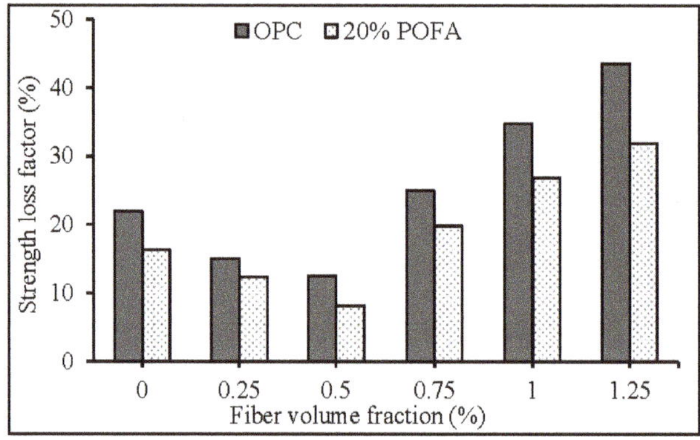

**Figure 10.** Strength loss factors of concrete mixes exposed to $MgSO_4$ solution at 365 days.

To assess the influence of $MgSO_4$ solution on the microstructure of concrete specimen, the SEM test was carried out. As revealed in Figure 11, the nature of paste specimens has changed due to the sulphate attach as compared to that SEM image of the water-cured specimen at the age of 365 days (Figure 5). The SEM images showed that the voids in the OPC paste were slowly occupied with newly precipitated flakes as a consequence of $MgSO_4$ contact. Therefore, a weak microstructure in the concrete matrix was formed, which caused the reduction of strength values as well as the low durability performance [34,35]. However, with the presence of POFA in the concrete mixture, due to the superior pozzolanic action, mostly with longer curing periods, additional hydration products such as C-S-H gel were formed. These C-S-H gels then filled up the existing cavities in the matrix, and, consequently, there was no place for the new harmful products, resultant from the sulphate attack. As the $C_3A$ content is higher in the OPC specimens, theoretically, the matrix is more prone to attacks by sulphate solution. However, with the existence of POFA, and due to the pozzolanic reactions given in Equations (6) and (7), the $Ca(OH)_2$ in OPC chemically reacted with active $SiO_2$ present in POFA and formed additional C-S-H gels, which prevent $Ca(OH)_2$ reacting with $MgSO_4$ solution to create gypsum and ettringite. Therefore, with a lower amount of gypsum and ettringite, the performance of concrete enhanced and improved the resistance against sulphate attacks [36].

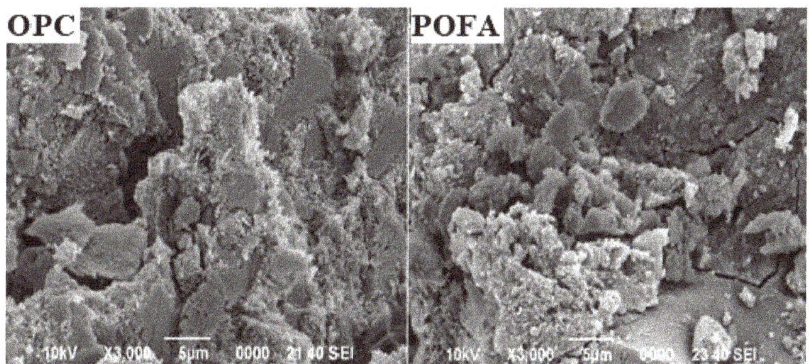


**Figure 11.** SEM of OPC and POFA paste exposed to sulphate solution for 365 days.

*3.4. Acid Resistance*

3.4.1. Visual Inspection

In this study, to assess the effects of $H_2SO_4$ solution on the concrete specimens, a visual inspection was performed. Generally, a visual observation which includes changes in colour, spalling, crack formation on the surface of specimens and deposit of new precipices on the surface is the first step to evaluate the effects of the acid solution on the concrete. Figure 12a,b illustrates the concrete specimens exposed to $H_2SO_4$ solution for one year. As shown in Figure 13, the colour of concrete offers comprehensive information on the influence of acid on concrete matrix; whether the colour implies the initially contacted surface or one resultant acid attack. It was detected that the shape and surface texture of plain specimens without fibres deteriorated as a result of the acid attack. However, slight changes in the shape and colour of specimens reinforced with WMP fibres were detected. The observation made indicates that the size and shape of the POFA-based specimens with WMP fibres remained same as the original due to the high resistance of POFA particles against acid attack, which prevented spalling and washed out the specimens, while the size of the OPC-based specimens were reduced by spalling and washing out as a result of acid attack. The visual assessment also revealed that with the addition and increase in the fibre dosage, the resistance against acid increased significantly. This superior resistance against acid attack was more evident in the POFA-based specimens. In general, lower spalling and changes in the shape were observed in the concrete mixes containing WMP fibres compared with those of plain mixes. This might be owing to the linking action of WMP fibres, which prevented the washing out and spalling of concrete attacked by acid, and thus, enhanced the performance of concrete in aggressive environments [37,38].

**Figure 12.** Concrete specimens of (**a**) OPC-based and (**b**) POFA-based mixes exposed to $H_2SO_4$ solution for 365 days.

**Figure 13.** Cross-section of concrete specimen exposed to $H_2SO_4$ solution.

### 3.4.2. Mass Loss

The changes in the mass of concrete cubes submerged in an acid solution for one year are shown in Figure 14. A similar trend of mass loss was observed for all mixes after one year of exposure to $H_2SO_4$ solution. The results show that the rate of mass losses in OPC-based specimens was comparatively higher than those of POFA-based specimens. It indicates that the OPC specimens washed out at a higher rate when immersed in an acid solution. This higher rate of leaching in the OPC specimens might be due to the existence of CaO at a higher volume in the OPC than POFA, which quickly washed out when coming into contact with acid [39,40]. However, by reinforcing OPC specimens with WMP fibres, the rate of mass loss decreased significantly. For instance, by adding fibres of 0.25%, 0.5%, 0.75%, 1% and 1.25%, the mass losses of 14.1%, 10.5%, 8.4%, 7.1% and 5.9% were recorded, respectively, which are comparatively lower than that of 19.6% mass loss noted for plain OPC mix without any fibre. Moreover, the rate of mass losses in the POFA mixes containing WMP fibres was considerably lower than those of OPC reinforced mixes. The mass losses of 11.3%, 8.7%, 6.4%, 4.2% and 3.7% were recorded for the fibre dosages of 0.25%, 0.5%, 0.75%, 1% and 1.25%, respectively, which are all lower than that of 16.8% recorded for plain POFA-based mix. With the addition of WMP fibres, through the bridging action of fibres the aggregates and cement past connected with a strong bond, and therefore, degradation and spalling of concrete specimens significantly reduced.

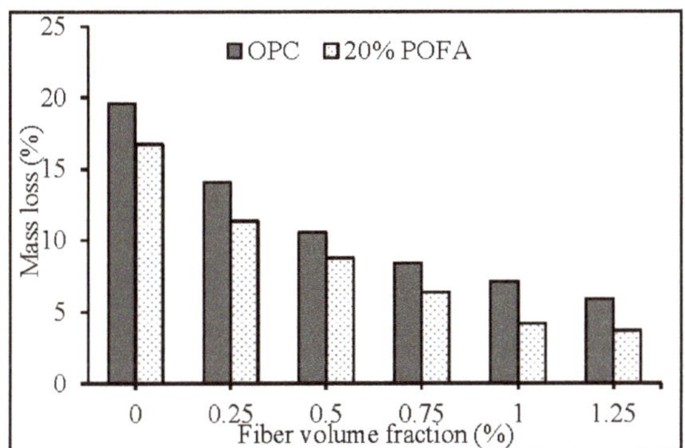


**Figure 14.** Effect of WMP fibres on the mass loss of concrete specimens exposed to the acid solution for 365 days.

### 3.4.3. Residual Strength

The outcomes of the residual compressive strength after 365 days of immersion in $H_2SO_4$ solution are shown in Figure 15 and compared with those mixes cured in water. Due to the long-term exposure to the acid solution, strength loss was revealed in all mixes. The results revealed that the losses in compressive strength loss, which is the difference between the strength of the water-cured specimens and those exposed to the acid solution, decreased with addition and a further increase in the WMP fibres. Moreover, the strength losses were observed to be higher for the OPC-based specimens than those of the POFA mixes. For the OPC mixes, the addition of WMP fibre at the dosages of 0.25%, 0.5%, 0.75%, 1% and 1.25% caused the strength losses of 28.9, 26.1, 24.5, 22.8 and 21.9 MPa, respectively. It can be seen that all of the recorded values were lower than that of 33 MPa obtained for plain concrete mix, which indicates the positive effect of fibres in controlling the strength loss. Besides, comparatively lower strength losses of 26.7, 24, 22.7, 21.5 and 20.9 MPa were noted for POFA mixes with the same fibre dosages, which are all lower than that of 29.8 MPa recorded for plain POFA mix. As revealed in Figure 15, those mixes with both WMP fibres and POFA obtained lower strength losses. This improvement was predictable due to the existence of POFA as a partial cement replacement, which contains a lower CaO content compared to OPC. The lower amount of CaO, which quickly dissolve in acid, results in the reduction of washing out and spalling of concrete specimens [41–43]. Besides, the existence of WMP fibres prevented the development of cracks by bridging action and reduced the spalling of concrete samples. Therefore, the mutual effects of POFA and WMP fibre lead to enhancements in the strength and durability performance of concrete mixtures exposed to acidic environments.

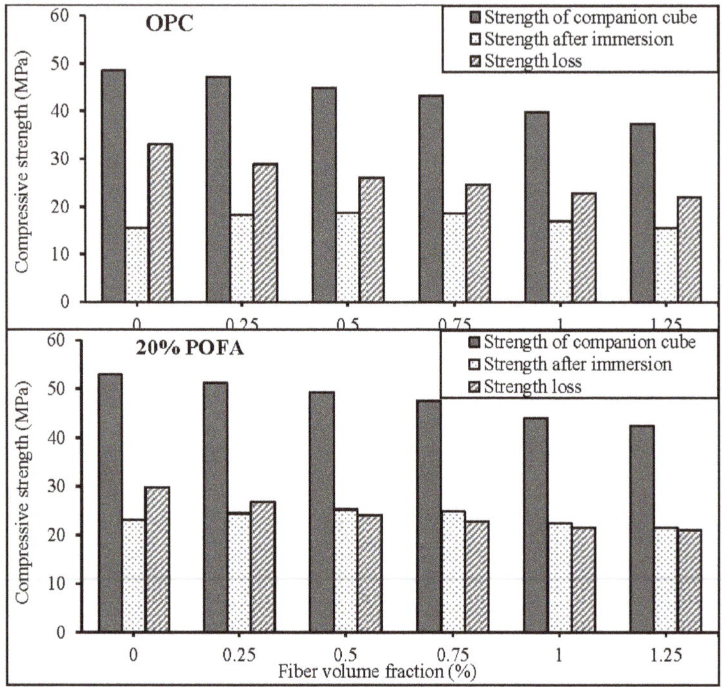



**Figure 15.** Effect of WMP fibres on the compressive strength of water-cured specimens and those exposed to $H_2SO_4$ solution after 365 days.

The SLF values of concrete mixes after one year of contact with acid were also evaluated, and the outcomes are shown in Figure 16. It can be perceived that the SLF values of POFA mixes were relatively lower than those values recorded for OPC mixes. As mentioned earlier, the better performance of POFA-based specimens in the acid environment could be due to the superior pozzolanic nature of POFA, which produced additional hydration products of C-S-H gel by consumption of $Ca(OH)_2$. These additional gels then filled up the porosities in the matrix and prevented the entrance of harmful particles into the matrix and, therefore, enhanced the performance of concrete [23].

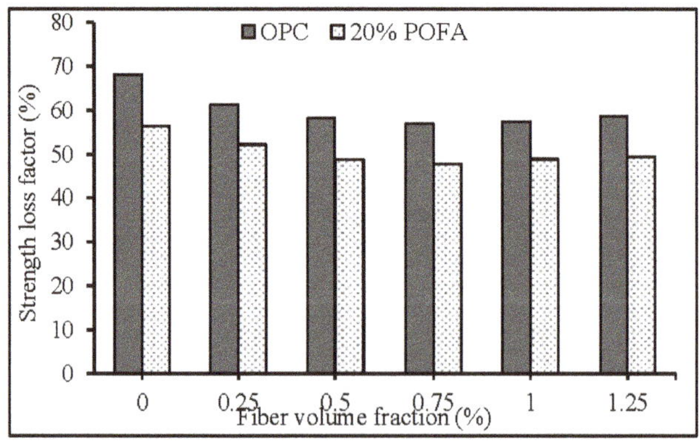



**Figure 16.** The strength loss factor of concrete mixes after 365-day immersion in the acid solution.

The microstructure of concrete mixes exposed $H_2SO_4$ solution for the one-year period are demonstrated in Figure 17. As revealed in the SEM images, the gypsum ($Ca_2SO_4.2H_2O$) was formed in large quantities during the exposure period. However, the gypsum particles were found to be lower in the POFA mix compared to that of OPC mix. Generally, soft and soluble gypsum is one of the main products of the chemical reactions between cement particles and sulphuric acid [42,43]. Owing to the lesser quantity of $Ca(OH)_2$ in POFA mixes which chemically reacts in the pozzolanic reaction, the formation of gypsum reduced. Consequently, the crack development and spalling reduced and results in the more durable and denser microstructure of POFA specimens against acid attack.

**Figure 17.** SEM images of OPC and POFA paste exposed to the acid solution for 365 days.

## 4. Conclusions

Considering the growth in the green production by using solid wastes in the production of concrete, which is vulnerable in the aggressive environment, the current study critically investigated the performance of concrete containing POFA and WMP fibres exposed to sulphuric acid and magnesium sulphate solutions. The following are the conclusions made based on a visual inspection and experimental results:

(1) The inclusion of WMP fibre reduced the workability of all concrete mixes, with lower slump values and higher VeBe times compared with those of the plain mix.

(2) The cube compressive strength of water-cured specimens was reduced by adding WMP fibres. However, the compressive strength significantly developed with the presence of POFA after 365 days.

(3) In sulphate solution, enhanced performance in the resistance of concrete mixes to sulphate attacks was attained by utilising WMP fibres and POFA. The main consequence of fibres was attributed to the formation of the grid structure to avoid the entering of destructive particles into the matrix and crack formation. Besides, the lower rate of mass gain and strength losses against sulphate attacks were detected for reinforced mixes.

(4) A satisfactory level of resistance to the sulphuric acid attacks was observed for the fibre reinforced specimens. The primary effect of POFA was attributed to the dilution effect of pastes owing to the replacement of OPC by POFA, which decreased the existing amount of $Ca(OH)_2$ content for reaction with acid and formation of gypsum. Besides, lower mass and strength losses were noted for the mixes containing POFA and WMP fibres.

(5) The microstructural analysis of paste specimens revealed changes in the nature of hydration products as a result of acid and sulphate attacks. However, the POFA-based specimens performed better against chemical attacks compared with those of OPC mixes owing to the development of supplementary C-S-H gels as a result of the pozzolanic reaction, which filled up the voids in the

matrix and provided a denser microstructure. It, therefore, enhanced the performance of concrete in terms of strength and durability under chemical attacks.

(6) The utilisation of plastic waste fibres and POFA has potential in retaining the integrity of the concrete matrix under chemical attacks. Additionally, conservation of waste materials and lower environmental contamination would also be achieved.

**Author Contributions:** Conceptualisation: M.M.T., H.M.; Methodology: F.A.; R.A. and H.A.; Validation: A.M.M. and H.A.; Formal analysis: H.M. and F.A.; Investigation: H.M. and R.A.; Resources: F.A. and R.A.; Data curation: H.A. and A.M.M.; Writing, original draft preparation: H.M.; Writing, review, and editing: H.M., F.A., and R.A.; Visualisation: A.M.M.; Supervision: M.M.T.; Project administration: H.A. All authors have read and agreed to the published version of the manuscript.

**Funding:** The project was supported by the Deanship of Scientific Research at Prince Sattam bin Abdulaziz University (Saudi Arabia) under the research project No. 2020/01/16810.

**Acknowledgments:** The authors gratefully acknowledge the technical and financial support received from the Universiti Teknologi Malaysia (UTM).

**Conflicts of Interest:** The authors declare no conflict of interest.

## References

1. Mohammadhosseini, H.; Tahir, M.M.; Alyousef, R.; Alabduljabbar, H.; Samadi, M. Effect of elevated temperatures on properties of sustainable concrete composites incorporating waste metalized plastic fibres. *SN Appl.* **2019**, *1*, 1520. [CrossRef]

2. Prem, P.R.; Verma, M.; Ambily, P.S. Sustainable cleaner production of concrete with high volume copper slag. *J. Clean. Prod.* **2018**, *193*, 43–58. [CrossRef]

3. Abdel-Gawwad, H.A.; Rashad, A.M.; Heikal, M. Sustainable utilisation of pretreated concrete waste in the production of one-part alkali-activated cement. *J. Clean. Prod.* **2019**, *232*, 318–328. [CrossRef]

4. Evi Aprianti, S. A huge number of artificial waste material can be supplementary cementitious material (SCM) for concrete production e a review part II. *J. Clean. Prod.* **2017**, *142*, 4178–4194. [CrossRef]

5. Jittin, V.; Bahurudeen, A.; Ajinkya, S.D. Utilisation of rice husk ash for cleaner production of different construction products. *J. Clean. Prod.* **2020**, *263*, 121578. [CrossRef]

6. Gu, L.; Ozbakkaloglu, T. Use of recycled plastics in concrete: A critical review. *Waste Manag.* **2016**, *51*, 19–42. [CrossRef]

7. Aryan, Y.; Yadav, P.; Samadder, S.R. Life Cycle Assessment of the existing and proposed plastic waste management options in India: A case study. *J. Clean. Prod.* **2019**, *211*, 1268–1283. [CrossRef]

8. Longo, F.; Cascardi, A.; Lassandro, P.; Sannino, A.; Aiello, M.A. Mechanical and thermal characterisation of FRCM-Matrices. *Key Eng. Mater.* **2019**, *817*, 189–194. [CrossRef]

9. Bhogayata, A.C.; Arora, N.K. Fresh and strength properties of concrete reinforced with metalized plastic waste fibres. *Constr. Build. Mater.* **2017**, *146*, 455–463. [CrossRef]

10. Bhogayata, A.C.; Arora, N.K. Impact strength, permeability and chemical resistance of concrete reinforced with metalized plastic waste fibres. *Constr. Build. Mater.* **2018**, *161*, 254–266. [CrossRef]

11. Khankhaje, E.; Hussin, M.W.; Mirza, J.; Rafieizonooz, M.; Salim, M.R.; Siong, H.C.; Warid, M.N.M. On blended cement and geopolymer concretes containing palm oil fuel ash. *Mater. Des.* **2016**, *89*, 385–398. [CrossRef]

12. Lim, N.H.A.S.; Mohammadhosseini, H.; Tahir, M.M.; Samadi, M.; Sam, A.R.M. Microstructure and strength properties of mortar containing waste ceramic nanoparticles. *Arab. J. Sci. Eng.* **2018**, *43*, 5305–5313. [CrossRef]

13. Mastali, M.; Dalvand, A.; Sattarifard, A. The impact resistance and mechanical properties of the reinforced self-compacting concrete incorporating recycled CFRP fibre with different lengths and dosages. *Compos. Part B Eng.* **2017**, *112*, 74–92. [CrossRef]

14. Brown, P.W.; Badger, S. The distributions of bound sulfates and chlorides in concrete subjected to mixed NaCl, MgSO4, Na2SO4 attack. *Cem. Concr. Res.* **2000**, *30*, 1535–1542. [CrossRef]

15. Santhanam, M.; Cohen, M.; Olek, J. Differentiating seawater and groundwater sulfate attack in Portland cement mortars. *Cem. Concr. Res.* **2006**, *36*, 2132–2137. [CrossRef]

16. Bulatović, V.; Melešev, M.; Radeka, M.; Radonjanin, V.; Lukić, I. Evaluation of sulfate resistance of concrete with recycled and natural aggregates. *Constr. Build. Mater.* **2017**, *152*, 614–631. [CrossRef]

17. Sotiriadis, K.; Nikolopoulou, E.; Tsivilis, S. Sulfate resistance of limestone cement concrete exposed to combined chloride and sulfate environment at low temperature. *Cem. Concr. Compos.* **2012**, *34*, 903–910. [CrossRef]

18. Monteny, J.; Vincke, E.; Beeldens, A.; De Belie, N.; Taerwe, L.; Van Gemert, D.; Verstraete, W. Chemical, microbiological, and in situ test methods for biogenic sulfuric acid corrosion of concrete. *Cem. Concr. Res.* **2000**, *30*, 623–634. [CrossRef]

19. Palankar, N.; Shankar, A.U.R.; Mithun, B.M. Durability studies on eco-friendly concrete mixes incorporating steel slag as coarse aggregates. *J. Clean. Prod.* **2016**, *129*, 437–448. [CrossRef]

20. Afroughsabet, V.; Biolzi, L.; Ozbakkaloglu, T. High-performance fibre-reinforced concrete: A review. *J. Mater. Sci.* **2016**, *51*, 6517–6551. [CrossRef]

21. Mastali, M.; Dalvand, A. The impact resistance and mechanical properties of self-compacting concrete reinforced with recycled CFRP pieces. *Compos. Part B Eng.* **2016**, *92*, 360–376. [CrossRef]

22. Ranjbar, N.; Behnia, A.; Alsubari, B.; Birgani, P.M.; Jumaat, M.Z. Durability and mechanical properties of self-compacting concrete incorporating palm oil fuel ash. *J. Clean. Prod.* **2016**, *112*, 723–730. [CrossRef]

23. Hossain, M.M.; Karim, M.R.; Hasan, M.; Hossain, M.K.; Zain, M.F.M. Durability of mortar and concrete made up of pozzolans as a partial replacement of cement: A review. *Constr. Build. Mater.* **2016**, *116*, 128–140. [CrossRef]

24. Zollo, R.F. Fibre-reinforced concrete: An overview after 30 years of development. *Cem. Concr. Compos.* **1997**, *19*, 107–122. [CrossRef]

25. Sideris, K.K.; Savva, A.E.; Papayianni, J. Sulfate resistance and carbonation of plain and blended cements. *Cem. Concr. Compos.* **2006**, *28*, 47–56. [CrossRef]

26. Rahman, M.R.; Hamdan, S.; Jayamani, E.; Kakar, A.; Bakri, M.K.B.; Yusof, F.A.B.M. The effect of palm oil fuel ash (POFA) and polyvinyl alcohol (PVA) on the physico-mechanical, thermal and morphological properties of hybrid bio-composites. *Polym. Bull.* **2019**, *77*, 3523–3535. [CrossRef]

27. Salami, B.A.; Johari, M.A.M.; Ahmad, Z.A.; Maslehuddin, M. Durability performance of palm oil fuel ash-based engineered alkaline-activated cementitious composite (POFA-EACC) mortar in sulfate environment. *Constr. Build. Mater.* **2017**, *131*, 229–244. [CrossRef]

28. Awal, A.S.M.A.; Mohammadhosseini, H. Green concrete production incorporating waste carpet fibre and palm oil fuel ash. *J. Clean. Prod.* **2016**, *137*, 157–166. [CrossRef]

29. Medina, N.; Barluenga, G.; Hernández-Olivares, F. Enhancement of durability of concrete composites containing natural pozzolans blended cement through the use of Polypropylene fibres. *Compos. Part B Eng.* **2014**, *61*, 214–221. [CrossRef]

30. Zuquan, J.; Wei, S.; Yunsheng, Z.; Jinyang, J.; Jianzhong, L. Interaction between sulfate and chloride solution attack of concretes with and without fly ash. *Cem. Concr. Res.* **2007**, *37*, 1223–1232. [CrossRef]

31. Söylev, T.A.; Özturan, T. Durability, physical and mechanical properties of fibre-reinforced concretes at low-volume fraction. *Constr. Build. Mater.* **2014**, *73*, 67–75. [CrossRef]

32. Mohammadhosseini, H.; Tahir, M.M.; Alaskar, A.; Alabduljabbar, H.; Alyousef, R. Enhancement of strength and transport properties of a novel preplaced aggregate fiber reinforced concrete by adding waste polypropylene carpet fibers. *J. Build. Eng.* **2020**, *27*, 101003. [CrossRef]

33. Alyousef, R.; Alabduljabbar, H.; Mohammadhosseini, H.; Mohamed, A.M.; Siddika, A.; Alrshoudi, F.; Alaskar, A. Utilisation of sheep wool as potential fibrous materials in the production of concrete composites. *J. Build. Eng.* **2020**, *30*, 101216. [CrossRef]

34. Al-Rousan, R.; Haddad, R.; Al-Sa'di, K. Effect of sulfates on bond behavior between carbon fibre reinforced polymer sheets and concrete. *Mater. Des.* **2013**, *43*, 237–248. [CrossRef]

35. Mohammadhosseini, H.; Alyousef, R.; Lim, N.H.A.S.; Tahir, M.M.; Alabduljabbar, H.; Mohamed, A.M. Creep and drying shrinkage performance of concrete composite comprising waste polypropylene carpet fibres and palm oil fuel ash. *J. Build. Eng.* **2020**, *30*, 101250. [CrossRef]

36. Mohammadhosseini, H.; Alyousef, R.; Lim, N.H.A.S.; Tahir, M.M.; Alabduljabbar, H.; Mohamed, A.M.; Samadi, M. Waste metalized film food packaging as low cost and ecofriendly fibrous materials in the production of sustainable and green concrete composites. *J. Clean. Prod.* **2020**, *258*, 120726. [CrossRef]

37. Miao, C.; Mu, R.; Tian, Q.; Sun, W. Effect of sulfate solution on the frost resistance of concrete with and without steel fibre reinforcement. *Cem. Concr. Res.* **2002**, *32*, 31–34. [CrossRef]

38. Mohammadhosseini, H.; Lim, N.H.A.S.; Tahir, M.M.; Alyousef, R.; Samadi, M.; Alabduljabbar, H.; Mohamed, A.M. Effects of waste ceramic as cement and fine aggregate on durability performance of sustainable mortar. *Arab. J. Sci. Eng.* **2020**, *45*, 3623–3634. [CrossRef]

39. Koushkbaghi, M.; Kazemi, M.J.; Mosavi, H.; Mohseni, E. Acid resistance and durability properties of steel fibre-reinforced concrete incorporating rice husk ash and recycled aggregate. *Constr. Build. Mater.* **2019**, *202*, 266–275. [CrossRef]

40. Alrshoudi, F.; Mohammadhosseini, H.; Tahir, M.M.; Alyousef, R.; Alghamdi, H.; Alharbi, Y.; Alsaif, A. Drying shrinkage and creep properties of prepacked aggregate concrete reinforced with waste polypropylene fibers. *J. Build. Eng.* **2020**, *32*, 101522. [CrossRef]

41. Hinchcliffe, S.A.; Hess, K.M.; Srubar III, W.V. Experimental and theoretical investigation of prestressed natural fibre-reinforced polylactic acid (PLA) composite materials. *Compos. Part B Eng.* **2016**, *95*, 346–354. [CrossRef]

42. Zhang, Y.; Mi, C. Strengthening bonding strength in NiTi SMA fibre-reinforced polymer composites through acid immersion and Nanosilica coating. *Compos. Struct.* **2020**, *239*, 112001. [CrossRef]

43. Alrshoudi, F.; Mohammadhosseini, H.; Alyousef, R.; Alghamdi, H.; Alharbi, Y.R.; Alsaif, A. Sustainable Use of Waste Polypropylene Fibers and Palm Oil Fuel Ash in the Production of Novel Prepacked Aggregate Fiber-Reinforced Concrete. *Sustainability* **2020**, *12*, 4871. [CrossRef]

*Article*

# A Comparative Study on Blast-Resistant Performance of Steel and PVA Fiber-Reinforced Concrete: Experimental and Numerical Analyses

Le Chen [1], Weiwei Sun [1,*], Bingcheng Chen [1], Sen Xu [1], Jianguo Liang [2], Chufan Ding [3] and Jun Feng [3,*]

[1] Department of Civil Engineering, Nanjing University of Science and Technology, Nanjing 210094, China; lechen.njust@outlook.com (L.C.); chenbingcheng912@163.com (B.C.); xs9427@njust.edu.cn (S.X.)

[2] College of Mechanical and Vehicle Engineering, Taiyuan University of Technology, Taiyuan 030024, China; liangjianguo20@hotmail.com

[3] National Key Laboratory of Transient Physics, Nanjing University of Science and Technology, Nanjing 210094, China; ding15195772018@163.com

* Correspondence: sww717@163.com (W.S.); jun.feng@njust.edu.cn (J.F.)

Received: 15 July 2020; Accepted: 14 August 2020; Published: 16 August 2020

**Abstract:** This paper deals with the blast-resistant performance of steel fiber-reinforced concrete (SFRC) and polyvinyl alcohol (PVA) fiber-reinforced concrete (PVA-FRC) panels with a contact detonation test both experimentally and numerically. With 2% fiber volumetric content, SFRC and PVA-FRC specimens were prepared and comparatively tested in comparison with plain concrete (PC). SFRC was found to exhibit better blast-resistant performance than PVA-FRC. The dynamic mechanical responses of FRC panels were numerically studied with Lattice Discrete Particle Model-Fiber (LDPM-F) which was recently developed to simulate the meso-structure of quasi-brittle materials. The effect of dispersed fibers was also introduced in this discrete model as a natural extension. Calibration of LDPM-F model parameters was achieved by fitting the compression and bending responses. A numerical model of FRC contact detonation was then validated against the blast test results in terms of damage modes and crater dimensions. Finally, FRC panels with different fiber volumetric fractions (e.g., 0.5%, 1.0% and 1.5%) under blast loadings were further investigated with the validated LDPM-F blast model. The numerical predictions shed some light on the fiber content effect on the FRC blast resistance performance.

**Keywords:** fiber-reinforced concrete; blast resistance; lattice discrete particle model-fiber; damage mode; contact detonation

---

## 1. Introduction

Recent terrorist attacks in Boston (2013), Moscow (2011), London (2005), Madrid (2004), New York (2001) and explosion accidents in Beirut (2020), Tianjin (2015) and San Juan Nico (1984) indicate great vulnerability of concrete material civil infrastructures to possible explosive loadings [1–3]. To protect civilian lives from these threats and casualties, civil infrastructures should provide good resistant performance against extreme loadings such as impact and blast. As the most widely used construction material, plain concrete exhibits high compressive strength but relatively low tensile strength, resulting in poor energy absorption capacity under extreme dynamic loadings. Incorporating randomly distributed fibers into the cementitious matrix is one effective way to overcome this defect [4–6]. The dispersed fibers in the matrix resist the crack initiation and delay its propagation, which attribute to the composite material ductility. Hence, fiber-reinforced concrete (FRC) shows novel tension and bending performance, leading to enhanced energy absorption capacity under blast loadings [7–9].

Many different types of fibers have been chosen as reinforcements added into the cementitious matrix [10]. Some are high strength fibers, including steel, carbon and glass fibers, which can effectively improve the strength of concrete. While others are low strength fibers like polyvinyl alcohol (PVA) and polypropylene fibers which are more effective in ductility improvement and cracking-resistance [11–13]. In recent literatures, it could be found that steel fibers and PVA fibers are the most widely investigated in the field of construction materials, mainly due to their good mechanical properties, perfect bond with cementitious matrix and relative low cost. Steel fiber-reinforced concrete (SFRC), consisting of cementitious matrix and steel fibers with high tensile strength in the range of 850–2000 MPa, usually exhibits greatly boosted mechanical performance (i.e., compressive strength, tensile strength, toughness, ductility) than normal plain concrete [14–16]. As a special type of SFRC, ultra-high performance fiber-reinforced concrete (UHPFRC) presents great compressive strength over 160 MPa [17]. Owing to its superior qualities, SFRC provides a perfect solution for construction material for infrastructure or protective shelter, where high strength and durability properties are required. On the other hand, PVA fiber-reinforced concrete (PVA-FRC), made of cementitious matrix and oiled PVA fibers [18], shows unique tensile strain-hardening mechanical behavior, i.e., high energy absorption capacity. Therefore, PVA-FRC is also believed to resist impact and blast loadings [19]. This work chooses to comparatively investigate the blast resistance of SFRC and PVA-FRC panels.

Previously, most fiber reinforcement researches have dealt with the concrete mechanical properties influenced by the fiber incorporation [20–23]. Naaman and Najm et al. have conducted extensive works on the mechanisms of physiochemical bonding and mechanical anchorage in SFRC, while Lee et al. focused on the pullout behavior of steel fiber in SFRC [24–27]. For PVA-FRC, many studies were been carried out by using cost-effective materials or improving the material strain-hardening behavior [18,28] to keep the balance between mechanical properties and cost. Since blast tests are highly costive and even dangerous, only few works handled the blast resistance performance of FRC by analyzing the effects of fiber content and fiber type [29–31]. It is widely acknowledged that numerical simulation method could greatly increase research efficiency and reduce research cost. For extreme loading tests, numerical simulation could also avoid the danger from the blast experiments. Consequently, numerical modeling has become the main method studying the concrete behavior under extreme dynamic loadings [32,33].

The recently developed Lattice Discrete Particle Model-Fiber (LDPM-F) was introduced herein to investigate blast-resistant performance of FRC with different types of fiber reinforcements. LDPM was designed to simulate the meso-structure of quasi-brittle materials through a three dimensional (3D) assemblage of sphere particles[34]. As a natural extension for this discrete model to add the effect of dispersed fibers to the meso-structure, LDPM-F incorporated this effect by modeling individual fiber, randomly placed within the matrix according to a given fiber volume fraction [35]. LDPM was good at simulating various features of the mesoscale material response, including strain softening in tension, strain hardening in compression, cohesive fracturing and material compaction caused by pore collapse. Material behaviors under extreme dynamic loadings could also be well predicted by LDPM, e.g., the deep penetration in concrete, thick panel projectile perforation and Hopkinson bar tests [36].

In this work, SFRC, PVA-FRC and control plain concrete (PC) specimens were prepared and tested with compression and bending. Sequent contact detonation tests were conducted to characterize the blast resistance of SFRC and PVA-FRC. The blast simulation model with LDPM-F was established with parameters calibrated by fitting the compression and bending responses. The model validation was achieved by predicting the blast damage on the front and rear surface. Finally, behavior of FRC panels with different fiber volume ratio (e.g., 0.5%, 1.0% and 1.5%) under blast loadings were numerically investigated.

## 2. Experimental Program

In this section, SFRC, PVA-FRC and control PC specimens were prepared and tested with compression, bending and blast loading. For material characterization purpose, compression specimens with a dimension of 40 mm × 40 mm × 40 mm and bending specimens with a dimension of 40 mm × 40 mm × 160 mm of both FRC and PC were prepared and tested. In sequence, anti-blast specimens of FRC and PC panels were also prepared with a dimension of 600 mm × 600 mm × 100 mm, then contact detonation tests were carried out to identify their blast-resistant performance.

## 2.1. FRC Preparation

According to previous work [31], the mix proportion used in this work was detailed listed in Table 1. Portland cement (P.I 42.5) was used herein as cementitious material and fly ash was added as mineral active fine admixture. Ground fine quartz sand worked as fine aggregate with average diameter of 40 μm. To improve fluidity, high-performance water-reducing agent, polycarboxylate superplasticizer (DC-WR2), was also added. For both PVA-FRC and SFRC specimens, samples were prepared by incorporating 2% volume fraction steel or PVA fibers into the fresh cementitious matrix. The fiber volume content of 2% is believed to improve the tensile property with good fluidity [31] since higher fiber fraction usually causes flowability problem. The fibers used in this work were depicted in Figure 1, whose detailed geometric information and mechanical properties were listed in Table 2.

In the preparation process, the dispersion of fibers was an important step in the mixing process, which will directly affect the strengthening and toughening effect of the fiber on the cementitious matrix [37–40]. Uneven dispersion can easily lead to performance degradation. Therefore, the influence of mixing process on dispersibility was very important. The preparation process proposed in this work was as follows: (1) First, add half of the calculated amount of water and a certain amount of superplasticizer to the measuring cup, and stir manually until completely mixed solution is formed. (2) Add cement and fly ash to the mixing container, then pouring the solution into the container, followed by a stirring cycle of the automatic stirring for 240 s (first stirring at a low speed for 60 s, then stirring at a high speed for 30 s, then pausing for 90 s, and finally stirring at a high speed for 60 s), until the mixture appears in a clear slurry state. (3) Add the quartz sand to the slurry and stir a cycle. (4) After pre-dispersion, fibers should be added into the mixing container slowly. Stirring two or three cycles until the fibers are uniformly dispersed. (5) The mixed FRC was cast and compacted, sealed with transparent plastic film. The mold was removed after air curing for 2 days, and cured in the curing room for 25 days. The test was carried out after air curing for 1 day, and the surface treatment of the specimen should be carried out before testing.

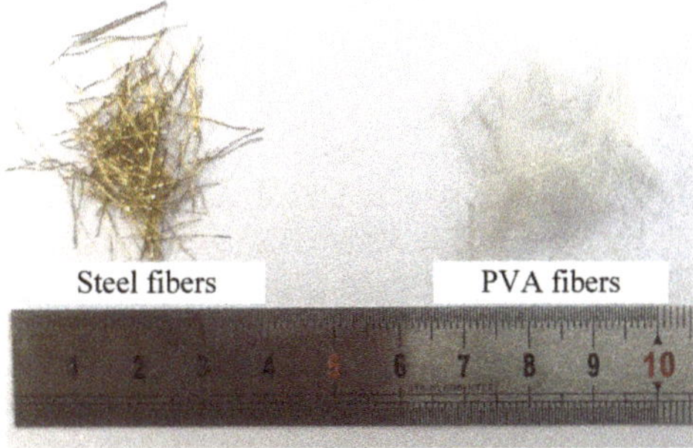


**Figure 1.** Fibers used for steel fiber-reinforced concrete (SFRC) and polyvinyl alcohol (PVA) fiber-reinforced concrete (PVA-FRC).

**Table 1.** Normalized mix proportion.

| Cement | Fly Ash | Water | Quartz Sand | Superplasticizer |
|---|---|---|---|---|
| 1.00 | 0.125 | 0.25 | 0.45 | 0.02 |

**Table 2.** Mechanical properties of fibers.

| Fiber Type | Diameter (μm) | Length (mm) | Density (g/cm$^3$) | Tensile Strength (MPa) | Elastic Modulus (GPa) |
|---|---|---|---|---|---|
| PVA | 30 | 12–15 | 1.30 | 1000 | 8 |
| Steel | 220 | 12–15 | 7.85 | 1200 | 200 |

*2.2. Compression and Bending Tests*

The quasi-static tests, including uniaxial compression (UC) and three-point bending test, were carried out herein to investigate the fiber reinforcement effect on the compressive strength and flexural strength of FRC specimens. Each specimen was cast with three samples with the average values of the experimental results recorded for further discussion.

Uniaxial compressive tests were carried out with cubic specimens with 40 mm edge. The dimension of the specimen and the test setup were shown in Figure 2. The compression tests were conducted using MTS machine with controlled speed rate of 0.5 mm/min. Before testing, abrasive paper was used to smooth the surface of the specimen for the purpose of specimen complete contact with the loading plates. Thereafter, the average peak strength of the specimens was captured as listed in Table 3.

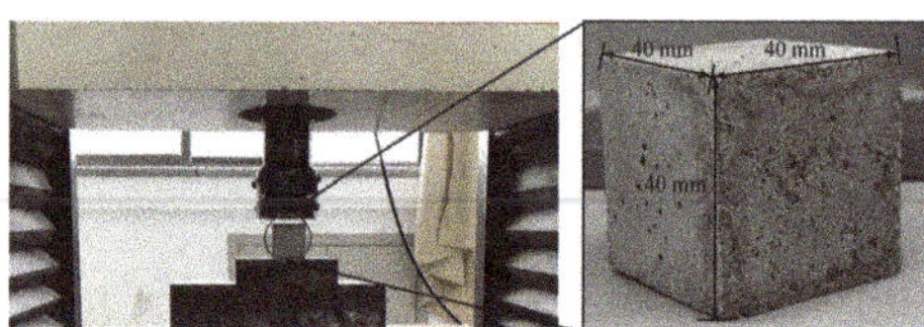


**Figure 2.** Setup of the uniaxial compression test.

Quasi-static three-point bending test were conducted on the 40 mm × 40 mm × 160 mm specimens and the test configuration was presented in Figure 3. The span was chosen as 120 mm, following the experimental standard of mechanical properties of cement mortar [7]. The PC and FRC specimens were tested under load speed of 0.1 mm/min with the peak load values recorded. The flexural strength was calculated with $f_f = 3F_p l/(2bd^2)$ [31], where $F_p$ is peak load, $l$ is span length, $d$ denotes the beam depth and $b$ is the beam width. The average flexure strength of the specimens was listed in Table 3 and the failure modes of the specimens after tests were depicted in Figure 4a.

The UC and 3PBT results, as listed in Table 3 and plotted in Figure 5, demonstrated that both SFRC and PVA-FRC showed boosted compressive and bending performance compared to PC. From the bending test results, it could be found that the flexural strength of PVA-FRC was increased to 12.34 MPa, improved by 42.3 % compared to PC specimen. With 26.95 MPa flexural strength, SFRC presented much more strong improvement in the bending performance. Similar results were also captured in the UC test, PVA-FRC compressive strength was 84.81 MPa while $f_c$ for SFCR was 109.66 MPa, improved by 20.2% and 55.4% respectively compared to control PC cubes. The reason might lie on the fact that with the stronger and stiffer steel fibers, the first crack strength and ultimate strength of SFRC were greatly improved. Besides, the properties of bond between fiber and cementitious matrix also plays an important role on the performance of FRC. According to literature[26,35], bond strength of steel fibers in the cementitious matrix is usually within the range between 5–8 MPa, while this value could be 2–5 MPa for PVA fibers bond. What's more is that the snubbing effect of steel fibers is also stronger than that of PVA fibers. While PVA fiber sometimes exhibits good pullout hardening behavior, it doesn't contribute much to FRC overall strength.

**Table 3.** Static test results.

| Test Data | PC | PVA-FRC | SFRC |
|---|---|---|---|
| Flexural strength $f_f$(MPa) | 8.67 | 12.34 | 26.95 |
| Compressive strength $f_c$(MPa) | 70.57 | 84.81 | 109.66 |

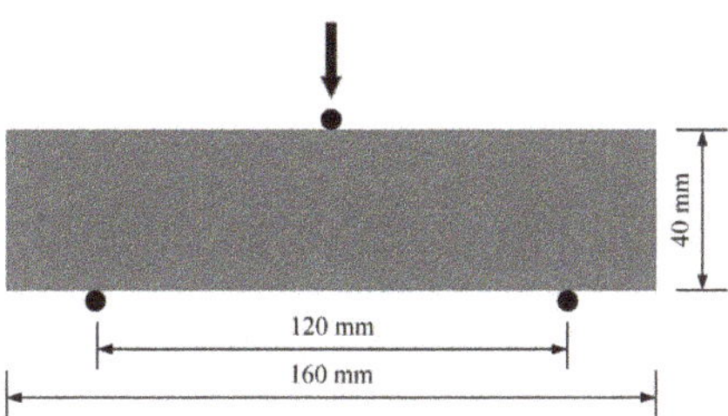


**Figure 3.** Setup of the three-point bending test.

**Figure 4.** Failure modes of tested and simulated specimens after 3PBT. (**a**) Test results, (**b**) Simulation results.

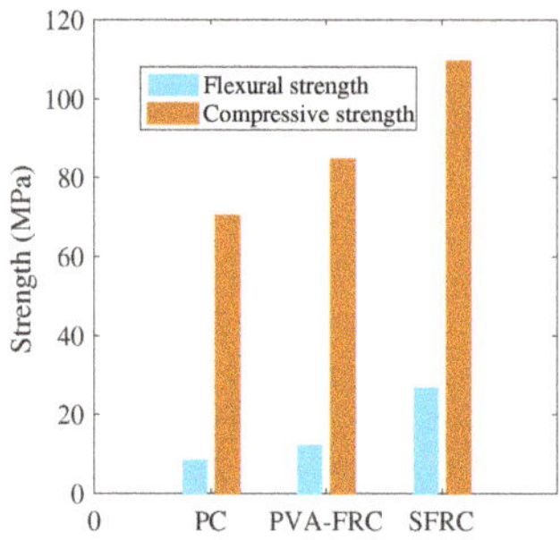



**Figure 5.** Fiber reinforcement effect on compressive and flexure strength.

## 2.3. Blast Test

The explosion tests via contact detonation were carried out in this section to simulate the bomb explosion scenario. Each mixture (i.e., PC, SFRC and PVA-FRC) was prepared with 2 identical panels with dimension of 600 mm × 600 mm × 100 mm for the blast tests. The labels of these specimens were listed in Table 4, whereas S1 represented the control plain concrete, S2 and S3 were the PVA-FRC and SFRC panels respectively. The test setup was shown in Figure 6 where the slab was placed on two bottom steel supporters whose length was 600 mm and section size was 50 mm × 50 mm. All the panels were tested with 100 g and 44 mm height cylinder TNT charge placed at the center of the panel upper surface. An electric detonator was used herein to trigger the TNT charge.

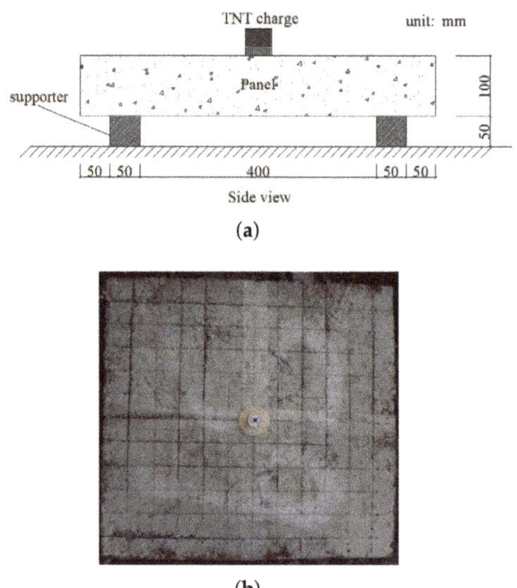



**Figure 6.** Blast test setup. (**a**) Test setup, (**b**) Top view.

The damage modes of the specimens after the blast tests were depicted in Figure 7. The detailed damage conditions were listed in Table 5 and depicted in Figure 8 for the sake of visualization. It should be noted that the crater depth of only SFRC panels was given because that only SFRC panels structure stayed intact after blast, PC and PVA-FRC panels were all perforated. In Figure 7, the plain concrete specimens S1 were totally torn apart under blast loadings. The natural brittleness makes the PC panels could not bear the extreme dynamic loading. While the FRC specimens S2 and S3 behaved much greater blast-resistance than control PC specimens. The PVA-FRC slabs (No.2-1, No.2-2) showed the boost blast-resistant performance. Panel No.2-1 was broken into 3 main blocks and 9 smaller pieces, 28 cracks were observed on the front surface and 21 cracks on the rear surface. Panel No.2-2 had similar damage mode, featured with 4 main blocks and 5 smaller pieces, and its front and rear surface had 19 and 15 main cracks. The SFRC specimens (No.3-1, No.3-2) behaved the greatest blast resistance, whose structure were remain intact after blast loading. Only craters and some fine cracks were found on the front and rear surface. There were about 20 fine cracks on the front and 26 cracks on the rear. Downward tendency was found in Figure 8 in terms of both front and rear crater sizes, which again proved its good blast resistance.

**Table 4.** Concrete mixture for blast test.

| No. | Quantity | PVA Fiber Volume Ratio (%) | Steel Fiber Volume Ratio (%) |
| --- | --- | --- | --- |
| S1 | 2 | - | - |
| S2 | 2 | 2.00 | - |
| S3 | 2 | - | 2.00 |

**Table 5.** Damaged crater for each specimen.

| No. | Front Crater Size (mm) | Front Crater Depth (mm) | Rear Crater Size (mm) |
| --- | --- | --- | --- |
| S1-1 | - | - | - |
| S1-2 | - | - | - |
| S2-1 | 245 × 320 | - | 336 × 390 |
| S2-2 | 234 × 289 | - | 207 × 273 |
| S3-1 | 139 × 156 | 20 | 230 × 290 |
| S3-2 | 152 × 204 | 33 | 241 × 280 |

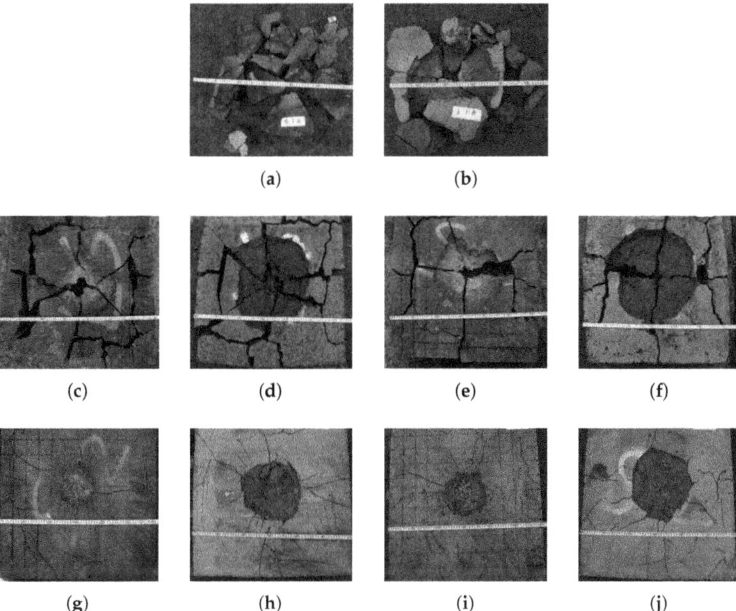

**Figure 7.** Damage modes of specimens after contact detonation. (**a**) S1-1, (**b**) S1-2, (**c**) S2-1 front, (**d**) S2-1 rear, (**e**) S2-2 front, (**f**) S2-2 rear, (**g**) S3-1 front, (**h**) S3-1 rear, (**i**) S3-2 front, (**j**) S3-2 rear.

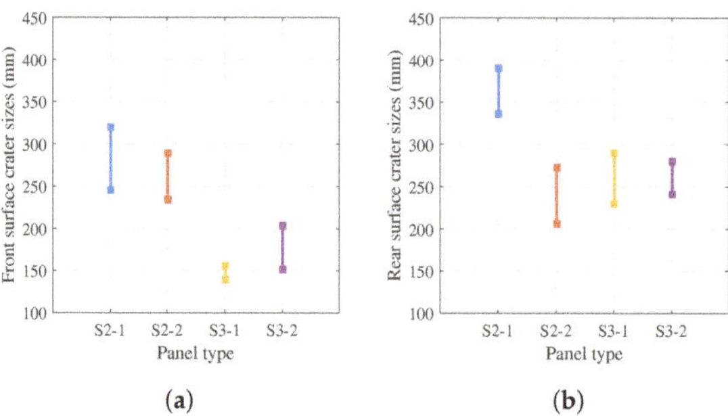

**Figure 8.** Crater sizes of SFRC and PVA-FRC panels. (**a**) Front surface, (**b**) Rear surface.

It could be concluded that plain concrete had the least blast-resistant capability, i.e., the natural brittleness make PC could seldom bear the blast loadings. FRC showed good blast resistance mainly because the fibers embedded into the matrix enhanced the strength and toughness of the concrete. Under blast loadings, the fiber-bridging effect resisted and limited the cracking, the energy absorption capacity and strength of the composite were greatly improved. With the stronger and stiffer embedded steel fibers, the SFRC exhibited the best resistance against blast loadings. Figure 9 showed the steel fiber recovered from damaged panels under scanning electron microscope (SEM) test. Magnified by 1000 times, it could be found that in contrast to smooth surface of pre-test steel fiber in Figure 9a, clear scratches due to friction were observed on the pulled out steel fiber surface. Moreover, some cementitious micro particles were adhered to the fiber surface in Figure 9b. These phenomena indicated that most of the steel fibers were pulled out during the blast process and there was a perfect bond between the steel fiber and the matrix. However, most PVA fibers found raptured after the test, that's another reason why PVA-FRC panels were perforated while SFRC panels were still intact. It was worth noting the FRC rear crater dimensions are much larger than the front crater. The reason for this phenomenon may lie in the fact that rear surface suffers reflected tensile stress wave, which could cause severer damage to the concrete panel since its tensile behavior is much weaker compared to its compressive behavior.

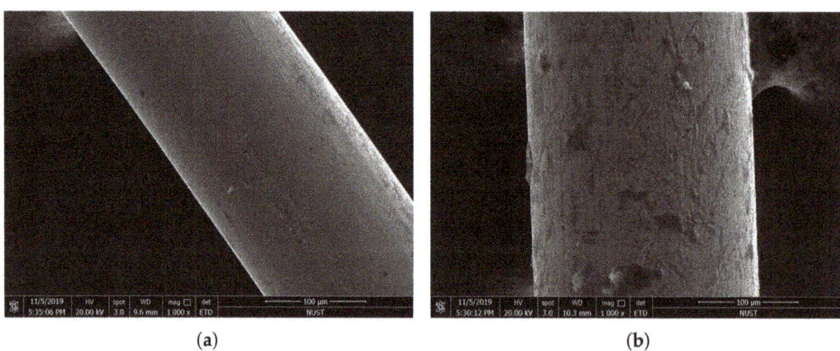

**Figure 9.** SEM images for steel fiber surface. (**a**) Original steel fiber, (**b**) Fiber after blast.

## 3. Review of LDPM-F

The LDPM model was selected as the simulation tool for FRC blast modeling. Prior to numerical study, the basic information of LDPM for PC and LDPM-F for FRC were reviewed briefly. LDPM model generation procedure and governing constitutive equations are explained in the following parts, in accordance with [34].

### 3.1. LDPM Model Construction

LDPM simulates concrete meso-structure through the following steps [41,42]:

The first step is the aggregate generation, which is carried out assuming that each aggregate can be approximated as a sphere. Then, the spherical aggregate size distribution function proposed by Stroeven [43] is considered

$$f(d) = \frac{qd_0^q}{[1-(d_0/d_a)^q]d^{q+1}} \tag{1}$$

in which $d_0$ and $d_a$ are the minimum and maximum spherical aggregate diameter respectively, and $q$ is a material parameter. Reference [34] shows that Equation (1) is associated with a sieve curve in the form

$$f(d) = (\frac{d}{d_a})^{n_f} \tag{2}$$

where $n_f = 3 - q$. When $n_f = 0.5$, Equation (2) corresponds to the classical Fuller curve [34], which is extensively used in concrete technology. For a given cement content $c$, water-to-cement ratio $w/c$, specimen volume $V$, minimum particle diameter $d_0$ and maximum aggregate diameter $d_a$ along with the considered Equation (1), the spherical aggregate system can be generated using a random number generator [34], which is depicted in Figure 10a. Delaunay tetrahedralization of the spherical aggregate center is utilized to define the interactions of the spherical aggregate system, as shown in Figure 10b. Figure 10c shows the final polyhedral particle discretization of the notched beam specimen.

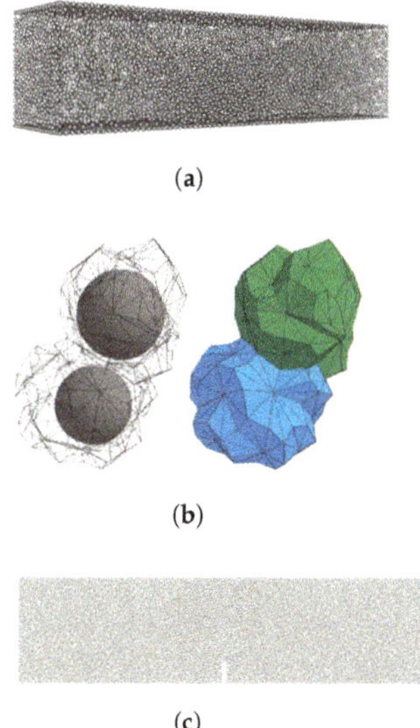

(a)

(b)

(c)

**Figure 10.** Lattice Discrete Particle Model (LDPM) polyhedral particles and cell discretization for a typical notched beam specimen. (**a**) Aggregate system, (**b**) Particles with interaction facets, (**c**) LDPM cell discretization.

### 3.2. LDPM Kinematics

The triangular facets forming the rigid polyhedral particles are assumed to be the potential material failure locations. Each facet is shared between two polyhedral particle and is characterized by a unit normal vector **n** and two tangential vectors **m** and **l**. Accordingly, three strain components are defined on each triangular facet using Equations (1) and (2), which for LDPM gives:

$$\varepsilon_N = \frac{\mathbf{n}^T[\![\mathbf{u}_C]\!]}{\ell}; \quad \varepsilon_L = \frac{\mathbf{l}^T[\![\mathbf{u}_C]\!]}{\ell}; \quad \varepsilon_M = \frac{\mathbf{m}^T[\![\mathbf{u}_C]\!]}{\ell} \tag{3}$$

where $\mathbf{u}_c$ is the displacement vector calculated at the facet centroid, $\ell$ is the length of the tetrahedron edge which means the distance between two particles. It was demonstrated that the meso-scale strain defined in Equation (3) corresponds to the local reference system projection of Green-Lagrange strain tensor of continuum mechanics [34,44,45].

When a facet strain increases beyond the tensile elastic limit, the mesoscale crack opening can be computed as $\mathbf{w} = w_N\mathbf{n} + w_M\mathbf{m} + w_L\mathbf{l}$, in which $w_N = \ell(\varepsilon_N - \sigma_N/E_N)$, $w_M = \ell(\varepsilon_M - \sigma_M/E_T)$ and $w_L = \ell(\varepsilon_L - \sigma_L/E_T)$, $E_N$ and $E_T$ are the elastic normal and tangential stiffness respectively.

### 3.3. LDPM Constitutive Equations

In the elastic regime, the normal and tangential meso-scale stress are proportional to the corresponding strains:

$$\sigma_N = E_N\varepsilon_N; \quad \sigma_M = E_T\varepsilon_M; \quad \sigma_L = E_T\varepsilon_L; \tag{4}$$

where $E_N = E_0$, $E_T = \alpha E_0$, $E_0$ is the normal modulus and $\alpha$ is the shear-normal coupling parameter.

For fracture behavior, through the relationship between equivalent strain, $\varepsilon = \sqrt{\varepsilon_N^2 + \alpha(\varepsilon_M^2 + \varepsilon_L^2)}$, and equivalent stress, $\sigma = \sqrt{\sigma_N^2 + (\sigma_M + \sigma_L)^2/\alpha}$, the fracture response is demonstrated as

$$\sigma_N = \varepsilon_N(\sigma/\varepsilon); \quad \sigma_M = \alpha\varepsilon_M(\sigma/\varepsilon); \quad \sigma_L = \alpha\varepsilon_L(\sigma/\varepsilon); \tag{5}$$

Equivalent stress corresponding to the tensile boundary is considered as

$$\sigma_{bt} = \sigma_0(\omega)exp[-H_0(\omega)\langle\varepsilon_{max} - \varepsilon_0(\omega)\rangle/\sigma_0(\omega)] \tag{6}$$

where $\varepsilon_{max}$ is the maximum equivalent strain, $\langle*\rangle = max(*,0)$, $\omega$ is a coupling variable representing the interaction degree between shear and normal loading and defined as $tan\omega = \frac{\varepsilon_N}{\sqrt{\alpha}\varepsilon_T}$, $\varepsilon_T = \sqrt{\varepsilon_M^2 + \varepsilon_L^2}$. Until the maximum equivalent strain reaches the elastic limit, the fracture damage begins to decline the boundary $\sigma_{bc}$ with the softening modulus $H_0(\omega)$, which governs the post-peak slope and is assumed as $H_0(\omega) = H_t(\frac{2\omega}{\pi})^{n_t}$, where $n_t$ is the softening exponent and $H_t = 2E_0/(l_t/l - 1)$ is the softening modulus in pure tension. $l_t = 2E_0G_t/\sigma_t^2$, $l_e$ is the length of the tetrahedron edge and $G_t$ is the meso-scale fracture energy. LDPM assumes $\sigma_0(\omega)$ as a variation providing a transition between pure tension and pure shear and the variation is denoted as

$$\sigma_0(\omega) = \sigma_t\gamma_{st}^2(-sin(\omega) + \sqrt{sin^2(\omega) + 4\alpha cos^2(\omega)/\gamma_{st}^2})/(2\alpha cos^2\omega) \tag{7}$$

where $\gamma_{st} = \sigma_s/\sigma_t$ is the shear-tensile strength ratio.

The second physical phenomenon simulated in LDPM is pore collapse from compression and compaction. For compressive loading($\varepsilon_N < 0$), the normal stress is computed by meeting the inequality $-\sigma_{bc}(\varepsilon_D, \varepsilon_V) \le \sigma_N \le 0$, where $\sigma_{bc}$ is a strain-dependent boundary associated with the volumetric strain $\varepsilon_V$ and the deviatoric strain $\varepsilon_D$. The volumetric strain $\varepsilon_V = (V - V_0)/3V_0$ ($V$ and $V_0$ are the current and original volumes of a tetrahedron), is calculated at the tetrahedron level. The limitation of the elastic regime is defined as: $\varepsilon_{c0} = \sigma_{c0}/E_0$, where $\sigma_{c0}$ is the meso-scale compressive yielding strength. Beyond the limitation, pore collapse begins, where the compressive boundary is assumed to have an initial linear evolution, up to a volumetric strain value $\varepsilon_{c1} = k_{c0}\varepsilon_{c0}$, $k_{c0}$ is a material parameter. For axial compression with lateral confinement, $\gamma_{DV} = \varepsilon_D/\varepsilon_V$ can significantly affect the response. The compressive boundary for the pore collapse phase is denoted as

$$\sigma_{bc} = \sigma_{c0} + \langle-\varepsilon_V - \varepsilon_{c0}\rangle H_c(\gamma_{DV}) \tag{8}$$

Following the pore collapse, modeling compaction starts with the compressive boundary experiencing an exponential evolution. In this case, the boundary can be expressed as:

$$\sigma_{bc} = \sigma_{c1}(\gamma_{DV})exp[(-\varepsilon_V - \varepsilon_{c1})H_c(\gamma_{DV})/\sigma_{c1}(\gamma_{DV})] \tag{9}$$

where $\sigma_{c1}(\gamma_{DV}) = \sigma_{c0} + (\varepsilon_{c1} - \varepsilon_{c0})H_c(\gamma_{DV})$

*3.4. Formulation of LDPM-F*

For the simulation of fiber-matrix interaction, LDPM-F adopts some assumptions, including that fiber is totally straight, ignoring the bending stiffness of fibers. Before the complete frictional pull-out stage, the embedded segment of a fiber was completely debonded from the surrounding matrix. The debonding stage is characterized by two main factors: The bond fracture energy $G_d$ and the frictional stress $\tau_0$ [46]. The critical slippage value is considered as $v_d$, which represents full debonding. For a given embedment length $L_e$, it can be expressed as [47]

$$v_d = \frac{2\tau_0 L_e^2}{E_f d_f} + (\frac{8G_d L_e^2}{E_f d_f})^{1/2} \tag{10}$$

where $E_f$ is the elastic modulus of the fiber. For the debonding stage ($v < v_d$), fiber load is given as [47]

$$P(v) = [\frac{\pi^2 E_f d_f^3}{2}(\tau_0 v + G_d)]^{1/2} \tag{11}$$

For the pull-out stage ($v > v_d$), the pull-out resistance is entirely frictional and considered as [47]

$$P(v) = P_0(1 - \frac{v - v_d}{L_e})[1 + \frac{\beta(v - v_d)}{d_f}] \tag{12}$$

where $P_0 = \pi L_e d_f \tau_0$, and $\beta$ is a dimensionless coefficient.

In most situations, it is quite different between the orientation of the embedded segment and the free segment of a fiber under crack-bridging force $P_f$. As shown in Figure 11 [35,48], the original angle between the embedded segment and the free segment is denoted as $\varphi_f$. At the point where the fiber exists the matrix, bearing stresses are created in the matrix [49,50]. When the bearing stresses reached a critical value, spalling occurs, causing that the embedment length of the fiber is reduced by a corresponding length $s_f$. The crack-bridging force in the fiber experiences a sudden drop, along with the angle between two fiber segments is reduced to $\varphi'_f$. The spalling length is obtained by [51] and is considered as

$$S_f = \frac{P_{fN}sin(\theta/2)}{k_{sp}\sigma_t d_f cos^2(\theta/2)} \tag{13}$$

where $P_{fN}$ is the normal component of the total pullout crack-bridging force $P_f$, $\sigma_t$ is the matrix tensile strength, $\theta = arccos(n_f^T n)$ and $k_{sp}$ is a material spalling parameter.

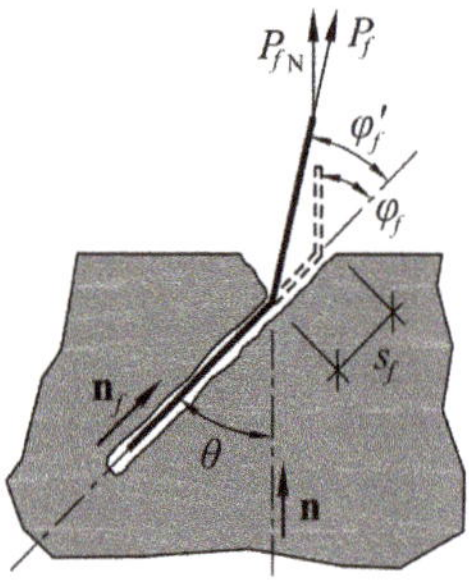

**Figure 11.** Spalling effect of fiber pullout.

The snubbing effect is also taken into account. When a fiber is pulled out from the matrix, at the point where the fiber exits the crack, LDPM-F assumes that the fiber wraps around the surrounding matrix in a totally flexible manner. The summation of the crack-bridging force paralleled to the embedded segment is denoted as $P$. The additional bearing force caused by the snubbing effect is considered as $P_f$, and $P_f > P$. The $P_f$ is obtained as

$$P_f = exp(k_{sn}\varphi'_f)P(v) \tag{14}$$

where $P$ is a function of relative fiber slippage $v$ and $k_{sn}$ is a material snubbing parameter. From Equations (13) and (14), it can be seen that the influence of fiber deflection angle on the fiber pullout response was introduced in LDPM-F, based on which LDPM-F could capture a more comprehensive effect that crack-bridging fiber has on the surrounding matrix. Due to the spalling and snubbing effect that was naturally introduced in LDPM-F, it is reasonable to effectively simulate the fiber orientation effect through LDPM-F.

In order to determine whether the crack-bridging force would cause the fiber rupture during the fiber pull-out process, $k_{rup}$ is introduced herein. Then, it is necessary to check for each fiber

$$\sigma_f = 4P_f / \pi d_f^2 \leq \sigma_{uf}e^{-k_{rup}\varphi'_f} \tag{15}$$

where $\sigma_{uf}$ is the ultimate strength of the fiber and $k_{rup}$ is a material rupture parameter.

Consider a fiber, with initial orientation $\mathbf{n}_f$, subject to a crack opening vector $\mathbf{w}$. The tangential component of $\mathbf{w}$ is defined as $w_T = \sqrt{w_M^2 + w_L^2}$. Assuming that the spalling length is the same on the both sides, the crack-bridging segment vector is computed as $\mathbf{w}' = \mathbf{w} + 2s_f\mathbf{n}_f$. The crack-bridging force vector is assumed to be paralleled to the crack-bridging fiber and is denoted as $\mathbf{P}_f = P_f\mathbf{n}'_f$, where $\mathbf{n}'_f = \mathbf{w}' / \left\|\mathbf{w}'\right\|$.

As discussed above, the realistic LDPM-F simulation response depends on two sets of parameters: (1) The LDPM material parameters which govern the behavior of plain concrete; (2) the parameters which govern the fiber-matrix interaction. All the parameters should be identified by fitting experimental data [16,52].

## 4. Numerical Modeling of FRC Contact Detonation

Numerical investigation on the blast-resistant performance of steel and PVA FRC was carried out in this section with LDPM-F. Figure 12 depicted the fiber distribution in a panel specimen simulated by LDPM-F. The geometry of an individual fiber can be characterized by two primary parameters: Diameter $d_f$ and length $L_f$, for a given fiber volume fraction $V_f$, individual fibers will be inserted in the matrix with randomly generated positions and orientations. In this work, $L_f$ and $d_f$ were consistent with the actual fiber dimensions.

**Figure 12.** Fiber distribution in a panel.

For the calibration purpose, uniaxial compression and 3-point bending test were firstly simulated with LDPM-F to calibrate the parameters controlling plain concrete and fiber properties. Furthermore, the validated LDPM-F blast model was further applied to investigate the fiber content effect on the blast-resistant performance. SFRC and PVA-FRC panels with fiber content of 0.5%, 1.0%, 1.5% were modeled with LDPM-F and the simulation results were compared to FRC with fiber content of 2.0%.

### 4.1. LDPM-F Parameter Calibration

For calibrating the parameters governing the plain concrete mechanical properties, the UC and 3PBT simulations of PC were carried out herein. The simulation constrains of UC and 3PBT simulations were depicted in Figure 13. The simulated response curve were shown in Figure 14 and the failure modes of the simulated 3PBT results were depicted in Figure 4b. It could be found that the simulation results were fitted well to the test results in terms of both mechanical responses and damage modes. The calibrated parameters were listed in Table 6.

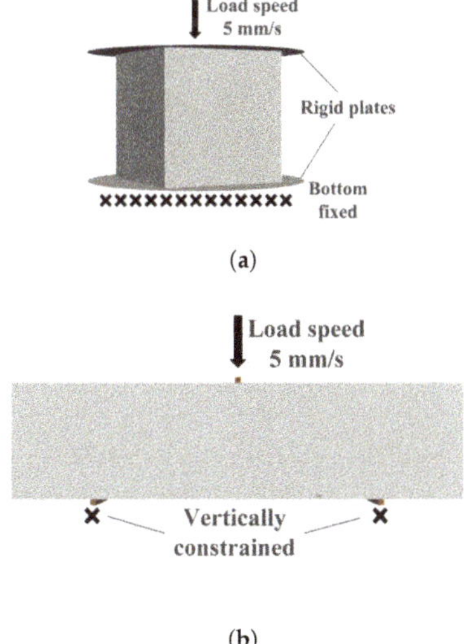


**Figure 13.** Simulation constrains. (**a**) Uniaxial compression, (**b**) 3-point bending test.

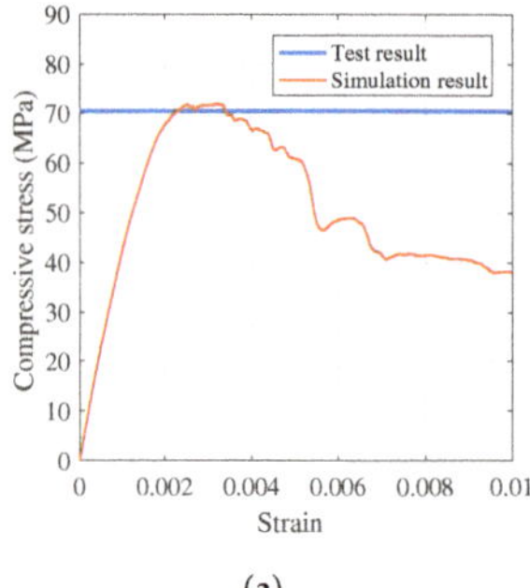

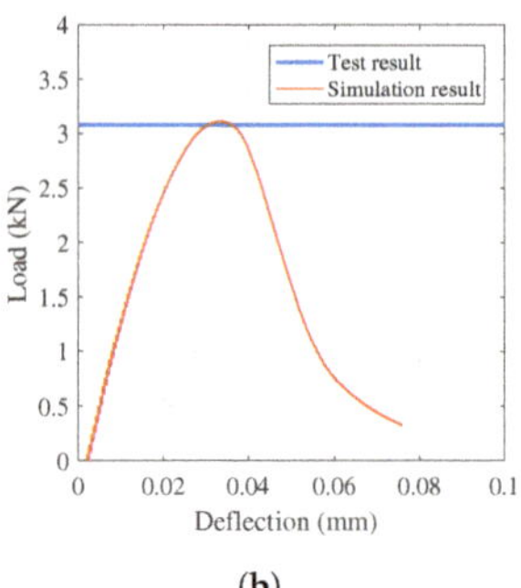

(a)

(b)

**Figure 14.** LDPM parameter calibration with test data. (a) Uniaxial compression, (b) 3-point bending test.

**Table 6.** LDPM parameters.

| $E_0$ [MPa] | $\alpha$ [-] | $\sigma_t$ [MPa] | $\sigma_{c0}$ [MPa] | $\sigma_s/\sigma_t$ [-] | $l_t$[mm] | $n_t$ [-] | $H_{c0}/E_0$ [-] | $\sigma_{N0}$ [MPa] | $k_{c0}$ [-] |
|---|---|---|---|---|---|---|---|---|---|
| 80,610 | 0.25 | 4.55 | 150 | 5.55 | 200 | 0.1 | 0.4 | 600 | 4 |

The fiber parameters were also calibrated herein by fitting the simulation results to the test results. The UC and 3PBT simulation constrains of FRC were the same with that of plain concrete. Both PVA-FRC and SFRC were simulated with the response curves plotted in Figures 15 and 16. Both PVA-FRC and SFRC presented similar compressive strength and flexural strength with test data. The predicted failure modes of the FRC specimens after 3PBT were depicted in Figure 4b, good consistency was also presented compared to test results, which further verified the calibrated parameters. The calibrated PVA and steel fiber parameters were listed in Table 7. After the calibration of LDPM parameters with plain concrete and LDPM-F parameters with FRC, the FRC panel blast model can be established for validation.

**Table 7.** Lattice Discrete Particle Model-Fiber (LDPM-F) parameters.

| Fiber Type | $E_f$ [MPa] | $k_{sp}$ [-] | $\sigma_{uf}$ [MPa] | $k_{rup}$ [-] | $k_{sn}$ [-] | $\tau_0$ [MPa] | $G_d$ [N/m] | $\beta$ [-] | $\gamma_f$ [-] |
|---|---|---|---|---|---|---|---|---|---|
| Steel fiber | 210,000 | 500 | 2800 | 0.2 | 0.2 | 6.0 | 0.0 | 0.0 | 0.6 |
| PVA fiber | 30,000 | 300 | 1000 | 0.0 | 0.2 | 2.5 | 3 | 0.05 | 1.0 |

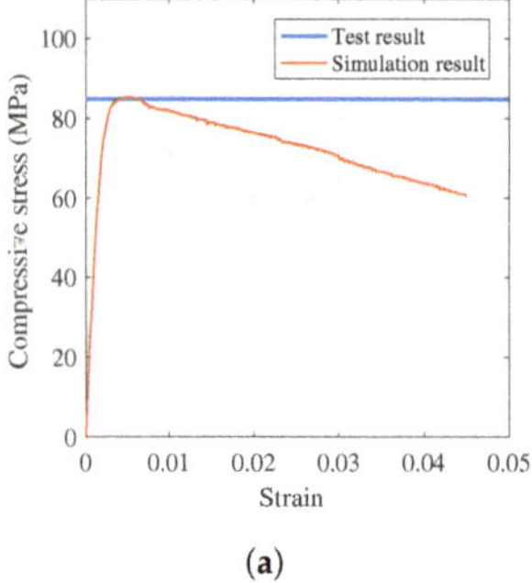

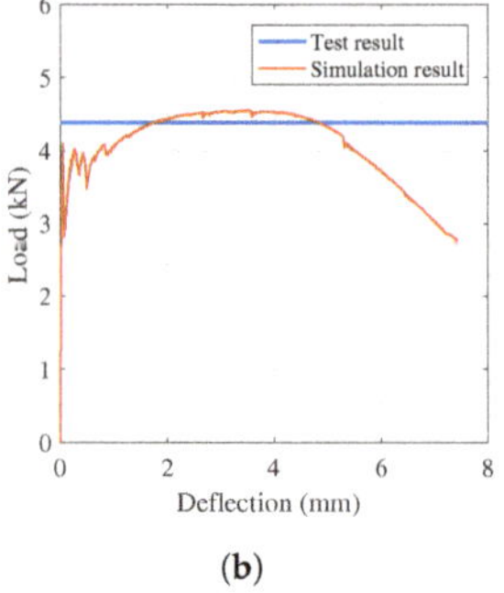

(a)

(b)

**Figure 15.** Comparison between simulation and PVA-FRC tests. (a) Uniaxial compression, (b) 3-point bending test.

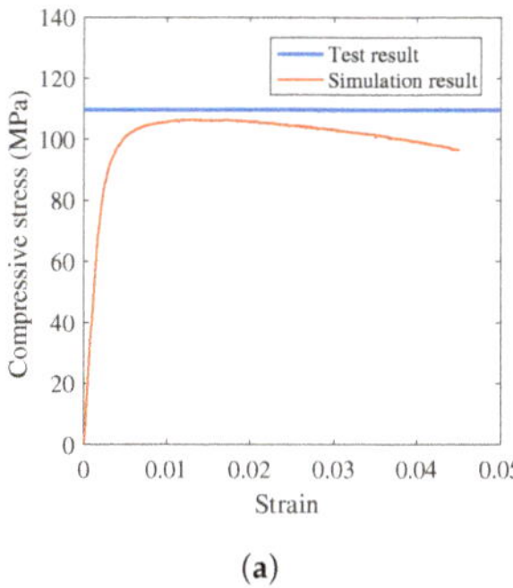

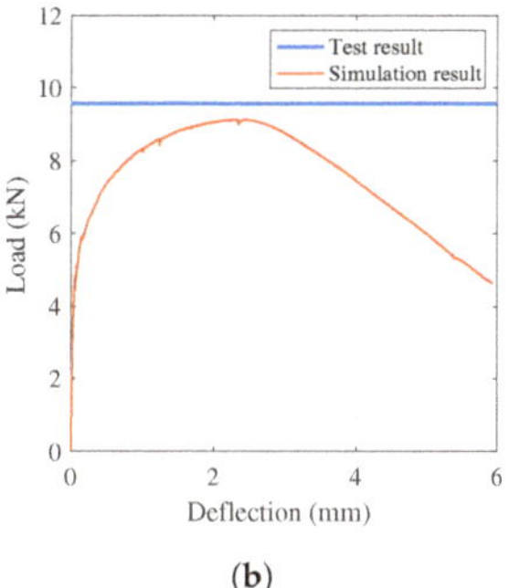

(a)                (b)

**Figure 16.** Comparison between simulation and SFRC tests. (**a**) Uniaxial compression, (**b**) 3-point bending test.

### 4.2. FRC Blast Simulation

Simulations of FRC panels under blast loadings were carried out with LDPM-F describing the FRC material properties. The FRC blast simulation model was plotted in Figure 17, wherethe blast model setup was consistent with the contact detonation test. The FRC panel was incorporated into 2% volume ratio fibers and the bottom supporters were set as rigid bodies, whose vertical freedom was constrained. The interaction between the FRC panel and the rigid supporters was set as penalty contact. The explosion source was set vertically 22 mm away from the center of the panel front surface.

After 5 ms simulation termination time, the numerically derived damage modes of PC and FRC were shown in Figure 18, in which the sizes of the craters on the front and rear surfaces were also presented in detail. The comparison of damage modes between test and simulation results were demonstrated in Figure 19, and detailed data was listed in Tables 8 and 9.

As can be seen in Figure 18, the PC panel was blown into pieces under blast loadings, just like the test results that the brittleness of the plain concrete was completely demonstrated. The PVA-FRC panel was also perforated by the blast loadings. Compared to PC specimen, the entire panel was greatly enhanced by the PVA fibers in terms of crater sizes on the front and rear surface and the panel integrity. The fiber-bridging effect brought by PVA fibers improved the concrete energy absorption capacity to some extent, as a result the PVA-FRC could withstand a certain amount of blast loadings. However, the steel FRC panels had the least damage, e.g., only craters and a few fine cracks occurred on the surface suggesting that the SFRC panel has the best blast-resistant performance.

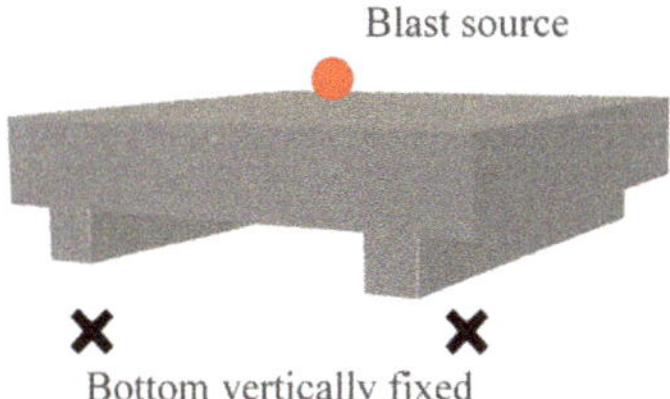


**Figure 17.** Blast simulation model.

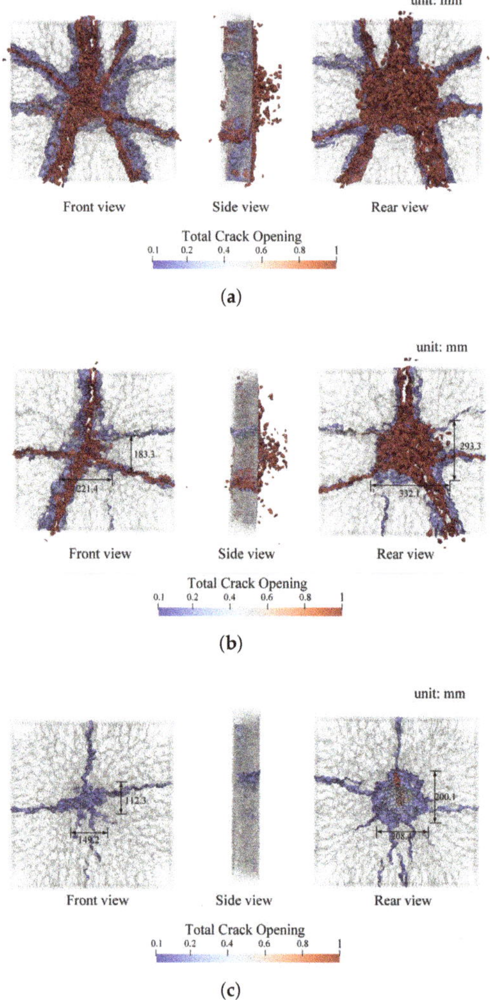



**Figure 18.** The damage modes of plain concrete (PC) and FRC under blast loadings. (**a**) Plain concrete (PC), (**b**) PVA-FRC, (**c**) SFRC.

**Table 8.** The comparison of the PVA-FRC crater size between test and simulation results.

| No. | Front Surface (mm) | Rear Surface (mm) |
|---|---|---|
| S2-1 | $245 \times 320$ | $336 \times 390$ |
| S2-2 | $234 \times 289$ | $207 \times 273$ |
| Simulation prediction | $183 \times 221$ | $293 \times 332$ |

**Table 9.** The comparison of the SFRC crater size between test and simulation results.

| No. | Front Surface (mm) | Rear Surface (mm) |
|---|---|---|
| S3-1 | $139 \times 156$ | $230 \times 290$ |
| S3-2 | $152 \times 204$ | $241 \times 280$ |
| Simulation prediction | $112 \times 149$ | $200 \times 208$ |

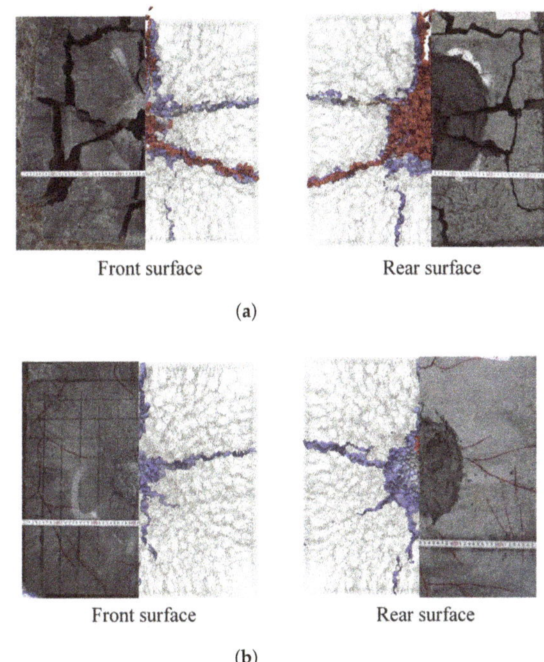

Front surface       Rear surface

(a)

Front surface       Rear surface

(b)

**Figure 19.** The comparison between damage modes of test and simulation results. (**a**) PVA-FRC, (**b**) SFRC.

The crater sizes of PVA-FRC and SFRC predicted by LDPM-F were listed in Tables 8 and 9. The comparison of damage modes between test and simulation results were shown in Figure 19. Good agreements were achieved in terms of both crater sizes and crack distribution, which strongly validated the LDPM-F blast model. As a result, the validated LDPM-F blast model was extended to investigate on the fiber content effect on the blast resistance of FRC for further discussion.

### 4.3. Fiber Content Effect

In this section, for the purpose of investigating on the blast resistance and damage modes of FRC with fiber content lower than 2.0%, the fiber content effect on the blast resistance was carried out herein through the validated LDPM-F blast model. For both PVA-FRC and SFRC, specimens embedded into fibers with fiber volume fraction of 0.5%, 1.0% and 1.5% under blast loadings were numerically simulated. The simulation setup was totally same with the contact detonation test. Thereafter, the simulation results were depicted in Figures 20 and 21 in front, side and rear view.

As can be seen in Figure 20, PVA-FRC with fiber content less than 2%, had limited blast resistance improvement since all panels were torn into some smaller blocks due to the reduction of PVA fibers amount. With less PVA fibers bridging cracking gaps, PVA-FRC exhibits reduced ductility and thus losing energy absorption capacity. It can be argued that 2% fiber volume fraction was a proper content for PVA-FRC shelter construction material since lower content leads to significant blast resistance degradation.

On the contrary, SFRC still behaved substantial blast-resistant performance even for the panel with only 0.5% steel fiber content. It was vividly suggested in the Figure 21 that all the SFRC panels maintain good integrity. With the increase of steel fiber content, less cracks with smaller cracking opening occurred on the panel surface. The strong bond between steel fiber and cementitous matrix contribute to the enhanced bending mechanical behavior even with much less fiber content. The crater sizes of SFRC with different fiber content were denoted in Figure 22 and listed in Table 10. It was demonstrated that as fiber content increases, the front surface crater has some oscillation while the rear surface crater sizes gradually decreases. The front crater mainly caused by compressive stress wave was less affected by the steel fiber content. On the other hand, the reflected tensile stress wave results in rear crater forming which was more sensitive to incorporated fiber dosage.

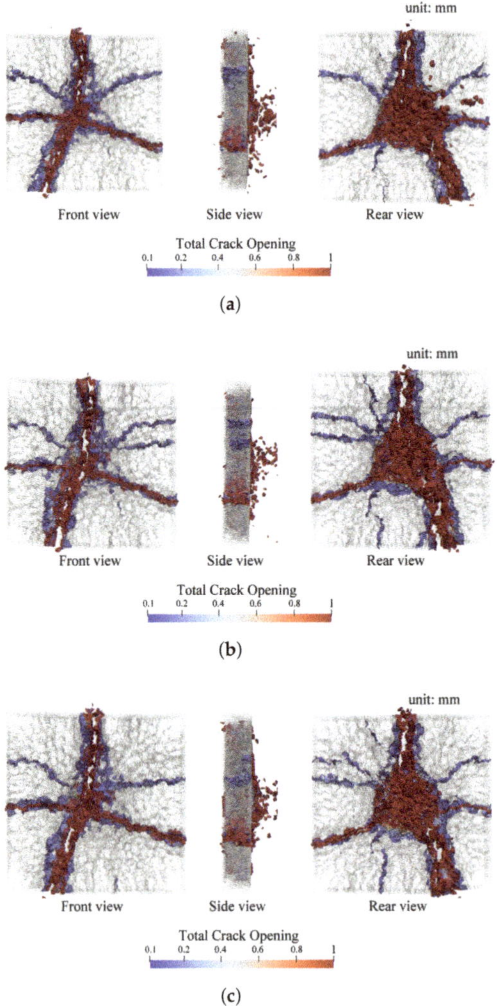

**Figure 20.** Damage modes of PVA-FRC with different fiber content under blast loadings. (**a**) 0.5% fiber volume fraction, (**b**) 1.0% fiber volume fraction, (**c**) 1.5% fiber volume fraction.

**Table 10.** The crater sizes of SFRC with different fiber content.

| Fiber Content | Front Surface (mm) | Rear Surface (mm) |
| --- | --- | --- |
| 0.5% | 139 × 210 | 280 × 299 |
| 1.0% | 120 × 161 | 272 × 281 |
| 1.5% | 200 × 144 | 265 × 241 |
| 2.0% | 112 × 149 | 200 × 208 |

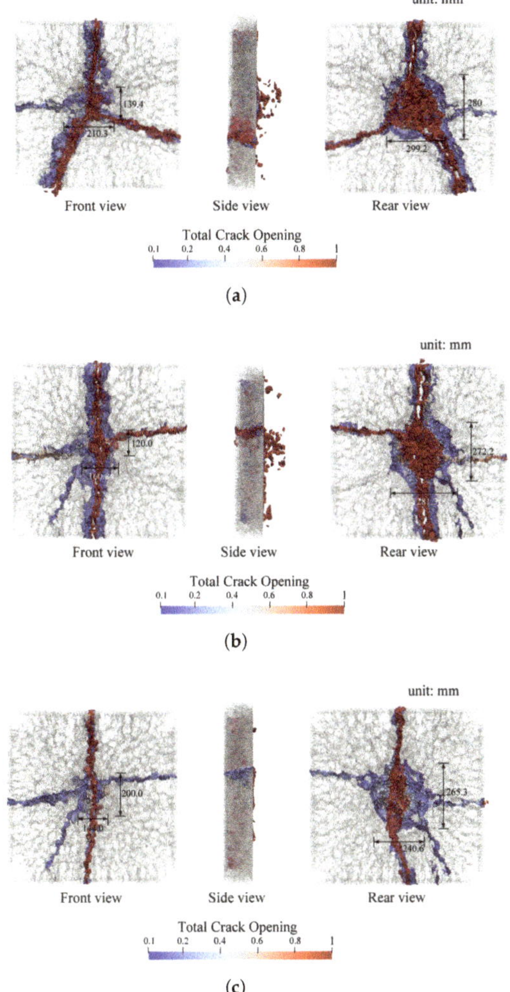



**Figure 21.** Damage modes of SFRC with different fiber content under blast loadings. (**a**) 0.5% fiber volume fraction, (**b**) 1.0% fiber volume fraction, (**c**) 1.5% fiber volume fraction.

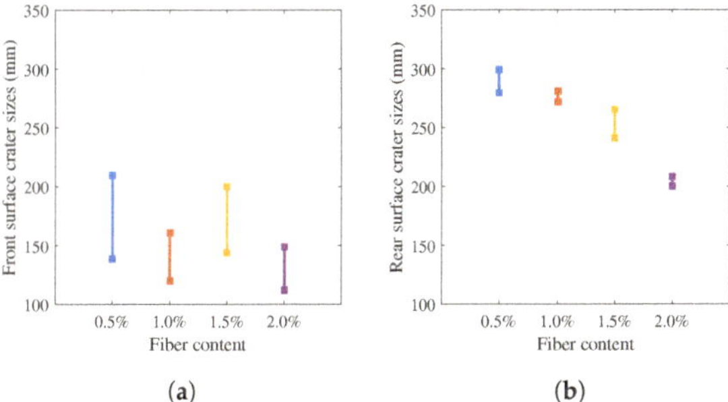



**Figure 22.** Crater sizes prediction of SFRC panels with different fiber content. (**a**) Front surface, (**b**) Rear surface.

## 5. Conclusions

This work aimed at comparative investigating on the blast-resistant performance of steel fiber-reinforced concrete and PVA fiber-reinforced concrete. Experimental program was conducted to characterize the mechanical properties and blast resistance. LDPM-F was introduced to model FRC contact detonation. The validated model was further applied for fiber content effect study. Several conclusions were drawn as follows: (1) The plain concrete could hardly bear the blast loadings. SFRC, PVA-FRC with fiber volume fraction of 2% showed much better blast resistance due to the improved ductility resulting from fiber incorporation. (2) Compared with PVA-FRC, SFRC exhibits higher compressive strength and bending ductility, thus providing better blast resistance against contact detonation. (3) The established LDPM-F blast model can predict the damage modes as well as crater sizes for both PVA-FRC and SFRC panels. (4) For fiber content lower than 2%, the blast-resistant performance of PVA-FRC was greatly degraded, while SFRC specimens still showed good integrity under blast even for panel with only 0.5% volume fraction steel fiber. (5) SFRC panel front surface crater is less affected by the steel fiber content while the rear surface crater is more sensitive to incorporated fiber dosage.

**Author Contributions:** Data curation, L.C.; Formal analysis, J.F.; Funding acquisition, J.L.; Investigation, B.C. and S.X.; Methodology, C.D.; Project administration, W.S. and J.L.; Software, L.C. and J.F.; Supervision, W.S.; Writing—original draft, L.C.; Writing—review & editing, J.F. All authors have read and agreed to the published version of the manuscript.

**Funding:** This research was funded by the Natural Science Foundation of Jiangsu Province (No. BK20170824).

**Acknowledgments:** This research was financial supported by the National Natural Science Foundation of China (No. 11902161). Weiwei Sun thanks the Foundation Strengthening Plan Technology Fund (No. 2019-JCJQ-JJ-371). Bingcheng Chen expresses gratitude to the Postgraduate Research & Practice Innovation Program of Jiangsu Province (No. KYCX20_0377). Le Chen was sponsored by the Postgraduate Research & Practice Innovation Program of Jiangsu Province via Grant No. SJCX19_0118.

**Conflicts of Interest:** The authors declare no conflict of interest.

## References

1. Yoo, D.; Banthia, N. Mechanical and structural behaviors of ultra-high-performance fiber-reinforced concrete subjected to impact and blast. *Constr. Build. Mater.* **2017**, *149*, 416–431. [CrossRef]
2. Feng, J.; Sun, W.; Wang, L.; Chen, L.; Xue, S.; Li, W. Terminal ballistic and static impactive loading on thick concrete target. *Constr. Build. Mater.* **2020**, *251*, 118899. [CrossRef]
3. Wang, Z.L.; Shi, Z.M.; Wang, J. On the strength and toughness properties of sfrc under static-dynamic compression. *Compos. Part B Eng.* **2011**, *42*, 1285–1290. [CrossRef]
4. Liu, J.; Han, F.; Cui, G.; Zhang, Q.; Lv, J.; Zhang, L.; Yang, Z. Combined effect of coarse aggregate and fiber on tensile behavior of ultra-high performance concrete. *Constr. Build. Mater.* **2016**, *121*, 310–318. [CrossRef]
5. Simoes, T.; Octavio, C.; Valenca, J.; Costa, H.; Diasdacosta, D.; Julio, E. Influence of concrete strength and steel fibre geometry on the fibre/matrix interface. *Compos. Part B Eng.* **2017**, *122*, 156–164. [CrossRef]
6. Li, P.P.; Brouwers, J.; Yu, Q.Q. Influence of key design parameters of ultra-high performance fibre reinforced concrete on in-service bullet resistance. *Int. J. Impact Eng.* **2020**, *136*, 103434. [CrossRef]
7. Zhang, W.; Chen, S.; Zhang, N.; Zhou, Y. Low-velocity flexural impact response of steel fiber reinforced concrete subjected to freeze–thaw cycles in nacl solution. *Constr. Build. Mater.* **2015**, *101*, 522–526. [CrossRef]
8. Hong, K.; Lee, S.; Han, S.; Yeon, Y. Evaluation of fe-based shape memory alloy (fe-sma) as strengthening material for reinforced concrete structures. *Appl. Sci.* **2018**, *8*, 730. [CrossRef]
9. Picazo, A.; Alberti, M.G.; Galvez, J.C.; Enfedaque, A.; Vega, A.C. The size effect on flexural fracture of polyolefin fibre reinforced concrete. *Appl. Sci.* **2019**, *9*, 1762. [CrossRef]
10. Feldman, D. Fibre reinforced cementitious composites. *Can. J. Civ. Eng.* **1993**, *20*, 341–341. [CrossRef]
11. Soe, K.; Zhang, Y.X.; Zhang, L. Material properties of a new hybrid fibre-reinforced engineered cementitious composite. *Constr. Build. Mater.* **2013**, *43*, 399–407. [CrossRef]
12. Halvaei, M.; Jamshidi, M.; Latifi, M. Investigation on pullout behavior of different polymeric fibers from fine aggregates concrete. *J. Ind. Text.* **2016**, *45*, 995–1008. [CrossRef]

13. Halvaei, M.; Jamshidi, M.; Pakravan, H.R.; Latifi, M. Interfacial bonding of fine aggregate concrete to low modulus fibers. *Constr. Build. Mater.* **2015**, *95*, 117–123. [CrossRef]

14. Sim, J.; Park, C.; Moon, D.Y. Characteristics of basalt fiber as a strengthening material for concrete structures. *Compos. Part B Eng.* **2005**, *36*, 504–512. [CrossRef]

15. Martinezperez, I.; Valivonis, J.; Salna, R.; Coboescamilla, A. Experimental study of flexural behaviour of layered steel fibre reinforced concrete beams. *J. Civ. Eng. Manag.* **2017**, *23*, 806–813. [CrossRef]

16. Jin, C.; Buratti, N.; Stacchini, M.; Savoia, M.; Cusatis, G. Lattice discrete particle modeling of fiber reinforced concrete: Experiments and simulations. *Eur. J. Mech. A Solids* **2016**, *57*, 85–107. [CrossRef]

17. Hassan, A.M.T.; Jones, S.W.; Mahmud, G.H. Experimental test methods to determine the uniaxial tensile and compressive behaviour of ultra high performance fibre reinforced concrete (uhpfrc). *Constr. Build. Mater.* **2012**, *37*, 874–882. [CrossRef]

18. Pan, Z.; Wu, C.; Liu, J.; Wang, W.; Liu, J. Study on mechanical properties of cost-effective polyvinyl alcohol engineered cementitious composites (pva-ecc). *Constr. Build. Mater.* **2015**, *78*, 397–404. [CrossRef]

19. Feng, J.; Sun, W.; Zhai, H.; Wang, L.; Dong, H.; Wu, Q. Experimental study on hybrid effect evaluation of fiber reinforced concrete subjected to drop weight impacts. *Materials* **2018**, *11*, 2563. [CrossRef]

20. Olivito, R.S.; Zuccarello, F.A. An experimental study on the tensile strength of steel fiber reinforced concrete. *Compos. Part B Eng.* **2010**, *41*, 246–255.[CrossRef]

21. Habel, K.; Gauvreau, P. Response of ultra-high performance fiber reinforced concrete (uhpfrc) to impact and static loading. *Cem. Concr. Compos.* **2008**, *30*, 938–946. [CrossRef]

22. Nili, M.; Afroughsabet, V. Combined effect of silica fume and steel fibers on the impact resistance and mechanical properties of concrete. *Int. J. Impact Eng.* **2010**, *37*, 879–886. [CrossRef]

23. Nia, A.A.; Hedayatian, M.; Nili, M.; Sabet, V.A. An experimental and numerical study on how steel and polypropylene fibers affect the impact resistance in fiber-reinforced concrete. *Int. J. Impact Eng.* **2012**, *46*, 62–73.

24. Naaman, A.E.; Najm, H. Bond-slip mechanisms of steel fibers in concrete. *Materials* **1991**, *88*, 135–145.

25. Lee, Y.; Kang, S.; Kim, J. Pullout behavior of inclined steel fiber in an ultra-high strength cementitious matrix. *Constr. Build. Mater.* **2010**, *24*, 2030–2041. [CrossRef]

26. Abdallah, S.; Fan, M.; Rees, D.W.A. Bonding mechanisms and strength of steel fiber–reinforced cementitious composites: Overview. *J. Mater. Civ. Eng.* **2018**, *30*, 04018001. [CrossRef]

27. Marcalikova, Z.; Cajka, R.; Bilek, V.; Bujdos, D.; Sucharda, O. Determination of mechanical characteristics for fiber-reinforced concrete with straight and hooked fibers. *Crystals* **2020**, *10*, 545. [CrossRef]

28. Ma, H.; Cai, J.; Lin, Z.; Qian, S.; Li, V.C. CaCO$_3$ whisker modified engineered cementitious composite with local ingredients. *Constr. Build. Mater.* **2017**, *151*, 1–8. [CrossRef]

29. Yao, W.; Sun, W.; Shi, Z.; Chen, B.; Chen, L.; Feng, J. Blast-resistant performance of hybrid fiber-reinforced concrete (hfrc) panels subjected to contact detonation. *Appl. Sci.* **2019**, *10*, 241. [CrossRef]

30. Yusof, M.A.; Nor, N.M.; Ismail, A.; Peng, N.C.; Sohaimi, R.M.; Yahya, M.A. Performance of hybrid steel fibers reinforced concrete subjected to air blast loading. *Adv. Mater. Sci. Eng.* **2013**, *2013*, 420136. [CrossRef]

31. Feng, J.; Gao, X.; Li, J.; Dong, H.; He, Q.; Liang, J.; Sun, W. Penetration resistance of hybrid-fiber-reinforced high-strength concrete under projectile multi-impact. *Constr. Build. Mater.* **2019**, *202*, 341–352. [CrossRef]

32. Huang, S.; Yuan, Z.; Fish, J. Computational framework for short-steel fiber-reinforced ultra-high performance concrete (cor-tuf). *Int. J. Multiscale Comput. Eng.* **2019**, *17*, 551–562. [CrossRef]

33. Yang, F.; Xie, W.; Meng, S. Impact and blast performance enhancement in bio-inspired helicoidal structures: A numerical study. *J. Mech. Phys. Solids* **2020**, *142*, 104025. [CrossRef]

34. Cusatis, G.; Pelessone, D.; Mencarelli, A. Lattice discrete particle model (ldpm) for failure behavior of concrete. I: Theory. *Cem. Concr. Compos.* **2011**, *33*, 881–890. [CrossRef]

35. Schauffert, E.A.; Cusatis, G. Lattice discrete particle model for fiber-reinforced concrete. I: Theory. *J. Eng. Mech.* **2011**, *138*, 826–833. [CrossRef]

36. Feng, J.; Yao, W.; Li, W.; Li, W. Lattice discrete particle modeling of plain concrete perforation responses. *Int. J. Impact Eng.* **2017**, *109*, 39–51. [CrossRef]

37. Yang, S.; Millard, S.; Soutsos, M.; Barnett, S.; Le, T.T. Influence of aggregate and curing regime on the mechanical properties of ultra-high performance fibre reinforced concrete (uhpfrc). *Constr. Build. Mater.* **2009**, *23*, 2291–2298. [CrossRef]

38. Cao, M.; Zhang, C.; Li, Y.; Wei, J. Using calcium carbonate whisker in hybrid fiber-reinforced cementitious composites. *J. Mater. Civ. Eng.* **2015**, *27*, 04014139. [CrossRef]
39. Cao, M.; Zhang, C.; Wei, J. Microscopic reinforcement for cement based composite materials. *Constr. Build. Mater.* **2013**, *40*, 14–25. [CrossRef]
40. Kanda, T.; Li, V.C. Practical design criteria for saturated pseudo strain hardening behavior in ecc. *J. Adv. Concr. Technol.* **2006**, *4*, 59–72. [CrossRef]
41. Feng, J.; Song, M.; Sun, W.; Wang, L.; Li, W.; Li, W. Thick plain concrete targets subjected to high speed penetration of 30crmnsini2a steel projectiles: Tests and analyses. *Int. J. Impact Eng.* **2018**, *122*, 305–317. [CrossRef]
42. Shen, L.; Li, W.; Zhou, X.; Feng, J.; Di Luzio, G.; Ren, Q.; Cusatis, G. Multiphysics lattice discrete particle model for the simulation of concrete thermal spalling. *Cem. Concr. Compos.* **2020**, *106*, 103457. [CrossRef]
43. Stroeven, P. A stereological approach to roughness of fracture surfaces and tortuosity of transport paths in concrete. *Cem. Concr. Compos.* **2000**, *22*, 331–341. [CrossRef]
44. Cusatis, G.; Rezakhani, R.; Schauffert, E.A. Discontinuous cell method (dcm) for the simulation of cohesive fracture and fragmentation of continuous media. *Eng. Fract. Mech.* **2017**, *170*, 1–22. [CrossRef]
45. Rezakhani, R.; Cusatis, G. Asymptotic expansion homogenization of discrete fine-scale models with rotational degrees of freedom for the simulation of quasi-brittle materials. *J. Mech. Phys. Solids* **2016**, *88*, 320–345. [CrossRef]
46. Yang, E.-H.; Yang, Y.; Li, V.C. Use of high volumes of fly ash to improve ecc mechanical properties and material greenness. *ACI Mater. J.* **2007**, *104*, 620.
47. Lin, Z.; Kanda, T.; Li, V.C. On interface property characterization and performance of fiber reinforced cementitious composites. *J. Concr. Sci. Eng.* **1999**, *1*, 173–184.
48. Feng, J.; Sun, W.; Liu, Z.; Cui, C.; Wang, X. An armour-piercing projectile penetration in a double-layered target of ultra-high-performance fiber reinforced concrete and armour steel: Experimental and numerical analyses. *Mater. Des.* **2016**, *102*, 131–141. [CrossRef]
49. Leung, C.K.; Ybanez, N. Pullout of inclined flexible fiber in cementitious composite. *J. Eng. Mech.* **1997**, *123*, 239–246. [CrossRef]
50. Leung, C.K.; Chi, J. Crack-bridging force in random ductile fiber brittle matrix composites. *J. Eng. Mech.* **1995**, *121*, 1315–1324. [CrossRef]
51. Yang, E.-H.; Wang, S.; Yang, Y.; Li, V.C. Fiber-bridging constitutive law of engineered cementitious composites. *J. Adv. Concr. Technol.* **2008**, *6*, 181–193. [CrossRef]
52. Feng, J.; Song, M.; He, Q.; Sun, W.; Wang, L.; Luo, K. Numerical study on the hard projectile perforation on rc panels with ldpm. *Constr. Build. Mater.* **2018**, *183*, 58–74. [CrossRef]

 *crystals*

*Article*

# Use of Flue Gas Desulfurization Gypsum, Construction and Demolition Waste, and Oil Palm Waste Trunks to Produce Concrete Bricks

**Lalitsuda Phutthimethakul [1], Park Kumpueng [1] and Nuta Supakata [1,2,***

[1] Department of Environmental Science, Faculty of Science, Chulalongkorn University, Bangkok 10330, Thailand; adustilal@gmail.com (L.P.); park.kumpueng@gmail.com (P.K.)

[2] Waste Utilization and Ecological Risk Assessment Research Group, the Ratchadaphiseksomphot Endowment Fund, Chulalongkorn University, Bangkok 10330, Thailand

* Correspondence: nuta.s@chula.ac.th

Received: 30 July 2020; Accepted: 14 August 2020; Published: 18 August 2020

**Abstract:** This research aims to study the utilization of waste from power plants, construction and demolition, and agriculture by varying the ratios of flue-gas desulfurization (FGD) gypsum, construction and demolition waste (CDW), and oil palm trunks (OPT) in concrete production. This research used these as the raw materials for the production of concrete bricks of $15 \times 15 \times 15$ cm. There were 12 ratios of concrete brick, fixing 5.5 wt% of FGD gypsum to replace Portland cement and substituting coarse sand with 0 wt%, 25 wt%, 50 wt%, or 75 wt% of CDW, and gravel with 0 wt%, 0.5 wt%, and 1 wt% of OPT. The initial binder:fine aggregate:coarse aggregate ratio was 1:2:4 and the water to cement ratio was 0.5, curing in water at room temperature for 28 days. Then, all concrete brick specimens were tested for compressive strength and water absorption. From the experiment, it was found that the highest compressive strength of concrete brick specimens was 45.18 MPa, which was produced from 5.5% gypsum without CDW and OPT, while 26.84 MPa was the lowest compressive strength obtained from concrete bricks produced from 5.5% FGD gypsum, 75% CDW, and 1% OPT. In terms of usage, all proportions can be applied in construction and building work because the compressive strength and water absorption were compliant with the Thai Industrial Standard TIS 57-2530 and TIS 60-2516.

**Keywords:** concrete brick; FGD gypsum; construction and demolition waste; oil palm trunks

---

## 1. Introduction

Concrete is a mixture of ordinary Portland cement (OPC), aggregates (coarse and fine), and water; it is a common material in building construction [1]. The demand for concrete is increasing because it is very durable and cost-efficient [2]. Social and economic factors also affect the demand for concrete. Every ton of cement production releases approximately one ton of carbon dioxide into the atmosphere; this is one of the major causes of global warming [3,4]. Cement production generates up to 7% of the world's total $CO_2$ emissions [5]. Sometimes, mineral admixtures are used as a binder in the concrete formulation to reduce the cement consumption; mostly, these come from industrial waste, which could account for 50%–70 wt% of the formula. This is one of the most effective methods of green concrete production [6–8]. The Krabi power plant is a thermal plant that uses fuel oil to generate electricity. One of the byproducts of this process is sulfur dioxide, which is generated from the sulfur contained in fuel oil and combines with oxygen during combustion [9]. Flue gas desulfurization (FGD) is the treatment used to eliminate sulfur dioxide; it produces FDG gypsum as a by-product [8]. FGD gypsum takes up a lot of space in landfills and may contaminate the surface and groundwater [8]. The main ingredient of FGD gypsum is the same as in natural gypsum [10,11], so it can be used as a substitute for natural gypsum in concrete production [11]. In addition, some research has found that FGD gypsum increases the hydration reaction of Portland cement [12].

The aggregate used in concrete production consists of sand, gravel, and stone [1], which are limited natural resources; there may not be enough to use them in concrete production in the future, according to the increase in demand. Therefore, some recent studies have explored alternative materials that can be substituted for natural resources in concrete production.

Construction activities increase along with population growth [13]. They generate construction and demolition waste (CDW), which is often handled inappropriately because the waste has unique characteristics, properties, and compositions that require special handling. There has been a lack of proper and comprehensive management [14]. In 2012, the generation of CDW in 40 countries worldwide reached more than three billion tons; this figure is predictedto increase rapidly [5,15]. However, this waste can be reused in the place of natural aggregates in construction [13], which not only reduces the amount of waste to be handled by landfills and recycling centers, but also can contribute to a circular economy. Previous studies have found that recycled CDW can be used to replace natural aggregates in concrete, replacing about 30%–50% of the total mass [5].

Thailand produces large amounts of agricultural waste. In 2017, Thailand had a palm oil production of around 14 million tons, of which southern Thailand contributed around 12 million tons [16]. Most of these oil palm plantations are located in four provinces in the southern part of Thailand: Krabi, Suratthani, Chumphon, and Trang, which produce approximately four million tons of palm oil per year [17]. Palm oil cultivation has also expanded from the south to other regions of Thailand. The main residues of palm oil from mills and plantations are palm kernel shells, mesocarp fibers, and empty fruit bunches [18]. When the oil palm trees are 16 years old or more, they have a decreased yield and have to be felled and replanted [19]. The felling of palm trees generates major residues such as oil palm leaves and trunks [18]. These materials can often be reused, e.g., oil palm shells are used to make a lightweight concrete [20] and empty palm bunches can be used for biomass briquettes [21], but managing oil palm trunks is difficult because they biodegrade slowly. There is research about oil palm trunk management methods such as anaerobic biodegradation [22] or use as fillers in gypsum composite [23], but there have been few studies on how to use them as an aggregate in concrete bricks. This research aims to study the utilization of waste from power plants (FGD gypsum), construction and demolition waste (CDW), and oil palm trunks (OPT) as raw materials for producing concrete bricks for the economic development of the local community in Krabi province.

## 2. Materials and Methods

### 2.1. Raw Materials

FGD gypsum was collected from the Electricity Generating Authority of Thailand (EGAT) (the Krabi power plant) and OPT was obtained from Krabi province, Thailand. CDW was obtained from the On Nut garbage disposal plant in Bangkok. Then, FGD gypsum and OPT were sun dried for five and 14 days, respectively. After drying, OPT was cut into small pieces by a grinding machine. For CDW, we reduced the size using a jaw crusher and roll crusher, and the particles were sieved through a vibration screen machine to ensure that they were the correct size, 720 µm to 4.75 mm. The chemical composition of the raw materials including OPC, FGD gypsum, coarse sand, CDW, gravel, and OPT were analyzed using an X-ray fluorescence spectrometer (S8 Tiger, Bruker, Karlsruhe, Germany). The mineralogical phase was analyzed by X-ray diffraction (XRD) (D8 Discover, Bruker, Karlsruhe, Germany). The microstructure was characterized using a scanning electron microscope (JSM-IT300, JEOL, Tokyo, Japan). The particle size of CDW, OPT, coarse sand, and gravel was verified using a laser particle distribution analyzer (Mastersizer 3000, Malvern, Malvern, UK).

### 2.2. Production of Concrete Bricks

According to IS:456-2000 and IS:1343-1980, M15 concrete having the specified characteristic strength for 28 days at 15 MPa [24] is the minimum grade for use as reinforced concrete. Therefore, this research used the ratio of concrete mix according to M15 concrete. The percentage replacement of fine aggregates (coarse sand) by CDW is 0%, 25%, 50%, and 75%, and the percentage replacement of coarse aggregates (gravel) by OPT is 0%, 0.5%, and 1%. Every proportion used FGD gypsum 5.5% to replace OPC type I. To produce concrete bricks, the mixture of each proportion has OPC and FGD gypsum as the binder, coarse sand and CDW as the fine aggregate, gravel, and OPT as the coarse aggregate, and water. The ratio of binder:fine aggregate:coarse aggregate was 1:2:4 by weight, and the ratio of binder:water was 1:0.5 by weight. All dried ingredients were mixed, then 1.5 L of water was added, and the mixture was stirred for 2 min. After that, the mixture was compacted in a cube mold ($15 \times 15 \times 15$ cm) and wrapped in plastic film for 24 h. After removing the specimen from the mold, all specimens were cured by immersion in water for 28 days at room temperature. For each proportion, we made five replicates for the compressive strength test and five replicates for the water absorption test. The proportions of concrete bricks are shown in Table 1.

**Table 1.** Proportions of concrete bricks. Abbreviations: CDW, construction and demolition waste; FGD, flue-gas desulfurization; OPC, ordinary Portland cement; OPT, oil palm trunks.

| Treatments | Binder (%) | | Fine Aggregate (%) | | Coarse Aggregate (%) | | Water/Cement Ratio |
|---|---|---|---|---|---|---|---|
| | OPC | FGD Gypsum | Coarse Sand | CDW | Gravel | OPT | |
| S01 | | | 100 | 0 | 100 | 0 | |
| S02 | 94.5 | 5.5 | 75 | 25 | 100 | 0 | |
| S03 | | | 50 | 50 | 100 | 0 | |
| S04 | | | 25 | 75 | 100 | 0 | |
| S05 | | | 100 | 0 | 99.5 | 0.5 | |
| S06 | 94.5 | 5.5 | 75 | 25 | 99.5 | 0.5 | |
| S07 | | | 50 | 50 | 99.5 | 0.5 | 0.5 |
| S08 | | | 25 | 75 | 99.5 | 0.5 | |
| S09 | | | 100 | 0 | 99 | 1 | |
| S10 | 94.5 | 5.5 | 75 | 25 | 99 | 1 | |
| S11 | | | 50 | 50 | 99 | 1 | |
| S12 | | | 25 | 75 | 99 | 1 | |
| Control | 100 | 0 | 100 | 0 | 100 | 0 | |

Then, the compressive strength and water absorption of the concrete bricks were determined according to ASTM C39 [25] and [26], respectively. The microstructure of the concrete bricks was characterized using a scanning electron microscope (JSM-IT300, JEOL, Tokyo, Japan). The mineralogical phases of the concrete bricks were identified using an X-ray diffractometer (D8 Discover, Bruker, Karlsruhe, Germany).

### 2.3. Environmental and Economic Analyses

For the environmental analysis, a comparison was made among all concrete brick treatments from S1 to S12 and the control. The functional unit in this study was one brick of $15 \times 15 \times 15$ cm. The system boundary was cradle-to-gate covering the raw materials acquisition, materials preparation, and concrete brick manufacturing. The life cycle assessment study was carried out using SimaPro 8.0.5.13, ReCiPe Midpoint (H) V1.06, and the World ReCiPe H method from the National Metal and Materials Technology Center (MTEC), Thailand. The results were expressed in six impact categories: abiotic depletion, acidification, global warming, ozone layer depletion, human toxicity and terrestrial ecotoxicity.

## 3. Results and Discussion

### 3.1. Characteristics of OPC, FGD Gypsum, Coarse Sand, CDW, Gravel, and OPT

#### 3.1.1. Chemical Composition

The chemical compositions of raw materials were analyzed by X-ray Fluorescence Spectrometry (S8 Tiger, Bruker, Karlsruhe, Germany), as shown in Table 2. The results showed that the major component of OPC was CaO, while FGD gypsum had $SO_3$ as the main component. $SiO_2$ was the major component of coarse sand and CDW. The largest components of gravel and OPT were CaO and $SiO_2$, respectively.

**Table 2.** Chemical properties of raw materials.

| Oxides | Content (wt%) | | | | | |
|---|---|---|---|---|---|---|
| | OPC | FGD Gypsum | Coarse Sand | CDW | Gravel | OPT |
| CaO | 68.78 | 43.84 | 1.34 | 27.90 | 68.98 | 20.32 |
| $SiO_2$ | 16.85 | 2.36 | 85.82 | 55.12 | 17.32 | 26.90 |
| $Al_2O_3$ | 4.10 | 0.88 | 7.25 | 7.74 | 5.47 | 2.22 |
| $SO_3$ | 3.94 | 49.45 | 0.03 | 0.67 | 0.11 | 4.05 |
| $Fe_2O_3$ | 3.60 | 0.82 | - | 3.34 | 3.36 | 8.75 |
| MgO | 1.30 | 1.79 | 0.21 | 1.56 | 2.20 | 3.09 |
| $K_2O$ | 0.54 | 0.14 | 4.47 | 2.67 | 1.88 | 22.75 |
| $TiO_2$ | 0.28 | 0.03 | 0.16 | 0.36 | 0.30 | 0.47 |

### 3.1.2. Particle Sizes

The particle size of the fine aggregate and coarse sand was more evenly distributed than in CDW but all passed through a 4.75 mm sieve and were retained on a 0.075 mm sieve, which was the size range specified for fine aggregate. The most common particle size of coarse sand was 0.850 mm, while for CDW it was 1.00 mm. The particle size distribution of the coarse aggregate was ostensibly different. The particle size of gravel was 5.60 mm and for OPT it was 0.71 mm, as shown in Figure 1.

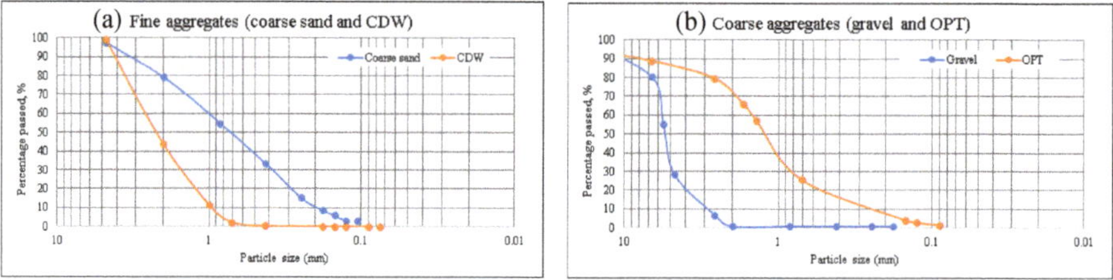


**Figure 1.** The particle size of the raw materials: (**a**) fine aggregates; (**b**) coarse aggregates.

### 3.1.3. The Mineralogical Phase

The results showed that the major crystalline phases of OPC and FGD gypsum were calcium silicate oxide ($Ca_2SiO_4$) and calcium sulfate dihydrate ($CaSO_4 \cdot 2H_2O$), respectively. Quartz ($SiO_2$) was the major crystalline phase of coarse sand, CDW, gravel, and OPT, as shown in Figure 2.

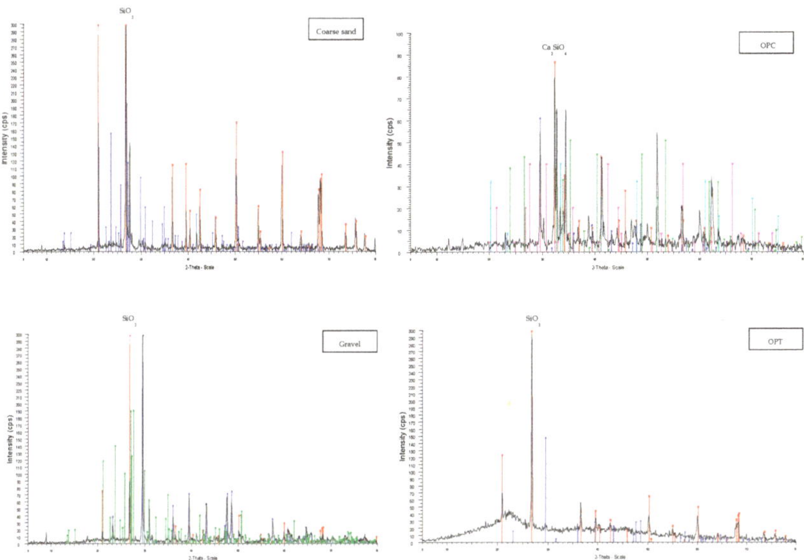

**Figure 2.** X-ray diffractometer (XRD) pattern of raw materials.

### 3.1.4. Microstructures of the Raw Materials

As shown in Figure 3, the particle morphologies of OPC were plate-shaped and smaller than FGD gypsum, while FGD gypsum had a columnar shape and smooth surface appearance. The main components of OPC and FGD gypsum were oxygen and calcium. The particle morphologies of coarse sand were an irregular angular shape and had a smoother surface than CDW, while CDW had a large size, highly irregular angles, and rough surface. Oxygen was the main element in both coarse sand and CDW, but the minor element of coarse sand was silicon, while for CDW it was carbon. The particle morphologies of gravel and OPT had a rough surface, but OPT also had a layer stacked on the surface. OPT had only two elements, carbon and oxygen, while oxygen was the major element and carbon was a minor element of gravel.

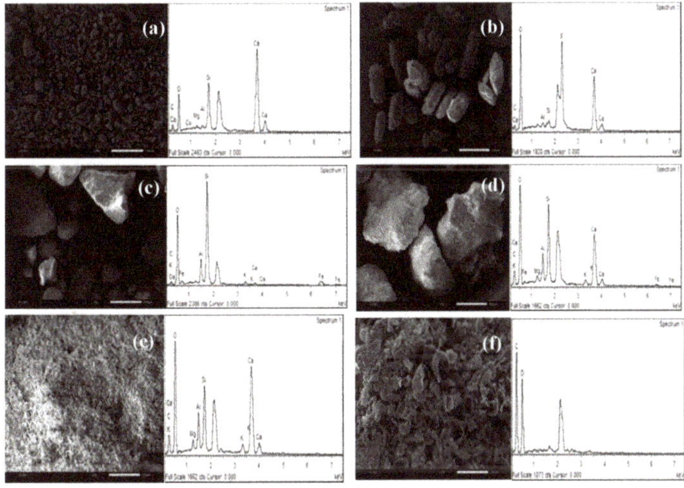

**Figure 3.** Microstructures of (**a**) OPC; (**b**) FGD gypsum; (**c**) coarse sand; (**d**) CDW; (**e**) gravel; and (**f**) OPT.

*3.2. Characteristics of Concrete Bricks Made from OPC, FGD Gypsum, Coarse Sand, CDW, Gravel, and OPT*

The concrete brick can be divided into three groups; Group 1 was a representative of concrete bricks without OPT including S01, S02, S03, and S04; Group 2 was S05, S06, S07, and S08, which were the concrete bricks with OPT 0.5%. The concrete bricks that used OPT 1% were S09, S10, S11, and S12, together called Group 3. In each group, there was a difference in the amount of replacement of fine aggregate by 0%, 25%, 50%, and 75% of CDW.

### 3.2.1. General Appearance of Concrete Bricks

The general appearances of concrete bricks were similar in all proportions, as shown in Figure 4. The surface of concrete bricks was not smooth due to the connection between cement paste and aggregate being incomplete. More voids were found in concrete bricks with a higher amount of OPT due to the different size of coarse aggregate particles between gravel and OPT.

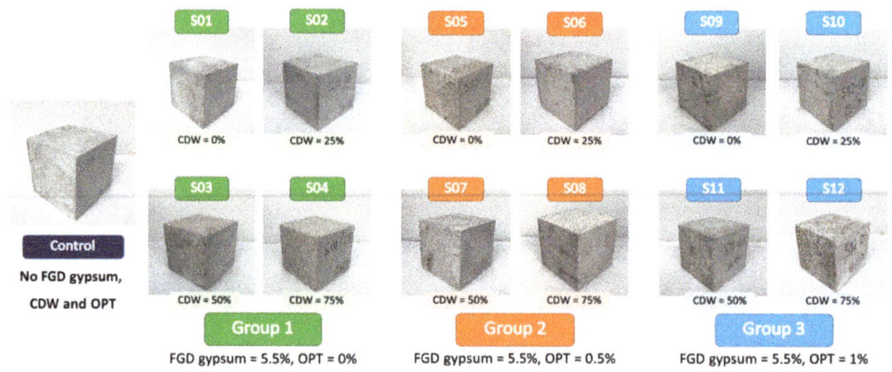


**Figure 4.** The appearance of the concrete bricks.

### 3.2.2. Compressive Strength of Concrete Bricks

Concrete brick specimen S01 produced from 5.5% FGD gypsum without CDW and OPT cured at 28 days obtained the highest compressive strength (45.18 MPa), while concrete brick specimen S12 produced from 5.5% FGD gypsum, 75% CDW, and 1% OPT obtained the lowest compressive strength (26.84 MPa). As shown in Figure 5, the compressive strength of concrete bricks decreased when higher amounts of CDW and OPT were added to the concrete mix. From the results of this study, the compressive strength of all specimens was higher than the compressive strength required by the standards of hollow load-bearing concrete masonry units (TIS 57–2530 all types, ≥5 MPa) [27] and solid load-bearing concrete masonry units (TIS 60–2516 all types, ≥8.34 MPa) [28]. Thus, the results of this study can be applied in production of concrete masonry blocks especially in masonry wall construction of 15 to 20 MPa [29–31].

### 3.2.3. Water Absorption

Concrete brick specimen S12 produced from 5.5% FGD gypsum, 75% CDW, and 1% OPT (with the highest amount of CDW and OPT) cured at 28 days obtained the highest water absorption (111.4 kg/m$^3$), while concrete brick specimen S01 produced from 5.5% FGD gypsum without CDW and OPT (excluding control) obtained the lowest water absorption (78 kg/m$^3$). As shown in Figure 6, the more CDW and OPT that were added, the more the water absorption increased. In addition, all concrete brick specimens were water absorption compliant, as defined by both standards (160 kg/m$^3$) [27,28].

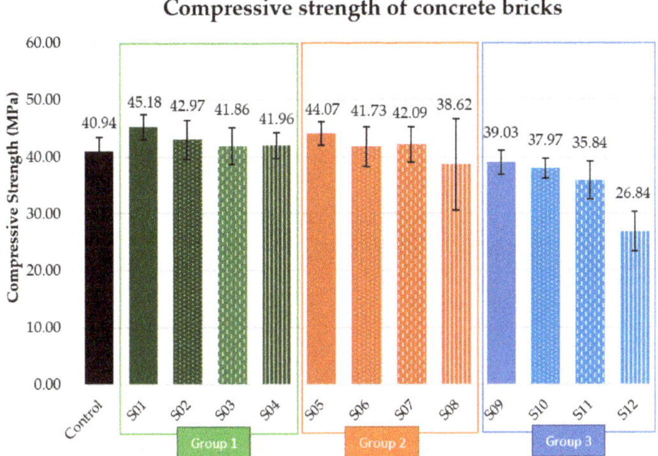



**Figure 5.** The compressive strength of concrete bricks.

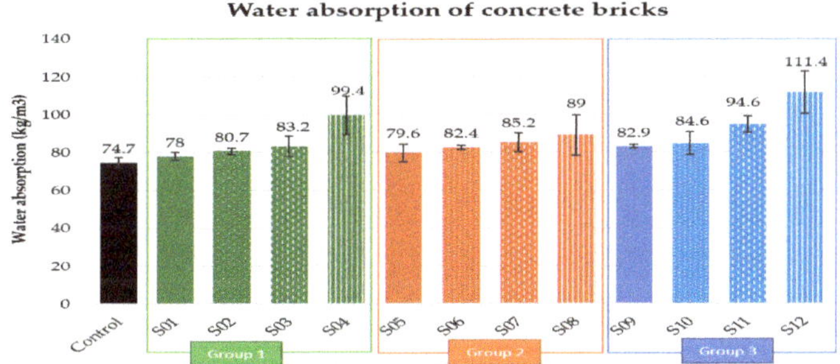



**Figure 6.** The water absorption of concrete bricks.

According to the compressive strength and water absorption analysis, concrete bricks with higher amounts of CDW and OPT had a lower compressive strength and higher water absorption, especially concrete brick specimen S12 (75% CDW and 1% OPT), which had the lowest compressive strength and the highest water absorption, while concrete brick specimen S01, produced from 5.5% FGD gypsum without CDW and OPT, had the highest compressive strength and the lowest water absorption. Figure 7 shows a cross section of the concrete brick specimen; the left side of Figure 7 represents a control concrete brick without CDW and OPT, having coarse sand (0.85 mm) as the fine aggregate and gravel (5.60 mm) as the coarse aggregate, with the same particle size as coarse sand as the fine aggregate can fill the gaps in gravel as coarse aggregate. The center image in Figure 7 represents a concrete brick having CDW (1.00 mm) replacing 25%–100% of coarse sand; the particle size of CDW is larger than that of coarse sand, and the particles were arranged in a disorderly manner in concrete brick specimens, causing more porosities and large voids that resulted in decreasing compressive strength and increasing water absorption when a higher amount of CDW was added. The right side of Figure 7 represents a concrete brick having CDW replacing 25%–100% of coarse sand and OPT replacing 0.5%–1% of gravel; vegetal particles like OPT tend to detach from the OPC binder because of the deficient cohesion, chemical incompatibility, and dissolution of hydration products, causing an increase in water absorption [32] and a decrease in compressive strength when a higher amount of OPT is added. According to Diquélou et al. (2015) and Dinh (2014) [33,34], the addition of plant aggregates (corn stalk, lavender stalk, and hemp shiv) decreased the products of hydration, including calcium silicate hydrate (CSH) and calcium hydroxide (CH). Moreover, the different size of coarse aggregate between gravel (5.60 mm) and OPT (0.71 mm) created porosity in concrete bricks, resulting in decreased compressive strength and increased water absorption.

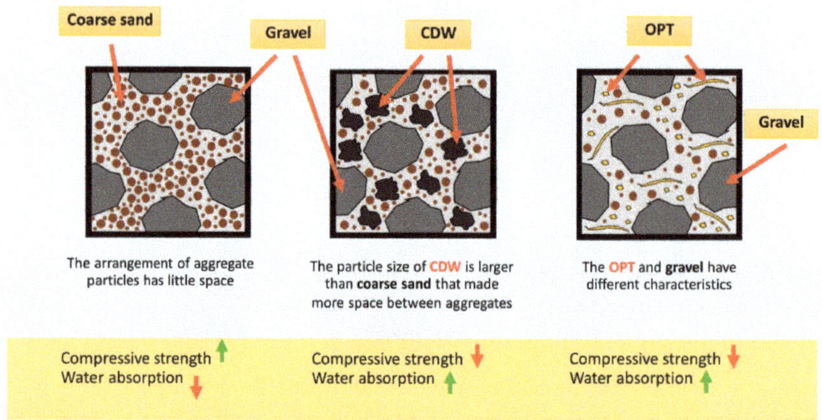


**Figure 7.** Cross sections of concrete brick specimens.

### 3.2.4. Mineralogical Phases of Concrete Bricks

The X-ray diffraction patterns of concrete brick after 28 days of curing were analyzed (D8 Discover, Bruker, Karlsruhe, Germany). The results are shown in Figure 8. The mineralogical phases of all concrete bricks were similar. All of the samples had intense peaks around 29.2° (2θ), which is both CSH and calcite. Other peaks of CSH were indicated at 42.7° and 55.3° (2θ) [35].

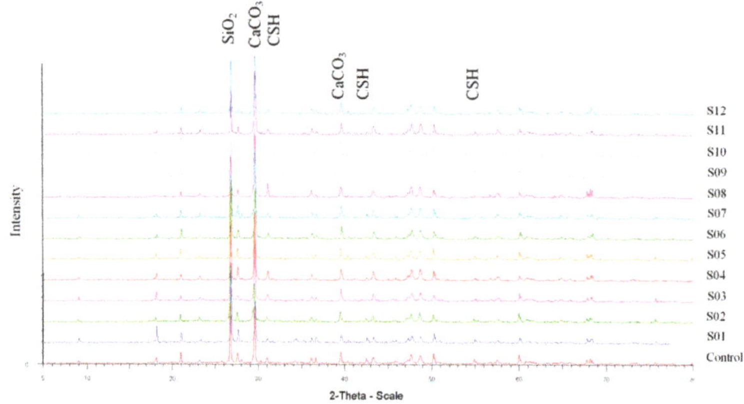


**Figure 8.** The X-ray diffraction patterns of concrete bricks. CSH: calcium silicate hydrate.

### 3.2.5. Microstructure of Concrete Bricks

The microstructure of concrete bricks was identified by a Scanning Electron Microscope (JSM-IT300, JEOL, Tokyo, Japan) (×1500 magnification). The element compositions of concrete brick were identified by an Energy-Dispersive X-ray Spectrometer (JSM-IT300, JEOL, Tokyo, Japan). As shown in Figure 9, ettringite, the insoluble calcium layer formed by the reaction of gypsum with tricalcium aluminate (C$_3$A), was obviously found in S04 and S09 on the aggregate surface due to the hydration reaction. The ettringite was a needle-shaped crystal caused by sulfate ions of gypsum reacting with C$_3$A and water. Most hydration reaction products consisted of calcium silicate hydrate (CSH), calcium hydroxide (CH), and ettringite, which filled the gap. CSH had a jelly-like function, to combine the aggregate.

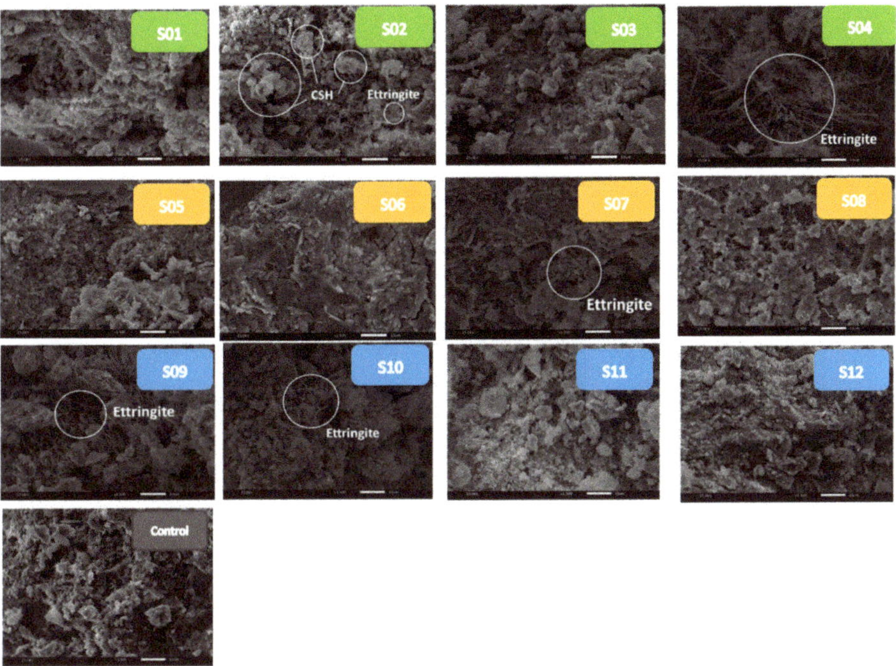


**Figure 9.** The microstructures of concrete bricks.

Considering the utilization of agricultural and industrial wastes as raw materials for concrete brick production, all proportions can be applied in construction and building activities because the compressive strength and water absorption of all concrete brick specimens in this study were compliant with the requirements. Furthermore, the X-ray diffraction pattern, microstructure, and element compositions showed that all of the concrete brick proportions had similar characteristics. To provide recommendations for how concrete brick manufacturing can be more eco-friendly, the environmental impact of the concrete brick made from FGD gypsum, construction and demolition waste, and oil palm trunks was analyzed in terms of abiotic depletion, acidification, global warming, ozone layer depletion, human toxicity and terrestrial ecotoxicity. In addition, the economic feasibility of the concrete bricks was studied and the results are given in Section 3.3.

*3.3. Life Cycle Assessment and Economic Feasibility of Concrete Bricks Made from OPC, FGD Gypsum, Coarse Sand, CDW, Gravel, and OPT*

The processes of concrete brick production consisted of raw material acquisition, raw material preparation, and brick manufacturing. The impact was calculated and converted into an equivalent factor to calculate the total impact in each category. Concrete brick production of 12 ratios (S01–S12) was divided into three groups differentiated by percent of OPT: S01–S04 used OPT as gravel substitution by 0%, S05–S08 by 0.5%, and S09–S12 by 1%, respectively. Within the group, the concrete brick CDW ratios also varied between 25%, 50%, and 75%, while the amount of FGD gypsum was fixed at 5.5% for all ratios. Control samples of concrete brick production were produced from OPC, sand, and gravel, which are the common materials used in construction work. The impact category of concrete brick production was compared between each ratio using S12 concrete brick production to represent 100% of emissions. The impacts of the concrete production tended to be higher due to the substitution of fine aggregates by construction and demolition waste at 25%, 50%, or 75% in S04, S08, and S12, respectively, as shown in Figure 10. The impacts from CDW were caused by the raw materials acquisition and preparation, e.g., diesel as a fuel for transportation and electricity for crushing, grinding, and screening the CDW [36]. Moreover, the substitution of coarse aggregate by oil palm trunks (OPT) had an impact on concrete brick production due to the OPT being transported to the experiment site by truck and the preparation needing shredding, which used diesel and electricity, respectively.

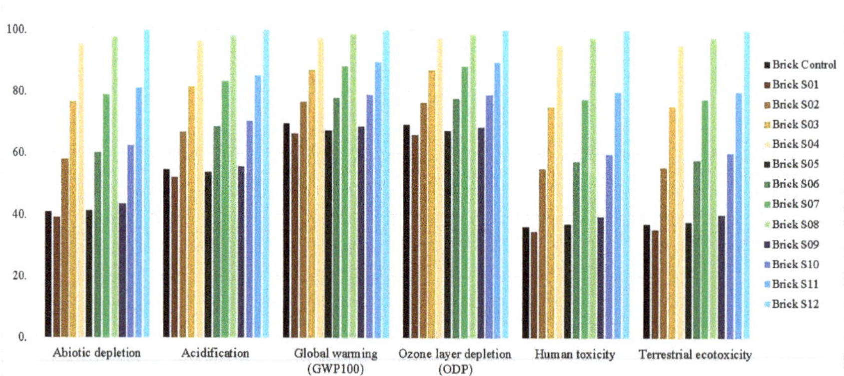


**Figure 10.** The comparative environmental impacts of concrete brick production of each ratio.

According to the results, most of the environmental impacts were directly related to the volume of aggregates being substituted. The more construction and demolition waste and oil palm trunks that were used, the more the environmental impacts of the concrete brick production increased due to the preparation of CDW and OPT consuming a high amount of electricity.

In this research, economic feasibility is divided into two case studies. First, the S01 concrete brick production produced from 5.5% FGD gypsum without CDW and OPT had the best physical properties including compressive strength, water absorption, and density. Second was S12 produced from 5.5% FGD gypsum, 75% CDW, and 1% OPT, which used the highest amount of recycled waste as a substitution in brick production. The comparison of the economic feasibility of concrete brick production between S01 and S12 showed that S12 concrete brick production (fixed cost: 226,000 Baht and variable cost: 17.57 Baht/piece) was higher than the S01 concrete brick production (fixed cost: 48,000 Baht and variable cost: 16.56 Baht/piece) due to the raw material preparation, which required machinery and high electricity consumption for material preparation, as shown in Figure 11.

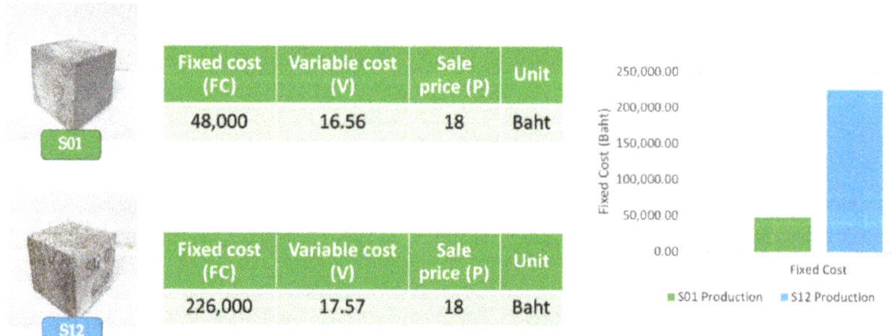


**Figure 11.** Cost comparison of the brick production between S01 and S12.

The breakeven point of S01 concrete brick production is 33,333 pieces, which is lower than the S12 concrete brick production of 528,581.4 pieces due to the lower capital from the production causing the payback period of S12 concrete brick production (3.383 years) to be higher than that of S01 concrete brick production (0.213 years) at the same sale price (18 Baht/piece), as shown in Figure 12.

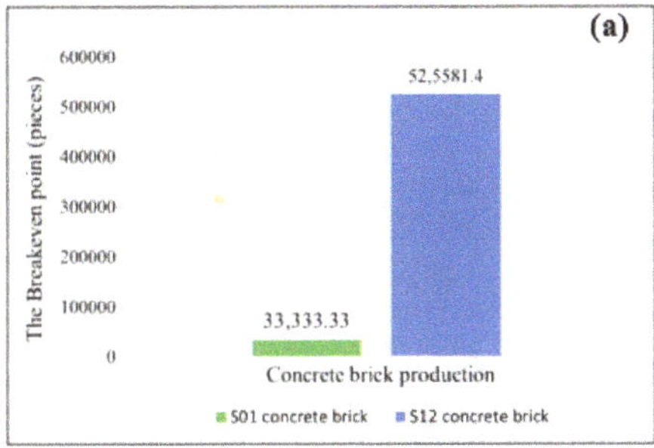

**Figure 12.** (a) Breakeven point comparison of brick production between S01 and S12; (b) payback period comparison of brick production between S01 and S12.

## 4. Conclusions

This study focused on the utilization of waste from power plants, construction and demolition waste, and agricultural wastes. The experiment was conducted to determine the utilization of FGD gypsum, construction and demolition waste (CDW), and oil palm trunks (OPT) for concrete brick production. The results found that the optimal ratio of concrete brick leading to the highest compressive strength was in concrete brick specimen S01. The main conclusions are summarized as follows:

1. Concrete brick specimen S01 produced from 5.5% FGD gypsum without CDW and OPT cured at 28 days obtained the highest compressive strength (45.18 MPa), while concrete brick specimen S12 produced from 5.5% FGD gypsum, 75% CDW, and 1% OPT obtained the lowest compressive strength (26.84 MPa).
2. Adding only FGD gypsum in the production of concrete brick specimens showed that the compressive strength of these specimens increased with significant differences when compared to the compressive strength of the control.
3. Using CDW and OPT led to similar effects: increasing CDW and OPT reduced the compressive strength and increased the water absorption of concrete bricks.
4. All of the concrete brick proportions had compressive strength and water absorption that were compliant with the requirements of TIS 57-2530 and TIS 60-2516; thus, CDW, OPT, and FGD gypsum can be used as raw materials for bearing concrete masonry units.
5. A comparison of the environmental impacts of concrete brick production showed that the production of S01 concrete bricks made with 5.5% FGD gypsum without CDW and OPT involved the lowest emissions. On the other hand, S12 concrete brick production led to the highest emissions due to the concrete bricks being made with 5.5% FGD gypsum, 75% CDW, and 1% OPT.
6. The S01 concrete brick production had the shortest payback period due to the lower cost of the raw material preparation process, which does not require machinery for grinding and sieving.

The authors would like to recommend that the size of concrete bricks/blocks made from these wastes can be differed to suit the construction practices. The compressive strength of concrete bricks in this study is high which can be used in the production of hollow blocks used to accommodate reinforcement in the walls for increased earthquake and lateral load resistance. To apply this in building and construction works, mechanical characteristics such as tensile strength, elastic modulus, stress–strain curves should be examined for future study.

**Author Contributions:** Conceptualization, N.S.; methodology, N.S., L.P., and P.K.; validation, N.S.; formal analysis, L.P. and P.K.; investigation, L.P. and P.K.; resources, N.S.; data curation, N.S., L.P., and P.K.; writing—original draft preparation, L.P. and P.K.; writing—review and editing, N.S.; visualization, N.S.; supervision, N.S.; project administration, N.S.; funding acquisition, N.S. All authors have read and agreed to the published version of the manuscript.

**Funding:** This research was funded by the Electricity Generating Authority of Thailand, grant number 62-F107000-11-IO.SS03F3008435. APC was funded by the Ratchadaphiseksomphot Endowment Fund.

**Acknowledgments:** The authors would like to thank Damrong Saiya, Chaiwat Boonroeang Kaow, Prapakorn Thachapradit, Sirikan Sawangsawai, and Awika Numnual from Electricity Generating Authority of Thailand (EGAT) (The Krabi power plant) for their support. The authors gratefully acknowledge the Department of Environmental Science in the Faculty of Science, the Department of Mining Engineering, the Department of Civil Engineering, and the Center of Laboratory Engineering in the Faculty of Engineering, the Environmental Research Institute and the Scientific and Technological Research Equipment Centre (STREC) for providing mechanical equipment, Electricity Generating Authority of Thailand (EGAT) (The Krabi power plant) for providing FGD gypsum and OPT, and On Nut garbage disposal plant in Bangkok for providing the CDW used in this investigation. The authors gratefully acknowledge Seksan Papong and the National Metal and Materials Technology Center for the database and software used in this investigation. The authors would also like to thank Sarawut Srithongouthai, Vorapot Kanokantapong, Chidsanuphong Chart-asa, and Nuttakorn Intaravicha for their kind suggestions.

**Conflicts of Interest:** The authors declare no conflict of interest.

# References


1. Verian, K.P.; Ashraf, W.; Cao, Y. Properties of recycled concrete aggregate and their influence in new concrete production. *Resour. Conserv. Recycl.* **2018**, *133*, 30–49. [CrossRef]
2. Medina, C.; Zhu, W.; Howind, T.; Frías, M.; De Sánchez Rojas, M.I. Effect of the constituents (asphalt, clay materials, floating particles and fines) of construction and demolition waste on the properties of recycled concretes. *Constr. Build. Mater.* **2015**, *79*, 22–33. [CrossRef]
3. Ahmari, S.; Ren, X.; Toufigh, V.; Zhang, L. Production of geopolymeric binder from blended waste concrete powder and fly ash. *Constr. Build. Mater.* **2012**, *35*, 718–729. [CrossRef]
4. Mo, K.H.; Johnson, A.U.; Jumaat, M.Z.; Yap, S.P. Feasibility study of high volume slag as cement replacement for sustainable structural lightweight oil palm shell concrete. *J. Clean. Prod.* **2015**, *91*, 297–304. [CrossRef]
5. Akhtar, A.; Sarmah, A.K. Construction and demolition waste generation and properties of recycled aggregate concrete: A global perspective. *J. Clean. Prod.* **2018**, *186*, 262–281. [CrossRef]
6. Van Den Heede, P.; De Belie, N. Environmental impact and life cycle assessment (LCA) of traditional and "green" concretes: Literature review and theoretical calculations. *Cem. Concr. Compos.* **2012**, *34*, 431–442. [CrossRef]
7. Igliński, B.; Buczkowski, R. Development of cement industry in Poland—History, current state, ecological aspects. A review. *J. Clean. Prod.* **2017**, *141*, 702–720. [CrossRef]
8. Jiang, L.; Li, C.; Wang, C.; Xu, N.; Chu, H. Utilization of flue gas desulfurization gypsum as an activation agent for high-volume slag concrete. *J. Clean. Prod.* **2018**, *205*, 589–598. [CrossRef]
9. Nikhom, S. Characterization of Gypsum Paste Using Flue Gas Desulfurization Sludge, Rubber Wood Fly Ash and Kaolin. Master's Thesis, Department of Civil Engineering (Structural Enginner), Prince of Songkla University, Songkhla, Thailand, 2010.
10. Lei, D.Y.; Guo, L.P.; Sun, W.; Liu, J.P.; Miao, C.W. Study on properties of untreated FGD gypsum-based high-strength building materials. *Constr. Build. Mater.* **2017**, *153*, 765–773. [CrossRef]
11. Caillahua, M.C.; Moura, F.J. Technical feasibility for use of FGD gypsum as an additive setting time retarder for Portland cement. *J. Mater. Res. Technol.* **2018**, *7*, 190–197. [CrossRef]
12. Poon, C.S.; Qiao, X.C.; Lin, Z.S. Effects of flue gas desulphurization sludge on the pozzolanic reaction of reject-fly-ash-blended cement pastes. *Cem. Concr. Res.* **2004**, *34*, 1907–1918. [CrossRef]
13. Fatemi, S.; Imaninasab, R. Performance evaluation of recycled asphalt mixtures by construction and demolition waste materials. *Constr. Build. Mater.* **2016**, *120*, 450–456. [CrossRef]
14. Kongchuenak, S. Legal Measures for Recycling of Construction and Demolition Waste: A Comparative Study between Thailand and Japan. Master's Thesis, Department of Natural Resources and Environmental Law, Faculty of Law, Thammasat University, Pathumthani, Thailand, 2016.
15. Hu, K.; Chen, Y.; Naz, F.; Zeng, C.; Cao, S. Separation studies of concrete and brick from construction and demolition waste. *Waste Manag.* **2019**, *85*, 396–404. [CrossRef] [PubMed]

16. Office of Agricultural Economics. Summary of Oil Palm Production in 2017. Available online: http://www.oae.go.th/assets/portals/1/fileups/prcaidata/files/oilpalm60_dis.pdf (accessed on 10 June 2019).

17. Agricultural Research Development Agency. Oil Palm Production in Southern Thailand. Available online: http://www.arda.or.th/kasetinfo/south/palm/controller/01-12.php (accessed on 9 June 2019).

18. Abnisa, F.; Arami-Niya, A.; Wan Daud, W.M.A.; Sahu, J.N.; Noor, I.M. Utilization of oil palm tree residues to produce bio-oil and bio-char via pyrolysis. *Energy Convers. Manag.* **2013**, *76*, 1073–1082. [CrossRef]

19. Oil Palm Agronomical Research Center. *Oil Palm Tree: The Propagation, Plantation Management*; Knowledge book Series; Prince of Songkla University: Songkhla, Thailand, 2010; pp. 71–73.

20. Traore, Y.B.; Messan, A.; Hannawi, K.; Gerard, J.; Prince, W.; Tsobnang, F. Effect of oil palm shell treatment on the physical and mechanical properties of lightweight concrete. *Constr. Build. Mater.* **2018**, *161*, 452–460. [CrossRef]

21. Srisang, N.; Srisang, S.; Wongpitithawat, P.; The-Eye, K.; Wongkeaw, K.; Sinthoo, C. Production of Biomass Briquette from Residual Bleaching Earth and Empty Palm Bunch. *Energy Procedia* **2017**, *138*, 1079–1084. [CrossRef]

22. Widyasti, E.; Shikata, A.; Hashim, R.; Sulaiman, O.; Sudesh, K.; Wahjono, E.; Kosugi, A. Biodegradation of fibrillated oil palm trunk fiber by a novel thermophilic, anaerobic, xylanolytic bacterium Caldicoprobacter sp. CL-2 isolated from compost. *Enzym. Microb. Technol.* **2018**, *111*, 21–28. [CrossRef]

23. Selamat, M.E.; Hashim, R.; Sulaiman, O.; Kassim, M.H.M.; Saharudin, N.I.; Taiwo, O.F.A. Comparative study of oil palm trunk and rice husk as fillers in gypsum composite for building material. *Constr. Build. Mater.* **2019**, *197*, 526–532. [CrossRef]

24. Gambhir, M.L. Grades of Concrete. In *Concrete Technology*, 3rd ed.; McGraw-Hill Publishing Co. Ltd.: New York, NY, USA, 2004; pp. 5–6.

25. ASTM C39/39M-18. *Standard Test Method for Compressive Strength of Cylindrical Concrete Specimens*; ASTM International: West Conshohocken, PA, USA, 2016.

26. ASTM C642-13. *Standard Test Method for Density, Absorption, and Voids in Hardened Concrete*; ASTM International: West Conshohocken, PA, USA, 2013.

27. TIS 57-2530. Standard for hollow load-bearing concrete masonry units. In *Thailand Industrial Standard*; Ministry of Industry: Bangkok, Thailand, 1990.

28. TIS 60-2516. Standard for solid load-bearing concrete masonry units. In *Thailand Industrial Standard*; Ministry of Industry: Bangkok, Thailand, 1973.

29. Li, D.; Zhuge, Y.; Gravina, R.; Mills, J.E. Compressive stress strain behavior of crumb rubber concrete (CRC) and application in reinforced CRC slab. *Constr. Build. Mater.* **2018**, *166*, 745–759. [CrossRef]

30. Zahra, T.; Dhanasekar, M. Prediction of masonry compressive behaviour using a damage mechanics inspired modelling method. *Constr. Build. Mater.* **2016**, *109*, 128–138. [CrossRef]

31. Mohamad, G.; Lourenço, P.B.; Roman, H.R. Mechanics of hollow concrete block masonry prisms under compression: Review and prospects. *Cem. Concr. Compos.* **2007**, *29*, 181–192. [CrossRef]

32. Ahmad, M.R.; Chen, B. Influence of type of binder and size of plant aggregate on the hygrothermal properties of bio-concrete. *Constr. Build. Mater.* **2020**, *251*, 1–14. [CrossRef]

33. Diquélou, Y.; Gourlay, E.; Arnaud, L.; Kurek, B. Impact of hemp shiv on cement setting and hardening: Influence of the extracted components from the aggregates and study of the interfaces with the inorganic matrix. *Cem. Concr. Compos.* **2015**, *55*, 112–121. [CrossRef]

34. Dinh, T.M. *Contribution to the Development of Precast Hempcrete Using Innovative Pozzolanic Binder*; University of Toulouse: Toulouse, France, 2014.

35. Hunnicutt, W. Characterization of Calcium-Silicate-Hydrate and Calcium-Alumino-Silicate-Hydrate. Master's Thesis, Department of Civil Engineering. Graduate College, University of Illinois at Urbana-Champaign, Urbana, IL, USA, 2013.

36. Afkhami, B.; Akbarian, B.; Beheshti, A.N.; Kakaee, A.H.; Shabani, B. Energy consumption assessment in a cement production plant. *Sustain. Energy Technol. Assess.* **2015**, *10*, 84–89. [CrossRef]

*Article*

# Applications of Gene Expression Programming and Regression Techniques for Estimating Compressive Strength of Bagasse Ash based Concrete

**Muhammad Faisal Javed [1], Muhammad Nasir Amin [2,*], Muhammad Izhar Shah [1,*], Kaffayatullah Khan [2], Bawar Iftikhar [1], Furqan Farooq [1], Fahid Aslam [3], Rayed Alyousef [3,*] and Hisham Alabduljabbar [3]**

[1]  Department of Civil Engineering, COMSATS University Islamabad, Abbottabad Campus, Abbottabad 22060, KP Pakistan; arbabfaisal@cuiatd.edu.pk (M.F.J.); bawar@cuiatd.edu.pk (B.I.); furqan@cuiatd.edu.pk (F.F.)

[2]  Department of Civil and Environmental Engineering, College of Engineering, King Faisal University (KFU), P.O. 380, Al-Hofuf, Al Ahsa 31982, Saudi Arabia; kkhan@kfu.edu.sa

[3]  Department of Civil Engineering, College of Engineering in Al-Kharj, Prince Sattam Bin Abdulaziz University, Al-Kharj 11942, Saudi Arabia; f.aslam@psau.edu.sa (F.A.); h.alabduljabbar@psau.edu.sa (H.A.)

*  Correspondence: mgadir@kfu.edu.sa (M.N.A.); mizhar@cuiatd.edu.pk (M.I.S.); r.alyousef@psau.edu.sa (R.A.); Tel.: +966-13-589-5431 (M.N.A.); +92-301-8040796 (M.I.S.)

Received: 25 July 2020; Accepted: 18 August 2020; Published: 21 August 2020

**Abstract:** Compressive strength is one of the important property of concrete and depends on many factors. Most of the concrete compressive strength predictive models mainly rely on available literature data, which are too simple to consider all the contributing factors. This study adopted a new approach to predict the compressive strength of sugarcane bagasse ash concrete (SCBAC). A vast amount of data from the literature study and fifteen laboratory tested concrete samples with different dosage of bagasse ash, were respectively used to calibrate and validate the models. The novel Gene Expression Programming, Multiple Linear Regression and Multiple Non-Linear Regression were used to model SCBAC compressive strength. The water cement ratio, bagasse ash percent replacement, quantity of fine and coarse aggregate and cement content were used as an input for models development. Various statistical indicators, i.e., NSE, $R^2$ and RMSE were used to assess the performance of the models. The results indicated a strong correlation between observed and predicted values with NSE and $R^2$ both above 0.8 during calibration and validation for the Gene Expression Programming (GEP). The outcomes from GEP outclassed all the models to predict SCBAC compressive strength. The validity of the model is further verified using data of fifteen tests conducted in the laboratory. Moreover, the cement content in the mix was revealed as the most sensitive parameter followed by water cement ratio form sensitivity analysis. The GEP fulfilled all the criteria for external validity. The simple formulae derived in this study could be used reliably for the prediction of SCBAC compressive strength.

**Keywords:** sugarcane bagasse ash; sensitivity analysis; compressive strength; gene expression programming; multiple linear and non-linear regression; green concrete

---

## 1. Introduction

A large amount of waste is generated due to rapid industrial development. Generation and disposal of large amounts of waste poses serious environmental issues [1,2]. The construction industry is responsible for 30% of global carbon dioxide emissions and also consumes maximum energy [3]. Concrete is the most consumed material and production of 1 ton of concrete releases approximately 0.05–0.13 tons of $CO_2$ into the environment [4–6]. Therefore, researchers are focusing on green technologies for sustainability [7–9]. Meanwhile, different types of wastes and by-products of industries were used as supplementary cementitious material for the production of green concrete. Such materials are considered as low-carbon alternatives to ordinary portland cement [10,11]. Incorporating these materials in concrete production can lead to durable and sustainable concrete. Hence, the production and use of green concrete is indispensable for reducing adverse environmental impacts [12,13], Sugarcane Bagasse Ash (SCBA) is agricultural waste produced in the sugar industry from the cogeneration process [14]. The use of SCBA as an alternative cementitious material in a concrete mix is considered to be advantageous to condense the environmental impact of the construction industry and preserve the natural environment [1].

Compressive strength is the fundamental property of concrete and indicates the ability to resist compression loads [15]. Concrete properties mainly depend on many factors such as mix design, type of material, testing procedure and mainly the concrete ingredients [16,17]. The properties and amount of supplementary material, curing techniques and concrete density [18] also effect the mechanical properties [19,20]. Therefore, knowing the relationship between mechanical properties, percentage replacement and water cement ratio used is indispensable before the frequent practice. For the assessment of concrete properties, the laboratory experiments are quite time-consuming, labor-intensive and costly. Therefore, models could be employed to reduce the experimental workload for assessment of mechanical properties [21]. Moreover, because of the small test record and restrictions in the parametric ranges, the accessible models could not provide accurate and desired results. Such a database, which covers a variety of parameters, is needed to create accurate equations which can easily predict multiple properties [1]. Developing strong prediction models for some key properties leads to time saving and making an effective mixture [22]. For this purpose, statistical regression techniques were employed by many researchers [23,24]. Some notable drawbacks were observed while using regression for empirical modeling. Firstly, the structure of the model should be definite and well defined by a linear or nonlinear equation for executing regression analysis. Secondly, the assumption of normality of residuals is another problem in regression analysis [25,26]. There are also concerns regarding the available equations in codes and standards for forecasting compressive strength because such equations are developed based on concrete tests without supplementary cementitious material [26].

Recent progress in the field of artificial intelligence (AI) techniques has resulted in the development of accurate and reliable models for the structural engineering problems [1]. The advancement and progress in the field of AI made it easy to produce models in order to adapt to difficulties related to concrete properties [1]. For solving problems specifically in Civil Engineering, AI has been in focus in both empirical and academic fields over the last decade [27]. The Artificial Neural Network (ANN) is the most widely used AI method to assess different concrete properties [28–30], although ANN is considered to be a black box model since it doesn't provide information about the anticipated prediction principles. The formulations based on neural networks are often too complex to be used because of the empiric formulization [1]. Moreover, the ANN is unable to provide practical prediction equations [22]. Genetic programming (GP) is another type of AI technique that automatically generates models which are totally based on genetic evaluation. GP is an influential optimization technique based on neural and regression techniques. However, GP is much more advantageous than conventional techniques. Without assuming the base form, GP has the ability to produce simple expressions. Some functions should be defined for regression and then analysis of the defined function is performed. There is no need to predefine the functions in GP, i.e., considering the formulations that fit the experimental outcomes, and GP itself adds or deletes various combinations of parameters. In this regard, Gene Expression Programming (GEP) is considered superior to other techniques [16]. The output of GEP is characterized by simple mathematical equations which are more applicable with high accuracy. GEP has a unique and multi-genic nature which permits the development of more complex programs. Recently, GEP has been used for commonly predicted method explicitly in the field of Civil Engineering [31–33].

Various researchers have modelled different properties of concrete using GEP. Abdollahzadeh et al. [16] predicted the 28-day compressive strength of concrete with the help of two GEP models considering the amount of cement, silica fume, CA, FA, water and superplasticizers. From literature, data of 159 concrete mixes were taken. The authors concluded that the GEP models can predict the compressive strength of high-strength concrete for each testing phase rendered to different statistical parameters. Gholampour A., et al. [1] used GEP to model the mechanical properties of recycled aggregate concrete utilizing data from literature. The authors reported a close agreement of the model results with the test results. Saridemir, M., and Bilir, T. [19] studied the elastic modulus of fly ash concrete using GEP models. The data for models were collected from 132 concrete mixes available in literature. The elastic modulus was predicted form the compressive strength of fly ash concrete. In the second model, prediction of the elastic modulus was made through the amount and strength of fly ash concrete. Iqbal, M.F., et al. [34] predicted the mechanical properties of waste foundry sand based concrete with the help of GEP. Extensive data from literature was used and four different parameters were used as input. They confirmed the high accuracy and prediction capacity of the GEP model. Mousavi, S.M., et al., [22] modeled the compressive strength of high performance concrete with GEP. A database was generated from literature study, and they observed that GEP is an effective technique to predict compressive strength of high performance concrete.

However, most of the above mentioned studies mainly depends on data collected from literature and it is very difficult to consider all the factors responsible for compressive strength. No organized and detailed study was conducted that considered both the combination of available literature data for model calibration and experimental testing of concrete for models validation. This study is devoted to apply the novel Gene Expression Programming (GEP), Multiple Linear Regression (MLR) and Multiple Non-linear Regression (MNLR) techniques to predict the compressive strength of Sugarcane Bagasse Ash Concrete (SCBAC). The literature data was used for models calibration and compressive strength results derived from laboratory tests were used to validate the model. SCBA was used to replace Portland cement in quantities of 0% 10%, 20%, 30% and 40% by weight of cement. The formulation of such models that accurately estimate the concrete compressive strength by utilizing the minimum number of parameters and then validating the model results through practical laboratory test results significantly enhances the utility and building trust in the forecasting models in many research areas.

## 2. Methods, Datasets and Experimental Procedure

### 2.1. Genetic Algorithm and Gene Expression Programming (GEP)

The genetic algorithm (GA) was first introduced by J. Holland [35] which was inspired by Darwin's theory of evolution. The biological evolution process is simply shown in the form of GAs and the solution is signified in fixed chromosome form. Similarly, Genetic programming (GP) was first proposed by Cramer [36] and further enhanced by Koza [37,38]. GP is basically an extension of GA, a type of machine learning that automatically generates models which are totally based on genetic evaluation. GP is an influential optimization technique based on neural and regression techniques. The computer-based program grows into a problem-independent solution based on the Darwinian reproduction principle.

The modified version of GP was proposed by Ferreira, C. [39] based on the population evolutionary algorithm and is named gene expression programming (GEP). A simple linear fixed array of chromosomes is encoded in GEP which outclasses the GP where a tree-like structure with variable length was used [40]. In the evolutionary GEP algorithm, the linear fixed length chromosomes and the nonlinear expression tree were inherited from GA and GP, respectively. GEP is an excellent method because of the linear fixed width of genetic programming as well as the genetic algorithm. Moreover, due to the genetic process on the chromosome level, GEP uses the simplest criteria and further permits the development of complex and nonlinear programs due to multi-genic behavior. The whole GEP comprised of five sets—Function set, terminal set, fitness measure set, parameters set and the criteria set—To terminate the functions. Each individual is set as a fixed-size linear string in GEP, which is called a genome. Furthermore, during the reproduction stage, the modification in the chromosomes takes place by genetic operators [41,42]. The schematic diagram of GEP algorithm is presented in Figure 1. Considering the formulations that fit the experimental outcomes, GEP itself adds or deletes various combinations of parameters [16].

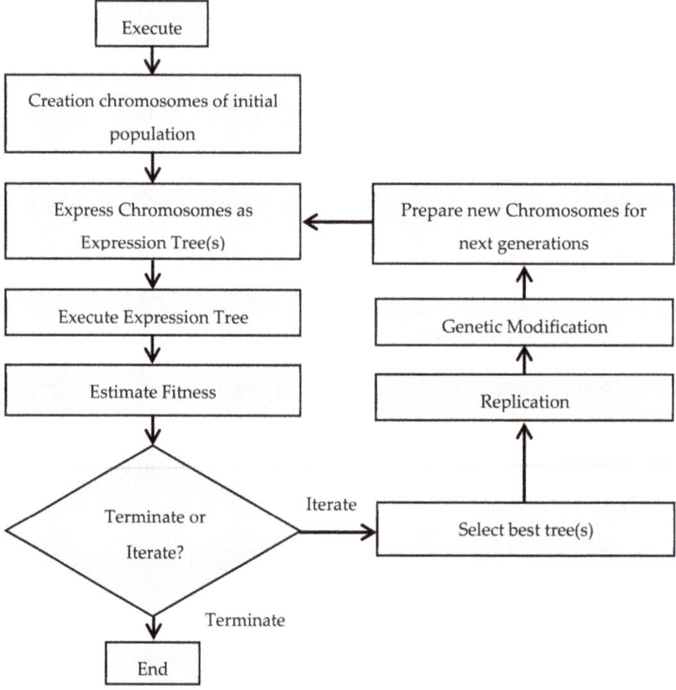


**Figure 1.** Schematic diagram of the Gene Expression Programming (GEP) algorithm [43].

In this paper, the GEP algorithm was applied with the help of GeneXproTools 5.0 [44]. Firstly, a chromosome of fixed length was randomly created for each individual. In the second step, the expression trees are created representing the chromosomes and the fitness of each expression tree is evaluated in terms of statistical indicators. And finally, the reproduction process is initiated for each individual assessed by a fitness function (which was RMSE for the developed model).

*2.2. Multiple Linear and Non-Linear Regression*

The multiple linear regression (MLR) model developed the relationship between many independent and dependent variables. The model expresses the value of a predicted variable as a linear function of one or more predictor variables. The basic assumption of MLR is that the association between $Y_i$ and the p vector of regression $X_i$ is linear. In the multiple non-linear regression (MNLR) model, the relationship between the dependent and independent variable is considered to be non-linear. The MNLR technique estimates the model by creating a random nonlinear connection between dependent and independent variables. The fundamental difference in MNLR is that estimating the equation nonlinearly depends on input variables [45]. The following equations, Equation (1) and Equation (2), represent the MLR [46,47] and MNLR [48], respectively.

$$Y = a + \beta_1 X_1 + \beta_2 X_2 + \cdots\cdots + \beta_i X_i \tag{1}$$

$$Y = a + \beta_1 X_i + \beta_2 X_j + \beta_3 X_i^2 + \beta_4 X_j^2 + \cdots\cdots + \beta_k X_i X_j \tag{2}$$

Where a is the intercept, $\beta$ is the slope or coefficient and k is the number of observations.

In this study, the MLR and MNLR models were developed using statistical analysis software called Statistical Package for Social Sciences (SPSS). It allows the user to perform a set of statistical analyses from basic to complex predictive functionalities. All the datasets were carefully analyzed before performing the linear and non-linear regression in SPSS.

### 2.3. Datasets

The dataset of SCBAC was collected from 21 published papers available in literature [14,49–66]. The final database form literature comprised of 65 data points. Each record of the collected data contained information about water/cement ratio (W/C), SCBA percentage replacement (SCBA%), fine aggregate in the mix (FA), cement content (CC), slump value (S), coarse aggregate in the mix (CA), water absorption (WA), specific gravity of the mix (SG) and compressive strength ($f'_c$). Numerous trials were performed to assess the consistency and validity of the data. The data points with up to 20% deviation from the global trend were not considered in the development and model performance evaluation [34]. The description of the collected data is shown in Table 1. The overall data from literature was used to develop and calibrate the GEP and regression models. For models validation, concrete specimens with varying dosages of SCBA were casted and then tested for compressive strength.

**Table 1.** Statistics of the Available Collected Data.

| Parameter | Unit | Range | Min | Max | Mean | SD |
|---|---|---|---|---|---|---|
| W/C | - | 0.3 | 0.3 | 0.6 | 0.47 | 0.074 |
| SCBA% | % | 50 | 0 | 50 | 14.41 | 11.16 |
| FA | Kg/m$^3$ | 625 | 240 | 865 | 609.1 | 239.1 |
| CC | Kg/m$^3$ | 444 | 116 | 560 | 341.2 | 101.2 |
| CA | Kg/m$^3$ | 770 | 490 | 1260 | 891.8 | 390.6 |
| $f'_c$ | MPa | 67.35 | 14.65 | 82 | 37.61 | 15.5 |

### 2.4. Model Development and Performance Evaluation

Prior to model development, the first and essential step is to choose the input parameters which can have an effect on the SCBAC properties. All the available parameters in the database were studied in detail and several preliminary runs were conducted in order to determine the most influential parameters for SCBAC to establish a generalized relationship. As a result, the compressive strength of SCBAC is considered to be a function of the following input parameters.

$$f'_c = f\left(\frac{W}{C}, \ SCBA\%, \ FA, \ CC, \ CA\right) \tag{3}$$

In order to assess the performance of the developed models, different indicators were used, such as Nash Sutcliff efficiency (NSE), correlation coefficient ($R^2$) and Root Mean Squared Error (RMSE) [67,68]. These statistical indicators were used to check the model simulation with the actual data. The NSE ranges between - $\infty$ to 1, where 1 is a perfect match. An NSE value greater than 0.65 depicts a very good correlation [69,70]. The $R^2$ value lies between 0 and 1 and the higher values indicate fewer errors, while value 1 represents completely matched data [70]. The RMSE is an error index type indicator commonly used in modelling studies. A lower value for RMSE is optimum. The mathematical expressions for RMSE, NSE and $R^2$ are shown below as Equations (4)–(6), respectively.

$$RMSE = \sqrt{\frac{\sum_{i=1}^{n}(P_i - M_i)^2}{N}} \tag{4}$$

$$NSE = 1 - \frac{\sum_{i=1}^{n}(M_i - P_i)^2}{\sum_{i=1}^{n}\left(M_i - \overline{M_i}\right)^2} \tag{5}$$

$$R^2 = \frac{\sum_{i=1}^{n}\left(M_i - \overline{M_i}\right)(P_i - \overline{P_i})}{\sqrt{\sum_{i=1}^{n}\left(M_i - \overline{M_i}\right)^2 \sum_{i=1}^{n}\left(P_i - \overline{P_i}\right)^2}} \tag{6}$$

where n is the number of input samples; $M_i$ and $P_i$ are the measured and predicted values, respectively. $\overline{M}_i$ and $\overline{P}_i$ are the average of measured and predicted values, respectively.

*2.5. Experimental Investigation*

2.5.1. Mix Proportions and Specimen Designation

In order to validate the robustness of the developed GEP and regression models, fifteen concrete specimens were casted. Mixes were abbreviated as CM and BC. CM represents control mix without addition of ash; whereas BC represents cement replaced with bagasse ash. The specific designation 10BC indicates that 10 percent cement has been replaced with SCBA. Target design strength of 21 MPa was designed. SCBA was used to replace OPC in quantities of 0% 10%, 20%, 30% and 40% by weight of cement. Each mix had the same water to cementitious material ratio of 0.5 and overall cementitious contents of 366 Kg/m$^3$ were kept constant. Furthermore, the X-ray Fluorescence (XRF) was conducted for chemical composition of sugarcane bagasse ash. Results of XRF are shown in Table 2. It can be seen that the sum of silica ($SiO_2$), alumina ($Al_2O_3$) and Iron oxide ($Fe_2O_3$) is 77.47 % (>70%), which meets one of the requirements of pozzolan as per ASTM C618-05.

**Table 2.** Chemical Composition of Bagasse Ash.

| Composition | Percentage |
|---|---|
| $SiO_2$ | 66.70 |
| $Al_2O_3$ | 9.24 |
| $Fe_2O_3$ | 1.53 |
| CaO | 10.07 |
| MgO | 4.60 |
| $P_2O_5$ | 1.98 |
| $K_2O$ | 4.32 |
| $Na_2O$ | 1.30 |
| $TiO_2$ | 0.22 |
| MnO | 0.02 |
| LOI | 1.9 |
| Moisture content | 1.1 |

Mix proportions of the concrete mixes are shown in Table 3. Each mix was tested in fresh as well as in hardened state. The fresh concrete tests, i.e., slump test and fresh concrete density were determined as per ASTM C143 and ASTM C138M-01, respectively. At the curing age of 28 days, the compressive strength of the SCBAC and control concrete were determined according to BS 1881: Part 116: 1983 standard.

**Table 3.** Mix Proportions of the Concrete Mix.

| Mix Design. | Cement (Kg/m$^3$) | Bagasse Ash (Kg/m$^3$) | W/C | Fine Aggregate (Kg/m$^3$) | Coarse Aggregate (Kg/m$^3$) | Water (Kg/m$^3$) |
|---|---|---|---|---|---|---|
| 0BC(CM) | 366 | 0 | 0.5 | 742.3 | 1013.5 | 183 |
| 10BC | 329.4 | 36.6 | 0.5 | 742.3 | 1013.5 | 183 |
| 20BC | 292.8 | 73.2 | 0.5 | 742.3 | 1013.5 | 183 |
| 30BC | 256.2 | 109.8 | 0.5 | 742.3 | 1013.5 | 183 |
| 40BC | 219.6 | 146.4 | 0.5 | 742.3 | 1013.5 | 183 |

### 2.5.2. Fresh Concrete Properties

Slump and fresh concrete densities of the concrete mixes at different replacement levels of SCBA (0, 10, 20, 30 & 40%) are presented in Figure 2a,b, respectively. Result shows that with the addition of further ash, the consistency of concrete increases. The maximum and minimum slump values of 58mm and 29mm were found for 40BC and control mix, respectively. Similarly, the maximum fresh density achieved by the CM was 2308.4 Kg/m$^3$. The fresh density decreased with the introduction of SCBA as cement replacement. In comparison to control concrete, the decrease in fresh concrete densities for 10BC, 20BC, 30BC and 40BC mixes was found to be 0.85%, 1.82%, 2.34% and 2.62%, respectively. The decrease in fresh density is due to the fact that the density is a function of specific gravity. Since the specific gravity of cement is more as compared to bagasse ash, therefore, the density of the CM mix is highest.

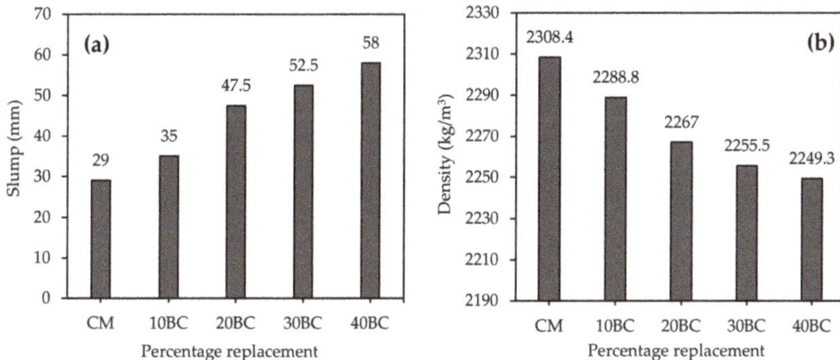

**Figure 2.** Comparison of different concrete mixes for (**a**) slump density and (**b**) fresh concrete density.

### 2.6. Sensitivity Analysis

The behavior of the developed models and its relationship with the input variables was measured using sensitivity analysis. It is a procedure to observe the most sensitive parameters as there are uncertainties related to model input values. The models can provide acceptable results during calibration and validation but it is not confirmed that they will also provide the same results on unknown data sets. Therefore, it is necessary to perform the sensitivity analysis to evaluate the relative contribution of each input parameter on model output. In the current study, the method developed by Gandomi et al. [71] was adopted. The authors used Equation (7) and Equation (8) to find out the contribution to output by each input variable.

$$N_i = f_{max}(x_i) - f_{min}(x_i) \tag{7}$$

$$S_i = \frac{N_i}{\sum_{j=i}^{n} N_j} \times 100 \tag{8}$$

The $f_{max}(x_i)$ and $f_{min}(x_i)$ are the max and min of the estimated output over the $i^{th}$ output.

## 3. Results and Discussion

### 3.1. SCBAC Mixes Compressive Strength from Laboratory Tests

For GEP and regression models validation, concrete specimens were casted with different percentages of SCBA and tested for compressive strength at the curing age of 28 days. The results are tabulated in Table 4. The results show the SCBAC (10BC) has a higher strength than all of the mixes. The difference in strength between 10BC and CM is found to be 3.58%. The compressive strength consistently decreases with the addition of SCBA. This may be attributed to the fact that cement is the primary binding material. The cement replacement of 30% and 40% reduces the lime content required for pozzolanic reaction. The optimum replacement level of SCBA with cement is found to be 10%.

**Table 4.** Compressive Strength of the Normal and Sugarcane Bagasse Ash Concrete (SCBAC) Mixes for Models Validation.

| MIX | 28 Day Compressive Strength (MPa) | | | |
| --- | --- | --- | --- | --- |
| | Specimen 1 | Specimen 2 | Specimen 3 | Average |
| 0BC (CM) | 23.14 | 23.80 | 24.10 | 23.68 |
| 10BC | 23.94 | 24.76 | 24.90 | 24.53 |
| 20BC | 21.43 | 20.70 | 20.92 | 21.01 |
| 30BC | 21.0 | 19.78 | 19.65 | 20.14 |
| 40BC | 15.30 | 14.65 | 14.8 | 14.91 |

*3.2. GEP Based Formulation for Compressive Strength of SCBAC Mixes*

The model to formulate the compressive strength of SCBAC, Equation (9), is proposed to predict the $28^{th}$ day compressive strength of SCBAC. The GEP model developed for compressive strength of SCBAC is shortlisted after running a set of GEP algorithms initiating from fundamental set function, small head size and a single gene chromosome. The important parameters involved for GEP setting are tabulated in Table 5. All the structural association of chromosomes and function sets were chosen before running the GEP algorithm. The capability of GEP is highly affected by parameters selection. Basic mathematical operators were used to assess the best and simplest model. Due to the number of potential outcomes and the complexity of estimating the test model, the three best populations of 50, 100 or 150 were considered.

$$f'_c = \frac{A}{D} \times B \times C \tag{9}$$

where,

$$A = (SCBA\% - 31)\frac{2.6}{W/C} \tag{10}$$

$$B = \frac{-0.014 \times FA \times (SCBA\% + CC)}{CC + SCBA\% + 110.6 - CA} \tag{11}$$

$$C = 0.27\,CC - \frac{9.6}{W/C} - 25.38 - 0.005FA + 11.16\frac{CC}{FA} \tag{12}$$

$$D = \frac{W/C}{7.2} + 5 \tag{13}$$

**Table 5.** Optimal Parameter Setting for GEP.

| Parameter | | Setting |
|---|---|---|
| General | Chromosomes | 30 |
| | Genes | 4 |
| | Head size | 10 |
| | Gene size | 26 |
| | Linking function | Addition |
| | Function set | $+, -, \times, \div, \hat{\ }2, \sqrt[3]{\ }$ |
| Genetic operators | Mutation rate | 0.0138 |
| | Inversion rate | 0.00546 |
| | IS transposition rate | 0.00546 |
| | RIS transposition rate | 0.00546 |
| | One point recombination rate | 0.00277 |
| | Two point recombination rate | 0.00277 |
| | Gene recombination rate | 0.00755 |
| | Gene transposition rate | 0.00277 |
| Numerical Constants | Constants per gene | 10 |
| | Data type | Floating type |
| | Lower bound | −10 |
| | Upper bound | 10 |

For GEP model simulation, five parameters, i.e., W/C, SCBA%, FA, CC, CA were used as an input. It is evident from Figure 3a,b that the developed GEP model effectively considered the influence of all the input parameters for both model calibration and validation. The statistical indicators, i.e., NSE, $R^2$ and RMSE, are 0.83, 0.83, 6.67 during model calibration and 0.87, 0.85, and 4.57 for model validation, respectively. An $R^2$ value of more than 0.8 is adequate [31]. Moreover, lower values of RMSE and higher values of $R^2$ and NSE show that the developed models adequately simulated the results [40].

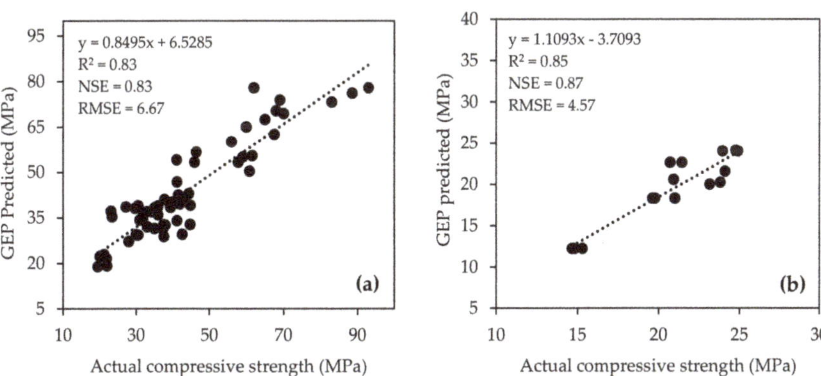


**Figure 3.** Comparison of actual and GEP predicted compressive strength during (**a**) calibration (**b**) validation.

### 3.3. MLR and MNLR Based Formulation for Compressive Strength of SCBAC Mixes

Figure 4a,b graphically demonstrates the MLR-simulated compressive strength of SCBAC mixes during model calibration and validation, respectively. The statistical indicators show a strong relationship between actual and predicted data. During MLR model calibration, the NSE, $R^2$ and RMSE, are 0.45, 0.45 and 12.21, respectively. Similarly, the values of NSE, $R^2$ and RMSE are 0.46, 0.44, and 24.74, respectively, for the validation period. It is obvious from the results that performance of the MLR during calibration is better than model validation. This could be considered as one of the limitations of MLR.

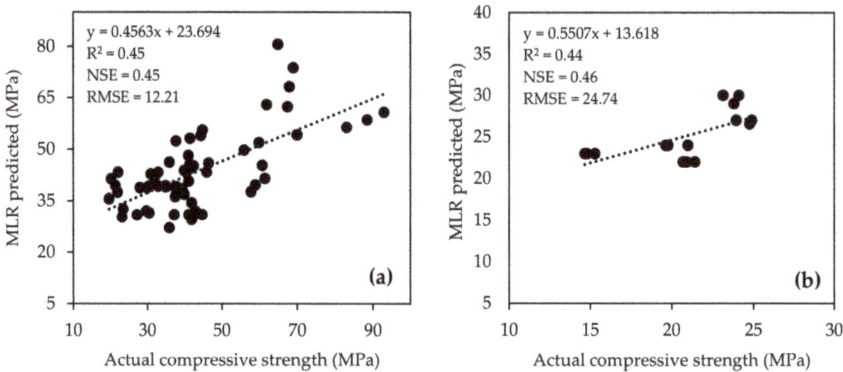

**Figure 4.** Comparison of actual and Multiple Linear Regression (MLR) predicted compressive strength during (**a**) calibration (**b**) validation.

The MNLR model simulation results for SCBAC compressive strength are shown in Figure 5a,b for calibration and validation, respectively. The NSE, $R^2$ and RMSE are found to be 0.55, 0.58 and 11.05, respectively, for calibration and 0.58, 0.56 and 5.52 for validation of the MNLR model. The absolute error between actual and predicted data for all the models is shown in Figure 6.

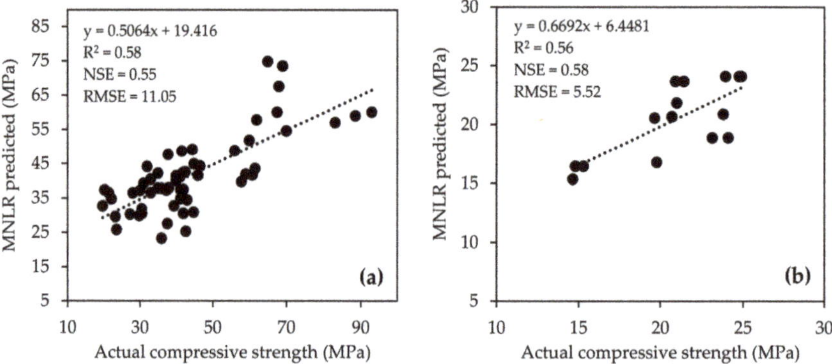

**Figure 5.** Comparison of actual and Multiple Non-Linear Regression (MNLR) predicted compressive strength during (**a**) calibration (**b**) validation.

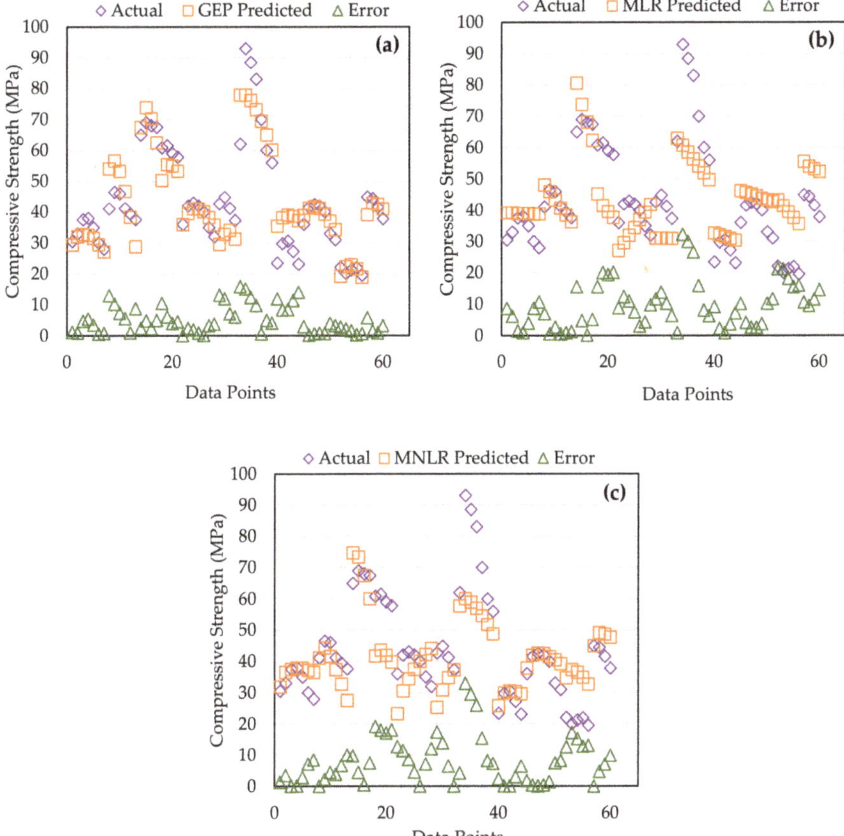

**Figure 6.** Absolute error in the actual and predicted results (**a**) GEP (**b**) MLR (**c**) MNLR.

### 3.4. Sensitivity Analysis

The contribution of the most relevant parameters (W/C, SCBA%, CC, FA, and CA) in the model was evaluated through sensitivity analysis. Various sensitive parameters were identified that are essential for modelling the compressive strength of SCBAC. The results are graphically shown in Figure 7. The results revealed that the cement content (CC) in the mix is the most sensitive parameter for SCBAC compressive strength with a relative contribution of 55.73%. The same results were observed during experimental testing of SCBAC where the strength of the concrete decreased consistently with the addition of SCBA up to 40% (reduction in cement content). So it is obvious that the model results are much in line with experimental results which further enhances the robustness of the developed model. The second and third most sensitive parameters for compressive strength of SCBAC turned out to be W/C with 17.14% and the CA aggregate content in the mix with 16.98% relative contribution to compressive strength. Furthermore, the compressive strength is least affected by FA content in the mix.

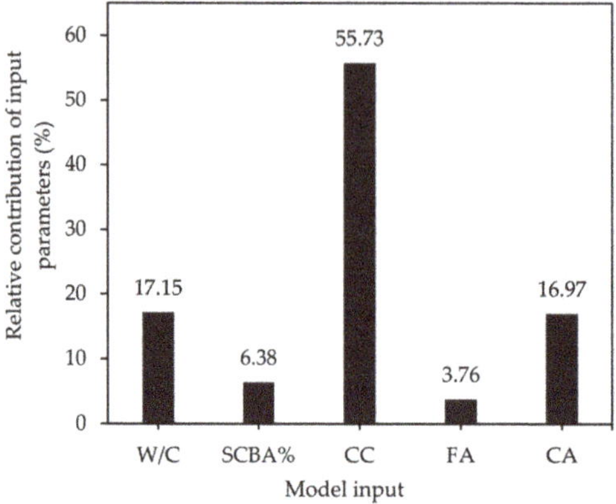


**Figure 7.** Relative contribution of input parameters to model output.

## 3.5. External Validation of the Models

Various researchers have suggested that the performance of the proposed model considerably depends on the ratio of data points to the number of inputs [72,73]. In order to check the suitability of data points for attaining the connection between selected variables, the ratio should be more than 5 for the ideal model [73]. In the current study, the suggested ratio is 13 which fulfill the criteria set by researchers. Furthermore, other criteria for external verification of the models proposed by Golbraikh and Tropsha [74] suggested that the slope of the regression line should be approximately near to 1. Roy and Roy [75] introduced a new indicator ($R_m$) for external predictability of the proposed models and suggested the $R_m$ value of more than 0.5 for a satisfied model. Alavi et al. [76] suggested that the squared correlation coefficients $R_o^2$ and $R_o'^2$ between the observed and predicted values should be near to 1. In this paper, the performances of the developed models were evaluated by the above mentioned criteria and the results are summarized in Table 6. The novel GEP model strongly fulfills all the aforementioned criteria.

**Table 6.** Statistical Parameters of the Models for the External Validation.

| S.No. | Equation | Condition | Models | Value | Suggested by |
|---|---|---|---|---|---|
| 1. | $R = \dfrac{\sum_{i=1}^{n}\left(M_i-\overline{M_i}\right)\left(P_i-\overline{P_i}\right)}{\sqrt{\sum_{i=1}^{n}\left(M_i-\overline{M_i}\right)^2 \sum_{i=1}^{n}\left(P_i-\overline{P_i}\right)^2}}$ | $R > 0.8$ | GEP | 0.838 | [74] |
| | | | MLR | 0.456 | |
| | | | MNLR | 0.580 | |
| 2. | $k = \dfrac{\sum_{i=1}^{n}(M_i-P_i)}{M_i^2}$ | $0.85 < k < 1.15$ | GEP | 0.992 | [75] |
| | | | MLR | 1.021 | |
| | | | MNLR | 0.954 | |
| 3. | $k' = \dfrac{\sum_{i=1}^{n}(M_i-P_i)}{P_i^2}$ | $0.85 < k' < 1.15$ | GEP | 1.008 | [75] |
| | | | MLR | 0.978 | |
| | | | MNLR | 1.047 | |
| 4. | $R_m = R^2 \times \left(1 - \sqrt{\left|R^2 - R_0^2\right|}\right)$ | $R_m > 0.5$ | GEP | 0.510 | [76] |
| | | | MLR | 0.119 | |
| | | | MNLR | 0.120 | |
| | $R_0^2 = \dfrac{\sum_{i=1}^{n}\left(P_i-M_i^0\right)^2}{\sum_{i=1}^{n}\left(P_i-\overline{P_i^0}\right)^2},\ M_i^0 = k \times P_i$ | $R_0^2 \cong 1$ | GEP | 0.999 | |
| | | | MLR | 0.999 | |
| | | | MNLR | 0.997 | |
| | $R_0'^2 = \dfrac{\sum_{i=1}^{n}\left(M_i-P_i^0\right)^2}{\sum_{i=1}^{n}\left(M_i-\overline{M_i^0}\right)^2},\ P_i^0 = k' \times M_i$ | $R_0'^2 \cong 1$ | GEP | 0.999 | |
| | | | MLR | 0.999 | |
| | | | MNLR | 0.997 | |

### 3.6. Comparison of GEP, MLR and MNLR Models

To the authors' best knowledge, no detailed study was conducted that considered the regression as well as novel GEP technique to predict the compressive strength of SCBAC and also to validate the model results by laboratory concrete testing. Therefore, GEP, MLR and MNLR models were developed to predict the compressive strength of SCBAC and the results are compared. The NSE and $R^2$ values are highest for GEP followed by MNLR. Similarly, the RMSE is lowest for the GEP model during both calibration and validation. Summarizing the results and performance of all the models, the GEP outclassed all other models during calibration and validation. Moreover, the GEP-developed equation can easily be used to predict the compressive strength of SCBAC. The performance of both MLR and MNLR reduced (in terms of statistical indicators) during validation. This may be considered as one of the shortcomings of the regression-based models. The average errors between actual and predicted data are shown in Figure 6 and were found to be 0.79%, 2.12% and 4.52% for GEP, MLR and MNLR models, respectively.

## 4. Conclusions

The present study reported the applications of the novel gene expression programming and regression methods, i.e., multiple linear and non-linear regression for modeling and predicting the compressive strength of sugarcane bagasse ash concrete. A new method was proposed where the already available data from literature study was used to develop and calibrate the proposed models. To validate the models, laboratory specimens of concrete with different dosages of bagasse ash were casted and simultaneously the test results were used for validation. Regardless of the large number of factors responsible for variation in the compressive strength, the developed models were successfully calibrated and validated. The goodness of fit for the models was estimated in terms of NSE, $R^2$ and RMSE. A good correlation was observed in both the calibration and validation period between actual and predicted values. However, it was observed that gene expression programming outperformed all the other techniques and is capable of forecasting the compressive strength of sugarcane bagasse ash concrete for a known set of input parameters. The empirical equation developed by gene expression programming could be used frequently to estimate the compressive strength. Gene expression programming played a major role in evaluating the suitable relations required for the qualitative depiction of the processes responsible for compressive strength. This technique does not assume any prior solution and establishes a suitable relation between dependent and independent variables, thus making the techniques superior to others. The performance of the multiple linear and non-linear regression models reduced during model validation and may be considered as one of the drawbacks of the regression based models. The novel gene expression programming applied in this study could serve as a baseline to model and predict the mechanical properties of concrete with high accuracy and the minimum number of parameters.

**Author Contributions:** Conceptualization, data analysis, review and editing, M.F.J.; Formal analysis and visualization, M.N.A.; conceptualization, data collection, writing original draft preparation, M.I.S.; Investigation and review, K.K.; Methodology and write up, B.I.; visualization and manuscript revision, F.F.; validation check and supervision, F.A.; Writing—Review & editing, R.A.; funding acquisition and supervision, H.A. All authors have read and agreed to the published version of the manuscript.

**Funding:** This research project was supported by the deanship of scientific research at prince sattam bin Abdulaziz University under the research project number 2020/01/16810.

**Acknowledgments:** The authors acknowledge the support of lab staff for providing access to materials lab and helped in casting and testing concrete specimens.

**Conflicts of Interest:** The authors declare no conflict of interest.

## References


1. Gholampour, A.A.; Gandomi, A.H.; Ozbakkaloglu, T. New formulations for mechanical properties of recycled aggregate concrete using gene expression programming. *Constr. Build. Mater.* **2017**, *130*, 122–145. [CrossRef]
2. Batayneh, M.; Marie, I.; Asi, I. Use of selected waste materials in concrete mixes. *Waste Manag.* **2007**, *27*, 1870–1876. [CrossRef] [PubMed]
3. Li, H.; Deng, Q.; Xia, B.; Zhang, J.; Skitmore, M. Assessing the life cycle CO2 emissions of reinforced concrete structures: Four cases from China. *J. Clean. Prod.* **2019**, *210*, 1496–1506. [CrossRef]
4. Mao, L.-X.; Hu, Z.; Xia, J.; Feng, G.-L.; Azim, I.; Yang, J.; Liu, Q.-F. Multi-phase modelling of electrochemical rehabilitation for ASR and chloride affected concrete composites. *Compos. Struct.* **2019**, *207*, 176–189. [CrossRef]
5. He, Z.; Zhu, X.; Wang, J.; Mu, M.; Wang, Y. Comparison of CO2 emissions from OPC and recycled cement production. *Constr. Build. Mater.* **2019**, *211*, 965–973. [CrossRef]
6. Obla, K.H. What is green concrete? *Indian Concr. J.* **2009**, *24*, 26–28.
7. Pan, S.-Y.; Fan, C.; Lin, Y.-P. Development and Deployment of Green Technologies for Sustainable Environment. *Environments* **2019**, *6*, 114. [CrossRef]
8. Sagotra, A.K.; Errandonea, D.; Cazorla, C. Mechanocaloric effects in superionic thin films from atomistic simulations. *Nat. Commun.* **2017**, *8*, 1–7. [CrossRef]
9. Shafiei, M.W.M.; Abadi, H. The Importance of Green Technologies and Energy Efficiency for Environmental Protection. *Int. J. Appl. Environ. Sci.* **2017**, *12*, 937–951.
10. Chen, L.; Wang, L.; Cho, D.-W.; Tsang, D.C.W.; Tong, L.; Zhou, Y.; Yang, J.; Hu, Q.; Poon, C.S. Sustainable stabilization/solidification of municipal solid waste incinerator fly ash by incorporation of green materials. *J. Clean. Prod.* **2019**, *222*, 335–343. [CrossRef]
11. Wang, L.; Chen, L.; Cho, D.-W.; Tsang, D.C.W.; Yang, J.; Hou, D.; Baek, K.; Kua, H.W.; Poon, C.-S. Novel synergy of Si-rich minerals and reactive MgO for stabilisation/solidification of contaminated sediment. *J. Hazard. Mater.* **2019**, *365*, 695–706. [CrossRef] [PubMed]
12. Du, H.; Tan, K.H. Properties of high volume glass powder concrete. *Cem. Concr. Compos.* **2017**, *75*, 22–29. [CrossRef]
13. Tan, K.H.; Du, H. Use of waste glass as sand in mortar: Part I–Fresh, mechanical and durability properties. *Cem. Concr. Compos.* **2013**, *35*, 109–117. [CrossRef]
14. Bahurudeen, A.; Kanraj, D.; Dev, V.G.; Santhanam, M. Performance evaluation of sugarcane bagasse ash blended cement in concrete. *Cem. Concr. Compos.* **2015**, *59*, 77–88. [CrossRef]
15. Yehia, S.; Abdelfatah, A.; Mansour, D. Effect of Aggregate Type and Specimen Configuration on Concrete Compressive Strength. *Crystals* **2020**, *10*, 625. [CrossRef]
16. Abdollahzadeh, G.; Jahani, E.; Kashir, Z. Genetic Programming Based Formulation to Predict Compressive Strength of High Strength Concrete. *Civ. Eng. Infrastruct. J.* **2017**, *50*, 207–219.
17. Hwang, R.; Lee, I.W.; Pyo, S.; Kim, D.J. Influence of the Aggregate Surface Conditions on the Strength of Quick-Converting Track Concrete. *Crystals* **2020**, *10*, 543. [CrossRef]
18. Marcalikova, Z.; Cajka, R.; Bilek, V.; Bujdos, D.; Sucharda, O. Determination of Mechanical Characteristics for Fiber-Reinforced Concrete with Straight and Hooked Fibers. *Crystals* **2020**, *10*, 545. [CrossRef]
19. Sarıdemir, M.; Billir, T. Modeling of elastic modulus of concrete containing fly ash by gene expression programming. In Proceedings of the Fourth International Conference on Sustainable Construction Materials and Technologies, Las Vegas, NV, USA, 7–11 August 2016.
20. Kliszczewicz, A.; Ajdukiewicz, A. Differences in instantaneous deformability of HS/HPC according to the kind of coarse aggregate. *Cem. Concr. Compos.* **2002**, *24*, 263–267. [CrossRef]
21. Zhang, J.; Zhao, Y.; Li, H. Experimental Investigation and Prediction of Compressive Strength of Ultra-High Performance Concrete Containing Supplementary Cementitious Materials. *Adv. Mater. Sci. Eng.* **2017**, *2017*, 4563164. [CrossRef]
22. Mousavi, S.M.; Aminian, P.; Gandomi, A.H.; Alavi, A.H.; Bolandi, H. A new predictive model for compressive strength of HPC using gene expression programming. *Adv. Eng. Softw.* **2012**, *45*, 105–114. [CrossRef]
23. Domone, P.; Soutsos, M. Approach to the proportioning of high-strength concrete mixes. *Concr. Int.* **1994**, *16*, 26–31.
24. Yeh, I.C. Modeling of strength of high-performance concrete using artificial neural networks. *Cem. Concr. Res.* **1998**, *28*, 1797–1808. [CrossRef]

25. Gandomi, A.H.; Alavi, A.H.; Arjmandi, P.; Aghaeifar, A.; Seyednour, R. Genetic programming and orthogonal least squares: A hybrid approach to modeling the compressive strength of CFRP-confined concrete cylinders. *J. Mech. Mater. Struct.* **2010**, *5*, 735–753. [CrossRef]

26. Mousavi, S.M.; Gandomi, A.H.; Alavi, A.H.; Vesalimahmood, M. Modeling of compressive strength of HPC mixes using a combined algorithm of genetic programming and orthogonal least squares. *Struct. Eng. Mech.* **2010**, *36*, 225–241. [CrossRef]

27. Jepsen, M.T. Predicting concrete durability by using artificial neural network. In Proceedings of the Durability of Exposed Concrete Containing Secondary Cementitious Materials, Hirtshals, Denmark, 20 November 2002. Special NCR-Publication.

28. Yeh, I.C. Modeling slump flow of concrete using second-order regressions and artificial neural networks. *Cem. Concr. Compos.* **2007**, *29*, 474–480. [CrossRef]

29. Basma, A.A.; Barakat, S.A.; Al-Oraimi, S. Prediction of cement degree of hydration using artificial neural networks. *ACI Mater. J.* **1999**, *96*, 167–172.

30. Ji, T.; Lin, T.; Lin, X. A concrete mix proportion design algorithm based on artificial neural networks. *Cem.'Concr. Res.* **2006**, *36*, 1399–1408. [CrossRef]

31. Gandomi, A.H.; Alavi, A.H.; Mirzahosseini, M.R.; Nejad, F.M. Nonlinear genetic-based models for prediction of flow number of asphalt mixtures. *J. Mater. Civ. Eng.* **2011**, *23*, 248–263. [CrossRef]

32. Mahdinia, S.; Eskandari-Naddaf, H.; Shadnia, R. Effect of cement strength class on the prediction of compressive strength of cement mortar using GEP method. *Constr. Build. Mater.* **2019**, *198*, 27–41. [CrossRef]

33. Azimi-Pour, M.; Eskandari-Naddaf, H. ANN and GEP prediction for simultaneous effect of nano and micro silica on the compressive and flexural strength of cement mortar. *Constr. Build. Mater.* **2018**, *189*, 978–992. [CrossRef]

34. Iqbal, M.F.; Liu, Q.F.; Azim, I.; Zhu, X.; Yang, J.; Javed, M.F.; Rauf, M. Prediction of mechanical properties of green concrete incorporating waste foundry sand based on gene expression programming. *J. Hazard. Mater.* **2020**, *384*, 121322. [CrossRef] [PubMed]

35. Holland, J. *Adaptation in Natural and Artificial Systems*; University of Michigan Press: Ann Arbor, MI, USA, 1975.

36. Cramer, N.L. A representation for the adaptive generation of simple sequential programs. In Proceedings of the First International Conference on Genetic Algorithms, Pittsburg, PA, USA, 24–25 July 1985.

37. Koza, J.R. *Genetic Programming II: Automatic Discovery of Reusable Subprograms*; MIT Press: Cambridge, MA, USA, 1994; Volume 13, p. 32.

38. Koza, J.R. Genetic programming as a means for programming computers by natural selection. *Stat. Comput.* **1994**, *4*, 87–112. [CrossRef]

39. Ferreira, C. Gene expression programming: A new adaptive algorithm for solving problems. *arXiv* **2001**, arXiv:cs/0102027.

40. Azim, I.; Yang, J.; Javed, M.F.; Iqbal, M.F.; Mahmood, Z.; Wang, F.; Liu, Q.F. Prediction model for compressive arch action capacity of RC frame structures under column removal scenario using gene expression programming. *Structures* **2020**, *25*, 212–228. [CrossRef]

41. Lopes, H.S.; Weinert, W.R. A gene expression programming system for time series modeling. In Proceedings of the XXV Iberian Latin American Congress on Computational Methods in Engineering, Recife, Brazil, 10–12 November 2004.

42. Ferreira, C. Gene expression programming in problem solving. In *Soft Computing and Industry*; Springer: Berlin/Heidelberg, Germany, 2002; pp. 635–653.

43. Ferreira, C. *Gene Expression Programming: Mathematical Modelling by an Artificial Intelligence*, 2nd ed.; Springer: Berlin/Heidelberg, Germany, 2006; 480p.

44. *GeneXopro Tools 5.0*; GEPSOFT Limited: Bristol, UK, 2013.

45. Simões, A.; Costa, E. Prediction in evolutionary algorithms for dynamic environments using markov chains and nonlinear regression. In Proceedings of the 11th Annual Conference on Genetic and Evolutionary Computation, Montreal, QC, Canada, 8–12 July 2009.

46. Pedhazur, E.J. *Multiple Regression in Behavioral Research: Prediction and Explanation*, 2nd ed.; Holt, Rinehart, & Winston: New York, NY, USA, 1982; 822p.

47. Montgomery, D.C.; Peck, E.A.; Vining, G.G. *Introduction to Linear Regression Analysis*, 5th ed.; Wiley: New York, NY, USA, 2012.

48.  Ivakhnenko, A. Heuristic self-organization in problems of automatic control. *Automatica* **1970**, *3*, 207–219. [CrossRef]

49.  Srinivasan, R.; Sathiya, K. Experimental study on bagasse ash in concrete. *Int. J. Serv. Learn Eng. Humanit. Eng. Soc. Entrep.* **2010**, *5*, 60–66. [CrossRef]

50.  Patel, J.A.; Raijiwala, D. Experimental study on use of sugar cane bagasse ash in concrete by partially replacement with cement. *Int. J. Innov. Res. Sci. Eng. Technol.* **2015**, *4*, 2228–2232.

51.  Neeraja, D.; Jagan, S.; Kumar, S.; Mohan, P.G. Experimental Study on Strength Properties of Concrete by Partial Replacement of Cement with Sugarcane Bagasse Ash. *Nat. Environ. Poll. Technol.* **2014**, *13*, 629.

52.  Ganesan, K.; Rajagopal, K.; Thangavel, K. Evaluation of bagasse ash as supplementary cementitious material. *Cem. Concr. Compos.* **2007**, *29*, 515–524. [CrossRef]

53.  Subramani, T.; Prabhakaran, M. Experimental study on bagasse ash in concrete. *Int. J. Innov. Eng. Res. Manag.* **2015**, *4*, 163–172.

54.  Rerkpiboon, A.; Tangchirapat, W.; Jaturapitakkul, C. Strength, chloride resistance, and expansion of concretes containing ground bagasse ash. *Constr. Build. Mater.* **2015**, *101*, 983–989. [CrossRef]

55.  Rukzon, S.; Chindaprasirt, P. Utilization of bagasse ash in high-strength concrete. *Mater. Des.* **2012**, *34*, 45–50. [CrossRef]

56.  Cordeiro, G.C.; Toledo Filho, R.D.; Tavares, L.M.; Fairbairn, E.D.M.R. Ultrafine grinding of sugar cane bagasse ash for application as pozzolanic admixture in concrete. *Cem. Concr. Res.* **2009**, *39*, 110–115. [CrossRef]

57.  Kumar, T.S.; Balaji, K.; Rajasekhar, K. Assessment of Sorptivity and Water Absorption of Concrete with Partial Replacement of Cement by Sugarcane Bagasse Ash (SCBA) and Silica Fume. *Int. J. Appl. Eng. Res.* **2016**, *11*, 5747–5752.

58.  Amin, N.U. Use of bagasse ash in concrete and its impact on the strength and chloride resistivity. *J. Mater. Civ. Eng.* **2011**, *23*, 717–720. [CrossRef]

59.  Hailu, B.; Dinku, A. Application of sugarcane bagasse ash as a partial cement replacement material. *Zede J.* **2012**, *29*, 1–12.

60.  Mangi, S.A.; Jamaluddin, N.; Ibrahim, M.W.; Abdullah, A.H.; Awal, A.S.M.A.; Sohu, S.; Ali, N. Utilization of sugarcane bagasse ash in concrete as partial replacement of cement. *IOP Conf. Ser. Mater. Sci. Eng.* **2017**, *271*, 1–8. [CrossRef]

61.  Dhengare, S.W.; Raut, S.P.; Bandwal, N.V.; Khangan, A. Investigation into utilization of sugarcane bagasse ash as supplementary cementitious material in concrete. *Int. J.* **2015**, *3*, 109.

62.  Hussein, A.A.E.; Shafiq, N.; Nuruddin, M.F.; Memon, F.A. Compressive strength and microstructure of sugar cane bagasse ash concrete. *Res. J. Appl. Sci. Eng. Technol.* **2014**, *7*, 2569–2577. [CrossRef]

63.  Reddy, M.V.S.; Ashalatha, K.; Madhuri, M.; Sumalatha, P. Utilization of sugarcane bagasse ash (SCBA) in concrete by partial replacement of cement. *J. Mech. Civ. Eng.* **2015**, *12*, 12–16.

64.  Ganesan, K.; Rajagopal, K.; Thangavel, K. Evaluation of bagasse ash as corrosion resisting admixture for carbon steel in concrete. *Anti-Corros. Methods Mater.* **2007**, *54*, 230–236. [CrossRef]

65.  Yashwanth, M.K.; Avinash, G.B.; Raghavendra, A.; Kumar, B.G.N. An experimental study on alternative cementitious materials: Bagasse ash as partial replacement for cement in structural lightweight concrete. *Indian Concr. J.* **2017**, *91*, 51–58.

66.  Jagadesh, P.; Ramachandramurthy, A.; Murugesan, R. Evaluation of mechanical properties of Sugar Cane Bagasse Ash concrete. *Constr. Build. Mater.* **2018**, *176*, 608–617. [CrossRef]

67.  Ghorbani, M.A.; Khatibi, R.; Hosseini, B.; Bilgili, M. Relative importance of parameters affecting wind speed prediction using artificial neural networks. *Theor. Appl. Climatol.* **2013**, *114*, 107–114. [CrossRef]

68.  Montaseri, M.; Ghavidel, S.Z.Z.; Sanikhani, H. Water quality variations in different climates of Iran: Toward modeling total dissolved solid using soft computing techniques. *Stoch. Environ. Res. Risk Assess.* **2018**, *32*, 2253–2273. [CrossRef]

69.  Mandeville, A.N.; O'connell, P.E.; Sutcliffe, J.V.; Nash, J.E. River flow forecasting through conceptual models part III-The Ray catchment at Grendon Underwood. *J. Hydrol.* **1970**, *11*, 109–128. [CrossRef]

70.  Moriasi, D.N.; Arnold, J.G.; van Liew, M.W.; Bingner, R.L.; Harmel, R.D.; Veith, T.L. Model evaluation guidelines for systematic quantification of accuracy in watershed simulations. *Trans. ASABE* **2007**, *50*, 885–900. [CrossRef]

71.  Gandomi, A.H.; Yun, G.J.; Alavi, A.H. An evolutionary approach for modeling of shear strength of RC deep beams. *Mater. Struct.* **2013**, *46*, 2109–2119. [CrossRef]

72. Gandomi, A.H.; Roke, D.A. Assessment of artificial neural network and genetic programming as predictive tools. *Adv. Eng. Softw.* **2015**, *88*, 63–72. [CrossRef]

73. Frank, I.E.; Todeschini, R. *The Data Analysis Handbook*; Elsevier: Amsterdam, The Netherlands, 1994.

74. Golbraikh, A.; Tropsha, A. Beware of q2! *J. Mol. Graph. Model.* **2002**, *20*, 269–276. [CrossRef]

75. Roy, P.P.; Roy, K. On some aspects of variable selection for partial least squares regression models. *QSAR Comb. Sci.* **2008**, *27*, 302–313. [CrossRef]

76. Alavi, A.H.; Ameri, M.; Gandomi, A.H.; Mirzahosseini, M.R. Formulation of flow number of asphalt mixes using a hybrid computational method. *Constr. Build. Mater.* **2011**, *25*, 1338–1355. [CrossRef]

*Article*

# New Prediction Model for the Ultimate Axial Capacity of Concrete-Filled Steel Tubes: An Evolutionary Approach

Muhammad Faisal Javed [1,*], Furqan Farooq [1,*], Shazim Ali Memon [2,*], Arslan Akbar [3,*], Mohsin Ali Khan [4], Fahid Aslam [5], Rayed Alyousef [5], Hisham Alabduljabbar [5] and Sardar Kashif Ur Rehman [1]

[1] Department of Civil Engineering, COMSATS University Islamabad, Abbottabad Campus, Abbottabad 22060, Pakistan; skashif@cuiatd.edu.pk
[2] Department of Civil and Environmental Engineering, School of Engineering and Digital Sciences, Nazarbayev University, Nur-Sultan, Kazakhstan
[3] Department of Architecture and Civil Engineering, City University of Hong Kong, Kowloon 999077, Hong Kong
[4] Department of Structural Engineering, Military College of Engineering (MCE), National University of Science and Technology (NUST), Islamabad 44000, Pakistan; moak.pg18mce@student.nust.edu.pk
[5] Department of Civil Engineering, College of Engineering in Al-Kharj, Prince Sattam Bin Abdulaziz University, Al-Kharj 11942, Saudi Arabia; f.aslam@psau.edu.sa (F.A.); r.alyousef@psau.edu.sa (R.A.); h.alabduljabbar@psau.edu.sa (H.A.)
* Correspondence: arbabfaisal@cuiatd.edu.pk (M.F.J.); furqan@cuiatd.edu.pk (F.F.); shazim.memon@nu.edu.kz (S.A.M.); aakbar4-c@my.cityu.edu.hk (A.A.)

Received: 11 July 2020; Accepted: 20 August 2020; Published: 22 August 2020

**Abstract:** The complication linked with the prediction of the ultimate capacity of concrete-filled steel tubes (CFST) short circular columns reveals a need for conducting an in-depth structural behavioral analyses of this member subjected to axial-load only. The distinguishing feature of gene expression programming (GEP) has been utilized for establishing a prediction model for the axial behavior of long CFST. The proposed equation correlates the ultimate axial capacity of long circular CFST with depth, thickness, yield strength of steel, the compressive strength of concrete and the length of the CFST, without need for conducting any expensive and laborious experiments. A comprehensive CFST short circular column under an axial load was obtained from extensive literature to build the proposed models, and subsequently implemented for verification purposes. This model consists of extensive database literature and is comprised of 227 data samples. External validations were carried out using several statistical criteria recommended by researchers. The developed GEP model demonstrated superior performance to the available design methods for AS5100.6, EC4, AISC, BS, DBJ and AIJ design codes. The proposed design equations can be reliably used for pre-design purposes—or may be used as a fast check for deterministic solutions.

**Keywords:** concrete-filled steel tube (CFST); axial capacity; genetic engineering programming (GEP); Euler's buckling load; GEP-based model

---

## 1. Introduction

A concrete-filled steel tube (CFST), consists of a steel tube full of concrete. Over the last decade, their use in the building-construction industry as a column and has increased exponentially [1,2]. They have been used in various modern construction projects [3–6]. The CFST structure provides adamant structural advantages that include desirable ductility with high energy-absorption capacities, high strength and fire resistance [7–9]. During concrete construction, the use of shuttering is also not necessary for that reason, the concrete construction costs and the time is lowered. These advantages have been commonly exploited and contributed to the widespread use of CFST members in civil engineering structures [1]. The behavior of CFST members has been broadly examined in the last three decades. This study focuses on the CFST columns with a circular steel tube, as it offers more efficient stiffness and post-yield strength than those with a rectangular or square cross-section [10–12]. Many experimental studies are available on CFST circular columns, with a prime focus on strength of concrete [13–15], the diameter-to-thickness ratio of the tube [16–20] or bond action among steel tube and concrete [21–24].

In the last two decades, a variety of numerical and analytical studies on the behavior of CFST square columns under axial compression have been performed [13,14,16–20]. Nonetheless, the influence of confinement on the enhancement of concrete infilled strength has been held in different opinions. The effects of other variables—for example, the impact of dimension on the concrete strength—likewise varies among numerous researchers [25–27]. The empirical formulas developed for the post-buckling of the steel tube differ from study to study. [15]. Model expression of square CFST columns for the axial load capacity is available in ACI 318 (ACI 2014), Eurocode 4 (CEN 2004) and AISC 360 (AISC 2016). However, none of these equations agree with one another. Such models were derived from the pre-assumed stress–strain relationship of the steel tube or infilled concrete; thus, the validity of these models is doubtful. Moreover, experimental tests require much money, expensive testing equipment and human effort. The accuracy of experimental tests also depends on many factors including the type of equipment, skilled labor, proper casting of test specimens and proper instrumentation. In contrast, numeric studies require experimental tests for the validation. Moreover, numeric modeling demands high-performance workstations and high computational skills. Hence, an accurate empirical equation is required that is easy to use in most conditions and includes all important factors.

Researchers have suggested different methods and techniques for the prediction of the ultimate load-bearing capacity of long circular CFST columns [28,29]. For instance, least median, linear and nonlinear regression techniques are used by various researcher in different domains of civil engineering and have found profound effects [30–32]. This regression-based equation helps us in the prediction of structural domains—and even gives an adamant relation to target-based values. However, regression models are based on some assumptions, making them unrealistic in terms of prediction aspects [33,34]. To address this issue, deep learning in the field of machine-learning-based model algorithms have been developed that have had robust effects for model prediction [35]. In fact, this deep-learning models has been used by various researcher and proved its supremacy over traditional method [12,29]. Artificial neural network (ANN), gene expression programming with supervised learning algorithms are some method which helps in prediction of mechanical properties in the civil engineering domain [36,37]. Nguyen et al. used feed forward neural network (FNN) to predict the compressive strength of rectangular concrete steel filled tubes [38]. The author used invasion weed optimization (IWO) for tuning of parameters, and hence made a hybrid FNN–IWO model. This yielded strong correlation of 0.979 [38]. Hai et al. predict the strength of CFST by using surrogate models. The author used neuro-fuzzy inference system (ANFIS) with meta-based optimization methods to make hybrid algorithm [39]. Particle swarm optimization (PSO), genetic algorithm (GA) and biogeography-based optimization (BBO) are some techniques in prediction of CFST. The result reveals that use of PSO with ANFIS yield strong correlation of about 0.942 with less error [39]. Quang et al. used hybrid algorithm to predict the bearing capacity of rectangular concrete steel tube column [40]. One step secant (OSS) algorithm with FNN algorithm to make hybrid algorithm was developed. The result reveals a good strong model with minimum error between actual and predicted targets. Nguyen et al. used hybrid algorithm namely as GAP-BART which is based on Bayesian additive regression tree (BART) to predict the strength of CFST [41]. Genetic algorithm (GA), particle swarm optimization (PSO) was used in making hybrid approach. The author reveals that particle swarm optimization give adamant model performance with less error. These algorithms train the data to solve the desire problem and then test was conducted on testing set to give results. However, there exist some flaws in ANN modeling as it acts as a black box and does not give adamant relation to model in term of the equation. This reduces its chance of modeling perspective. Also, ANN parameters are based on several hit and trail cycles which in turn requires more time in prediction. In contrary, use of gene expression in supervised mechanism produces and gives a well-defined prediction model [42–44]. Ipek et al. predicted the axial capacity of concrete-filled double-skin steel column section by using gene expression [45]. The author achieved a strong correlation with actual and predicted one with minimum errors. The Gene expression model take the best input and optimize it and predict the outcome by minimizing its error and thus provide best prediction with adamant fitness. Numerous scholars' study and used GP in generating an accurate model for complex engineering domains. Different modifications were proposed to enhance the performance of GP. Genetic engineering programming (GEP) is the most advanced one. Yet, the use of GEP to address complex structural engineering problems has been limited [22]. Esra et al. estimated the axial carry capacity of concrete-filled tube by using GEP algorithm [46]. It is worth mentioning here, that the developed equation is two lengthy and cannot be used for practical implementation [47,48].

Experimental works is time consuming and thus required lot of resources to give a good justified strength. This tradition approach and misplacement of quantities during casting produces malignant effect to strength. Hence, use of supervised algorithms increases the efficiency of prediction by not only just taking data point, but can also help us in generating a hand-based equation. This equation can be then used to predict the overall efficiency of desired model. Moreover, supervised machine learning approaches just predict the strength by giving us the strong correlation but cannot give a relation-based equation. Hence, gene expression programming algorithm was used which can give a strong-based equation with stronger correlation with target and predicted values.

In this research, the GEP approach is exercised to evaluate the axial performance of CFST members. The developed model correlates the axial strength to a few affecting parameters. To effectively design the CFST members with lesser costs, it is essential to establish some models correlating the basic parameters with an axial ultimate capacity of CFST members. Special attention has been given in making a simplified equation that can predicts the strength of CFST even by hand calculation. The model proposed is built based on a huge number of published axial tests on CFST members. The results produced by the model developed are further than judged with those achieved through various codes of practice as several authors show their concerns about the existing design codes [23].

## 2. Comparison of Genetic Programming vs. Genetic Engineering Programming

Ferreira [24] proposed supervised learning machine algorithm ahead from GP which is based on the genetic human evolutionary algorithm. This modified form is also termed as gene expression programming (GEP). GEP develops computer supervised programs that are encrypted in fixed-length chromosomes whereas GP grows a solitary tree expression [49–51]. Gene expression programming (GEP) is like genetic algorithms (GA) and is an alternative form of traditional genetic programming (GP). It was proposed by Ferreira [24] and is used to predict the relationship between input and output data. In GEP the chromosome consist of linear, symbolic strings of genes and each gene in it is a code for object selection while expression tree (ET) is also used for the similar purpose. The parameters that are used by GEP are similar to the ones that were used in GP [52–54]. In these algorithms the computer programs consist of the characters of defined length comparing with the expression trees of length which varies in genetic programming. In computer programs each expression hide as cramped twine of rooted capacity and intentionally declared as the function in which entities are not affected by the change in their values. These types of programs are called complex tree structure or expression trees (ETs) [55–57]. GEP uses genotype and phenotype algorithms in which genotype is detached from phenotype and this programming results as an evolutionary advantage [24]. In GEP size of genome is defined clearly by the problems and is determined by hit and trial rule. For this purpose, a method that utilizes the capability of a system to choose a best possible mode of operation is adaptive control that is employed [58–60]. This approach uses the parameters that are same as of GP. Since all adaptation take place in simple linear structure because in overall structure mutation and structure replication is not required. Moreover, each chromosome comprises of genes which have two well-defined adjacent regions which is called head and the terminal symbols (nodes of leaf) called tail. In head the symbol are used to code internal on ET and in tail it is encoded in expression tree (ET) [61–63].

Figure 1 displays the GEP algorithm schematic layout. The procedure is started with the random formation of fixed dimension chromosome for each singular. Second, the genes are fetching as ETs and tested for their best fitness. Afterwards, the reproduction is applied to the individuals evaluated by the fitness function. The complete hierarchy is repetitive with newly produced gene until the obstinate solution is attained. In short, genetic procedures for example X-over, mutation and reproduction are used for the transformation in population.

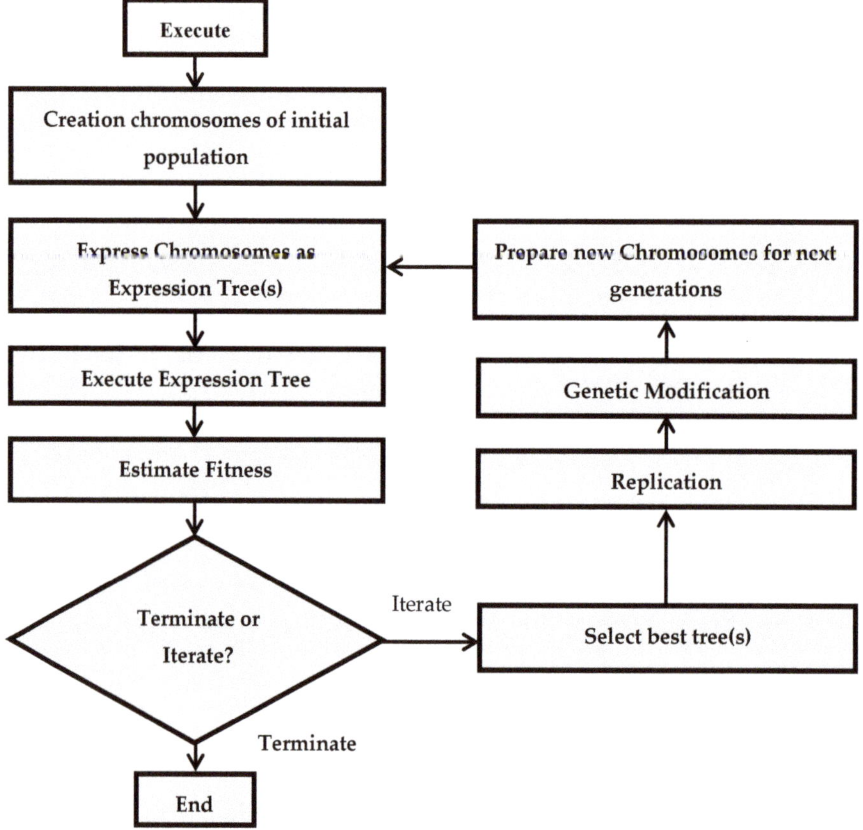


Figure 1. Simple illustration of the of the gene expression programming (GEP) algorithm [22].

## 3. Experimental Database

The model is built with the aid of 227 test results collected from more than 40 literature studies is attached in Appendix A. Only those results were included in the database in which no reinforcement in the infilled concrete is used. Frequency histograms are used for the visualization of the data distribution as shown in Figure 2. These distributions show the maximum parametric influence in total data points taken from literature. The maximum thickness of outer steel tube in CFST lies in the range of 3–5 mm. similarly, the maximum values of diameter, compressive strength, L/D lies in the range of 73–146 mm,

30–50 MPa and 7 to 20, respectively. This shows us the optimum variables values which when take in experimental work produce utmost effect with strength

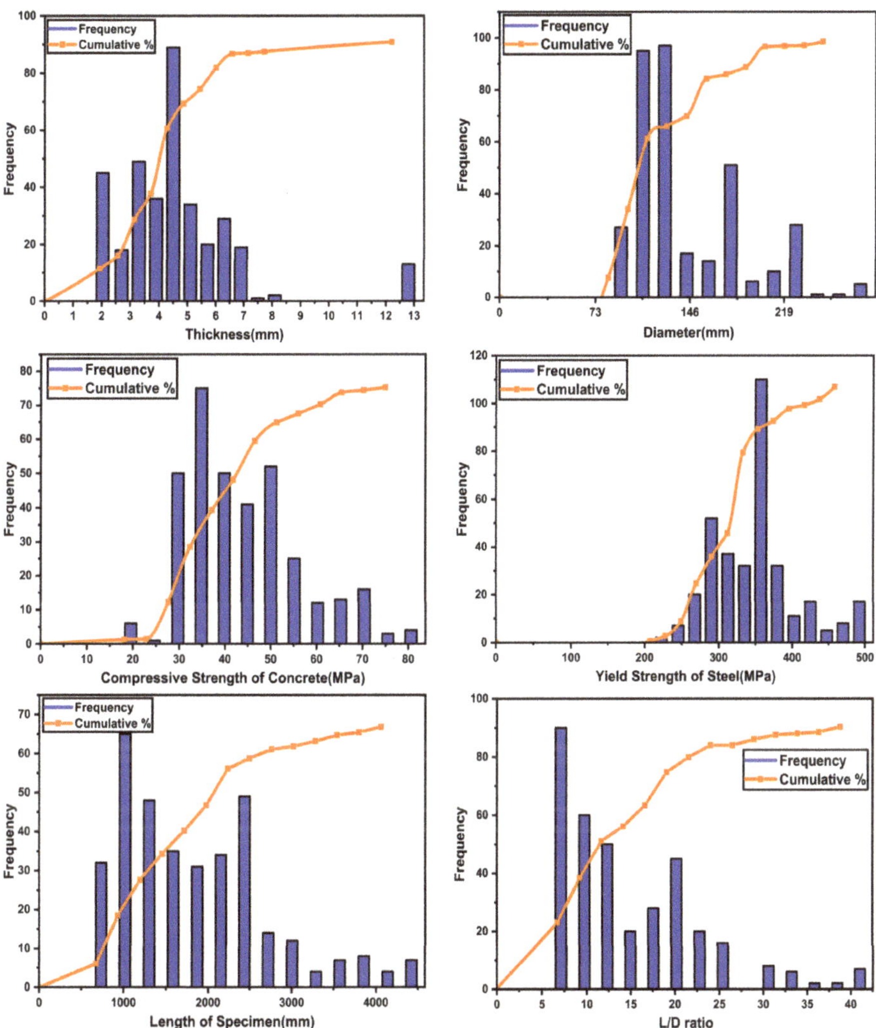

**Figure 2.** Histogram of the variables exercised in the establishment of the model.

The statistical parameters for the development of model including testing, training and validation set are shown in Table 1. Moreover, Figure 3 represents the relationship of individual variables with each other.

One major drawback comes in the supervised machine learning algorithms is the over fitting of data [64,65]. Abundant explanations have been recommended in the literature to evade this problem. Fulcher suggested to train and validate the data on different set of data [66]. In this study, this procedure is used by arbitrarily separating the obtainable data into three sets, namely as a validation set, learning set and testing set. First, the model is established created on the learning set or train set which is then validated by dividing set of data and finally test was conducted to evaluate the performance of model on test set [67]. The validated model is test on the data which is not used on train data.

**Table 1.** Descriptive variables statistics.

| Parameters | Diameter | Thickness | Yield Stress | Compressive Strength | Length | Length/Diameter | Test |
|---|---|---|---|---|---|---|---|
| **Training set data** | | | | | | | |
| Mean | 137.9 | 4.6 | 344.9 | 43.9 | 36.4 | 1651.8 | 13.0 |
| Standard error | 4.8 | 0.2 | 5.4 | 1.5 | 1.4 | 81.1 | 0.7 |
| Median | 114.3 | 4.1 | 338.9 | 40.5 | 33.4 | 1420.0 | 10.3 |
| Mode | 108.0 | 5.0 | 348.0 | 35.7 | 29.0 | 1040.0 | 20.0 |
| Standard deviation | 54.4 | 2.5 | 61.2 | 17.3 | 15.9 | 917.2 | 7.9 |
| Sample variance | 2956.3 | 6.2 | 3744.4 | 298.6 | 254.3 | 841,300.1 | 62.6 |
| Kurtosis | 4.6 | 3.8 | 0.2 | 3.4 | 4.3 | 2.6 | 3.9 |
| Skewness | 1.9 | 1.8 | 0.8 | 1.6 | 1.8 | 1.5 | 1.7 |
| Range | 279.6 | 11.4 | 271.8 | 91.6 | 86.0 | 4892.0 | 45.5 |
| Minimum | 76.0 | 1.4 | 233.2 | 14.4 | 10.0 | 508.0 | 4.5 |
| Maximum | 355.6 | 12.8 | 505.0 | 106.0 | 96.0 | 5400.0 | 50.0 |
| Sum | 17,646.2 | 592.2 | 44,143.2 | 5615.7 | 4656.1 | 211,432.9 | 1663.8 |
| Count | 128.0 | 128.0 | 128.0 | 128.0 | 128.0 | 128.0 | 128.0 |
| **Testing set data** | | | | | | | |
| Mean | 127.8 | 4.3 | 340.4 | 39.6 | 32.5 | 1806.0 | 15.0 |
| Standard error | 5.2 | 0.3 | 7.8 | 1.9 | 1.7 | 152.4 | 1.4 |
| Median | 114.0 | 4.0 | 340.0 | 35.7 | 29.0 | 1648.2 | 11.5 |
| Mode | 108.0 | 4.0 | 338.9 | 35.7 | 29.0 | 2700.0 | 6.0 |
| Standard deviation | 37.0 | 2.2 | 55.0 | 13.5 | 12.1 | 1077.3 | 9.9 |
| Sample variance | 1367.0 | 4.7 | 3021.6 | 182.2 | 147.1 | 1,160,607.4 | 97.2 |
| Kurtosis | 1.0 | 7.7 | 1.3 | 1.0 | 1.2 | −0.2 | 0.4 |
| Skewness | 1.4 | 2.3 | 0.7 | 1.1 | 1.2 | 0.8 | 1.1 |
| Range | 136.5 | 11.3 | 267.8 | 62.8 | 57.2 | 3812.0 | 35.5 |
| Minimum | 82.6 | 1.4 | 237.2 | 14.4 | 10.0 | 508.0 | 4.5 |
| Maximum | 219.0 | 12.7 | 505.0 | 77.2 | 67.2 | 4320.0 | 40.0 |
| Sum | 6388.6 | 214.0 | 17,017.5 | 1982.3 | 1626.9 | 90,302.2 | 751.0 |
| Count | 50.0 | 50.0 | 50.0 | 50.0 | 50.0 | 50.0 | 50.0 |
| **Validation set data** | | | | | | | |
| Mean | 137.3 | 4.5 | 347.2 | 42.2 | 34.8 | 1755.6 | 13.8 |
| Standard error | 5.7 | 0.3 | 10.2 | 1.7 | 1.5 | 127.2 | 1.2 |
| Median | 114.3 | 4.1 | 338.9 | 40.2 | 33.0 | 1572.0 | 10.9 |
| Mode | 108.0 | 4.0 | 486.0 | 35.7 | 29.0 | 1640.0 | 6.0 |
| Standard deviation | 43.9 | 2.3 | 79.1 | 13.3 | 11.9 | 985.3 | 9.0 |
| Sample variance | 1927.1 | 5.1 | 6256.5 | 175.8 | 141.8 | 970,885.0 | 81.2 |
| Kurtosis | 0.4 | 4.5 | 1.0 | 0.5 | 0.9 | 2.3 | 3.8 |
| Skewness | 1.1 | 1.8 | 1.0 | 1.0 | 1.1 | 1.4 | 1.7 |
| Range | 190.9 | 11.4 | 404.5 | 57.5 | 52.8 | 4892.0 | 45.2 |
| Minimum | 76.5 | 1.4 | 200.2 | 25.5 | 20.2 | 508.0 | 4.8 |
| Maximum | 267.4 | 12.8 | 604.7 | 83.0 | 73.0 | 5400.0 | 50.0 |
| Sum | 8236.8 | 271.1 | 20,831.6 | 2534.8 | 2089.2 | 105,337.0 | 830.1 |
| Count | 60 | 60 | 60 | 60 | 60 | 60 | 60 |

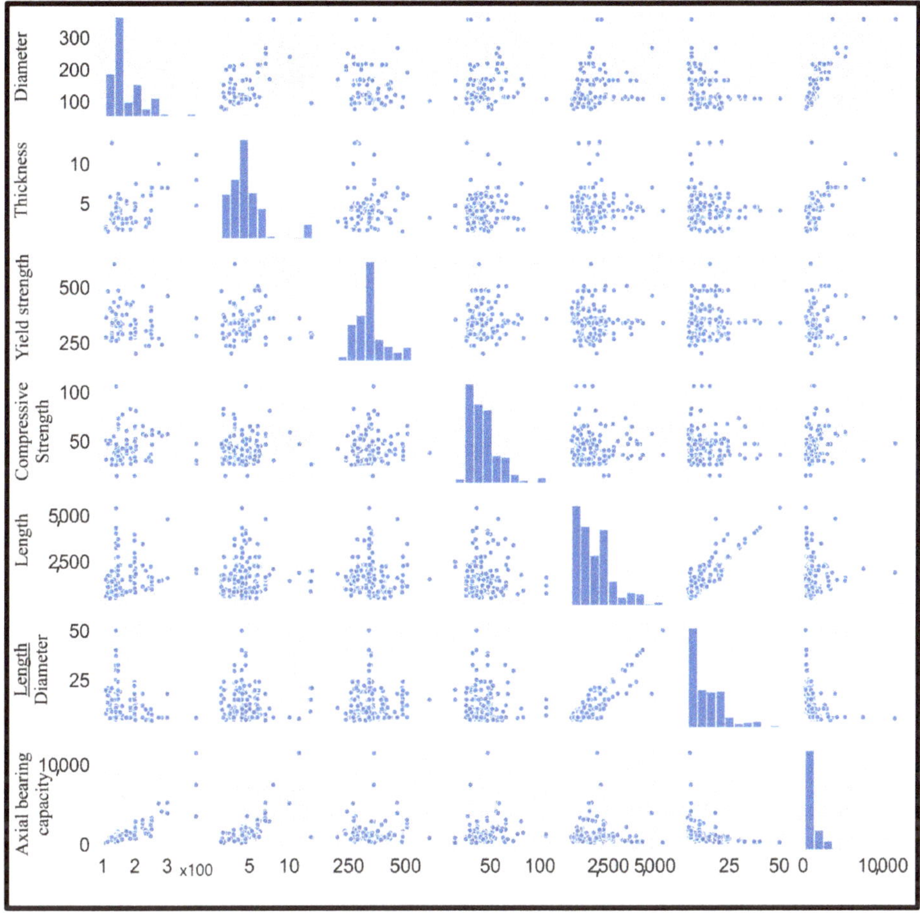

**Figure 3.** Relationship among individual variables.

Various parameters in designing long circular CFST members may be interdependent. Interdependency is needed to be check as it leads to difficulty in the interpretation of the model. In addition, the interdependency causes numerous problems during investigation as it upsurges the strength of relations between different parameters. This kind of problematic is often mentioned to as a "multicollinearity problem" [68]. Therefore, the association coefficients are calculated for all the possible mixtures among the parameters and are presented in Table 2. It can be detected that all the relation coefficients (both negative positive and) are not extraordinary, presentation no danger of "multicollinearity problem".

**Table 2.** Correlation coefficients for different variables.

| Variable | Diameter | Thickness | Steel Yield Strength | Compressive Strength | Length | Length/Diameter |
|---|---|---|---|---|---|---|
| Diameter | 1 | 0.367 | −0.197 | 0.123 | 0.246 | −0.293 |
| Thickness | 0.367 | 1 | 0.031 | −0.041 | 0.091 | −0.102 |
| Steel yield strength | −0.197 | 0.031 | 1 | 0.088 | −0.028 | 0.075 |
| Compressive strength | 0.123 | −0.041 | 0.088 | 1 | −0.016 | −0.102 |
| Length | 0.246 | 0.091 | −0.028 | −0.016 | 1 | 0.813 |
| Length/diameter | −0.293 | −0.102 | 0.075 | −0.102 | 0.813 | 1 |

## 4. Development of Model

The study aims in establishing a novel-based prediction equation for the axial compressive strength of CFST members using the GEP approach. The main variables frequently used in the earlier codes and analytical models were used as input variables. These parameters were evaluated based on the literature [15,21,69]. Therefore, the formulation of the axial ultimate strength of CFST member was assumed as follows:

$$N = f\left(D, t, f_y, f_c, L, \frac{L}{D}\right) \tag{1}$$

In the above equation, $N$ is the ultimate axial capacities of the long circular CFST column. $f_y, t, D$ and $L$ are the yield strength, thickness, outer diameter and outer steel tube length, respectively. Whereas $f_c$ is the 28-day compressive strength of concrete cylinder. The key input parameters used in the GEP algorithm are shown in Table 3. These variables have influence on model and thus importance should be given while selecting the governing one. Moreover, six basic mathematic operators ($+, -, \div$, $\times$, square, cubic root) were used in predication of model.

**Table 3.** GEP parameters settings.

| Parameter | Settings |
|---|---|
| **General** | |
| Chromosomes | 30 |
| Genes | 3 |
| Head size | 8 |
| Gene size | 26 |
| Linking function | Addition |
| Function set | $+, -, \times, \div, \sqrt{}, \sqrt[3]{}$ |
| **Genetic operators** | |
| Mutation rate | 0.0138 |
| Inversion rate | 0.00546 |
| IS Transposition rate | 0.00546 |
| RIS transposition rate | 0.00546 |
| One-point recombination rate | 0.00277 |
| Two-point recombination rate | 0.00277 |
| Gene recombination rate | 0.00755 |
| Gene transposition rate | 0.00277 |
| **Numerical constants** | |
| Constants per gene | 10 |
| Data type | Floating Point |
| Lower bound | −10 |
| Upper bound | 10 |

The model prediction and time required to model is completely dependent on the difficulty of the problems, the population size and the variables. The model gets stopped after best fitness. In addition, gene size and chromosomes of the model have influence on the prediction of properties. Each gene size consists of a unique expression tree. The number of chromosomes in the genes and head size describes the difficulty level of GEP-based model. The overall fitness of the new programs is calculated via the mean absolute error (MAE) function. The parameters values included are calculated using trial and error. GeneXproTools 5.0 by Gepsoft Lda- Portugal was used to implement the GEP algorithm [70].

To achieve a consistent distribution of data, numerous arrangements of testing and training sets were established. The distribution of data in term of learning set, validation set and the model which predicts the response was used in GEP model to select the best response, namely as testing set. An objective function presented by Babanajad, Gandomi [71] is used to measure the fitness of learning and validation set. The finest GEP model was obtained by reducing the objective function (Equation (2)).

$$f_{min} = \left(\frac{n_L - n_V}{n_L + n_V}\right)\left(\frac{m_L + rm_L}{R_L^2}\right) + \frac{2n_V}{n_L + n_V}\left(\frac{m_V + rm_V}{R_V^2}\right) \tag{2}$$

In the above equation, $n_V$ and $n_L$ are the test numbers in validation sets and learning sets, respectively. $R_L^2, m_L$ and $rm_L$ are the determination coefficient, mean absolute error and root mean square error of learning set, respectively. $R_V^2$, $m_V$, and $rm_V$ are the determination coefficient, mean absolute error and root mean square error of validation set, respectively. These all are calculated using the following equations. The mathematical forms of mean square error (MAE), root mean square error (RMSE) and determination coefficient are represented in Equations (3) and (4).

$$MAE = \frac{1}{n}\sum_{i=1}^{n}|x_i - y_i| \tag{3}$$

$$RMSE = \sqrt{\frac{1}{n}\sum_{i=1}^{n}(x_i - y_i)^2} \tag{4}$$

$$R = \frac{\sum_{i=1}^{n}(x_i - \overline{x_i})(y_i - \overline{y_i})}{\sqrt{\frac{1}{n}\sum_{i=1}^{n}(x_i - \overline{x_i})^2 \sum_{i=1}^{n}(y_i - \overline{y_i})^2}} \tag{5}$$

In the above equations, $x_i$ and $y_i$ are the actual output and calculated output for the $i$th output, respectively. It is worth noting that the objective function presented in Equation (2) considers $m$, $rm$ and $R$ together, which results in a more accurate model. Furthermore, the given objective function takes into consideration the effect of distinct data sets, i.e., learning and validation sets. Lower values of $m$ and $rm$ indicates higher accuracy of the model.

## 5. Results and Discussion

The equation obtained for the ultimate axial capacity of circular CFST members is specified in Equation (6). The objective function ($f_{min}$) value obtained for Equation (2) is 182.52. Equation (6) is obtained from the expression tree which is shown in Figure 3. In Figure 4, the c1–c9 represents different constant values tried by the GEP, d0–d6 are different variables explained in Equation (1), while the 3Rt represents the cubic root of the value. It can be seen that the capacity of a concrete-filled steel tube is dependent on the input variables, namely as diameter, thickness, length to diameter ratio, yield strength, compressive strength as shown in Equation (6). Moreover, every parameter has a key influence on capacity thus increasing one or decreasing another will sufficiently have a benignant and malignant effect on its strength.

$$N_{GEP}(kN) = D(3t - 1) - t^2 - 137.67t - (4t + 1)\frac{L}{D} + \frac{f_y}{t} + 6.72f_c' + \left(f_y - L\right)^{\frac{1}{3}} - 46.61 \tag{6}$$

where $N_{GEP}$ is the ultimate axial moment capacity of the column calculated from Equation (6) and $f_c'$ is the compressive strength of infilled concrete. $D, t$, L and $f_y$ are the diameter, thickness, length and yield strength of the steel tube, respectively.

The relationship between predicted values and experimental values is shown in Figure 5. The important statistical values of the proposed equation for learning, validation and testing sets are given in Table 4. It can be seen that the $R^2$ value was increased from 0.97 to 0.99 while MAE and RMSE decreases 134 to 124 and 210 to 173, respectively. Moreover, that the error value for testing is lesser as compared with other training and validation set. This illustrates that the present GEP model can accurately predict the axial capacity of CFST members and can be used for the generalization purpose [72].

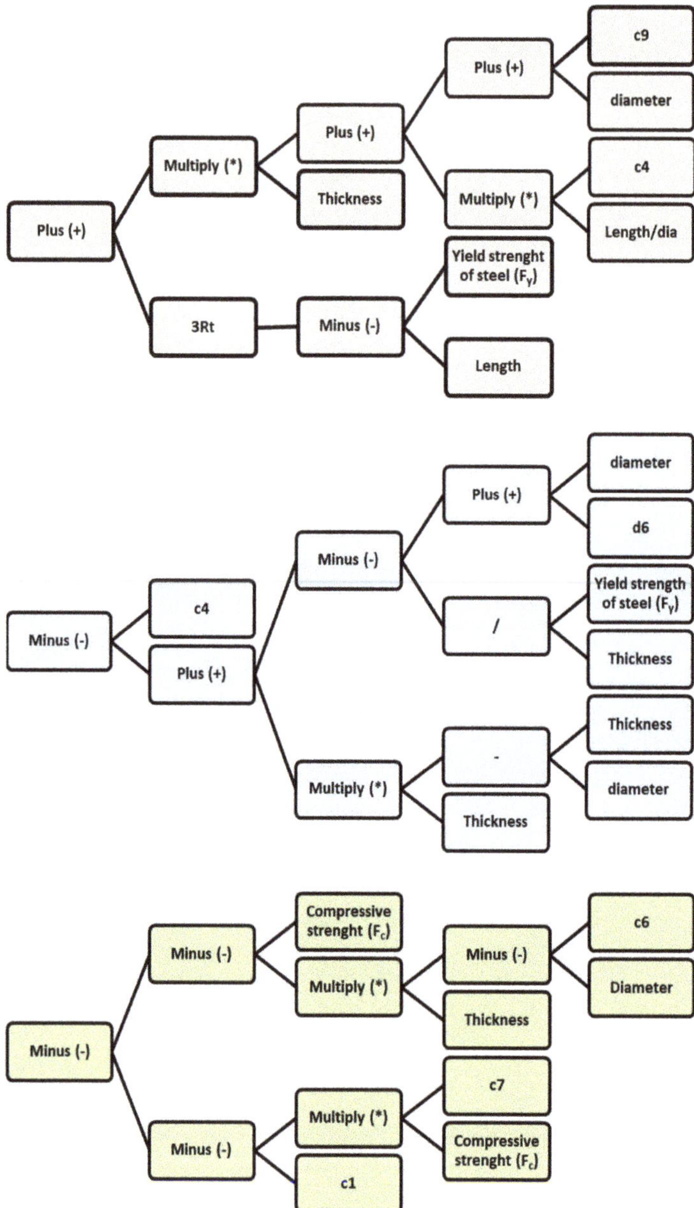



**Figure 4.** Expression tree for the GEP model.

**Table 4.** GEP model execution results.

| Model | Experimental Axial Capacity vs. Predicted Axial Capacity | | |
|---|---|---|---|
| | $R^2$ | MAE | RMSE |
| Learning | 0.97 | 134.8 | 210.3 |
| Validation | 0.98 | 153.9 | 226.1 |
| Testing | 0.99 | 124.3 | 173.7 |

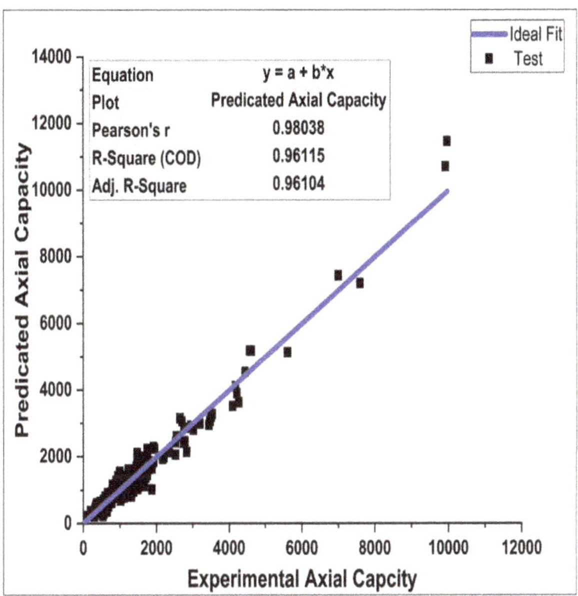



**Figure 5.** Predicted axial capacity vs. experimental results using the GEP model.

*Model Performance, Validity and Comparative Study*

The existing formulae provided by six different design codes (AS5100.6 (2004), EC4 (2004), AISC, BS, DBJ, AIJ) are utilized for the comparison of the suggested model. The process for the calculation of the axial load capacity of circular CFST columns is described in Table 5. The Australian standard (AS5100.6) counteract for the interaction effect of and steel tube concrete core. It also contains the effectiveness of concrete confinement. The relation presented by British standard (BS5400) contains an allocation for the eccentricity of the minor axis that does not surpass 0.03 times the composite column's least lateral dimension. It is improper as the engineer's preference may increase it. The equation of the American Institute of Steel Construction (AISC 2005) accounts for the effect of the restraining hoop that results from transverse confinement. This phenomenon increases the usable concrete stress. The relation provided by the Architectural Institute of Japan (AIJ 2001) involves a confinement factor that accounts for the reduction in the steel tube effective yield stress, caused by the hoop stresses. In the Eurocode 4 (EC4 2004), the equation accounts for the confinement effect in addition to the effect of steel tube and concrete core interaction. The concrete strength is increased by the triaxial state of stress conditions and the hoop stress that reduces the steel effective yield stress. The Chinese code (DBJ 1999) provides an equation ultimate axial moment capacity that cannot be used for ultra-high-strength concrete.

The comparison between the predicted values from the GEP model and different established codes is shown in Figure 6. In Figure 6, the model accuracy is highest for the value of 1. The frequency of 1 is highest for the GEP model while it is lowest for AS5100.6. In addition, it can be seen from the below Figures that the frequency of all the codes lies above 1. Thus, minimizing its practical implementation in calculation of strength. On the other side, GEP model show the distribution of its frequency in the range of 0 and 1. Thus, making it a safe approach in prediction. The statistical parameters for the comparison purpose are shown in Table 6. The R-value must approach to 1 for maximum accuracy. A value of R greater than 0.8 is deemed acceptable [73]. GEP model gives the best results than the available design codes. Furthermore, the MAE and RMSE are calculated for available design codes and the GEP model. Both MAE and RMSE should be minimum for higher accuracy. Based on MAE and RMSE, GEP gives the most accurate results followed by AIJ and BS, respectively.

**Table 5.** Details of the codes.

| Equation No: | Code Specification | Ultimate Axial Moment Capacity ($N_u$) | Limitations |
|---|---|---|---|
| 1 | AS5100.6 (2004) | $N_u = \alpha_c \left[ \eta_a A_s f_y + \left( 1 + \frac{\eta_c t f_y}{d_o f_c'} \right) A_c f_c' \right]$ <br> $\alpha_c = \xi \left( 1 - \sqrt{1 - \left( \frac{90}{\xi \lambda} \right)^2} \right)$ ; $\xi = \frac{\left( \frac{\lambda}{90} \right)^2 + 1 + \eta}{2 \left( \frac{\lambda}{90} \right)^2}$ <br> $\lambda = \lambda_n + \alpha_a \alpha_b$ ; $\eta = 0.00326(\lambda_n - 13.5) \geq 0$ ; $\lambda_n = 90 \lambda_r$ <br> $\lambda_r = \sqrt{\frac{N_s}{N_{cr}}}$ ; $N_s = A_s f_y + A_c f_c'$ <br> $N_{cr} = \frac{\pi^2 (EI)_{eff}}{l^2}$ ; $(EI)_{eff} = E_s I_s + E_c I_c$ <br> $\alpha_a = \frac{2100(\lambda_n - 13.5)}{\lambda_n^2 - 13.5\lambda_n + 2050}$ ; $\alpha_b = Presented\ in\ code$ <br> $\eta_2 = 0.25(3 + 2\lambda_r) \geq 0$ ; $\eta_1 = 4.9 - 18.5\lambda_r + 17.5\lambda_r^2 \geq 0$ | |
| 2 | AISC (2005) | $N_u = \phi_c N_n$ ; $\phi_c = 0.75\ (LRFD)$ <br> $If\ N_e \geq 0.44 N_o$ ; $N_n = N_o \left[ 0.658^{\left( \frac{N_o}{N_e} \right)} \right]$ <br> $If\ N_e < 0.44 N_o$ ; $N_n = 0.877 N_e$ <br> $N_o = A_s f_y + 0.95 A_c f_c'$ <br> $N_e = \frac{\pi^2 (EI)_{eff1}}{(KL)^2}$ ; $EI_{eff1} = E_s I_s + C_1 E_c I_c$ <br> $C_1 = 0.1 + 2 \left( \frac{A_s}{A_c + A_s} \right) \leq 0.3$ ; $E_c = (f_c')^{\frac{1}{2}}\ (MPa)$ | $21\ MPa \leq f_c' \leq 70\ MPa$ <br> $f_y \leq 525\ MPa$ <br> $A_s \geq 0.01 A_g$ <br> $\frac{D}{t} \leq \sqrt{\frac{8 E_s}{f_y}}$ |
| 3 | BS5400 | $\alpha_c = \frac{0.45 f_{cc} A_c}{N_u}$ ; $0.1 < \alpha_c < 0.8$ <br> $N_u = 0.91 f_y' A_s + 0.45 f_{cc} A_c$ <br> $f_y' = C_2 f_y$ ; $f_{cc} = f_{cu} + C_1 \frac{t}{D} f_y$ <br> $C_1\ and\ C_2\ are\ constants\ depends\ on\ \frac{l_e}{D}$ | $f_{cu} \geq 20\ MPa$ <br> $f_y = Grade\ 43\ or\ 50$ <br> $\frac{D}{t} \leq \sqrt{\frac{8 E_s}{f_y}}$ <br> $Nominal\ aggregate\ size \leq 20\ mm$ |
| 4 | DBJ (1999) | $N_u = \gamma_m f_{scy} W_{scm}$ <br> $f_{scy} = (1.18 + 0.85 \xi) f_{ck}$ <br> $W_{scm} = \frac{\pi}{32} D^3$ <br> $\xi = \frac{A_s f_{yk}}{A_c f_{ck}}$ <br> $\gamma_m = 1.04 + 0.48 \ln(\xi + 0.1)$ | $100\ mm \leq D \leq 2000\ mm$ <br> $200\ MPa \leq f_{scy} \leq 500\ MPa$ <br> $20\ MPa \leq f_{ck} \leq 80\ MPa$ |

**Table 5.** *Cont.*

| Equation No: | Code Specification | Ultimate Axial Moment Capacity ($N_U$) | Limitations |
|---|---|---|---|
| 5 | AIJ (2001) | $N_{u1} = 0.85 A_c f'_c + (1+\eta) A_s f_y \; ; \left(\frac{l}{D} \le 4\right)$ <br> $N_{u2} = N_{u1} - 0.125\{N_{u1} - N_{u3}\}\left(\frac{l}{D} - 4\right) ; \left(4 < \frac{l}{D} \le 12\right)$ <br> $N_{u3} = A_c \sigma_{cr} + N_{crs} \; ; \left(\frac{l}{D} > 12\right)$ <br> $\sigma_{cr} = \frac{1.7 f'_c}{1 + \sqrt{\lambda_1^4 + 1}} \; ; \; \lambda_1 \le 1.0$ <br> $\sigma_{cr} = 0.83 \exp\{(0.568 + 0.00612 f'_c)(1 - \lambda_1)\}0.85 f'_c \; ; \; \lambda_1 > 1$ <br> $\lambda_1 = \frac{\lambda}{\pi}\sqrt{0.93(0.85 f'_c)^{\frac{1}{4}} \times 10^{-3}}$ <br> $N_{crs} = A_s f_y \; ; \; \lambda_1 \le 0.3$ <br> $N_{crs} = 1 - 0.545(\lambda_1 - 0.3) \; ; \; 0.3 \le \lambda_1 < 1.3$ <br> $N_{crs} = \frac{N_{Es}}{1.3} \; ; \; \lambda_1 \ge 1.3$ <br> $\lambda_1 = \frac{\lambda}{\pi}\sqrt{\frac{f_y}{E_s}}$ <br> $N_{Es} = \frac{\pi^2 E_s I_s}{l^2}$ <br> $\lambda = $ *slenderness ratio of concrete column* | $\frac{D}{t} \le \frac{35250}{f_y}$ |
| 6 | EC4 (2004) | $N_u = \eta_a A_s f_y + \left(1 + \eta_c \frac{t}{D}\frac{f_y}{f'_c}\right)A_c f'_c$ <br> $\eta_a = 0.25\left(3 + 2\overline{\lambda}\right) \le 1.0 \; ; \; \eta_c = 4.9 - 18.5\overline{\lambda} + 17\overline{\lambda}^2 \ge 0$ <br> $\overline{\lambda} = \frac{N_{plR}}{N_{cr}} ; \; N_{plR} = A_s f_y + A_c f'_c$ <br> $N_{cr} = \frac{\pi^2 (EI)_{eff2}}{l_2} ; \; (EI)_{eff2} = E_s I_s + K_c E_c I_c ; \; K_c = 0.6$ <br> $E_c = 22,000\left[\frac{(f'_c + 8)}{10}\right]^{0.3} (MPa)$ | $20\,MPa \le f'_c \le 60\,MPa$ <br> $f_y \le 460\,MPa$ <br> $\frac{D}{t} \le \frac{0.15 E_s}{f_y}$ |

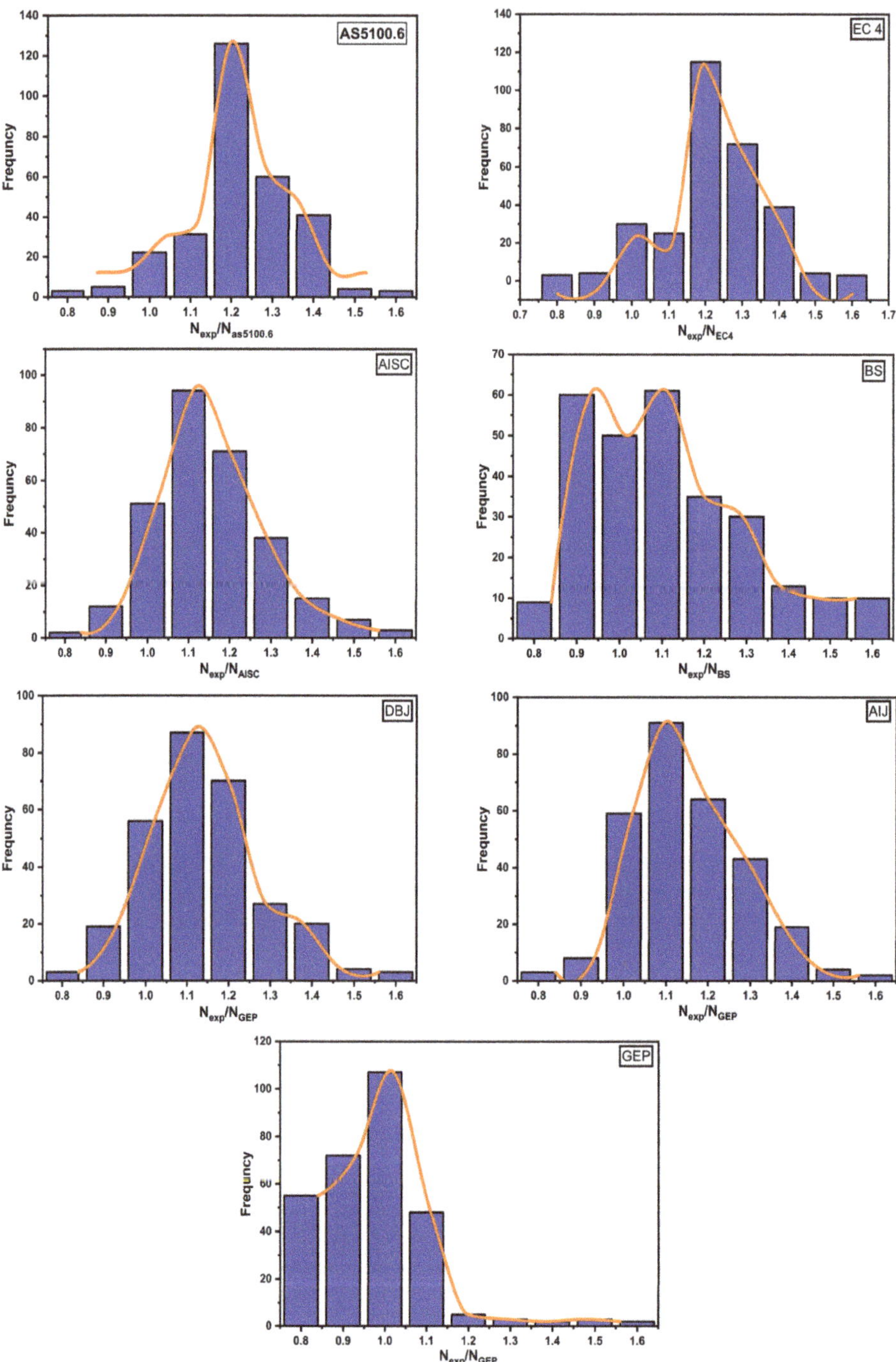



**Figure 6.** Evaluation of the concrete-filled steel tubes (CFST) columns experimental and predicted axial bearing capacity.

**Table 6.** Axial strength prediction models overall performance.

| Statistical Parameters | GEP | AS5100.6 | EC4 | AISC | BS | DBJ | AIJ |
|---|---|---|---|---|---|---|---|
| Rsq | 0.98 | 0.98 | 0.98 | 0.97 | 0.96 | 0.97 | 0.97 |
| MAE | 138.7 | 249.4 | 220.6 | 333.5 | 205.0 | 228.0 | 194.4 |
| RMSE | 258.0 | 484.7 | 452.9 | 701.4 | 352.9 | 512.0 | 408.4 |
| Row ($\rho$) | 0.1 | 0.2 | 0.1 | 0.2 | 0.3 | 0.2 | 0.2 |
| Average | 1.2 | 1.1 | 1.2 | 1.0 | 1.0 | 0.9 | 1.2 |
| Maximum | 1.2 | 1.6 | 1.7 | 2.0 | 1.7 | 1.5 | 1.2 |
| Minimum | 0.7 | 0.7 | 0.8 | 0.6 | 0.5 | 0.6 | 0.8 |
| SD | 0.1 | 0.1 | 0.1 | 0.1 | 0.3 | 0.2 | 0.1 |
| COV | 0.1 | 0.1 | 0.1 | 0.1 | 0.2 | 0.1 | 0.1 |

The model evaluation between errors and performance coefficient is measured by performance index ($\rho$) [74]. $\rho$ is used successfully by numerous researchers and is calculated by using Equation (7):

$$\rho = \frac{Rr_m}{1+R} \tag{7}$$

where $Rr_m$ is the relative $r_m$. Higher value of $\rho$ shows bad achievement of the model and vice versa. From Table 6, it is determined that the GEP model outperforms the available design codes by huge margin.

The model accuracy can also be checked by several statistical measures. Frank and Todeschini [74] proposed that the accuracy of model is based on the number of testing set and the numbers of parameters used in modeling. He suggested and equation in which the ratio of both aforementioned should be greater than or equal to 5 as presented in Equation (8):

$$\frac{No.\ of\ experimental\ tests}{No.\ of\ variables\ used} \geq 05 \tag{8}$$

In this research, the ratio is 44. Furthermore, external verification is also suggested by researcher [75]. The test recommended that the slope of one of the regression lines moving through the origin should be approximately 1 [76]. In addition, test recommended by Roy [77] was also conducted for the given model. Table 7 outlines the acceptance benchmarks and the results of the built GEP model. The model developed based on GEP adamantly fulfils the criteria of all the above-mentioned tests. It is therefore inferred that the GEP model established is accurate and is not a simple correlation.

**Table 7.** GEP model statistical parameters for external validation.

| Sr. No | Formula | Condition | GEP |
|---|---|---|---|
| 1 | Equation (5) | $R > 0.8$ | 0.973 |
| 2 | $K = \dfrac{\sum_{i=1}^{n}(x_i \times y_i)}{x_i^2}$ | $0.85 < K < 1.15$ | 0.983 |
| 3 | $K' = \dfrac{\sum_{i=1}^{n}(x_i \times y_i)}{y_i^2}$ | $0.85 < K' < 1.15$ | 1.003 |
| 4 | $R_m = R^2 \times \left(1 - \sqrt{\left|R^2 - R_0^2\right|}\right)$ | $R_m > 0.5$ | 0.838 |
| $R_0^2$ is squared correlation coefficient between predicted and experimental values | | | 0.999 |

Simplicity is the utmost advantages in prediction of mechanical properties based on GEP algorithm. This adamant advantage helps in calculation of ultimate axial capacity by hand calculations using GEP-based formula. GEP model is completely independent and does not depend on the previous equations and design models. Moreover, increasing the training and validation set data enhance the overall accuracy of the model.

A comparison of GEP model with equations suggested by various authors was made on all data set [78–80]. It can be seen in Figure 7 that GEP model give an adamant $R^2$ accuracy of about

0.94 as compared to other models. This is due to simplified nature of GEP in prediction. Moreover, Glakoumelis et al. [80] predict the compressive nature of CFST by giving an empirical relation with a strong correlation value $R^2$ of about 0.895. Also, Goode et al. [79] and Lu et al. [78] give same empirical equation with some modification with $R^2$ value of 0.807 and 0.903, respectively as illustrated in Figure 7. This study show us that GEP-based empirical equations can be used in prediction of different variables.

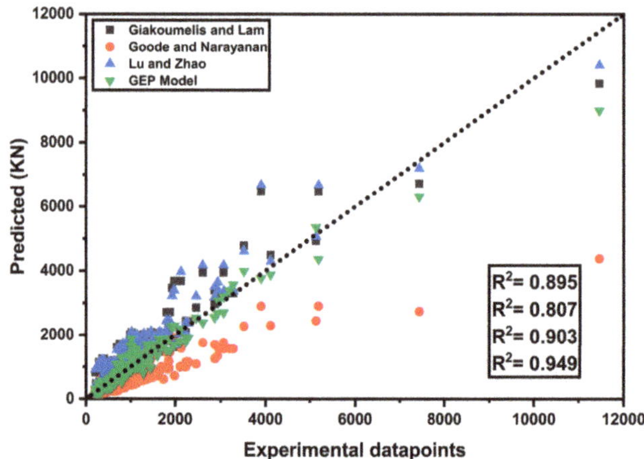


**Figure 7.** Comparison of GEP model with other published equations [78–80].

## 6. Conclusions

This study represents a novel and dominant method for the derivation of the expression to compute the ultimate axial capacity of CFST long circular columns by genetic engineering programming (GEP) for the first time. The resulting equation is empirical, and is formed by previous experimental data published in literatures. The suggested equation is simplest and CFST axial capacity can be determined by hand calculations. All the model outcomes show outstanding consent to the experimental results. Different statistical parameters such as RMSE, MAE and $R^2$ proved the accuracy and reliability of GEP-based derived equations. In addition, this supervised machine learning algorithm can be used in many other domains. As they help us in making the forecast prediction by training and testing of data. This artificial intelligence-based algorithm then helps scientific community by taking measures and overcome the issues associated in mechanical work or in experimental work. Though, the comparison between the MAE, RMSE and $R^2$ of GEP model, AS5100.6, EC4, AISC, BS, DBJ and AIJ shows that GEP model performs best for all sets (learning, training and validation) of data. Even though the GEP-based model can calculate short CFST shear strength, it is restricted to long circular columns. The findings from this new research will give civil engineers and structural designers some useful information and can be used as a modern and powerful method to help decision-making in concrete construction fields.

**Author Contributions:** Fahid Aslam (F.A.); visualization and Conceptualization, Muhammad Faisal Javed (M.F.J.); Conceptualization, Furqan Farooq (F.F.); software and Investigation, Arslan Akbar (A.A) and Mohsin Ali Khan (M.A.K.); writing—review and editing; Shazim Ali Memon (S.A.M.); writing—review & editing, Project administration, Funding acquisition", Rayed Alyousef (R.A.); review & editing; Hisham Alabduljabbar (H.A.); Review and Graphs; Sardar Kashif Ur Rehman (S.K.U.R.); review & editing. All authors have read and agreed to the published version of the manuscript.

**Funding:** This research was supported by Nazarbayev University Faculty development competitive research grants (090118FD5316) and deanship of scientific research at Prince Sattam Bin Abdulaziz University under the research project number 2020/01/16810.

**Acknowledgments:** The authors acknowledge the support of COMSATs University Islamabad, Abbottabad Campus and Engr. Kashif Nazir for providing fruitful environment in completing this research.

**Conflicts of Interest:** The authors declare no conflict of interest.

## Nomenclature

CFST = Concrete-filled steel tube
ANN = Artificial neutron network
GP = Genetic programming
GEP = Genetic engineering programming
$ETs$ = Expression trees
$MAE$ = Mean absolute error
$RMSE$ = Root mean square error
$N_u$ = Ultimate axial moment capacity
$N_n$ = Nominal axial moment capacity
$N_e$ = Euler's bucking load
$N_o$ = Nominal axial compressive strength exclusive of length effects
$A_s$ = Steel section areas
$A_c$ = Concrete area
$A_g$ = Total composite cross-section area
$D$ = Diameter of concrete core
$E_c$ = Concrete elastic modulus = $0.043\omega_c^{1.5}\sqrt{f_c'}$ MPa
$E_s$ = Steel elastic modulus = $200,000$ MPa
$f_c'$ = Concrete compressive strength
$f_y$ = Steel section minimum yield strength
$I_c$ = Concrete section moment of inertia
$I_s$ = Steel section moment of inertia
$K$ = Length effectiveness factor
$L$ = Length of laterally braced member
$(EI)_{eff}$ = Composite section effective stiffness
$N_e$ = Elastic bucking load
$\alpha_c$ = Concrete contribution factor
$f_{cu}$ = 28-day characteristic strength of concrete cube
$f_{cc}$ = Triaxially contained concrete improved characteristic strength
$f_{scy}$ = Steel-tube nominal yield strength
$f_{ck}$ = Concrete characteristic strength
$f_y'$ = Reduced nominal yield strength of the steel casing
$l_e$ = Effective length = $0.7l$
$l$ = Actual length
$\eta_c$ = Concrete confinement coefficient
$\eta_a$ = Steel tube confinement coefficient
$\bar{\lambda}$ = Relative slenderness
$(EI)_{eff2}$ = Effective flexural stiffness
$K_c$ = Correction factor
$\eta$ = Confinement factor = $0.27$
$\xi$ = Confinement factor
$W_{scm}$ = Section modulus of composite cross section
$\gamma_m$ = Flexural strength index
$f_{min}$ = Objective function

## Appendix A

**Table A1.** Data used to model concrete-filled steel tube.

| S. No | Diameter | Thickness | Yield Strength | Compressive Strength | Length | Length/Diameter | Axial Capacity |
|---|---|---|---|---|---|---|---|
| 1 | 120.9 | 3.73 | 312 | 30.22 | 2311 | 19.11 | 725 |
| 2 | 166 | 5 | 288.1 | 63.70 | 1040 | 6.27 | 1862 |
| 3 | 88.9 | 5.842 | 406 | 50.50 | 1117.6 | 12.57 | 715.56 |
| 4 | 114.3 | 3.1 | 348 | 62.78 | 670 | 5.86 | 898 |
| 5 | 95 | 3.68 | 392 | 31.44 | 860 | 9.05 | 686 |
| 6 | 166 | 5 | 288.1 | 36.55 | 1040 | 6.27 | 1495 |
| 7 | 168.2 | 4.52 | 302 | 52.80 | 813 | 4.83 | 2113 |
| 8 | 114.3 | 3.1 | 348 | 62.78 | 670 | 5.86 | 904 |
| 9 | 219 | 7 | 273 | 46.50 | 1200 | 5.48 | 3200 |
| 10 | 114 | 6.34 | 486 | 45.00 | 850 | 7.46 | 1608 |
| 11 | 100 | 2.5 | 433.2 | 54.78 | 600 | 6.00 | 750 |
| 12 | 108 | 4 | 338.88 | 35.71 | 5400 | 50.00 | 210.7 |
| 13 | 219 | 7 | 273 | 46.50 | 1420 | 6.48 | 3070 |
| 14 | 215.9 | 4.08 | 292 | 28.67 | 2220 | 10.28 | 1650 |
| 15 | 152.4 | 1.55 | 294 | 43.25 | 914 | 6.00 | 721.5 |
| 16 | 114 | 6 | 486 | 45.00 | 850 | 7.46 | 1334 |
| 17 | 114.3 | 3.1 | 340 | 73.10 | 3370 | 29.48 | 379 |
| 18 | 95 | 3.66 | 338 | 30.00 | 2032 | 21.39 | 463 |
| 19 | 216 | 4.04 | 293 | 36.89 | 2220 | 10.28 | 2289 |
| 20 | 114.3 | 3.19 | 414 | 35.44 | 838 | 7.33 | 734 |
| 21 | 95 | 12.4 | 277 | 26.22 | 1420 | 14.95 | 907 |
| 22 | 108 | 4 | 337.6 | 43.12 | 756 | 7.00 | 785 |
| 23 | 152.4 | 1.55 | 330 | 32.11 | 1499 | 9.84 | 734 |
| 24 | 166 | 5 | 288.1 | 65.17 | 1040 | 6.27 | 1852 |
| 25 | 166 | 5 | 289.1 | 34.68 | 2700 | 16.27 | 1117.2 |
| 26 | 110 | 1.9 | 350 | 14.44 | 2200 | 20.00 | 252 |
| 27 | 120.9 | 3.76 | 312 | 26.78 | 1049 | 8.68 | 721 |
| 28 | 216 | 4.11 | 291 | 36.89 | 2220 | 10.28 | 2239 |
| 29 | 108 | 4.5 | 348.1 | 46.87 | 4023 | 37.25 | 318 |
| 30 | 190.7 | 6 | 505 | 57.40 | 3450 | 18.09 | 2130 |
| 31 | 166 | 5 | 288.1 | 53.11 | 1040 | 6.27 | 1695 |
| 32 | 108 | 4 | 338.88 | 35.71 | 864 | 8.00 | 766.36 |
| 33 | 95 | 12.75 | 277 | 26.22 | 1420 | 14.95 | 938 |
| 34 | 114 | 5.94 | 486 | 45.00 | 1750 | 15.35 | 1138 |
| 35 | 216 | 4.11 | 304 | 29.11 | 2220 | 10.28 | 1834 |
| 36 | 152.4 | 3.17 | 415 | 26.56 | 2271 | 14.90 | 939 |
| 37 | 114 | 4.68 | 332 | 45.00 | 850 | 7.46 | 1049 |
| 38 | 108 | 4.5 | 259.7 | 25.48 | 1620 | 15.00 | 524 |
| 39 | 108 | 4 | 338.88 | 35.71 | 3240 | 30.00 | 478.24 |
| 40 | 110 | 1.9 | 350 | 14.44 | 2200 | 20.00 | 219 |
| 41 | 76.48 | 1.73 | 369 | 32.56 | 609.45 | 7.97 | 330.04 |
| 42 | 166 | 5 | 274.4 | 36.43 | 1100 | 6.63 | 1985 |
| 43 | 127.1 | 2.95 | 376 | 77.20 | 711 | 5.59 | 1305 |
| 44 | 114.3 | 3.1 | 348 | 62.67 | 1020 | 8.92 | 888 |
| 45 | 110 | 1.9 | 350 | 40.50 | 2200 | 20.00 | 437 |
| 46 | 355.6 | 11.18 | 361 | 47.00 | 1880 | 5.29 | 11,460 |
| 47 | 88.9 | 5.85 | 400 | 49.75 | 508 | 5.71 | 992 |
| 48 | 127.3 | 1.63 | 334 | 77.20 | 711 | 5.59 | 1285 |
| 49 | 210 | 3 | 233.2 | 33.52 | 1040 | 4.95 | 1705 |
| 50 | 114.3 | 3.1 | 340 | 64.56 | 3720 | 32.55 | 293 |
| 51 | 355.6 | 4.72 | 281 | 27.00 | 1880 | 5.29 | 3517 |
| 52 | 114 | 3.41 | 291 | 43.75 | 2750 | 24.12 | 569 |
| 53 | 95 | 3.66 | 332 | 31.44 | 860 | 9.05 | 656 |
| 54 | 95 | 12.7 | 277 | 26.22 | 860 | 9.05 | 1034 |
| 55 | 108 | 4.5 | 358 | 106.00 | 1188 | 11.00 | 1194 |
| 56 | 121 | 3.73 | 333 | 27.11 | 1050 | 8.68 | 746 |
| 57 | 219 | 7 | 273 | 46.50 | 990 | 4.52 | 3278 |
| 58 | 168.4 | 4.52 | 302 | 52.80 | 813 | 4.83 | 2233 |
| 59 | 160 | 2.5 | 433.2 | 39.40 | 960 | 6.00 | 1426 |
| 60 | 121 | 3.71 | 313 | 30.67 | 2310 | 19.09 | 695 |
| 61 | 88.9 | 5.842 | 406 | 50.50 | 1422.4 | 16.00 | 712 |
| 62 | 88.9 | 5.72 | 400 | 48.25 | 1422 | 16.00 | 712 |
| 63 | 165.2 | 4.1 | 353 | 49.88 | 3965 | 24.00 | 1019 |
| 64 | 168.1 | 4.52 | 298 | 52.30 | 813 | 4.84 | 2233 |
| 65 | 92 | 3 | 260.7 | 26.07 | 1380 | 15.00 | 409 |
| 66 | 114 | 5.94 | 486 | 31.11 | 1280 | 11.23 | 1285 |
| 67 | 114 | 6.11 | 486 | 40.00 | 2750 | 24.12 | 941 |
| 68 | 168.8 | 5 | 302.4 | 40.50 | 2135 | 12.65 | 1130 |
| 69 | 108 | 4.5 | 348.1 | 31.91 | 4158 | 38.50 | 342 |

**Table A1.** *Cont.*

| S. No | Diameter | Thickness | Yield Strength | Compressive Strength | Length | Length/Diameter | Axial Capacity |
|---|---|---|---|---|---|---|---|
| 70 | 82.55 | 1.397 | 482.3 | 47.29 | 1422.4 | 17.23 | 294.59 |
| 71 | 114 | 3.23 | 290 | 36.67 | 1751 | 15.36 | 706 |
| 72 | 121 | 5.41 | 348 | 27.11 | 1050 | 8.68 | 1018 |
| 73 | 165.2 | 4.17 | 358.7 | 49.82 | 1321.6 | 8.00 | 1445 |
| 74 | 95 | 3.86 | 332 | 31.44 | 1420 | 14.95 | 567 |
| 75 | 95 | 12.6 | 279 | 26.22 | 860 | 9.05 | 1018 |
| 76 | 166 | 5 | 289.1 | 34.68 | 2700 | 16.27 | 1271.06 |
| 77 | 114.3 | 3.1 | 348 | 65.56 | 1335 | 11.68 | 794 |
| 78 | 108 | 4.5 | 358 | 106.00 | 1620 | 15.00 | 1018 |
| 79 | 114.3 | 3.1 | 348 | 67.22 | 2040 | 17.85 | 688 |
| 80 | 250 | 7 | 243 | 55.58 | 1480 | 5.92 | 4116 |
| 81 | 76.5 | 1.74 | 364 | 49.88 | 610 | 7.97 | 423 |
| 82 | 95 | 3.91 | 392 | 31.44 | 1420 | 14.95 | 606 |
| 83 | 108 | 4.5 | 358 | 106.00 | 756 | 7.00 | 1286 |
| 84 | 165 | 4.7 | 355 | 33.40 | 2475 | 15.00 | 1058 |
| 85 | 114 | 1.72 | 266 | 43.75 | 2750 | 24.12 | 353 |
| 86 | 95 | 12.6 | 275 | 25.89 | 1981 | 20.85 | 903 |
| 87 | 200 | 3 | 303.5 | 55.80 | 2002 | 10.01 | 1882 |
| 88 | 169 | 7.5 | 360 | 80.80 | 1768 | 10.46 | 2870 |
| 89 | 152.4 | 3.17 | 415 | 26.56 | 2271 | 14.90 | 881 |
| 90 | 121 | 3.86 | 332 | 30.67 | 2310 | 19.09 | 755 |
| 91 | 165.2 | 4.17 | 358.7 | 49.82 | 1982.4 | 12.00 | 1305 |
| 92 | 95 | 12.5 | 279 | 26.22 | 1420 | 14.95 | 947 |
| 93 | 108 | 4.5 | 348.1 | 31.91 | 3510 | 32.50 | 400 |
| 94 | 88.9 | 5.82 | 400 | 48.75 | 1727 | 19.43 | 614 |
| 95 | 88.9 | 5.842 | 406 | 50.50 | 812.8 | 9.14 | 918.925 |
| 96 | 166 | 5 | 288.1 | 52.90 | 1040 | 6.27 | 1764 |
| 97 | 121 | 5.44 | 327 | 30.67 | 2310 | 19.09 | 865 |
| 98 | 169.3 | 2.62 | 338.1 | 41.38 | 1830 | 10.81 | 689 |
| 99 | 121.01 | 3.66 | 300 | 27.11 | 1050 | 8.68 | 695 |
| 100 | 108 | 4 | 338.88 | 35.71 | 2160 | 20.00 | 672.28 |
| 101 | 166 | 5 | 284.2 | 51.24 | 870 | 5.24 | 1862 |
| 102 | 108 | 5 | 379.8 | 40.91 | 548 | 5.07 | 1084 |
| 103 | 114 | 3.35 | 291 | 45.00 | 850 | 7.46 | 785 |
| 104 | 108 | 4 | 338.88 | 35.71 | 1620 | 15.00 | 646.8 |
| 105 | 76.5 | 1.73 | 364 | 32.11 | 610 | 7.97 | 330 |
| 106 | 355.6 | 7.98 | 361 | 29.78 | 2083 | 5.86 | 7433 |
| 107 | 114 | 1.79 | 266 | 45.00 | 850 | 7.46 | 515 |
| 108 | 267.4 | 7 | 461 | 57.40 | 4800 | 17.95 | 3900 |
| 109 | 95 | 12.6 | 294 | 26.22 | 1980 | 20.84 | 917 |
| 110 | 114.3 | 3.1 | 348 | 62.67 | 1020 | 8.92 | 849 |
| 111 | 108 | 4.5 | 358 | 106.00 | 1188 | 11.00 | 1232 |
| 112 | 76 | 2 | 275 | 50.60 | 1556 | 20.47 | 330 |
| 113 | 216 | 6.3 | 411 | 36.89 | 2220 | 10.28 | 2932 |
| 114 | 114 | 5.73 | 486 | 40.00 | 2750 | 24.12 | 824 |
| 115 | 110 | 1.9 | 350 | 33.40 | 2200 | 20.00 | 374 |
| 116 | 219 | 4 | 325 | 61.44 | 1000 | 4.57 | 1980 |
| 117 | 267.4 | 7 | 461 | 57.40 | 1600 | 5.98 | 5190 |
| 118 | 88.9 | 5.81 | 400 | 47.62 | 1118 | 12.58 | 716 |
| 119 | 121 | 3.76 | 313 | 30.67 | 1050 | 8.68 | 837 |
| 120 | 108 | 4 | 338.88 | 35.71 | 2160 | 20.00 | 676.2 |
| 121 | 127 | 2.413 | 336 | 32.56 | 914 | 7.20 | 658.3 |
| 122 | 120.83 | 4.09 | 451.3 | 36.18 | 1050.04 | 8.69 | 1091.91 |
| 123 | 200 | 3 | 303.5 | 55.80 | 2001 | 10.01 | 1806 |
| 124 | 108 | 4 | 338.88 | 35.71 | 1080 | 10.00 | 783.02 |
| 125 | 82.55 | 1.397 | 482.3 | 47.29 | 1727.2 | 20.92 | 224.725 |
| 126 | 121.01 | 3.71 | 300 | 27.11 | 2310 | 19.09 | 641 |
| 127 | 140 | 2.5 | 433.2 | 47.43 | 840 | 6.00 | 1124 |
| 128 | 152.4 | 1.57 | 330 | 26.67 | 1499 | 9.84 | 681 |
| 129 | 120.65 | 4.09 | 451.3 | 41.72 | 1050.04 | 8.70 | 1155.7 |
| 130 | 215.9 | 6.02 | 350 | 36.44 | 2220 | 10.28 | 2869 |
| 131 | 121 | 5.49 | 348 | 27.11 | 2310 | 19.09 | 816 |
| 132 | 200 | 2 | 237.2 | 30.28 | 980 | 4.90 | 1411 |
| 133 | 82.55 | 1.397 | 482.3 | 47.29 | 812.8 | 9.85 | 400.5 |
| 134 | 108 | 4 | 338.88 | 35.71 | 2700 | 25.00 | 648.76 |
| 135 | 114.3 | 3.1 | 340 | 67.22 | 2700 | 23.62 | 516 |
| 136 | 108 | 4.5 | 348.1 | 31.91 | 3510 | 32.50 | 390 |
| 137 | 110 | 1.9 | 350 | 40.50 | 2200 | 20.00 | 368 |
| 138 | 114.3 | 3.19 | 414 | 35.44 | 838 | 7.33 | 756 |
| 139 | 114.3 | 3.1 | 340 | 67.22 | 2700 | 23.62 | 536 |

**Table A1.** *Cont.*

| S. No | Diameter | Thickness | Yield Strength | Compressive Strength | Length | Length/Diameter | Axial Capacity |
|---|---|---|---|---|---|---|---|
| 140 | 95 | 12.7 | 277 | 26.22 | 860 | 9.05 | 1008 |
| 141 | 108 | 4.5 | 348.1 | 46.87 | 3510 | 32.50 | 440 |
| 142 | 165 | 4.7 | 355 | 14.44 | 2477 | 15.01 | 800 |
| 143 | 140 | 5 | 378.3 | 37.53 | 840 | 6.00 | 1379 |
| 144 | 108 | 4.5 | 259.7 | 25.48 | 1994 | 18.46 | 495 |
| 145 | 152.4 | 1.55 | 330 | 32.11 | 1499 | 9.84 | 725 |
| 146 | 110 | 1.9 | 350 | 33.40 | 2200 | 20.00 | 368 |
| 147 | 219 | 4 | 325 | 56.60 | 1000 | 4.57 | 1931 |
| 148 | 114 | 4.44 | 332 | 45.00 | 850 | 7.46 | 902 |
| 149 | 108 | 4.5 | 348.1 | 46.87 | 4158 | 38.50 | 298 |
| 150 | 108 | 4 | 347.7 | 40.47 | 1620 | 15.00 | 672 |
| 151 | 152.7 | 3.15 | 421 | 26.89 | 1676.4 | 10.98 | 880.11 |
| 152 | 108 | 4 | 338.88 | 35.71 | 4320 | 40.00 | 294 |
| 153 | 108 | 4 | 338.88 | 35.71 | 1620 | 15.00 | 707.56 |
| 154 | 108 | 4.2 | 259.7 | 25.87 | 648 | 6.00 | 722 |
| 155 | 92 | 3 | 260.7 | 26.07 | 920 | 10.00 | 431 |
| 156 | 108 | 4 | 338.88 | 35.71 | 864 | 8.00 | 869.26 |
| 157 | 219 | 7 | 273 | 46.50 | 990 | 4.52 | 3278 |
| 158 | 108 | 4.5 | 344 | 40.91 | 548 | 5.07 | 917 |
| 159 | 107 | 4 | 379.8 | 38.32 | 542 | 5.07 | 889 |
| 160 | 108 | 4.5 | 259.7 | 25.48 | 648 | 6.00 | 665 |
| 161 | 219 | 7 | 273 | 46.50 | 990 | 4.52 | 3278 |
| 162 | 190.7 | 6 | 505 | 65.44 | 2300 | 12.06 | 2610 |
| 163 | 114 | 3.31 | 291 | 30.00 | 2320 | 20.35 | 535 |
| 164 | 95 | 12.6 | 275 | 25.89 | 861 | 9.06 | 1019 |
| 165 | 114.3 | 3.1 | 348 | 62.67 | 1020 | 8.92 | 845 |
| 166 | 140 | 5 | 378.3 | 42.63 | 840 | 6.00 | 1501 |
| 167 | 88.9 | 5.842 | 406 | 50.50 | 508 | 5.71 | 890 |
| 168 | 95 | 3.51 | 340 | 31.44 | 1980 | 20.84 | 488 |
| 169 | 108 | 4 | 338.88 | 35.71 | 4320 | 40.00 | 345.94 |
| 170 | 127.3 | 1.63 | 376 | 77.20 | 711 | 5.59 | 1285 |
| 171 | 95 | 3.76 | 332 | 31.44 | 1980 | 20.84 | 536 |
| 172 | 165 | 4.7 | 355 | 40.50 | 2476 | 15.01 | 1037 |
| 173 | 166 | 5 | 287.14 | 34.68 | 3700 | 22.29 | 958.44 |
| 174 | 127 | 2.413 | 336 | 27.11 | 914 | 7.20 | 627.2 |
| 175 | 114.3 | 3.1 | 340 | 73.10 | 3370 | 29.48 | 362 |
| 176 | 165 | 4.7 | 355 | 33.40 | 2475 | 15.00 | 1037 |
| 177 | 95 | 3.66 | 350 | 31.11 | 1981 | 20.85 | 529 |
| 178 | 114 | 1.73 | 266 | 40.00 | 1751 | 15.36 | 461 |
| 179 | 219 | 7 | 273 | 46.50 | 1640 | 7.49 | 2956 |
| 180 | 95 | 3.4 | 343 | 30.44 | 1980 | 20.84 | 473 |
| 181 | 210 | 2.5 | 237.2 | 32.93 | 1670 | 7.95 | 1323 |
| 182 | 95 | 3.78 | 392 | 31.44 | 1980 | 20.84 | 567 |
| 183 | 114 | 5.99 | 486 | 45.00 | 1750 | 15.35 | 1177 |
| 184 | 114 | 3.28 | 291 | 43.75 | 2750 | 24.12 | 667 |
| 185 | 108 | 4 | 338.88 | 35.71 | 5400 | 50.00 | 225.4 |
| 186 | 114.3 | 3.1 | 340 | 64.56 | 3720 | 32.55 | 305 |
| 187 | 108 | 4.5 | 259.7 | 25.48 | 1296 | 12.00 | 563 |
| 188 | 165 | 4.3 | 317.7 | 52.30 | 3640 | 22.06 | 987 |
| 189 | 114 | 3.29 | 291 | 30.00 | 2250 | 19.74 | 652 |
| 190 | 166 | 5 | 288.1 | 33.12 | 1040 | 6.27 | 1372 |
| 191 | 114.3 | 3.1 | 348 | 67.22 | 2040 | 17.85 | 617 |
| 192 | 168.3 | 4.47 | 302 | 29.33 | 813 | 4.83 | 1744 |
| 193 | 108 | 4 | 327.1 | 41.55 | 1188 | 11.00 | 686 |
| 194 | 140 | 5 | 378.3 | 51.25 | 840 | 6.00 | 1539 |
| 195 | 101.73 | 3.1 | 604.67 | 37.93 | 1524 | 14.98 | 800.1 |
| 196 | 152.4 | 1.55 | 294 | 43.25 | 914 | 6.00 | 733 |
| 197 | 219 | 7 | 273 | 46.50 | 1640 | 7.49 | 2956 |
| 198 | 168.8 | 2.64 | 200.2 | 42.13 | 1830 | 10.84 | 916 |
| 199 | 108 | 4 | 338.88 | 35.71 | 648 | 6.00 | 828.1 |
| 200 | 165.2 | 4.1 | 353 | 49.88 | 1322 | 8.00 | 1412 |
| 201 | 120.9 | 5.54 | 343 | 30.22 | 2311 | 19.11 | 867 |
| 202 | 114 | 6.14 | 486 | 34.44 | 2250 | 19.74 | 1000 |
| 203 | 166 | 5 | 313.6 | 51.24 | 1700 | 10.24 | 1460.2 |
| 204 | 219 | 7 | 273 | 46.50 | 1640 | 7.49 | 2956 |
| 205 | 76.5 | 1.73 | 364 | 31.11 | 1524 | 19.92 | 245 |
| 206 | 82.55 | 1.397 | 482.3 | 47.29 | 1117.6 | 13.54 | 356 |
| 207 | 95 | 12.8 | 283 | 26.22 | 1980 | 20.84 | 886 |
| 208 | 190.7 | 6 | 505 | 65.44 | 1150 | 6.03 | 3064 |
| 209 | 95 | 3.4 | 340 | 31.44 | 860 | 9.05 | 656 |

**Table A1.** *Cont.*

| S. No | Diameter | Thickness | Yield Strength | Compressive Strength | Length | Length/Diameter | Axial Capacity |
|---|---|---|---|---|---|---|---|
| 210 | 114 | 5.64 | 486 | 34.44 | 2250 | 19.74 | 902 |
| 211 | 240 | 10 | 269 | 58.80 | 1440 | 6.00 | 5135 |
| 212 | 152 | 1.65 | 270 | 83.00 | 900 | 5.92 | 1458 |
| 213 | 140 | 3 | 426.3 | 40.38 | 840 | 6.00 | 1208 |
| 214 | 216 | 6.05 | 395 | 29.11 | 2220 | 10.28 | 2462 |
| 215 | 108 | 4.5 | 348.1 | 46.87 | 4023 | 37.25 | 320 |
| 216 | 95 | 3.58 | 340 | 31.44 | 1420 | 14.95 | 576 |
| 217 | 169.3 | 2.62 | 338.1 | 45.13 | 1830 | 10.81 | 756 |
| 218 | 95 | 12.65 | 275 | 25.89 | 1420 | 14.95 | 930 |
| 219 | 114 | 6.21 | 486 | 40.00 | 2750 | 24.12 | 941 |
| 220 | 108 | 4.5 | 259.7 | 25.48 | 972 | 9.00 | 666 |
| 221 | 121 | 5.56 | 327 | 30.67 | 1050 | 8.68 | 1079 |
| 222 | 168.1 | 4.52 | 298 | 52.30 | 813 | 4.84 | 2113 |
| 223 | 216 | 4.06 | 289 | 29.11 | 2220 | 10.28 | 1023 |
| 224 | 114.3 | 3.1 | 340 | 73.10 | 3370 | 29.48 | 401 |
| 225 | 165.2 | 4.1 | 353 | 49.88 | 2974 | 18.00 | 1147 |
| 226 | 140 | 5.3 | 378.3 | 60.56 | 840 | 6.00 | 1664 |
| 227 | 127 | 2.39 | 289 | 42.75 | 1499 | 11.80 | 623 |

## References


1. El-Heweity, M.M. On the performance of circular concrete-filled high strength steel columns under axial loading. *Alex. Eng. J.* **2012**, *51*, 109–119. [CrossRef]
2. Elbakry, H.M.F. A numerical approach for evaluating the stiffness of steel tube–R.C. beam composite joint. *Alex. Eng. J.* **2014**, *53*, 583–589. [CrossRef]
3. Jayalekshmi, S.; Jegadesh, J.S.S.; Goel, A. Empirical approach for determining axial strength of circular concrete filled steel tubular columns. *J. Inst. Eng. Ser. A* **2018**, *99*, 257–268. [CrossRef]
4. Ekmekyapar, T.; AL-Eliwi, B.J.M. Concrete filled double circular steel tube (CFDCST) stub columns. *Eng. Struct.* **2017**, *135*, 68–80. [CrossRef]
5. Chen, D.; Montano, V.; Huo, L.; Fan, S.; Song, G. Detection of subsurface voids in concrete-filled steel tubular (CFST) structure using percussion approach. *Constr. Build. Mater.* **2020**, *262*, 119761. [CrossRef]
6. Chen, D.; Montano, V.; Huo, L.; Song, G. Depth detection of subsurface voids in concrete-filled steel tubular (CFST) structure using percussion and decision tree. *Meas. J. Int. Meas. Confed.* **2020**, *163*, 107869. [CrossRef]
7. Rong, B.; Liu, S.; Li, Z.; Liu, R. Experimental and numerical studies of failure modes and load-carrying capacity of through-diaphragm connections. *Trans. Tianjin Univ.* **2018**, *24*, 387–400. [CrossRef]
8. Alatshan, F.; Osman, S.A.; Mashiri, F.; Hamid, R. Explicit simulation of circular CFST stub columns with external steel confinement under axial compression. *Materials* **2019**, *13*, 23. [CrossRef]
9. Akbar, A.; Farooq, F.; Shafique, M.; Aslam, F.; Alyousef, R.; Alabduljabbar, H. Sugarcane bagasse ash-based engineered geopolymer mortar incorporating propylene fibers. *J. Build. Eng.* **2020**, *33*, 101492. [CrossRef]
10. Ren, Q.; Li, M.; Zhang, M.; Shen, Y.; Si, W. Prediction of ultimate axial capacity of square concrete-filled steel tubular short columns using a hybrid intelligent algorithm. *Appl. Sci.* **2019**, *9*, 2802. [CrossRef]
11. Zhang, T.; Lyu, X.; Yu, Y. Prediction and analysis of the residual capacity of concrete-filled steel tube stub columns under axial compression subjected to combined freeze–thaw cycles and acid rain corrosion. *Materials* **2019**, *12*, 3070. [CrossRef]
12. Ly, H.-B.; Le, T.-T.; Le, L.M.; Tran, V.Q.; Le, V.M.; Vu, H.-L.T.; Nguyen, Q.H.; Pham, B.T. Development of hybrid machine learning models for predicting the critical buckling load of I-shaped cellular beams. *Appl. Sci.* **2019**, *9*, 5458. [CrossRef]
13. Chen, J.B.; Chan, T.M.; Castro, J.M. Parametric study on the flexural behaviour of circular rubberized concrete-filled steel tubes. In *Tubular Structures XVI, Proceedings of the 16th International Symposium on Tubular Structures, ISTS 2017, Melbourne, Australia, 4–6 December 2017*; CRC Press/Balkema: Leiden, The Netherlands, 2018; pp. 51–59.
14. Farooq, F.; Akbar, A.; Khushnood, R.A.; Muhammad, W.L.B.; Rehman, S.K.U.; Javed, M.F. Experimental Investigation of Hybrid Carbon Nanotubes and Graphite Nanoplatelets on Rheology, Shrinkage, Mechanical, and Microstructure of SCCM. *Materials* **2020**, *13*, 230. [CrossRef] [PubMed]
15. Hu, H.S.; Liu, Y.; Zhuo, B.T.; Guo, Z.X.; Shahrooz, B.M. Axial compressive behavior of square CFST columns through direct measurement of load components. *J. Struct. Eng.* **2018**. [CrossRef]

16. Li, G.; Chen, B.; Yang, Z.; Feng, Y. Experimental and numerical behaviour of eccentrically loaded high strength concrete filled high strength square steel tube stub columns. *Thin-Walled Struct.* **2018**, *127*, 483–499. [CrossRef]

17. Mendoza, R.; Yamamoto, Y.; Nakamura, H.; Miura, T. Numerical simulation of compressive failure behaviors of concrete-filled steel tube using coupled discrete model and shell finite element. In *High Tech Concrete: Where Technology and Engineering Meet*; Springer International Publishing: Berlin, Germany, 2018; pp. 1353–1361.

18. Lai, Z.; Varma, A.H. Effective stress-strain relationships for analysis of noncompact and slender filled composite (CFT) members. *Eng. Struct.* **2016**, *124*, 457–472. [CrossRef]

19. Chen, C.C.; Ko, J.W.; Huang, G.L.; Chang, Y.M. Local buckling and concrete confinement of concrete-filled box columns under axial load. *J. Constr. Steel Res.* **2012**, *78*, 8–21. [CrossRef]

20. Mendoza, R.; Yamamoto, Y.; Nakamura, H.; Miura, T. Numerical evaluation of localization and softening behavior of concrete confined by steel tubes. *Struct. Concr.* **2018**, *19*, 1956–1970. [CrossRef]

21. Jumaa, G.B.; Yousif, A.R. Predicting shear capacity of frp-reinforced concrete beams without stirrups by artificial neural networks, gene expression programming, and regression analysis. *Adv. Civ. Eng.* **2018**. [CrossRef]

22. Javed, M.F.; Amin, M.N.; Shah, M.I.; Khan, K.; Iftikhar, B.; Farooq, F.; Aslam, F.; Alyousef, R.; Alabduljabbar, H. Applications of Gene Expression Programming and Regression Techniques for Estimating Compressive Strength of Bagasse Ash based Concrete. *Crystals* **2020**, *10*, 737. [CrossRef]

23. Javed, M.F.; Sulong, N.H.R.; Memon, S.A.; Rehman, S.K.U.; Khan, N.B. FE modelling of the flexural behaviour of square and rectangular steel tubes filled with normal and high strength concrete. *Thin-Walled Struct.* **2017**, *119*, 470–481. [CrossRef]

24. Ferreira, C. Gene Expression Programming: A Newadaptive Algorithm for Solving Problems. Available online: https://scholar.google.com/scholar?hl=en&as_sdt=0%2C5&q=Ferreira%2C+C.%2C+Gene+Expression+Programming%3A+A+New+Adaptive+Algorithm+for+Solving+Problems.+Complex+Systems%2C+2001.+13%282%29%3A+p.+87-129.&btnG= (accessed on 17 December 2019).

25. Sun, Y.; Li, G.; Zhang, J.; Qian, D.; Binh, G.E.; Pham, A. Prediction of the strength of rubberized concrete by an evolved random forest model. *Adv. Civ. Eng.* **2019**. [CrossRef]

26. Zhang, J.; Ma, G.; Huang, Y.; Aslani, F.; Nener, B. Modelling uniaxial compressive strength of lightweight self-compacting concrete using random forest regression. *Constr. Build. Mater.* **2019**. [CrossRef]

27. Kohavi, R. A study of cross-validation and bootstrap for accuracy estimation and model selection. In Proceedings of the International Joint Conference on Artificial Intelligence (IJCAI), Montreal, QC, Canada, 20 August 1995; pp. 1137–1145.

28. Le, L.M.; Ly, H.-B.; Pham, B.T.; Le, V.M.; Pham, T.A.; Nguyen, D.-H.; Tran, X.-T.; Le, T.-T. Hybrid artificial intelligence approaches for predicting buckling damage of steel columns under axial compression. *Materials* **2019**, *12*, 1670. [CrossRef]

29. Sarir, P.; Chen, J.; Asteris, P.G.; Armaghani, D.J.; Tahir, M.M. Developing GEP tree-based, neuro-swarm, and whale optimization models for evaluation of bearing capacity of concrete-filled steel tube columns. *Eng. Comput.* **2019**, *1*, 3. [CrossRef]

30. Elfahham, Y. Estimation and prediction of construction cost index using neural networks, time series, and regression. *Alex. Eng. J.* **2019**, *58*, 499–506. [CrossRef]

31. Diab, A.M.; Elyamany, H.E.; Abd Elmoaty, A.E.M.; Shalan, A.H. Prediction of concrete compressive strength due to long term sulfate attack using neural network. *Alex. Eng. J.* **2014**, *53*, 627–642. [CrossRef]

32. Mohamed, Y.S.; Shehata, H.M.; Abdellatif, M.; Awad, T.H. Steel crack depth estimation based on 2D images using artificial neural networks. *Alex. Eng. J.* **2019**, *58*, 1167–1174. [CrossRef]

33. Mai, S.H.; Seghier, M.E.; Nguyen, P.L.; Jafari-Asl, J.; Thai, D.K. A hybrid model for predicting the axial compression capacity of square concrete-filled steel tubular columns. *Eng. Comput.* **2020**, *1*, 3. [CrossRef]

34. Xu, J.; Wang, Y.; Ren, R.; Wu, Z.; Ozbakkaloglu, T. Performance evaluation of recycled aggregate concrete-filled steel tubes under different loading conditions: Database analysis and modelling. *J. Build. Eng.* **2020**, *30*, 101308. [CrossRef]

35. Zhou, C.; Chen, W.; Ruan, X.; Tang, X. Experimental study on axial compression behavior and bearing capacity analysis of high titanium slag CFST columns. *Appl. Sci.* **2019**, *9*, 2021. [CrossRef]

36. Dao, D.; Trinh, S.; Ly, H.-B.; Pham, B. Prediction of compressive strength of geopolymer concrete using entirely steel slag aggregates: Novel hybrid artificial intelligence approaches. *Appl. Sci.* **2019**, *9*, 1113. [CrossRef]

37. Van Dao, D.; Ly, H.-B.; Vu, H.-L.T.; Le, T.-T.; Pham, B.T. Investigation and optimization of the C-ANN structure in predicting the compressive strength of foamed concrete. *Materials* **2020**, *13*, 1072. [CrossRef] [PubMed]

38. Nguyen, H.Q.; Ly, H.B.; Tran, V.Q.; Nguyen, T.A.; Le, T.T.; Pham, B.T. Optimization of artificial intelligence system by evolutionary algorithm for prediction of axial capacity of rectangular concrete filled steel tubes under compression. *Materials* **2020**, *13*, 1205. [CrossRef]

39. Ly, H.-B.; Pham, B.T.; Le, L.M.; Le, T.-T.; Le, V.M.; Asteris, P.G. Estimation of axial load-carrying capacity of concrete-filled steel tubes using surrogate models. *Neural Comput. Appl.* **2020**. [CrossRef]

40. Nguyen, Q.H.; Ly, H.-B.; Tran, V.Q.; Nguyen, T.-A.; Phan, V.-H.; Le, T.-T.; Pham, B.T. A novel hybrid model based on a feedforward neural network and one step secant algorithm for prediction of load-bearing capacity of rectangular concrete-filled steel tube columns. *Molecules* **2020**, *25*, 3486. [CrossRef]

41. Luat, N.V.; Shin, J.; Lee, K. Hybrid BART-based models optimized by nature-inspired metaheuristics to predict ultimate axial capacity of CCFST columns. *Eng. Comput.* **2020**, *1*, 3. [CrossRef]

42. Han, Q.; Gui, C.; Xu, J.; Lacidogna, G. A generalized method to predict the compressive strength of high-performance concrete by improved random forest algorithm. *Constr. Build. Mater.* **2019**, *226*, 734–742. [CrossRef]

43. Nilsen, V.; Pham, L.T.; Hibbard, M.; Klager, A.; Cramer, S.M.; Morgan, D. Prediction of concrete coefficient of thermal expansion and other properties using machine learning. *Constr. Build. Mater.* **2019**, *220*, 587–595. [CrossRef]

44. Fawagreh, K.; Medhat Gaber, M.; Elyan, E.; Gaber, M.M. Random forests: From early developments to recent advancements. *Syst. Sci. Control Eng. Open Access J.* **2014**, *2*, 602–609. [CrossRef]

45. Ipek, S.; Güneyisi, E.M. Ultimate axial strength of concrete-filled double skin steel tubular column sections. *Adv. Civ. Eng.* **2019**. [CrossRef]

46. Güneyisi, E.M.; Gültekin, A.; Mermerdaş, K. Ultimate capacity prediction of axially loaded CFST short columns. *Int. J. Steel Struct.* **2016**, *16*, 99–114. [CrossRef]

47. Yang, C.; Gao, P.; Wu, X.; Chen, Y.F.; Li, Q.; Li, Z. Practical formula for predicting axial strength of circular-CFST columns considering size effect. *J. Constr. Steel Res.* **2020**, *168*, 105979. [CrossRef]

48. Avci-Karatas, C. Prediction of ultimate load capacity of concrete-filled steel tube columns using multivariate adaptive regression splines (MARS). *Steel Compos. Struct.* **2019**, *33*, 583–594.

49. Wang, X.Y. Optimal design of the cement, fly ash, and slag mixture in ternary blended concrete based on gene expression programming and the genetic algorithm. *Materials* **2019**, *12*, 2448. [CrossRef]

50. Alkroosh, I.S.; Sarker, P.K. Prediction of the compressive strength of fly ash geopolymer concrete using gene expression programming. *Comput. Concr.* **2019**, *24*, 295–302.

51. Iqbal, M.F.; Liu, Q.-F.; Azim, I.; Zhu, X.; Yang, J.; Javed, M.F.; Rauf, M. Prediction of mechanical properties of green concrete incorporating waste foundry sand based on gene expression programming. *J. Hazard. Mater.* **2020**, *384*, 121322. [CrossRef]

52. Bolandi, H.; Banzhaf, W.; Lajnef, N.; Barri, K.; Alavi, A.H. An intelligent model for the prediction of bond strength of FRP bars in concrete: A soft computing approach. *Technologies* **2019**, *7*, 42. [CrossRef]

53. Nour, A.I.; Güneyisi, E.M. Prediction model on compressive strength of recycled aggregate concrete filled steel tube columns. *Compos. Part B Eng.* **2019**, *173*, 106938. [CrossRef]

54. Leon, L.P.; Gay, D. Gene expression programming for evaluation of aggregate angularity effects on permanent deformation of asphalt mixtures. *Constr. Build. Mater.* **2019**, *211*, 470–478. [CrossRef]

55. Gunal, M.; Guven, A.; Asce, M. A genetic programming approach for prediction of local scour downstream hydraulic structures genetic programming approach for prediction of local scour downstream of hydraulic structures. *Artic. J. Irrig. Drain. Eng.* **2008**. [CrossRef]

56. Ferreira, C. *Gene Expression Programming Mathematical Modeling by an Artificial Intelligence*; Springer: Berlin, Germany, 2006; ISBN 3-540-32796-7.

57. Ferreira, C. Gene expression programming in problem solving. In *Soft Computing and Industry*; Springer: London, UK, 2002; pp. 635–653.

58. Tenpe, A.R.; Patel, A. Utilization of support vector models and gene expression programming for soil strength modeling. *Arab. J. Sci. Eng.* **2020**, *45*, 4301–4319. [CrossRef]

59. Nematzadeh, M.; Shahmansouri, A.A.; Fakoor, M. Post-fire compressive strength of recycled PET aggregate concrete reinforced with steel fibers: Optimization and prediction via RSM and GEP. *Constr. Build. Mater.* **2020**, *252*, 119057. [CrossRef]

60. Momeni, M.; Hadianfard, M.A.; Bedon, C.; Baghlani, A. Damage evaluation of H-section steel columns under impulsive blast loads via gene expression programming. *Eng. Struct.* **2020**, *219*, 110909. [CrossRef]

61. Gandomi, A.H.; Roke, D.A. Assessment of artificial neural network and genetic programming as predictive tools. *Adv. Eng. Softw.* **2015**, *88*, 63–72. [CrossRef]

62. Alavi, A.H.; Gandomi, A.H. A robust data mining approach for formulation of geotechnical engineering systems. *Eng. Comput. Int. J. Comput.-Aided Eng.* **2011**, *28*, 242–274.

63. Gandomi, A.H.; Babanajad, S.K.; Alavi, A.H.; Farnam, Y. Novel approach to strength modeling of concrete under triaxial compression. *J. Mater. Civ. Eng.* **2012**. [CrossRef]

64. Muduli, P.K.; Das, S.K.; Bhattacharya, S. CPT-based probabilistic evaluation of seismic soil liquefaction potential using multi-gene genetic programming. *Georisk* **2014**, *8*, 14–28. [CrossRef]

65. Das, S.K. Artificial neural networks in geotechnical engineering: Modeling and application issues. In *Metaheuristics in Water, Geotechnical and Transport Engineering*; Elsevier: Amsterdam, The Netherlands, 2013; pp. 231–270. ISBN 9780123982964.

66. Fulcher, J. Computational intelligence: An introduction. *Stud. Comput. Intell.* **2008**, *115*, 3–78.

67. Alabduljabbar, H.; Haido, J.H.; Alyousef, R.; Yousif, S.T.; McConnell, J.; Wakil, K.; Jermsittiparsert, K. Prediction of the flexural behavior of corroded concrete beams using combined method. *Structures* **2020**, *25*, 1000–1008. [CrossRef]

68. Dunlop, P.; Smith, S. Estimating key characteristics of the concrete delivery and placement process using linear regression analysis. *Civ. Eng. Environ. Syst.* **2003**, *20*, 273–290. [CrossRef]

69. Han, L.H.; Li, W.; Bjorhovde, R. Developments and advanced applications of concrete-filled steel tubular (CFST) structures: Members. *J. Constr. Steel Res.* **2014**, *100*, 211–228. [CrossRef]

70. Ferreira, C. Function finding and the creation of numerical constants in gene expression programming. In *Advances in Soft Computing*; Springer: London, UK, 2003; pp. 257–265.

71. Babanajad, S.K.; Gandomi, A.H.; Alavi, A.H. New prediction models for concrete ultimate strength under true-triaxial stress states: An evolutionary approach. *Adv. Eng. Softw.* **2017**, *110*, 55–68. [CrossRef]

72. Soleimani, S.; Rajaei, S.; Jiao, P.; Sabz, A.; Soheilinia, S. New prediction models for unconfined compressive strength of geopolymer stabilized soil using multi-gen genetic programming. *Meas. J. Int. Meas. Confed.* **2018**, *113*, 99–107. [CrossRef]

73. Kupper, L.L. Probability, statistics, and decision for civil engineers. *Technometrics* **1971**, *13*, 211. [CrossRef]

74. Frank, I.; Todeschini, R. The data analysis handbook. *Data Handl. Sci. Technol.* **1994**, *14*, 1–352.

75. Golbraikh, A.; Tropsha, A. Beware of q2! *J. Mol. Graph. Model.* **2002**. [CrossRef]

76. Gandomi, A.H.; Alavi, A.H.; Mirzahosseini, M.R.; Nejad, F.M. Nonlinear genetic-based models for prediction of flow number of asphalt mixtures. *J. Mater. Civ. Eng.* **2011**, *23*, 248–263. [CrossRef]

77. Roy, P.P.; Roy, K. On some aspects of variable selection for partial least squares regression models. *QSAR Comb. Sci.* **2008**, *27*, 302–313. [CrossRef]

78. Lu, Z.; Zhao, Y. Suggested empirical models for the axial capacity of circular CFT stub columns. *J. Constr. Steel Res.* **2010**. [CrossRef]

79. Goode, C.; Narayanan, R. Design of concrete filled steel tubes to EC4. In Proceedings of the ASCCS Seminar on Concrete Filled Steel Tubes—A Comparison of International Codes and Practice, Innsbruck, Austria, 18 September 1997; pp. 1–25.

80. Giakoumelis, G.; Lam, D. Axial capacity of circular concrete-filled tube columns. *J. Constr. Steel Res.* **2004**, *60*, 1049–1068. [CrossRef]

*Article*

# Numerical Simulation of Adsorption of Organic Inhibitors on C-S-H Gel

Zijian Song [1], Huanchun Cai [1], Qingyang Liu [2,*], Xing Liu [3], Qi Pu [4], Yingjie Zang [1] and Na Xu [1]

[1] College of Mechanics and Materials, Hohai University, Xikang Road 1#, Nanjing 210098, China; songzijian@hhu.edu.cn (Z.S.); 191308070001@hhu.edu.cn (H.C.); zyj062818@hhu.edu.cn (Y.Z.); 181308050003@hhu.edu.cn (N.X.)

[2] State Key Laboratory of Water Resources and Hydropower Engineering Science, Wuhan University, Donghu South Road 8#, Wuhan 430072, China

[3] State Key Laboratory of Solidification Processing, School of Materials Science and Engineering, Northwestern Polytechnical University, Youyi West Road 127#, Xian 710072, China; xingliu_lx@163.com

[4] Suzhou Concrete and Cement Products Research Institute Co., Ltd., Sanxiang Road 718#, Suzhou 215000, China; puqi0512@163.com

* Correspondence: liuqingyang@whu.edu.cn

Received: 18 July 2020; Accepted: 20 August 2020; Published: 23 August 2020

**Abstract:** Corrosion inhibitors are one of the most effective anticorrosion techniques in reinforced concrete structures. Molecule dynamics (MD) was usually utilized to simulate the interaction between the inhibitor molecules and the surface of Fe to evaluate the inhibition effect, ignoring the influence of cement hydration products. In this paper, the adsorption characteristics of five types of common alkanol-amine inhibitors on C-S-H gel in the alkaline liquid environment were simulated via the MD and the grand canonical Monte Carlo (GCMC) methods. It is found that, in the MD system, the liquid phase environment had a certain impact on the adsorption configuration of compounds. According to the analysis of the energy, the binding ability of MEA on the surface of the C-S-H gel was the strongest. In the GCMC system, the adsorption of MEA was the largest at the same temperature. Furthermore, for the competitive adsorption in the GCMC system, the adsorption characteristics of the inhibitors on the C-S-H gel were to follow the order: MEA > DEA > TEA > NDE > DETA. Both MD and GCMC simulations confirmed that the C-S-H gel would adsorb the organic inhibitors to a different extent, which might have a considerable influence on the organic inhibitors to exert their inhibition effects.

**Keywords:** calcium silicate hydrate; simulation; concrete; corrosion inhibitor; grand canonical Monte Carlo method; molecular dynamics; adsorption

## 1. Introduction

Corrosion inhibitors have been regarded as one of the most effective anticorrosion techniques applied in reinforced concrete structures [1,2]. Due to the relatively low biological toxicity and carcinogenicity, the organic inhibitors have attracted increasing attention of researchers in recent decades. However, the practical inhibition effects of many organic inhibitors are still unclear or disputable. Omellese et al. [3] tested 80 kinds of organic inhibitors and concluded that many amine inhibitors, e.g., DMEA (dimethylethanolamine), had almost no inhibition effect in concrete while the amino acid inhibitors had a certain effect, but they still could not meet the needs of industrial applications. E. Rakanta et al. [4] studied the effect of the organic corrosion inhibitor DMEA on the corrosion of steel bar in simulated concrete pore (SCP) solution, and pointed out that DMEA showed a good inhibition effect. Morris et al. [5] studied the inhibition effect of amine inhibitors on the corrosion of reinforcement in concrete and found that only when the concentration of chloride ions in concrete was less than 0.2% (mass fraction), this kind of corrosion inhibitor could exert its effect. Thus, it is still urgent to properly evaluate the inhibition effect of the organic inhibitors.

In recent years, molecular-scale simulations have been widely used in the field of anticorrosion towards metals in acid mediums. Many researchers utilized molecular dynamics (MD) methods to investigate the active sites, chemical reactivity and the interaction between inhibitors and the metal surface so as to evaluate the inhibition effectiveness [6–9]. Khaled et al. [7] simulated the adsorption of inhibitor molecules on the Fe (0 0 1) surface in an ideal aqueous liquid to investigate their anticorrosion behavior towards mild steel. Similarly, Dehghani et al. [8] adopted the ideal aqueous liquid environment to study the adsorption tendency, energy and affinity of selected inhibitors on the steel adsorbent (i.e., Fe (1 1 0)). Sulaiman et al. [9] made a further step and started to use an electrolyte with high ionic strength as a liquid environment to imitate the corrosion process that occurred in the practical acidic environment. It should be recognized that the simulations in the ideal aqueous liquid or electrolyte solutions can meet the corrosion situations in acid environments and possibly satisfy the requirements of the estimation of the inhibition effect. In the reinforced concrete, however, the corrosion situations largely differ. On the one hand, the exposure surface would be the passive layer due to the high pH concrete pore solution. On the other hand, the surrounding cement hydrated productions might also influence the exertion of the inhibition effect [10]. However, most of the related numerical studies merely considered the adsorption process between inhibitors and steel reinforcements. For instance, Alibakhshi et al. [11] built a setup to simulate the corrosion inhibition of mild steel occurred in the presence of solvent molecules. Ormellese et al. [12] studied the inhibition effect of five kinds of organic compounds on chloride-induced corrosion of concrete by theoretical methodology to discuss the influence of their interaction with passive film on the corrosion initiation. In these studies, the researchers usually simulated the adsorption process by placing inhibitor molecules above the surface of passive film in a water or SCP environment, ignoring the influence of hydration products in cement. However, the ignorance of the effect of cement hydrated productions may largely increase the gap between the simulation and practical results.

The hydration products of concrete can adsorb the nitrogen-containing organic inhibitors when these inhibitors transport in concrete [13]. Among these hydration products, the calcium-silicate-hydrate (C-S-H) gel was the main hydration product of concrete. Knowing this, this paper intended to adopt the calcium-silicate-hydrate (C-S-H) gel to construct the nanopore of concrete as a simulated adsorption circumstance [14]. Then the behavior and ability of C-S-H gel to adsorb different organic corrosion inhibitors were simulated by the method of grand canonical Monte Carlo (GCMC) [15,16]. Meanwhile, the adsorption characteristics of corrosion inhibitors on the C-S-H gel in alkaline liquid environment were studied by molecular dynamics. This paper aims to promote the understanding of the adsorption effect of organic inhibitors in cement-hydrated system, so as to make a contribution to the proper estimation or prediction of the inhibition effects of the common organic inhibitors.

## 2. Model Constructions and Computational Methods

### 2.1. Model Constructions

The calcium-silicate-hydrate (C-S-H) gel, the most important component in concrete, accounts for the largest proportion about 60% of hydration products in cement paste, which also plays a dominate role in the structural and mechanical properties of concrete [17]. However, C-S-H gel at the nanoscale level is still not completely understood as it has complex and diverse structures. C-S-H gel is a non-stoichiometric compound in which $CaO \cdot SiO \cdot H_2O$ is represented in different proportions [16]. In recent years, with the development of research technologies, small angle X-ray diffraction (SAXS), small angle neutron scattering (SANS), inelastic neutron scattering (INS), extended X-ray absorption microstructural spectroscopy (EXAFS) and atomic force microscopy (AFM) have been applied to the study of the C-S-H gel and some common characteristics have been identified [18–21]. Based on the properties of the C-S-H gel obtained by XRD, NMR and SANS [22,23], Pellenq et al. [24] established a realistic model of C-S-H gel, which accorded with the true $Q_n$ distribution, density, chemical composition and characteristics similar to the jennite structure.

Based on the model construction procedures proposed by Pellenq [24], the C-S-H model was constructed by the following steps. Firstly, this paper utilized the tobermorite 11 Å structure as the initial configuration of the C-S-H gel [25–27]. Secondly, an amount of bridging $SiO_2$ (/silicate chains) groups and silicate dimers were removed to satisfy the calcium-to-silicon ratio (C/S) of C-S-H gel and the $Q_n$ distribution of silicon chains [24,28]. Then the Ca/Si ratio of the configuration reached 1.44 and the model achieved $Q_0$, $Q_1$ and $Q_2$ percentages of 9.7%, 19.3% and 71% respectively, which was consistent with the test result of NMR and other MD models of the C-S-H gel. In this way, the silicate skeleton structure in the dry state was obtained. Finally, the GCMC method, which is a simulation method specially used to study the adsorption, was utilized to make the dry calcium silicate skeleton absorb water. The simulation process was similar to water adsorption of the microporous structure (such as zeolite) [29]. The GCMC simulation is a dynamic process including 100,000 circles. The Compass II force field was chosen, the Ewald addition method was used for electrostatic interaction, and the atom-based method was used for the van der Waals effect and hydrogen bond interaction [30]. In this study, the ratios of water molecules exchange, conformation isomerization, rotation, translation and regeneration were set as 2, 1, 1, 1 and 0.1 respectively. After water absorption, the relaxation configuration and energy minimization were obtained and the water-to-silicon ratio of C-S-H gel reached 1.98 while its density reached 2.43 $g/cm^3$. The C-S-H gel model construction process was shown in Figure 1.

Five types of common alkanol-amine compounds were selected as adsorbates. They all belong to the amino-alcohol based migrating corrosion inhibitors. The corrosion inhibitors migrate into the concrete through capillary and microcracks to adsorb and form films on the surface of steel bars so as to exert their inhibiting effect. Their molecular structures are given in Table 1. In this study, according to the molecular structure, the material visualizer module was used to draw the compounds and the compass II force field was utilized to optimize the geometry.

**Table 1.** The molecule structures of the compounds.

| Name | Abbreviation | Molecular Formula | Molecular Mass | Molecular Structure |
|---|---|---|---|---|
| Ethanolamine | MEA | $C_2HNO$ | 61.08 | 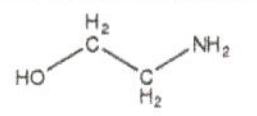 |
| Diethanolamine | DEA | $C_4H_{11}NO_2$ | 105.14 | 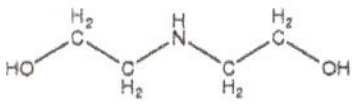 |
| Triethylolamine | TEA | $C_6H_{15}NO_3$ | 149.19 | 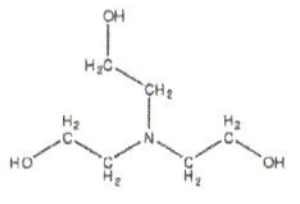 |
| N,N-dimethylethanolamine | NDE | $C_4H_{11}NO$ | 89.14 | 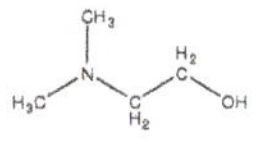 |
| Diethyleneteiamine | DETA | $C_4H_{13}N_3$ | 103.17 | 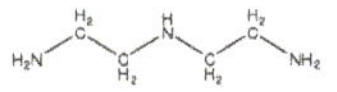 |

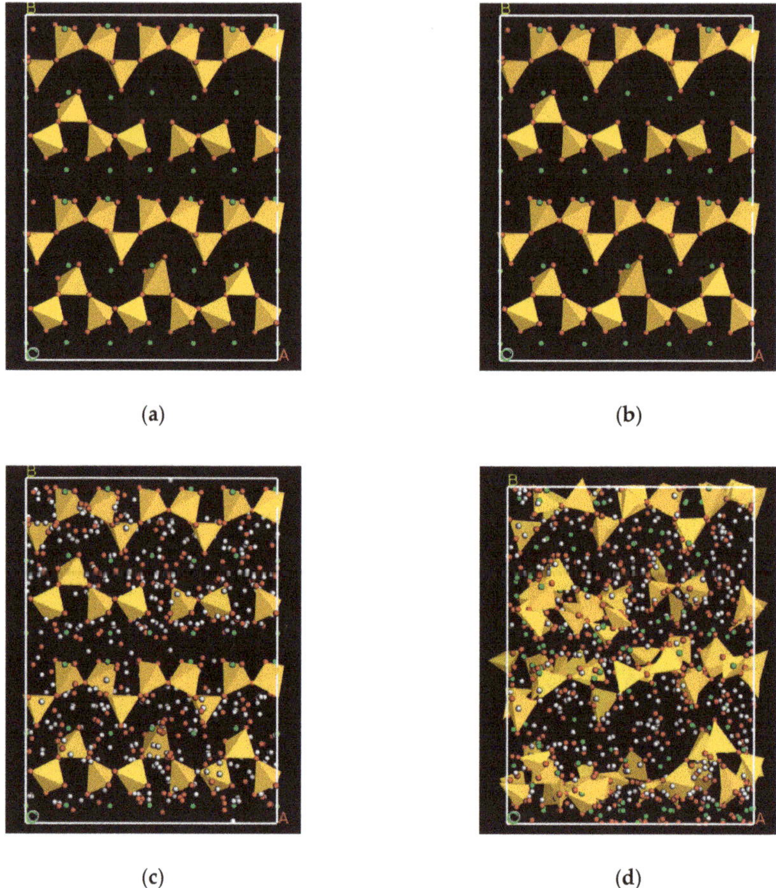

(a)  (b)

(c)  (d)

**Figure 1.** The C-S-H gel model construction process (Red ball: oxygen atom; green ball: calcium atom; white ball: hydrogen atom; yellow tetrahedral: silicate tetrahedral): (**a**) initial tobermorite 11 Å structure without water molecules. Cell parameters: a = 18.37 Å, b = 24.32 Å, c = 26.48 Å, $\alpha$, $\beta$, $\gamma$, equal to 90°; (**b**) removal of the bridging silicate tetrahedral, where $Q_n$ represents n species of silicate tetrahedral, n corresponds to the number of bridging oxygen atoms (BO) in one silicate tetrahedral; (**c**) grand canonical Monte Carlo (GCMC) water adsorption and (**d**) C-S-H structure finality achieved.

## 2.2. Surface Adsorption Test

In this study, the adsorption characteristics of the five types of alkanol-amine compounds on the C-S-H gel in alkaline liquid environment were studied by molecule dynamics. It is known that the carbonation and calcium leaching are two important processes that would occur on a concrete surface [31–33]. However, this paper is to study the adoption of organic inhibitors on the C-S-H gel, which usually occurs on the interface of cement hydration products and pore solution inside the concrete. Therefore, the effects of the carbonation and dissolution of calcium ions are preliminarily ignored in this study. Initially, the C-S-H gel was cleaved to show the surface along the (0 0 1) direction at 24.32 Å depth so as to simulate the non-bonded interaction between adsorbates and the surface of the C-S-H gel. Figure 2 shows the snapshots for the surface adsorption test. As shown in Figure 2, the lower part of the model consists of a fixed C-S-H gel with a thickness of 24.32 Å and 6 adsorbates were placed on the surface. The GCMC method was used to search for the best adsorption site in the range of 10 Å on the upper surface [34]. After that, 351 water molecules, 2 $Na^+$ and $OH^-$ were added, in order to simulate the concrete pore solution environment with pH 13.5 including 1.0 mol/L adsorbates. The crystal cell adopted three-dimensional periodic boundary conditions. A vacuum space thicker than 25 Å was laid on the top layers so as to avoid the influence of the periodic cell above [35]. Finally, the MD simulations were carried out using the compass II force field and 20 ps NVT (isothermal-isasteric ensemble) running for the system.

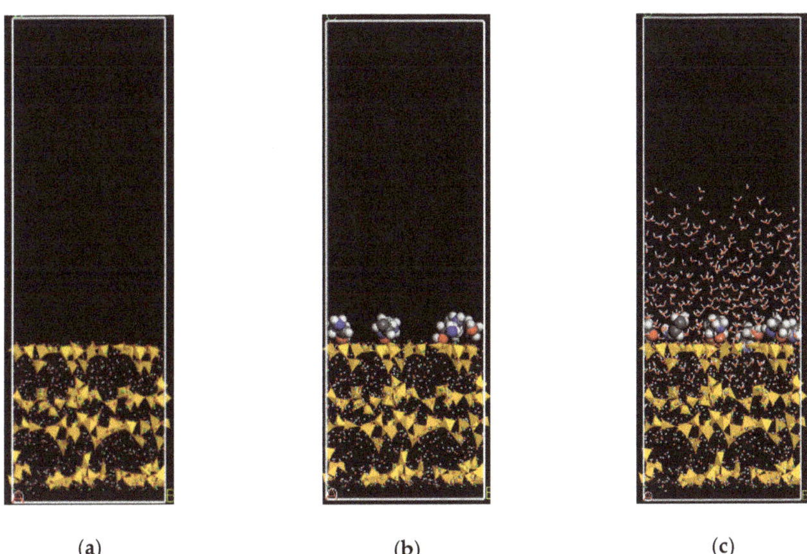

(a)        (b)        (c)

**Figure 2.** Snapshots for the surface adsorption test (red ball: oxygen atom; green ball: calcium atom; white ball: hydrogen atom; yellow tetrahedral: silicate tetrahedral; violet ball: sodium atom; blue ball: nitrogen atom and grey ball: carbon atom): (**a**) C-S-H surface. Cell parameters: a = 18.37 Å, b = 26.48 Å, c = 78.72 Å, $\alpha$, $\beta$, $\gamma$, equal to 90°; (**b**) GCMC adsorbents placed and (**c**) final balancing snapshots.

*2.3. Pore Adsorption Test*

In order to reduce the statistical errors caused by the size effect and make the simulation closer to the pore structure of concrete, a larger C-S-H gel channel was constructed. As depicted in Figure 3, under the condition of periodic boundary, the C-S-H pore consisted of two parallel layers of C-S-H gel with 24.32 Å thickness and the surface of C-S-H gel layer was 55.12 Å × 52.97 Å. The upper and lower gel surfaces were spaced 35 Å for simulating the concrete pore [36]. Subsequently, the GCMC simulations were applied into the structure and energy analysis [33]. The adsorbates and water molecules were regarded as stiff small molecules. According to the partial pressure of 1.0 mol/L adsorbate solution, the two-component adsorption simulations of five types of alkanol-amine compounds were carried out respectively. In order to observe the competitive strength of different component adsorption more intuitively, six-component adsorption simulation ($H_2O$ 110 kPa + MEA 2 kPa + DEA 2 kPa + TEA 2 kPa + NDE 2 kPa + DETA 2 kPa) were carried out. The minimum energy structure was output. In addition, this paper is to fundamentally study the adsorption of organic inhibitor molecules on C-S-H gel. In MD, we simulated the adsorption behavior of the single organic inhibitor molecule on C-S-H gel. In contrast, we simulated as many inhibitor molecules as possible to adsorb on the surface of C-S-H gel in GCMC to study the adsorption characteristics of organic inhibitor molecules.

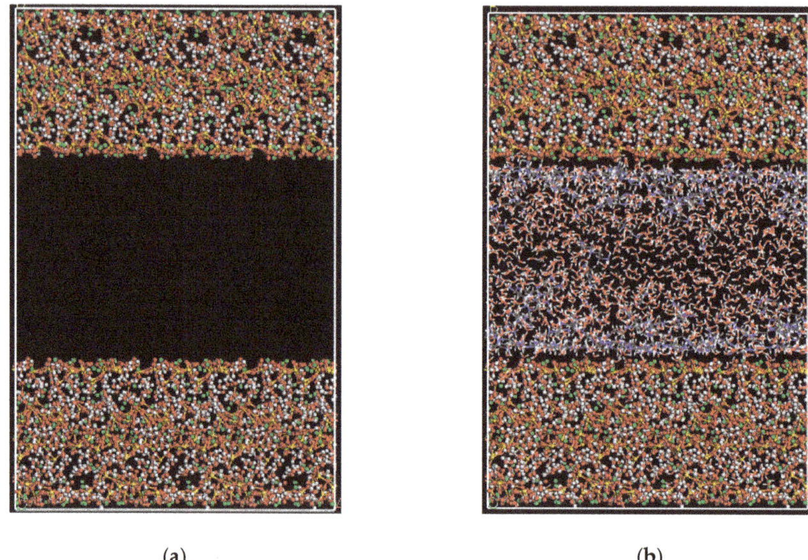

(a)             (b)

**Figure 3.** Snapshots for the pore adsorption test (red ball: oxygen atom; green ball: calcium atom; white ball: hydrogen atom; yellow ball: silicate atom; violet ball: sodium atom; blue ball: nitrogen atom; grey ball: carbon atom): (**a**) C-S-H pore. Cell parameters: a = 18.37 Å, b = 26.48 Å, c = 78.72 Å, $\alpha$, $\beta$, $\gamma$, equal to 90°; (**b**) GCMC adsorption.

## 3. Simulation Results of Surface Adsorption System

### 3.1. Adsorption Morphology

The changes of energy with time are shown in Figure 4. It can be seen that the energy of the system decreased rapidly within 2 ps. Finally, the energy of the system remained basically unchanged, and the change range of total energy was less than 1‰. Therefore, it can be considered that the system had reached the equilibrium state. At the equilibrium state of dynamic simulations, the final interfacial configuration of the surface adsorbing inhibitor could be obtained. Figure 5 shows the final balancing snapshots of five types of alkanol-amine compounds after molecular dynamics simulation. As shown in Figure 5, the five types of inhibitor molecules have parallel adsorption on the surface of C-S-H gel in the way of flat orientations. It is found that their structures all have deformation phenomenon in different degrees. Most of the compounds were almost planar and parallel to the C-S-H gel surface. However, in the liquid phase, the compounds were affected both by the adsorption force of the C-S-H gel and the solvent. As a result, some compounds did not adhere to the surface of the C-S-H gel. The final equilibrated mode and the changes of the energy were different from those under vacuum condition [37], which indicated that the liquid phase environment would have a certain impact on the adsorption configuration of compounds. Hence, for making a more precise simulation, it is necessary to construct a solvent environment closer to the pore liquid of concrete.

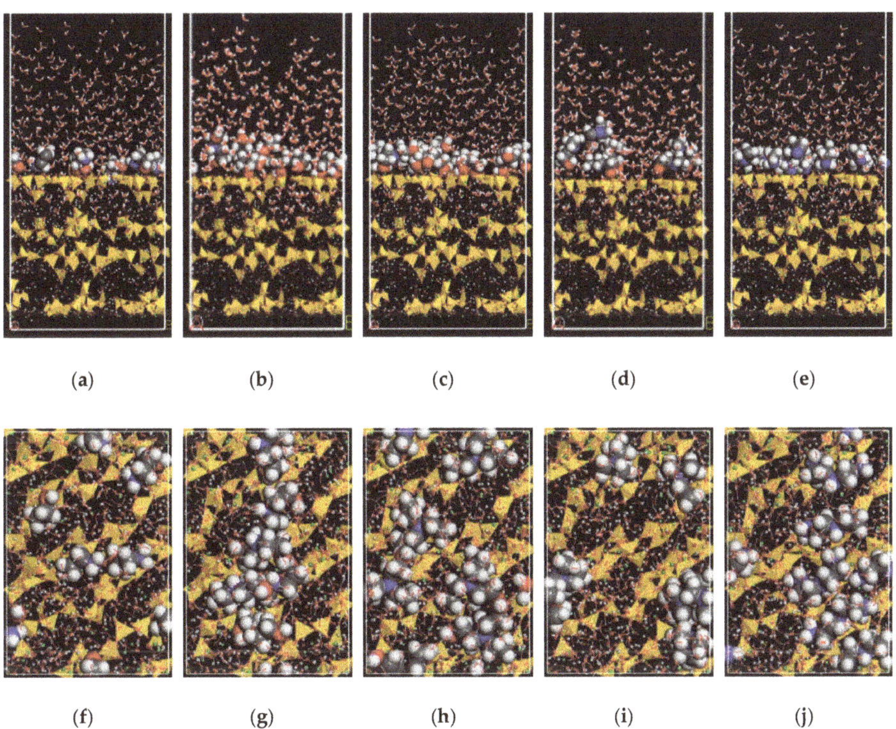

**Figure 4.** The change of energy with time in the molecular dynamics simulation.

**Figure 5.** The final equilibrated snapshots (red ball: oxygen atom; green ball: calcium atom; white ball: hydrogen atom; yellow tetrahedral: silicate tetrahedral; violet ball: sodium atom; blue ball: nitrogen atom and grey ball: carbon atom): (**a**) main view of MEA; (**b**) main view of DEA; (**c**) main view of TEA; (**d**) main view of NDE; (**e**) main view of DETA; (**f**) top view of MEA; (**g**) top view of DEA; (**h**) top view of TEA; (**i**) top view of NDE and (**j**) top view of DETA.

## 3.2. Energy Analysis

In order to quantitatively analyze the adsorption of the inhibitor molecules, the adsorption energy ($E_{Ads}$) is used to characterize the interaction strength between the molecules and the surface of the

C-S-H gel [6]. While the solvation energy ($E_{sol}$) is used to characterize the interaction between molecules and the concrete pore solution [38]. These three energies can be calculated by the following equations:

$$E_{Ads} = E_{inhi/csh} - \left(E_{inhi} + E_{csh}\right) \tag{1}$$

$$E_{sol} = E_{inhi/water} - \left(E_{inhi} + E_{water}\right) \tag{2}$$

where, $E_{inhi/CSH}$ is the total potential energy of the whole simulated system without aqueous solution, $E_{inhi}$ is the internal energy of the inhibitor molecules after adsorption, $E_{CSH}$ is the surface energy of the isolated C-S-H gel, $E_{inhi/water}$ is the total potential energy of the whole simulated system without the C-S-H gel and $E_{water}$ is the energy of the solvent system. The above parameters were calculated by the first-principles using the Forcite module. The calculated results of the energy at their most stable conformation are shown in Table 2. From Table 2, it can be seen that the adsorption energy of the five types of inhibitors was all negative, indicating that the inhibitors could adsorb on the solid surface by reducing the system energy in order to achieve the maximization of the adsorption energy of system and the lowest state of the energy of the system [39]. The molecules could be spontaneously adsorbed on the surface of the solid. The more negative adsorption energy embodied the adsorption between molecules and the surface of C-S-H gel were more prone to occur and more difficult to be destroyed after adsorption [40]. The solvation energy of the inhibitors was also negative, which indicated that the inhibitors tend to dissolve in alkaline aqueous solution. It is noticed that the solvation energy was obviously more positive than the corresponding adsorption energy, indicating the dissolve process could not forbid the spontaneous occurrence of the adsorption process [38]. However, the combination of the inhibitors and the solvent could weaken its binding ability to the C-S-H gel surface. Hence, the difference between the adsorption energy and solvent energy could represent the binding ability of the inhibitor molecules and the surface of the C-S-H gel in alkaline concrete pore solution environment. It is found that, although the adsorption capacity of TEA was the largest, the binding effect of the solvent was also stronger, which led to the difficulty of the removal of the binding of water molecules. Thus, its binding ability to the surface of the C-S-H gel was not the strongest under the solvent environment. In the alkaline concrete pore solution environment, the binding ability between the five types of inhibitor molecules and the surface of C-S-H gel could be ranged according to the difference as: MEA > DEA > TEA > NDE > DETA. According to the $E_{ads}$ value, the trend became TEA > MEA > DEA > DETA > NDE. Thus, the liquid environment truly impacted the adsorption of inhibitor molecules on the C-S-H gel.

**Table 2.** Adsorption and solvation energy of five molecules. (kcal/mol).

| Molecule | $E_{Ads}$ | $E_{sol}$ | $E_{Ads} - E_{sol}$ |
|---|---|---|---|
| MEA | −697.496064 | −92.595926 | −604.900138 |
| NDE | −629.069637 | −143.23066 | −485.838977 |
| TEA | −756.114049 | −199.036085 | −557.077964 |
| DEA | −671.377916 | −105.307953 | −566.069963 |
| DETA | −658.629254 | −188.678217 | −469.951037 |

## 4. The Grant Canonical Monte Carlo System

### 4.1. Adsorption Quantity Analysis

Mixing different components together for adsorption could make a more intuitive comparison of the adsorption capacity of different components. Set the partial pressure of each adsorption component as: $H_2O$ 110 kPa, MEA 2 kPa, DEA 2 kPa, TEA 2 kPa, NDE 2 kPa and DETA 2 kPa. At the same time, in order to test the effect of temperature on adsorption, the set of simulations was carried out every 5 K in the temperature range of 293–318 K. The adsorption capacities of five types of alkanol-amine compounds at different temperature are shown in Figure 6. Adsorption selectivity of a to b can be calculated by the following equations:

$$S = \frac{x_a / y_a}{x_b / y_b} \tag{3}$$

where, $S$ is adsorption selectivity, $x_a$ is the concentration of $a$ in the adsorbed phase, $y_a$ is the partial pressure of $a$ in the adsorbed phase (2 kPa adopted in this research), $x_b$ is the concentration of $b$ in the adsorbed phase and $y_b$ is the partial pressure of $b$ in the adsorbed phase (2 kPa adopted in this research) [41]. The calculated results are shown in Table 3. The adsorption selectivity of each inhibitor molecules to water was greater than 1, indicating that they were easily precipitated from aqueous solution and adsorbed on the C-S-H gel. At the same temperature, the adsorption of the inhibitor MEA was the most and more than twice of the other four inhibitors, which indicated the adsorption of the inhibitor MEA was the largest. With the increase of the temperature, the adsorption of water molecules increased while that of inhibitor molecules presented decreasing tendency, indicating they were exothermic physical adsorption process. The five types of the inhibitor molecules placed at further distances with respect to each other resulting in weaker interactions at a higher temperature [42]. Therefore, if temperature increased from 298 to 333 K, the adsorption effect between the inhibitor molecules and the surface of the C-S-H gel decreased. It is found that the adsorption processes of five types of the inhibitor molecules were all spontaneous, which was exothermic at a different temperature, indicating that a high temperature was not conducive to adsorption.

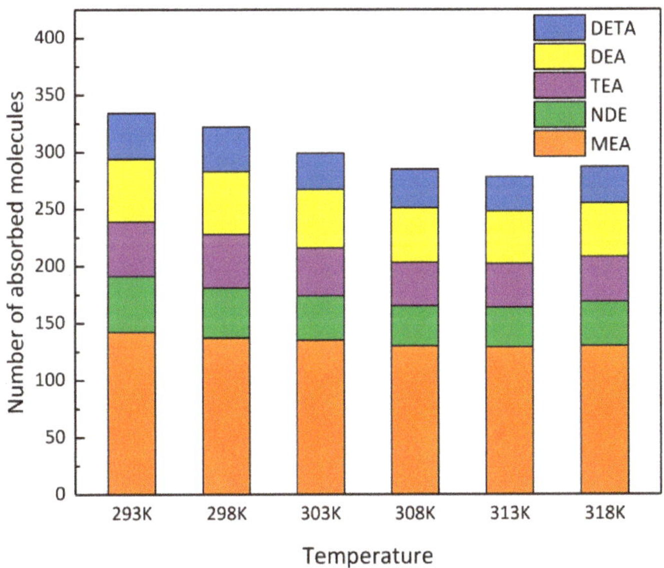


**Figure 6.** Adsorption quantity at a different temperature.

**Table 3.** Adsorption selectivity of alkanol-amine compounds to $H_2O$ at different temperature.

| Molecule | 293 K | 298 K | 303 K | 308 K | 313 K | 318 K |
|---|---|---|---|---|---|---|
| MEA to $H_2O$ | 12.93 | 12.54 | 12.19 | 11.68 | 11.46 | 11.12 |
| NDE to $H_2O$ | 4.46 | 4.03 | 3.52 | 3.15 | 3.11 | 3.16 |
| TEA to $H_2O$ | 4.37 | 4.30 | 3.79 | 3.42 | 3.38 | 3.34 |
| DEA to $H_2O$ | 5.01 | 5.03 | 4.61 | 4.31 | 4.09 | 4.02 |
| DETA to $H_2O$ | 3.64 | 3.57 | 2.89 | 3.06 | 2.67 | 2.65 |

Figure 6 presents the adsorption quantity at a different temperature. As shown in Figure 6, when the temperature was 318 K, the adsorption capacity increased slightly, this might be caused by the increase of the total amount of adsorbed materials. However, if we observe their adsorption selectivity to water in Table 3, it can be easily found that most of them decreased with the increase of temperature. According to the competitive adsorption quantity, the binding ability between the five types of inhibitor molecules and the surface of the C-S-H gel in the alkaline concrete pore fluid environment were to follow the order: MEA > DEA > TEA > NDE > DETA.

*4.2. Energy Analysis*

In order to verify the energy change trend in the relatively big system and get close to the structure of the internal channel in concrete, we used the method of GCMC to simulate the adsorption process in the larger C-S-H gel channel. Table 4 lists the changes of total energy, van der Waals energy, electrostatic energy and intramolecular energy before and after adsorption in each adsorbate system under a different temperature. As clearly shown in Table 4, with the increase of temperature, the absolute values of electrostatic force showed a downward trend. It also can be seen in the Table 4 that the negative value of the electrostatic energy is relatively high, which might be able to break the optimum state of the adsorption caused by the vdW (Van Der Waals) energy and the intramolecular energy. Then the values of the vdW energy and the intramolecular energy might become a little positive. The electrostatic energy was negative, which indicated that the adsorption process mainly depended on the electrostatic force, and the adsorption effect became stronger with the increase of the absolute values. Meanwhile, the absolute values of the total energies of five types of inhibitor molecules all decreased and the adsorption function diminished, which indicated the high temperature would restrain the physical adsorption. When adsorbing a single inhibitor molecule, van der Waals energy mainly came from dipole–dipole attraction, induction force and dispersion force caused by the reversal of permanent dipoles [43]. Van der Waals energy of inhibitor molecules MEA and TEA was strong. This might be because their spatial configurations were more conducive to their combination and adsorption on the C-S-H gel. For the intramolecular energy, the increase of the intramolecular energy reflected the amount of adsorption. It can be seen from Table 4 that with the increase of temperature, the intramolecular energy of each inhibitor molecule did not change much, which also showed that the adsorption of each inhibitor in Table 3 had little difference with the change of temperature. The magnitude of the total energy could reflect the level of stability of the equilibrium state of the system. Any system had a tendency to decrease energy. When the energy was growing smaller, the system became more stable. Meanwhile, the adsorption of the inhibitor molecules on the C-S-H gel was more likely to occur. At 298 K, the total energies of five types of inhibitor molecules MEA, NDE, TEA, DEA and DETA adsorbed by C-S-H gel were $-64690.743$ kcal·mol$^{-1}$, $-51813.292$ kcal·mol$^{-1}$, $-53083.227$ kcal·mol$^{-1}$, $-53338.167$ kcal·mol$^{-1}$ and $-53106.886$ kcal·mol$^{-1}$ respectively. It is found that the adsorption system of MEA was the most stable.

**Table 4.** Total energy, van der Waals energy, electrostatic energy and intramolecular energy of every system in different temperatures (kcal/mol).

| Molecule | Temperature | Total Energy | Van Der Waals Energy | Electrostatic Energy | Intramolecular Energy |
|---|---|---|---|---|---|
| MEA | 298 K | −64690.74324 | 2951.46269 | −78902.78028 | 11260.57435 |
|  | 303 K | −64530.30697 | 2898.92290 | −78303.21347 | 11073.98360 |
|  | 308 K | −64075.01570 | 2467.56340 | −77464.32230 | 10921.74313 |
|  | 313 K | −63639.92284 | 2366.96724 | −76879.99999 | 10873.10984 |
| NDE | 298 K | −51813.29218 | 2257.08450 | −65106.40411 | 11036.02744 |
|  | 303 K | −51011.18774 | 2247.36929 | −63938.87839 | 11080.32136 |
|  | 308 K | −49461.25152 | 2175.76475 | −62563.41327 | 10926.39701 |
|  | 313 K | −49008.36233 | 2068.31950 | −61393.39248 | 10316.71066 |
| TEA | 298 K | −53083.22706 | 2715.29208 | −67080.09967 | 11281.58053 |
|  | 303 K | −52678.75286 | 2680.23054 | −66935.38242 | 11206.39902 |
|  | 308 K | −52069.37783 | 2615.87605 | −65897.47767 | 11012.22379 |
|  | 313 K | −52092.91453 | 2677.50214 | −65938.40445 | 11067.98778 |
| DEA | 298 K | −53338.16669 | 2593.65200 | −66600.79470 | 11208.97600 |
|  | 303 K | −53020.39933 | 2603.99637 | −65289.85228 | 11065.45658 |
|  | 308 K | −52974.72470 | 2544.07300 | −64940.22606 | 10721.42836 |
|  | 313 K | −51087.92460 | 2507.73629 | −64016.20586 | 10830.54496 |
| DETA | 298 K | −53106.88604 | 2703.01003 | −64217.20122 | 11448.15528 |
|  | 303 K | −51976.23648 | 2604.64465 | −63585.48501 | 11004.60387 |
|  | 308 K | −508903.83110 | 2592.05362 | −62751.71009 | 11525.82537 |
|  | 313 K | −50057.24012 | 2562.15990 | −61100.04881 | 11021.49893 |

## 5. Conclusions

In this study, the adsorption characteristics of the five alkanol-amine compounds on the C-S-H gel in the alkaline liquid environment were studied by molecule dynamics and GCMC. Conclusions can be summarized as follows:

(1) In the MD system, the five types of inhibitor molecules were almost planar and parallel to the surface of the C-S-H gel. It is found that their molecular structures all had the deformation phenomenon in different degrees. In the liquid phase, the compounds were affected both by the adsorption force of the C-S-H gel and the solvent. The liquid phase environment would have a certain impact on the adsorption configuration of compounds.

(2) According to the analysis of the energy in the MD system, it can be concluded that the inhibitors tended to dissolve in the alkaline aqueous solution. The solvation energy was obviously more positive than the corresponding adsorption energy and the dissolve process could not forbid the spontaneous occurrence of the adsorption process. In the alkaline concrete pore solution environment, the binding ability of MEA on the surface of C-S-H gel was the strongest. Considering the effect of the solvent, the binding ability of TEA on the surface of C-S-H gel was the strongest.

(3) In the GCMC system, it is found that all the five types of inhibitor molecules were easily precipitated from aqueous solution and adsorbed on the C-S-H gel. At the same temperature, the adsorption of the inhibitor MEA was the largest and more than twice of the other four inhibitors. When the temperature increased from 298 to 333 K, the adsorption of water molecules increased while that of inhibitor molecules presented a decreasing tendency.

(4) For the competitive adsorption of different types of inhibitor molecules in the GCMC system, the binding ability among the five types of inhibitor molecules on the C-S-H gel were to follow the order: MEA > DEA > TEA > NDE > DETA.

**Author Contributions:** Conceptualization, Z.S. and Q.L.; Data curation, H.C. and X.L.; Formal analysis, H.C. and Y.Z.; Funding acquisition, Z.S. and Q.P.; Investigation, Q.L. and Y.Z.; Methodology, Z.S. and Q.L.; Project administration, Z.S. and X.L.; Resources, Z.S. and Q.P.; Software, X.L.; Writing—original draft, H.C. and N.X.; Writing—review and editing, Z.S. and Q.L. All authors have read and agreed to the published version of the manuscript.

**Funding:** This investigation is financed by the National Natural Science Foundation of China (Nos.51609075 and 51778209), the National Key Research and Development Program of China (2018YFC15087040), the Fundamental Research Funds for the Central Universities (B200202123), the Open Research Fund of Guangxi Key Laboratory of Water Engineering Materials and Structures (GXHRI-WEMS-2019-04, GXHRI-WEMS-2019-05).

**Acknowledgments:** We also wish to thank Jiajia Wang, College of mechanics and materials, Hohai University, for his guidance to some of our numerical work.

**Conflicts of Interest:** All authors of this article declare no conflicts of interest.

## References

1. Bertolini, L.; Elsener, B.; Pedeferri, P.; Redaelli, E.; Polder, R.B. *Corrosion of Steel in Concrete: Prevention, Diagnosis, Repair*; Wiley-VCH Verlag GmbH & Co. KGaA: Weinheim, Germany, 2013; Volume 49, pp. 4113–4133.

2. Liu, Q.Y.; Song, Z.J.; Cai, H.C.; Zhou, A.P.; Wang, W.Y.; Jiang, L.H.; Liu, Y.Q.; Zhang, Y.J.; Xu, N. Effect of Ultrasonic Parameters on Electrochemical Chloride Removal and Rebar Repassivation of Reinforced Concrete. *Materials* **2019**, *12*, 2774. [CrossRef]

3. Ormellese, M.; Lazzari, L.; Goidanich, S.; Fumagalli, G.; Brenna, A. A study of organic substances as inhibitors for chloride-induced corrosion in concrete. *Corros. Sci.* **2009**, *51*, 2959–2968. [CrossRef]

4. Rakanta, E.; Zafeiropoulou, T.; Batis, G. Corrosion protection of steel with DMEA-based organic inhibitor. *Constr. Build. Mater.* **2013**, *44*, 507–513. [CrossRef]

5. Morris, W.; Vico, A.; Vazquze, M. Corrosion of reinforcing steel by means of concrete resistively measurements. *Corros. Sci.* **2002**, *44*, 81–99. [CrossRef]

6. Liu, Q.Y.; Song, Z.J.; Han, H. A novel green reinforcement corrosion inhibitor extracted from waste Platanus acerifolia leaves. *Constr. Build. Mater.* **2020**, *260*, 119695. [CrossRef]

7. Khaled, K.F.; El-Maghraby, A. Experimental, Monte Carlo and molecular dynamics simulations to investigate corrosion inhibition of mild steel in hydrochloric acid solutions. *Arab. J. Chem.* **2014**, *7*, 319–326. [CrossRef]

8. Dehghani, A.; Bahlakeh, G.; Ramezanzadeh, B.; Karati, M.R. Potential role of a novel green eco-friendly inhibitor in corrosion inhibition of mild steel in HCl solution: Detailed macro/micro-scale experimental and computational explorations. *Constr. Build. Mater.* **2020**, *245*, 118464. [CrossRef]

9. Sulaiman, K.O.; Onawole, A.T.; Faye, O.; Shuaib, D.T. Understanding the corrosion inhibition of mild steel by selected green compounds using chemical quantum based assessments and molecular dynamics simulations. *J. Mol. Liq.* **2019**, *279*, 342–350. [CrossRef]

10. Zhu, Y.Y.; Ma, Y.W.; Hu, J.; Zhang, Z.M.; Huang, J.Z.; Wang, Y.Y.; Wang, H.; Cai, W.X.; Huang, H.L.; Yu, Q.J.; et al. Adsorption of organic core-shell corrosion inhibitors on cement particles and their influence on early age properties of fresh cement paste. *Cem. Concr. Res.* **2020**, *130*, 106000. [CrossRef]

11. Alibakhshi, E.; Ramezanzadeh, M.; Bahlakeh, G.; Ramezanzadeh, B.; Mahdavian, M.; Motamedi, M. Glycyrrhiza glabra leaves extract as a green corrosion inhibitor for mild steel in 1 M hydrochloric acid solution: Experimental, molecular dynamics, Monte Carlo and quantum mechanics study. *J. Mol. Liq.* **2018**, *255*, 185–198. [CrossRef]

12. Ormellese, M.; Perez, E.A.; Raffaini, G.; Ganazzoli, F.; Lazzari, L. Inhibition mechanism in concrete by organic substances: An experimental and theoretical study. *J. Mater. Sci. Eng.* **2010**, *13*, 66–77.

13. Zheng, H.; Li, W.; Ma, F.; Kong, Q. The performance of a surface-applied corrosion inhibitor for the carbon steel in saturated $Ca(OH)_2$ solutions. *Cem. Concr. Res.* **2014**, *55*, 102–108. [CrossRef]

14. Li, Z. Structure of concrete. In *Proceedings of the Advanced Concrete Technology*; John Wiley & Sons Inc.: Hoboken, NJ, USA, 2011; pp. 140–163.

15. Puibasset, J.; Pellenq, J.M. Water adsorption on hydrophilic mesoporous and plane silica substrates: A grand canonical Monte Carlo simulation study. *J. Chem. Phys.* **2003**, *118*, 5613–5622. [CrossRef]

16. Ma, H.; Li, Z. Realistic pore structure of Portland cementpaste: Experimental study and numerical simulation. *Comput. Concr.* **2013**, *11*, 317–336. [CrossRef]

17. Mehta, P.K.; Paulo, J.M. *Concrete: Microstructure, Properties, and Materials*; Prentice-Hall: Upper Saddle River, NJ, USA, 2013.

18. Allen, A.J.; Thomas, J.J. Analysis of C–S–H gel and cement paste by small-angle neutron scattering. *Cem. Concr. Res.* **2007**, *37*, 319–324. [CrossRef]

19. Ji, Q.; Pellenq, J.M.; Vliet, K.J.V. Comparison of computational water models for simulation of calcium-silicate-hydrate. *Comput. Mater. Sci.* **2012**, *53*, 234–240. [CrossRef]

20. Bonhoure, I.; Wieland, E.; Scheidegger, A.M.; Ochs, M.; Kunz, D. EXAFS study of Sn(IV) immobilization by hardened cement paste and calcium silicate hydrates. *Environ. Sci. Technol.* **2003**, *37*, 2184–2191. [CrossRef]

21. Gauffinet, S.; Finot, E.; Lesniewska, E.; Nonat, A. Studies of CSH growth on C3S surface during its early hydration. In Proceedings of the 20th International Conference on Cement Microscopy, Guadalajara, Mexico, 19–23 April 1998.

22. Grangeon, S.; Claret, F.; Linard, Y.; Chiaberge, C. X-ray diffraction: A powerful tool to probe and understand the structure of nanocrystalline calcium silicate hydrates. *Acta Crystallogr. B Struct. Sci. Cryst. Eng. Mater.* **2013**, *69*, 465–473. [CrossRef]

23. Cong, X.; Kirkpatrick, R. 29Si MAS NMR study of the structure of calcium silicate hydrate. *Adv. Cem. Based Mater.* **1996**, *3*, 144–156. [CrossRef]

24. Pellenq, R.J.; Kushima, A.; Shahsavari, R.; Van Vliet, K.; Buehler, M.J.; Yip, S.; Ulm, F.-J. A realistic molecular model of cement hydrates. *Proc. Natl. Acad. Sci. USA* **2009**, *106*, 16102–16107. [CrossRef]

25. Hamid, S. The crystal structure of the 11 Å natural tobermorite $Ca_{2.25}$ $Si_3$ $O_{7.5}$ $(OH)_{1.5}$ $1H_2$ O. *Z. Fur Krist.* **1981**, *154*, 189–198.

26. Murray, S.J.; Subramani, V.J.; Selvam, R.P.; Hall, K.D. Molecular dynamics to understand the mechanical behavior of cement Paste. *J. Transp. Res. Board* **2010**, *2142*, 75–82. [CrossRef]

27. Selvam, R.P.; Subramani, V.J.; Murray, S.; Hall, K.D. *Potential Application of Nanotechnology on Cement Based Materials*; The National Academies of Sciences, Engineering, and Medicine: Keck Center, DC, USA, 2009.

28. Manzano, H.; Moeini, S.; Marinelli, F.; van Duin, A.C.T.; Ulm, F.-J.; Pellenq, R.J.-M. Confined water dissociation in microporous defective silicates: Mechanism, dipole distribution, and impact on substrate properties. *J. Am. Chem. Soc.* **2011**, *134*, 2208–2215. [CrossRef] [PubMed]

29. Puibasset, J.; Pellenq, R. Grand canonical monte carlo simulation study of water adsorption in silicalite at 300 K. *J. Phys. Chem. B* **2008**, *112*, 6390–6397. [CrossRef]

30. Valleau, J.P.; Cohen, L.K. Primitive model electrolytes. I. Grand canonical Monte Carlo computations. *J. Chem. Phys.* **1980**, *72*, 5935–5941. [CrossRef]

31. Oelkers, E.H.; Golubev, S.V.; Chairat, C.; Pokrovsky, O.S.; Schott, J. The surface chemistry of multi-oxide silicates. *Geochim. Et Cosmochim. Acta* **2009**, *73*, 4617–4634. [CrossRef]

32. Longo, R.; Cho, K.; Brüner, P.; Welle, A.; Gerdes, A.; Thissen, P. Carbonation of wollastonite(001) competing hydration: Microscopic insights from ion spectroscopy and density functional theory. *ACS Appl. Mater. Interfaces* **2015**, *7*, 4706–4712. [CrossRef]

33. Giraudo, N.; Thissen, P. Carbonation Competing Functionalization on Calcium-Silicate-Hydrates: Investigation of Four Promising Surface-Activation Techniques. *ACS Sustain. Chem. Eng.* **2016**, *4*, 3985–3994. [CrossRef]

34. Sun, H. Simulation of organic adsorption on metallic materials. *J. Phys. Chem. B* **1998**, *102*, 7338. [CrossRef]

35. Musa, A.Y.; Jalgham, R.T.T.; Mohamad, A.B. Molecular dynamic and quantum chemical calculations for phthalazine derivatives as corrosion inhibitors of mild steel in 1M HCl. *Corros. Sci.* **2012**, *56*, 176–183. [CrossRef]

36. Hou, D.; Jia, Y.; Yu, J.; Wang, P.; Liu, Q.F. Transport Properties of Sulfate and Chloride Ions Confined between Calcium Silicate Hydrate Surfaces: A Molecular Dynamics Study. *J. Phys. Chem. C* **2018**, *122*, 28021–28032. [CrossRef]

37. Diamanti, M.V.; Pérez Rosales, E.A.; Raffaini, G.; Ganazzoli, F.; Brenna, A.; Pedeferri, M.; Ormellese, M. Molecular modelling and electrochemical evaluation of organic inhibitors in concrete. *Corros. Sci.* **2015**, *100*, 231–241. [CrossRef]

38. Mendonca, G.L.F.; Costa, S.N.; Freire, V.N.; Casciano, P.N.S.; Correia, A.N.; de Lima-Neto, P. Understanding the corrosion inhibition of carbon steel and copper in sulphuric acid medium by amino acids using electrochemical techniques allied to molecular modelling methods. *Corros. Sci.* **2017**, *115*, 41–55. [CrossRef]

39. Gustincic, D.; Kokalj, A. DFT Study of Azole Corrosion Inhibitors on $Cu_2O$ Model of Oxidized Copper Surfaces: I. Molecule-Surface and Cl-Surface Bonding. *Metals* **2018**, *8*, 311. [CrossRef]

40. Guo, L.; Tan, J.; Kaya, S.; Leng, S.; Li, Q.; Zhang, F. Multidimensional insights into the corrosion inhibition of 3,3-dithiodipropionic acid on Q235 steel in $H_2SO_4$ medium: A combined experimental and in silico investigation. *J. Colloid Interface Sci.* **2020**, *570*, 116–124. [CrossRef]

41. Zhu, Z.-S.; Li, L.-Q.; Gao, Y.-C.; Zhang, J.-F.; Song, J.-F.; Qin, Y.-X.; Liu, X.-Y. Analysis of Multi-component Organic Gas Adsorption Behaviors and Selectivity of Adsorbent. *Nat. Gas Chem. Ind.* **2008**, *33*, 47–51.
42. Abdollahi, F.; Razmkhah, M.; Moosavi, F. The role of hydrogen bond interaction on molecular orientation of alkanolamines through temperature and pressure variation: A mixed molecular dynamics and quantum mechanics study. *Comput. Mater. Sci.* **2017**, *131*, 239–249. [CrossRef]
43. Button, J.K.; Gubbins, K.E.; Tanaka, H.; Nakanishi, K. Molecular dynamics simulation of hydrogen bonding in monoethanolamine. *Fluid Phase Equilibria* **1996**, *116*, 320–325. [CrossRef]

*Article*

# Numerical Evaluation of the Perfobond (PBL) Shear Connector Subjected to Lateral Pressure Using Coupled Rigid Body Spring Model (RBSM) and Nonlinear Solid Finite Element Method (FEM)

**Muhammad Shoaib Karam [1],*, Yoshihito Yamamoto [2], Hikaru Nakamura [1] and Taito Miura [1]**

[1] Department of Civil and Environmental Engineering, Nagoya University, Furo-cho, Chikusa-ku, Nagoya 464-8603, Japan; hikaru@cc.nagoya-u.ac.jp (H.N.); t.miura@civil.nagoya-u.ac.jp (T.M.)

[2] Department of Civil and Environmental Engineering, Hosei University, 2-33 Ichigaya-Tamachi, Shinjuku, Tokyo 162-0843, Japan; y.yamamoto@hosei.ac.jp

* Correspondence: shoaib.karam@gmail.com

Received: 22 July 2020; Accepted: 23 August 2020; Published: 24 August 2020

**Abstract:** An analytical investigation focusing on the concrete damage progress of the PBL shear connector under the influence of various lateral pressures, employing a coupled RBSM and solid FEM model was carried out. The analytical model succeeded in simulating the test shear capacities and the failure modes adequately. The internal failure process was also clarified; the two horizontal cracks occurred near the top of the concrete dowels through the hole of the perforated steel plate, and afterward, the two vertical cracks also initiated and propagated along with the shear surface. In a low lateral pressure case, the shear strength was determined by the vertical cracks propagated along the shear surface. While as the amount of applied lateral pressure increased, the shear strength of the two vertical cracked surfaces was enhanced, and the shear strength of the PBL was characterized by the occurrence of the splitting cracks and caused the splitting failure into the side concrete blocks. Moreover, the combined effects of lateral pressure and hole diameters were also evaluated numerically, and it was found that the increase in shear strength was more in a large diameter case subjected to high lateral pressure because of the wide compressive regions generated around the concrete dowel.

**Keywords:** coupled RBSM and solid FEM model; PBL shear connector; shear strength; lateral pressures; failure mechanism

## 1. Introduction

In recent years, the use of concrete-steel composite construction has been adopted widely and extensively, contributing toward the superior structural response (strength, stiffness, resistance against seismic and monotonic loadings, and provision of reduced member sizes, etc.) and the ease of construction. The critical aspect of the concrete-steel composite structures involves the existence of a load-transferring element (shear connector) between steel and concrete. The composite action between concrete and steel in the composite construction heavily depends on the mechanical behavior of the shear connectors. Therefore, the efforts are always devised and proposed for the structural improvement of the shear connectors in concrete–steel composite construction.

The perfobond rib shear connector [Perfobond Leiste in German (PBL)], firstly introduced and developed by the German consultants, was being practiced in concrete–steel composite structures (hybrid girder joints, hybrid truss joints, the hybrid pylon joints, and the anchorage joints between the suspenders and the girders) [1–6]. It behaved as a key and transferred the large internal forces between concrete and steel; it provided advantages related to ease of installation, economic reliability, ductility, and the excellent bearing capacity and the anti-fatigue behavior, etc. [7–9]. A typical PBL shear connector is composed of a perforated steel plate, which is attached with a steel section with transverse reinforcement within the rib perforations and the concrete passing through the rib perforations and forming the concrete dowels. The shear capacity of a typical PBL shear connector is mainly comprised of: (1) the shear resistance of the concrete dowel, (2) the shear resistance of the transverse reinforcement, (3) the concrete end bearing resistance and (4) the frictional and the bond effects between concrete–steel interfaces [10,11], as shown in Figure 1.

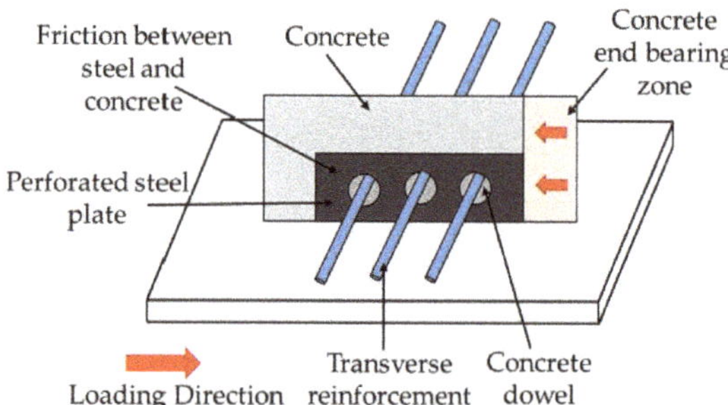


**Figure 1.** Representation of force-resisting components of a typical perfobond rib shear connector [Perfobond Leiste in German (PBL)] shear connector.

Over the years, the researchers performed numerous studies to investigate the shear resistance of the PBL shear connector under the influence of various connection parameters and structural conditions. It was mentioned that the shear resistance of the PBL shear connector was dependent on the diameter of transverse reinforcement inside the hole of the PBL, the diameter of the hole, the thickness of the perforated steel plate, and the stress state in the vicinity of the concrete of the PBL [12]. The experimental investigation [13] revealed that the shear resistance was influenced by the laterally applied varying amounts of pressures to surrounding concrete blocks of the PBL in the simple push-out test, and the shear capacity was increased with the increased amounts of the applied lateral pressures. Similarly, the influence of the lateral constraints, concrete compressive strength, the diameter of transverse steel in the rib hole, and the bond between concrete–steel plate interfaces on the bearing mechanism of the PBL in push-out tests were also investigated experimentally [14]. Furthermore, an experimental study was performed to evaluate the shear resistance capacity and the failure behavior of the PBL under the effect of mechanical properties of concrete and the different rib arrangements, and a shear capacity equation was proposed that took into account the influence of the rib height, the rib spacing, and the rib arrangement [15]. Similarly, experimental and parametric-based studies were also carried out to observe the effect of different types of rib patterns, rib hole configurations, including the varying number of drilled holes in the rib, and the transverse rebar effect inside the rib hole, etc. on the shear response, the modes of failure, and the slip response of the PBL [16,17]. Furthermore, it was reported that the shear resistance and the failure modes (local shear failure and splitting failure) of the PBL shear connector were also dependent on the concrete sizes of the adjoining concrete blocks of the PBL in the simple push-out test [18]. Considering the PBL shear connector relatively more effective concerning the structural integrity compared with the conventional shear connectors (head studs and group studs etc.), the shear resistance of the PBL was also studied in various concrete–steel composite structural elements, e.g., slab, slim floor steel beams, and the diaphragm walls, etc. [19–21].

There existed two major investigation techniques or approaches for the shear response evaluation of the PBL, firstly through the experimental studies by conducting model tests and secondly through the numerical analyses using the finite-element method (FEM). On the one hand, most of the past studies were experimental based, which evaluated the structural performance of the PBL under various loading conditions and test parameters as mentioned above [22–29].

On the other hand, the past numerical simulation studies utilized the finite element analyses mainly focused on capturing the test shear strength and the macroscopic load–displacement response under the influence of several test parameters that affected the shear capacity of the PBL and mostly addressed the shear resistance (deformation, stress contours, and strain distributions) of the steel plate [30–36]. In contrast to focusing on the steel plate, few simulation studies [37,38] were conducted aimed at verifying the reproducibility of crack propagation behaviors and the failure modes of concrete. The investigation and understanding regarding the detailed internal failure mechanism are important and essential for establishing rational design methods and reinforcement details. Therefore, it is required to analyze the internal failure behavior of concrete comprehensively using numerical simulation analyses.

The current research aims to investigate the effect of lateral pressures on the shear strength of the PBL shear connector and also highlights the detailed failure process of concrete, especially the internal crack propagation behavior and stress distributions, through simulation analyses, which has not been discussed efficiently in past research studies, using a coupled Rigid Body Spring Model (RBSM) and the nonlinear solid Finite Element Method (FEM). Furthermore, the current research primarily focuses on determining the shear resistance solely of the PBL shear connector without presenting the comparison related to shear performance for different types of shear connectors. The coupled RBSM and solid FEM model combines the use of 3D-RBSM and the nonlinear solid FEM that was proposed by authors [39,40]. The 3D-RBSM has been referred to as an effective numerical approach for the evaluation of nonlinear fracture behavior of concrete (internal crack initiation, propagation, and orientation), quantitatively [41]. In the current study, firstly, the analytical approach based on a coupled RBSM and solid FEM model is presented; then, the validation of the numerical model is carried out for the test shear capacities and the failure modes of the PBL shear connector specifically under the influence of the various amounts of the lateral pressures applied to the surrounding concrete of the PBL. After the validation of the numerical model, the detailed internal crack propagation process and the failure mechanism of concrete influenced by the varying amounts of the lateral pressures are highlighted and discussed. Furthermore, the combined effects of the varying lateral pressures and the diameters of the holes on the shear resistance of the PBL shear connector are also evaluated.

## 2. Numerical Modeling of Concrete, Steel, and Concrete–Steel Interface

In the coupled Rigid Body Spring Model (RBSM) and the nonlinear solid Finite Element Method (FEM) model, the concrete is modeled using 3D-RBSM, and the steel embedded in concrete is modeled using eight-noded nonlinear solid finite elements utilizing the actual geometrical features of the steel. The modeling of the concrete, the steel, and the concrete–steel interface is described in this chapter.

### 2.1. Modeling of Concrete Using 3D-RBSM

The 3D-RBSM is comprised of an assemblage of rigid elements. The rigid elements are interconnected by employing normal and tangential springs along with their interfaces of boundaries. The 3D-RBSM has been proved to be an effective and an efficient numerical approach for the quantitative evaluation of the nonlinear mechanical response of concrete, such as crack propagation behavior, shear transfer behavior of cracked surfaces, and compression failure assessment inclusive of localization and constraint pressure dependence [42,43], developed by the authors at the Concrete Engineering Laboratory of Nagoya University, Japan. The cracks initiate and propagate through the interfaces of boundaries of rigid elements and are strongly influenced by the mesh design of the concrete in numerical analyses. In order to overcome this fact, the random geometry of rigid particles has been adopted by using Voronoi diagram, as shown in Figure 2. The response of the spring model represents an insight into the interaction between the rigid particles. Each rigid particle involves the three rotational and the three translational degrees of freedom assigned at the geometric centroid that characterize the particles according to the Voronoi diagram, as illustrated in Figure 2.

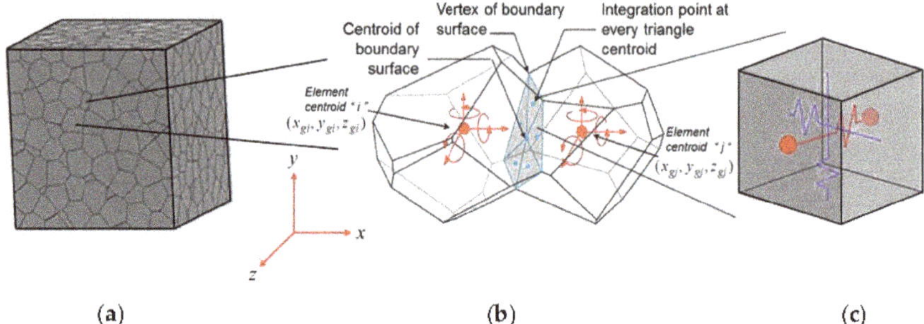

(a)           (b)           (c)

**Figure 2.** Depiction of 3D-RBSM (Rigid Body Spring Model) for modeling of concrete; (**a**) Discretization of concrete using Voronoi diagram; (**b**) Illustration of integration points on the boundary interface of rigid particles; (**c**) Representation of normal and shear springs [44].

The springs are defined on the integration points as depicted in Figure 2. The integration points are produced through the division of the surface of rigid particles into sets of triangles. As shown in Figure 2, the geometric centroid of each triangle is corresponding to one integration point. Furthermore, the induction and arrangement of integration points and springs in such manner as shown in Figure 2 automatically develops the effects of flexural and torsional behavior without the need to introduce any additional rotational springs. The complete detail regarding the formulation of the 3D-RBSM employed in the current study is available in previous research [44].

The numerical simulation analyses for the nonlinear behavior of concrete in 3D-RBSM greatly depend on the constitutive models assigned to the springs. The combination of the constitutive models as well as the distribution of the springs over the boundary surfaces together facilitates the reproduction of the nonlinear mechanical response of concrete. The formation of cracks in concrete is expressed by the failure of these springs. The models that can capture the localization and softening behavior under various stress states are proposed and applied to the constitutive models of the springs. The material constitutive models for the tension, compression, and shear of concrete are presented in Figure 3.

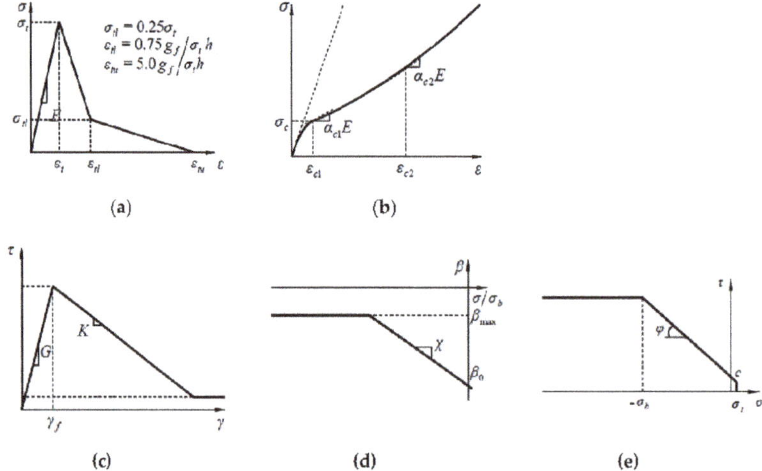

**Figure 3.** The constitutive models of normal and shear springs for concrete in 3D-RBSM: (**a**) Tension model for normal spring; (**b**) Compression model for normal spring; (**c**) Model for shear spring; (**d**) Softening coefficient for shear spring; (**e**) Mohr–Coulumb criterion for shear spring.

The calibrated material parameters of the constitutive models for normal and shear springs are shown in Tables 1 and 2, respectively. The calibration of material parameters of the constitutive models has been conducted through the parametric analyses (specimen size, shape, mesh size, and compressive strength of concrete) comparing with the macro stress–strain test relationships including the softening part subjected to uniaxial compression, hydrostatic compression, triaxial compression, and uniaxial tension [42,43]. After calibration of the material parameters as shown in Tables 1 and 2, the applicability of the calibrated model parameters in 3D-RBSM has been confirmed through the initiation and propagation of the cracks in the concrete under various structural and mechanical behaviors, e.g., shear response evaluation of reinforced concrete (RC) deep beams [45], RC wall panels subjected to cycling loading [46], numerical evaluation of localization and softening behavior of the concrete confined by steel tubes [44], the investigation of internal crack propagation behavior due to corrosion [47–50], the simulation of the bond behavior between steel and concrete [51], etc. Furthermore, the compression model of a normal spring does not include the softening behavior and the failure of the normal springs. However, the compressive failure response including the confinement effect and localized failure behavior can be reproduced by means of the combination of a normal and shear spring.

**Table 1.** Calibrated model parameters for normal spring.

| Elastic Modulus | Tensile Response | | Compressive Response | | | |
|---|---|---|---|---|---|---|
| $E$ $(N/mm^2)$ | $\sigma_t$ $(N/mm^2)$ | $g_f$ $(N/mm)$ | $\sigma_c$ $(N/mm^2)$ | $\varepsilon_{c2}$ | $\alpha_{c1}$ | $\alpha_{c2}$ |
| 1.4 $E^*$ | 0.65 $f_t{}^*$ | 0.5 $G_F{}^*$ | 1.5 $f_c{}'^*$ | −0.015 | 0.15 | 0.25 |

* The macroscopic material parameters obtained from the concrete specimen's test. $E^*$: Young's Modulus, $f_t{}^*$: Tensile Strength, $g_f{}^*$: Fracture Energy, $f_c{}'^*$: Compressive Strength.

**Table 2.** Calibrated model parameters for shear spring.

| Elastic Modulus | Fracture Criterion | | | Softening Behavior | | | |
|---|---|---|---|---|---|---|---|
| $G$ $(N/mm^2)$ | $c$ $(N/mm^2)$ | $\varphi$ $(degree)$ | $\sigma_b$ $(N/mm^2)$ | $\beta_0$ | $\beta_{max}$ | $\chi$ | $\kappa$ |
| 0.35 $E$ | 0.14 $f_c{}'^*$ | 37 | 1.00 $f_c{}'^*$ | −0.05 | −0.025 | −0.01 | −0.3 |

* The macroscopic material parameters obtained from the concrete specimen's test. $f_c{}'^*$: Compressive Strength.

## 2.2. Modeling of Steel Using Solid FEM

The steel is modeled using eight-noded nonlinear solid finite elements. The Von Mises plasticity model with strain hardening is used for the constitutive model of the steel for reproducing the elastoplasticity of the steel elements in the coupled RBSM and solid FEM model.

The perforated steel plate of the PBL shear connector between side concrete blocks is modeled considering the nonlinear solid finite elements, taking into account the actual geometrical features. The 3D model for the perforated steel plate of the PBL used in the current study is shown in Figure 4.

## 2.3. Concrete–Steel Interface

The concrete elements (3D-RBSM) and the steel elements (solid FEM) have been connected through the coupling of boundary interfaces of concrete and steel elements utilizing link elements. The coupling approach of boundary interfaces through link elements using springs on the boundary interfaces between constituent materials in 3D discrete macroscopic element modeling was also adopted by the other researchers. In the numerical evaluation of out-of-plane behavior of brick masonry infill wall panels [52–54], the nonlinear interaction of the frame and masonry infills was modeled by 3D discrete nonlinear interface link elements. The structural behavior of integrally attached timber plates [55] and the simulation of the progressive collapse of RC frame structures [56] were also performed through nonlinear springs and beam elements on the interfaces.

Similarly, in a coupled RBSM and solid FEM model, each link element on the interface between the RBSM element and nonlinear solid FEM element consists of two shear springs and one normal spring, as shown in Figure 5. The deformation of each spring of the link element is obtained by the relative displacement between the surfaces of the RBSM element and the nonlinear solid FEM element, subsequently converting it to the local coordinate system. Furthermore, the constitutive models of normal and shear springs on the boundary interface are assumed to be the same as that for modeling of the concrete in 3D-RBSM (Figure 2, Tables 1 and 2). However, the model parameters of constitutive models for normal and shear springs on the boundary interface have been changed considering the friction between steel and concrete on the boundary interface [57,58].

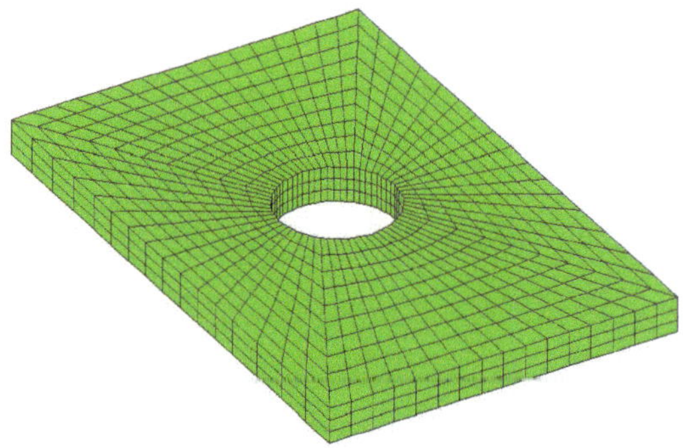

**Figure 4.** 3D model of perforated steel plate using eight-noded nonlinear solid FEM.

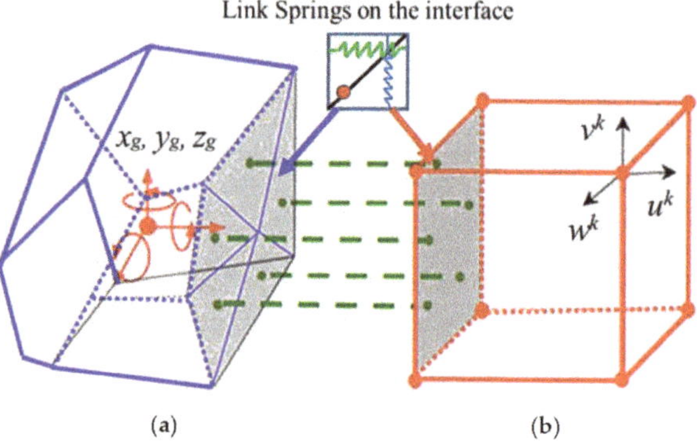


**Figure 5.** The boundary interface of a coupled RBSM and solid FEM model: (**a**) Representation of RBSM element; (**b**) Representation of eight-noded solid FEM element [39,40].

The numerical model couples the RBSM elements and the solid FEM elements regardless of the position of nodes of the FEM elements; additionally, the numerical model has the capability to reproduce the link elements arbitrarily on the boundary interface between concrete and steel elements, which consequently reduces the analytical computational cost. The numerical model employs the same coupling technique for combining the RBSM elements and the solid FEM elements as applied by the past researcher for coupling of the RBSM elements and the shell FEM elements [44,59], based on the Inverse-Mapping algorithm using Taylor expansion.

## 3. Validation of Coupled RBSM and Solid FEM Model

In this chapter, the numerical model is validated by utilizing the already published experimental investigations focusing on the shear resistance of the PBL under a simple push-out test. For validation of the model, the test investigations corresponding to the most fundamental-type PBL specimens with a single hole in the steel plate are selected so that the shear resistance, internal failure process, and mechanism of concrete can be discussed comprehensively and clearly. Specifically, the validation is performed for capturing the test shear capacities and the failure modes under the influence of the varying lateral pressures applied to the surrounding concrete of the PBL [13].

### 3.1. Test Overview and Numerical Modeling

The shear capacity of the PBL is influenced by the amounts of lateral pressures applied to the surrounding concrete of the PBL [12,13]. In steel–concrete hybrid construction, the presence of transverse prestressing tendons in the concrete causes the generation of a varied stress state around the PBL, and the concrete is loaded to varying levels of compressive forces. In the test, the mechanical response of a single PBL under simple push-out tests was investigated considering the influence of varying amounts of lateral pressures applied to surrounding concrete blocks. The main varying parameter in the test was the amount of the lateral pressures (1 MPa, 2 MPa, 3 MPa, 4 MPa, 6 MPa, 8 MPa, and 10 MPa) applied to two opposite faces of concrete blocks using mechanical jacks. The diameter of the hole and the thickness of the perforated steel plate were maintained constant. The test specimens were embedded with a perforated steel plate ($230 \times 150 \times 12$ mm$^3$) having a hole of 50 mm diameter between two side concrete blocks. All the test specimens had the rectangular geometry, with the size of $212 \times 200 \times 150$ mm$^3$. The geometrical details of the test specimens are presented in Figure 6.

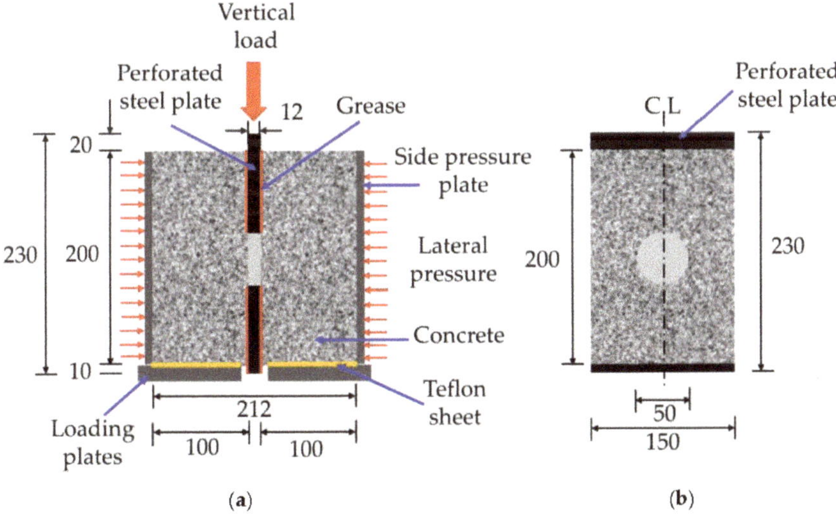


**Figure 6.** Geometrical details of the test specimen (unit: mm): (**a**) Front view; (**b**) Side view [13].

The Teflon sheet was placed at the bottom of the test specimens between steel loading plates and concrete, and the grease was used on the surface of the perforated steel plate in order to eliminate the frictional and bond effects between steel plates and surrounding concrete blocks. The simple push-out tests were performed by applying a vertical load on the top of the perforated steel plate loaded under the influence of the varying amounts of lateral pressures, which were applied to two opposite surfaces of the surrounding concrete blocks simultaneously. The specimens subjected to lateral pressures of 1 MPa, 2 MPa, 3 MPa and 4 MPa had a concrete compressive strength around 42.2 MPa, while the specimens with lateral pressures 6 MPa, 8 MPa, and 10 MPa had a concrete compressive strength around 45.2 MPa.

The numerical models corresponding to a single PBL shear connector, subjected to the varying amounts of lateral pressures, are illustrated in Figure 7. The proper numerical modeling of concrete in the circular region (concrete dowel) between two side concrete blocks requires adopting an average mesh size less than the thickness (12 mm) of the perforated steel plate, as the crack propagation behavior in the concrete dowel region depends on the selection of average mesh size. The crack propagation can occur inside the hole or along with the shear surface; therefore, the mesh size in the concrete dowel region is selected in such a manner so that the internal failure process of the concrete dowel can be captured effectively in numerical simulations, whereas a relatively large mesh size has been selected near the ends of concrete blocks to minimize the analytical computational cost, as shown in Figure 7a. Therefore, the average mesh size in the circular region (concrete dowel) is set as 3 mm (less than the thickness of the perforated steel plate: 12 mm), and then, the mesh size is gradually increased from the circular region to the side concrete blocks. In this regard, the average mesh size in numerical simulations incorporated for the numerical modeling of the concrete approximately ranges between 3 and 13 mm. The mesh size dependency of the numerical model is also investigated numerically through the mesh sensitivity analysis ranging between the mesh sizes selected for the numerical modeling. The difference of results is found to be negligible and also confirms that the numerical model does not have mesh size dependency.

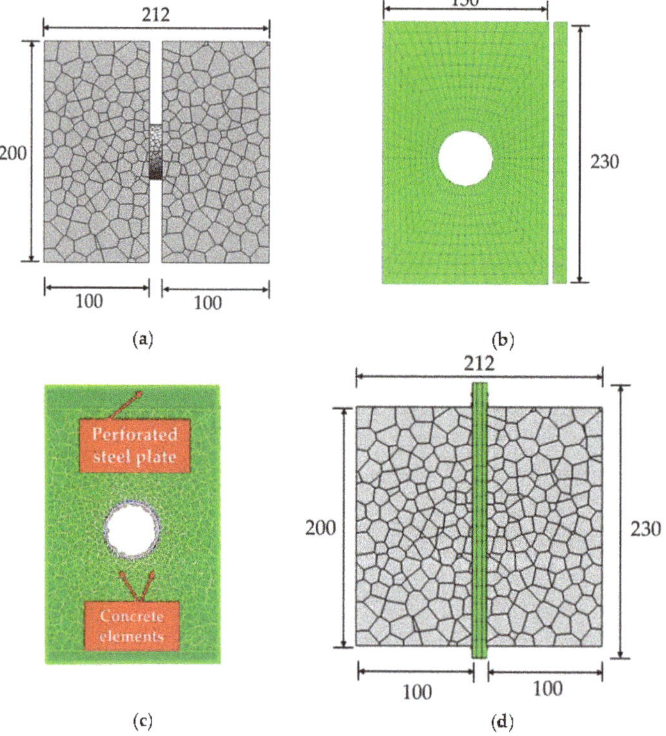


**Figure 7.** The analytical model of the test specimen (unit: mm): (**a**) Modeling of the concrete dowel and the side blocks; (**b**) Modeling of the perforated steel plate; (**c**) Representation of concrete and steel elements on the concrete–steel interface; (**d**) Analytical model of the PBL shear connector.

The numerical modeling of the test boundaries is elaborated here. It was reported that the shear capacity of the PBL shear connector was influenced by the friction between the concrete blocks and the test bed, which was loaded in the simple push-out test [60]. In the same manner, the preliminary numerical simulations also highlighted the dependency of shear response of PBL on the sensitivity of material parameters; tensile strength ($\sigma_t$), cohesion (c), and angle of internal friction ($\varphi$) were selected for the modeling of test boundaries. As mentioned earlier, the grease was applied on the surface of the perforated steel plate, and a Teflon sheet was placed between the bottom loading plates and the concrete blocks in the test. In this regard, the test boundary conditions were modeled in simulation analyses of the PBL by adopting the minimal material parameters of the constitutive models ($\sigma_t = 0.01$ MPa, c = 0.01 MPa, and $\varphi = 2$ degrees) for the normal and shear springs (Tables 1 and 2) between the steel plate and the concrete. The different amounts of lateral pressures to the surrounding concrete blocks were applied by the load control of a pressure plate consisting of one RBSM element. The vertical shear load was applied by the displacement control of solid FEM nodes on top of the perforated steel plate.

### 3.2. Results and Discussions

Since the reference test [13] only provided the information about the concrete compressive strength ($f_c'$), the tensile strength ($\sigma_t$), elastic modulus (E), and fracture energy ($g_f$) as the model parameters required in Tables 1 and 2 were assessed from the compressive strength of the concrete using the conversion formulae proposed in the standard specifications of Japan Society of Civil Engineers for concrete structures (JSCE) [61]. However, the experimental investigations reported that the tensile strength of concrete was influenced by many factors; e.g., water–cement ratio, type and size of aggregates, curing and storage (moisture) conditions of the specimen, size of the specimen, and conditions of testing (splitting test, uniaxial test, and flexural test), etc., and tensile strength might be reduced. In this regard, an approximately 30% reduction only for the tensile strength ($\sigma_t$) was incorporated in the numerical simulations, while all the other remaining material parameters (Tables 1 and 2) were kept the same for the modeling of concrete.

Figure 8 illustrates the analytical load and relative displacement (between the bottom of the steel plate and mid-height) relationships and the influence of lateral pressures on the shear capacity of the PBL. In the test, it was observed that the shear capacity increased as the applied lateral pressure increased, as shown in Figure 8b. In the same manner, the numerical simulation results also exhibit the increase in shear capacity with the increase in the lateral pressures applied to side concrete blocks. Figure 8b shows the slight deviation and overestimation of the numerical shear capacities compared with the test results; however, it captures the tendency for an increase in shear capacity against increased lateral pressure, which is consistent with the test investigations. In the test, mainly two types of failure modes were observed: (1) local shear failure along with the edge of the hole of the perforated steel plate, and (2) the splitting failure of side concrete blocks perpendicular to the thickness of the perforated steel plate, as shown in Figure 9.

The test photos show the damage and cracking at the steel–concrete contact surface, which was parallel to the thickness of the steel plate. Figure 9 also presents the analytical failure modes considering the cut section near the hole of steel plate parallel to the thickness of the plate, in the post-peak region for low (2 MPa) and high (10 MPa) lateral pressures applications corresponding to 1.2 mm and 2.1 mm displacements, as marked in Figure 8a, respectively. The comparison of the test and the analytical failure modes are found to be in good agreement. The dominant failure mode corresponding to low lateral pressure applications (1 MPa and 2 MPa) was shear failure, while the splitting failure mode was investigated for all the remaining pressures applications (3 MPa, 4 MPa, 5 MPa, 6 MPa, 8 MPa, and 10 MPa), as it can be confirmed through the load–displacement relationships in Figure 8a. The specimens subjected to low lateral pressure applications (1 MPa and 2 MPa) exhibit the ductile failure behavior, whereas all other specimens reproduce the splitting failure mode and load decreases after the peak due to the occurrence of the splitting crack. The validation of the numerical model is confirmed and highlighted for the test shear capacities and the failure modes, while the detailed failure process and mechanism are discussed in the next chapter.

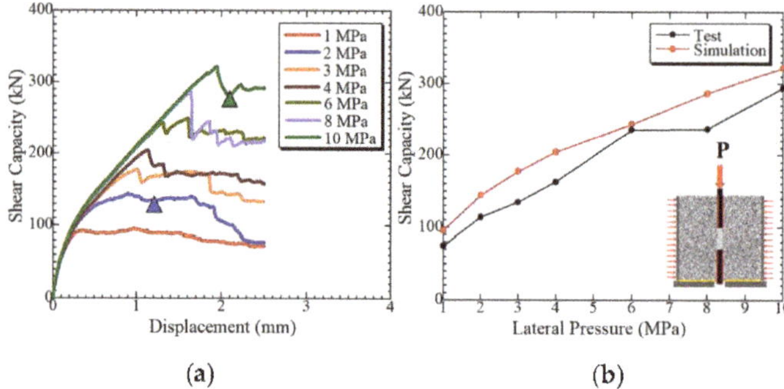

**Figure 8.** Influence of lateral pressures on the shear capacity of PBL: (**a**) Analytical load–displacement relations; (**b**) Comparison of test and analytical shear capacities.

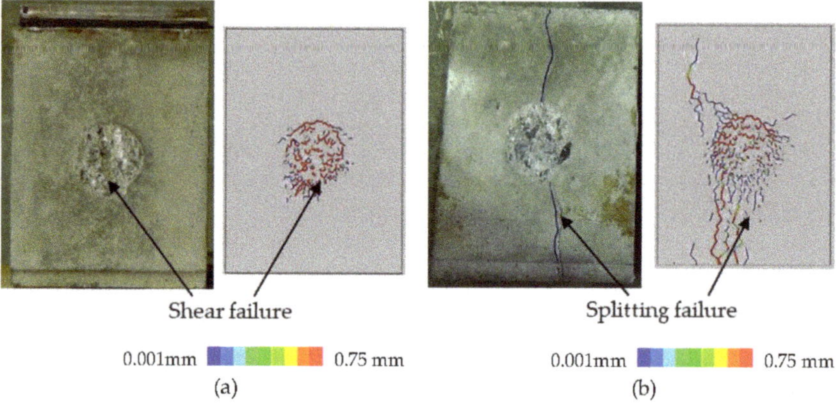



**Figure 9.** Comparison of test and analytical failure modes: (**a**) Low lateral pressure (2 MPa) case; (**b**) High lateral pressure (10 MPa) case.

## 4. Investigation of Failure Process and Mechanism

After validation of the numerical model, the detailed failure process and mechanism is evaluated. The deformation response including internal crack patterns and normal stress distribution along with height and width (direction of lateral pressure application) for low (2 MPa) and high (10 MPa) lateral pressures with different failure modes are investigated, as shown in Figure 9. In order to highlight the discussion, the quantitative shear response in numerical simulations at various stages is selected on load–displacement relationships, as marked in Figure 10.

The detailed failure process in a specimen subjected to low lateral pressure (2 MPa) corresponding to various slip stages is shown in Figure 11. Figure 11 presents the surface-deformed behaviors, internal crack patterns at the cut sections perpendicular and parallel to the thickness of steel plate, and the normal stress distribution along with height (y-axis: direction corresponding to the application of push-out force) and width (z-axis: direction corresponding to the application of lateral pressure) at a cut section perpendicular to the thickness of the steel plate. The crack widths indices for the internal cracks are highlighted through the variation of colors, from blue (0.001 mm) to red (0.75 mm). Similarly, the variation of normal stresses in concrete is shown by the indices, from 2 MPa (pink: tensile stress) to −50 MPa (red: compressive stress).

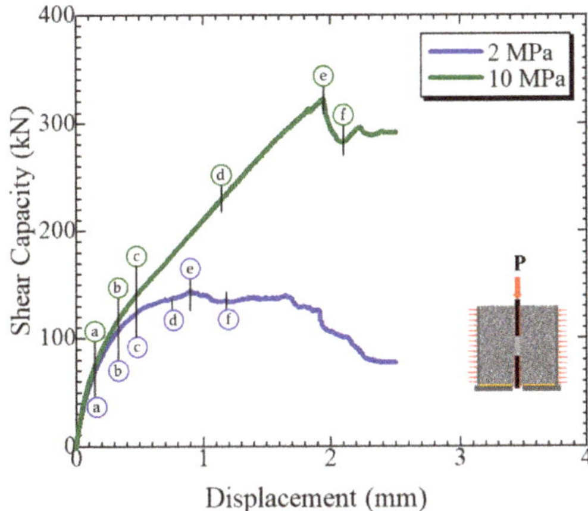


**Figure 10.** Representation of various stages for shear response evaluation.

From the internal crack pattern and normal stress distribution at the cut section defined near to mid-depth perpendicular to the thickness of the steel plate at slip stage (a), it can be observed that firstly, the two horizontal cracks initiate near the top of the concrete dowel through the hole of the perforated steel plate, and compressive stresses ($\sigma_y$) occur on the top of the concrete dowel. Afterward, at slip stage (b), the horizontal cracks further propagate, and the two vertical cracks also initiate and propagate around the shear surface against a slightly increased shear force, and the compressive stresses ($\sigma_y$) also transfer from top to almost mid-height of the concrete dowel. After slip stage (b), the slope of the load–displacement curve changes because of the propagation of the vertical crack along with the shear surface and tends to become milder after slip stage (c). From the internal crack patterns and the normal stress distribution along with the y-axis ($\sigma_y$), at slip stages (c) and (d), it can be recognized that the vertical crack penetrates completely to the bottom of the hole and causes the shear slip on the vertical cracked surface, and the compressive stresses are concentrated in the concrete dowel region.

In the case of low lateral pressure application, the shear strength of the PBL is characterized by the shear failure of two cracked surfaces along with the edge of the hole in the steel plate, consequently causing the failure of the concrete dowel, the same as observed in the test, as illustrated in Figure 9. The distribution of the normal stress along with the y-axis ($\sigma_y$) clearly depicts the transmission of large compressive stress (red color) from the top to bottom of the concrete dowel with the increase of shear force, while along with the z-axis ($\sigma_z$), the large compressive stresses are only concentrated on the top edge of the hole of the steel plate. The surface crack patterns shown in Figure 11 also reveal that no splitting crack occurs perpendicular to the thickness of the perforated steel in the surrounding concrete blocks in the low lateral pressure application case, the concrete dowel mainly resists the vertical push-out force, and the stresses along with the y-axis and z-axis are mainly localized in the circular region between the side concrete blocks.

Similarly, the detailed failure process and the mechanism for high lateral pressure (10 MPa) application captured by the numerical analyses is also discussed and presented in Figure 12. In high lateral pressure application, the initial internal crack propagation process near the steel plate hole follows the same failure mechanism as observed in the low lateral pressure case, i.e., the occurrence of two horizontal cracks near the top of concrete dowels. Afterward, the initiation and propagation of vertical cracks around the shear surface causes the transformation of compressive stresses from the top to bottom of the concrete dowel. However, the vertical crack propagation is restrained by the application of high lateral pressure, as shown in Figure 12. It can be confirmed from the comparison of internal crack patterns for low and high pressures applications corresponding to slip stages (a), (b), and (c) as shown in Figures 11 and 12, respectively.

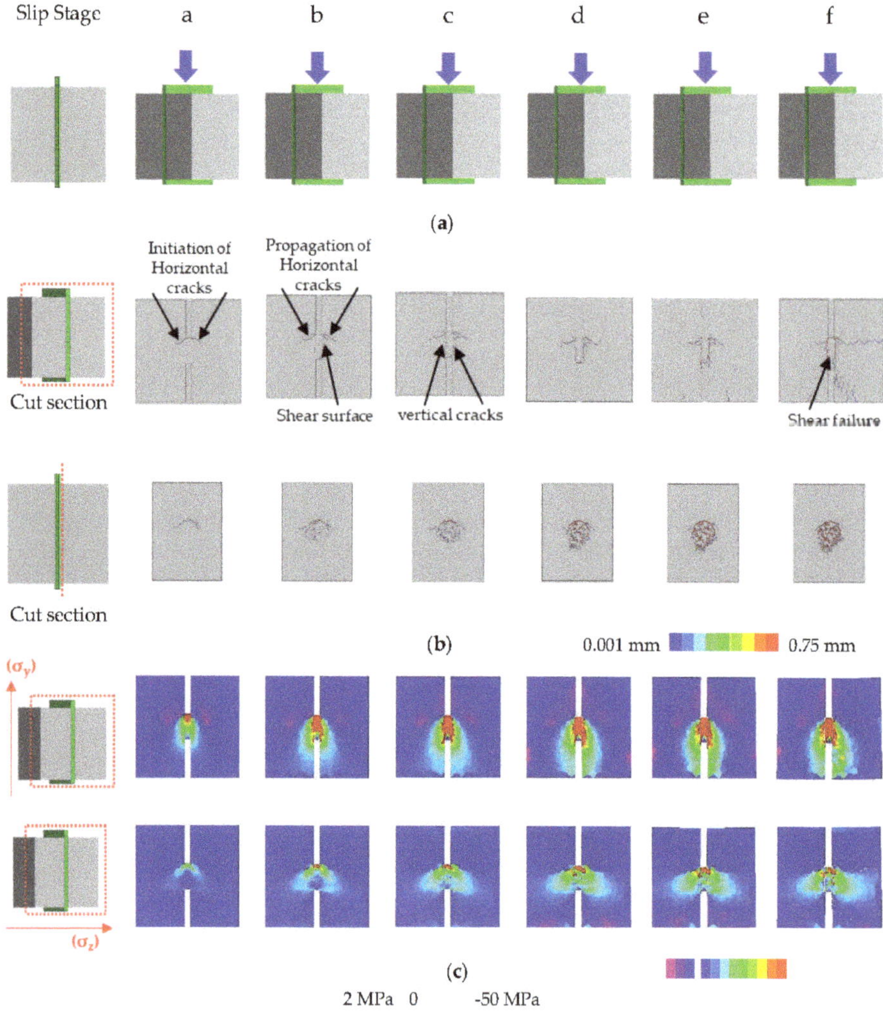


**Figure 11.** Detailed failure process and mechanism for low lateral pressure case: (**a**) Surface-deformed behaviors (×5); (**b**) Internal crack patterns at cut sections perpendicular and parallel to the thickness of the steel plate; (**c**) Normal stress distribution along with height (*y*-axis: direction corresponding to the application of push-out force) and width (*z*-axis: direction corresponding to the application of lateral pressure) at a cut section perpendicular to the thickness of the steel plate.

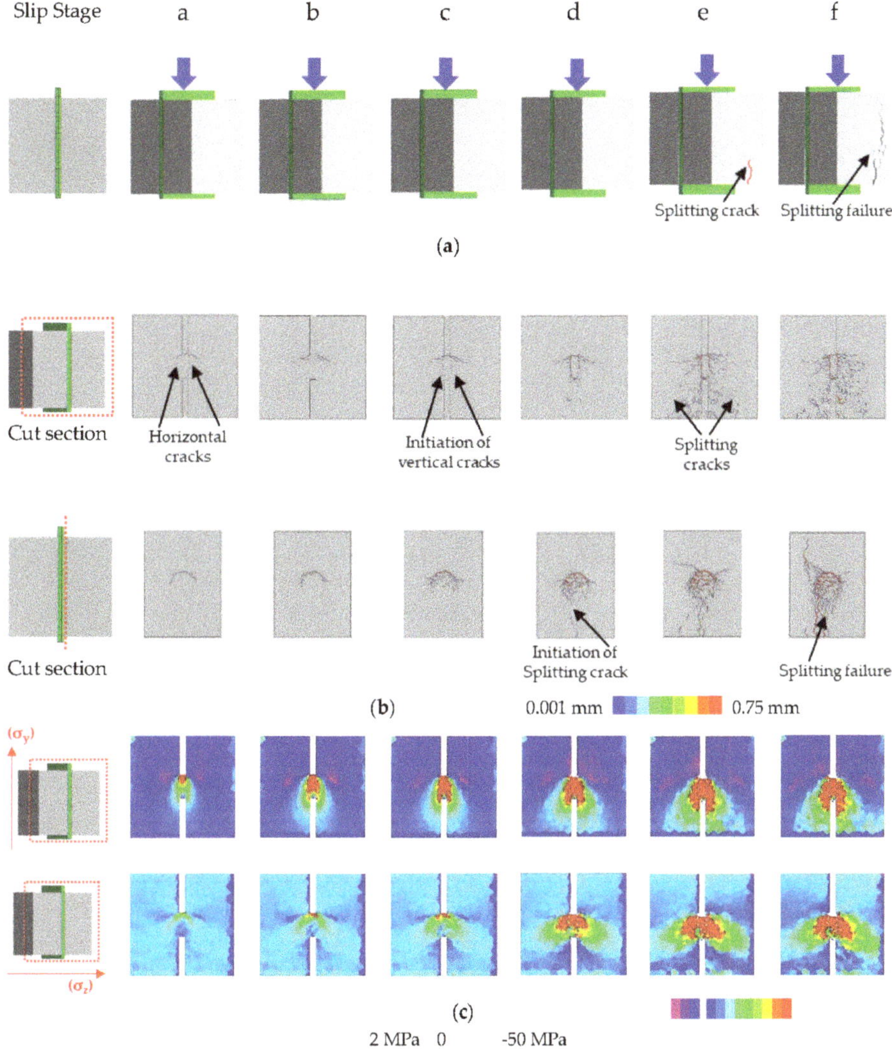



**Figure 12.** Detailed failure process and mechanism for high lateral pressure case: (**a**) Surface deformed behaviors (×5); (**b**) Internal crack patterns at cut sections perpendicular and parallel to the thickness of the steel plate; (**c**) Normal stress distribution along with height (y-axis: direction corresponding to the application of push-out force) and width (z-axis: direction corresponding to the application of lateral pressure) at the cut section perpendicular to the thickness of the steel plate.

At slip stage (b) in Figure 11, the vertical crack initiates along with the shear surface, while in Figure 12, there is no initiation of vertical crack. Similarly, the internal crack pattern corresponding to slip stage (c) in Figure 11 shows that the vertical crack completely propagates along the shear surface, whereas in high pressure application, only the initiation of a vertical crack occurs. After the initiation of a vertical crack at slip stage (c), the load tends to further increase (Figure 10) and exhibits an increasing slope of the load–displacement curve because of the restraining effect of the shear dilatancy behavior by the externally applied high lateral pressure, in contrast with the low lateral pressure application.

Afterward, at slip stage (d) in Figure 12, the splitting cracks initiate under the concrete dowel, into the inner surfaces of the surrounding concrete blocks, in the perpendicular direction to the thickness of the perforated steel plate, and the compressive stresses along the y-axis ($\sigma_y$) and z-axis ($\sigma_z$) are induced in the side concrete blocks. Furthermore, the normal stresses along with height ($\sigma_y$) and along with width ($\sigma_z$) in the side concrete blocks almost have the same damage zone as that of the splitting crack propagation region in the side concrete blocks in contrast with the low lateral pressure application case, where the normal stresses ($\sigma_y$ and $\sigma_z$) are only localized around the concrete dowel region, as shown in Figure 11c. At the peak stage, the splitting cracks propagate the surface of the surrounding concrete blocks, and they consequently cause the splitting failure behavior post-peak, which is consistent with that observed in the test. In high lateral pressure application, the shear strength of the PBL is determined by the occurrence of splitting failure behavior perpendicular to the thickness of the perforated steel plate in the surrounding concrete blocks.

It is pertinent to mention here that the distribution of normal stresses along with height ($\sigma_y$) and width ($\sigma_z$) in Figures 11 and 12 clearly highlights the difference of the internal resistance mechanism. The normal stresses along with height ($\sigma_y$), in the low lateral pressure application case, show that the compressive stresses are concentrated around the concrete dowel region because of shear failure, whereas in the high lateral pressure application case, the concrete in surrounding blocks and under the concrete dowels also undergoes compressive stresses. Moreover, the normal stresses along with height ($\sigma_y$) in Figures 11c and 12c also highlight the difference of the shear resistance behavior of the concrete dowel concerning the varying amounts of lateral pressures applied. The damage zone of the concrete dowel in the low lateral pressure application case experiencing large compressive stresses, as shown in Figure 11c, is less compared to the high lateral pressure application case, as shown in Figure 12c. The shear resistance of the concrete dowel is enhanced because of the high lateral pressure application, and the damage zone of the concrete dowel also extends in the side concrete blocks. Similarly, the normal stresses along with the width ($\sigma_z$) show that in the low lateral pressure application case, the large compressive stresses only concentrate on the top edge of the hole, while in high lateral pressure application case, the compressive stresses are observed in the complete concrete dowel region. Furthermore, the envelope of the compressive stresses induced in the side concrete blocks as shown in Figure 12c is more than the low lateral pressure application case shown in Figure 11c because of splitting crack propagation. It is also observed that as the amount of lateral pressure applied to the side concrete blocks increases, consequently, the envelope of the compressive stresses also expands more, and the shear resistance of the PBL shear connector is improved.

The detailed internal failure process and mechanism revealed in Figures 11 and 12 captured by the numerical model efficiently highlight the difference of the internal shear resistance mechanism of the PBL subjected to varying amounts of lateral pressures. The numerical model not only highlights the internal failure mechanism but also reproduces the transformation of failure modes (local shear failure in low lateral pressure and splitting failure in high lateral pressure) with respect to the amounts of lateral pressures applied to concrete surfaces, which is consistent as observed in the test.

## 5. Evaluation of Combined Effects for the Lateral Pressure and Hole Diameter

After the validation of the numerical model and clarifying the detailed failure process and the mechanism in low and high lateral pressures as shown in Figures 11 and 12 respectively, the numerical evaluation is performed to investigate the combined effects of the various lateral pressures and the holes' diameters on the shear capacity of the PBL shear connector. Although the diameter of the hole is one of the main connection parameters of the PBL shear connector, only the individual effect of each parameter on the shear capacity has been studied yet [12,13]. For numerical evaluation in this regard, the specimen with the same geometrical dimensions as those in Section 3 (Figure 6) is simulated, where only the hole diameter is changed from 50-mm to 75-mm and the lateral pressures of 2 MPa, 6 MPa, and 10 MPa are also applied. The analytical model of the PBL shear connector with a hole of 75-mm diameter is shown in Figure 13.

The analytical relationships of the shear capacities corresponding to the combined effects of the lateral pressures and the holes' diameters are shown in Figure 14. It is noticed through the numerical investigations that the specimen with the hole of 75-mm diameter subjected to a lateral pressure of 2 MPa reproduces the shear failure mode, while the specimens with applied lateral pressures of 6 MPa and 10 MPa exhibit the splitting failure modes, which are found to be consistent as obtained for the 50-mm hole diameter specimen (chapter 4).

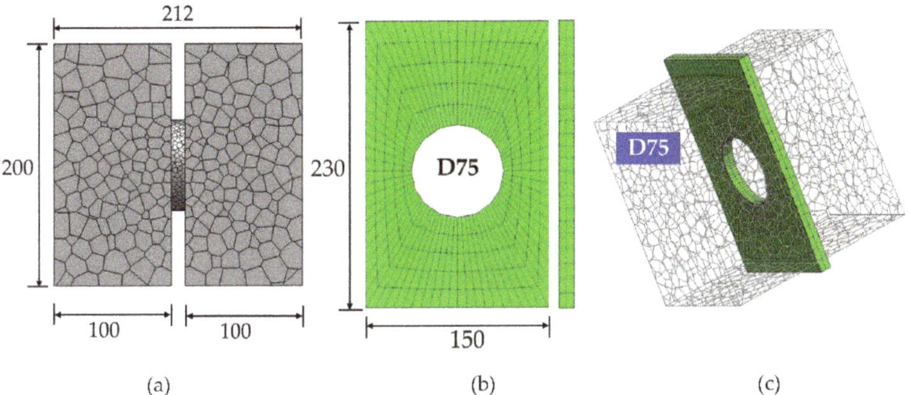


**Figure 13.** The analytical model of the specimen with a 75-mm diameter of the hole (unit: mm): (**a**) Modeling of the concrete dowel and the side blocks; (**b**) Modeling of the perforated steel plate; (**c**) Analytical model of the PBL shear connector.

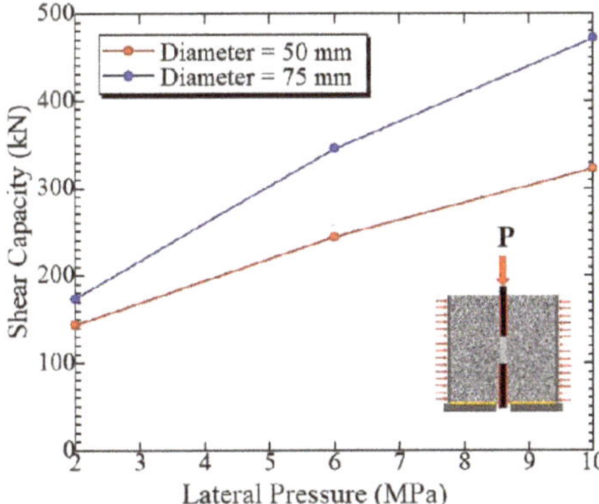


**Figure 14.** Illustration of the combined effects of the lateral pressures and the diameters of the holes on analytical shear capacities.

Figure 14 shows that the shear capacity increases with the increase in the lateral pressures and the holes' diameters, exhibiting the same trend against the individual effect of each parameter, as observed in the past research studies [12,13]. However, it should be emphasized that there is a clear existence of the combined effects against the shear capacities, as the relationship for the 50-mm hole diameter and 75-mm hole diameter are not parallel to each other, as shown in Figure 14. Specifically, the increase in the shear capacity for a specimen with a hole of 75-mm diameter was found to be more in high lateral pressure compared with a lateral pressure of 2 MPa. The combined effects of the lateral pressures and the holes' diameters can also be efficiently examined through the internal cracking behavior and the normal stress distribution shown in Figure 15, considering the advantage of the reliable numerical approach. Figure 15 shows the internal crack patterns and the normal stress distribution along with height ($\sigma_y$) for both diameters of the holes under lateral pressures of 2 MPa and 10 MPa at the peak stage. The normal stress distribution along with height ($\sigma_y$) in Figure 15 shows that as the diameter of the hole is increased from 50-mm to 75-mm, the envelope of compressive stresses and the region of the large compressive stresses around the concrete dowel and in the side concrete blocks are also expanded more for both lateral pressures (2 MPa and 10 MPa).

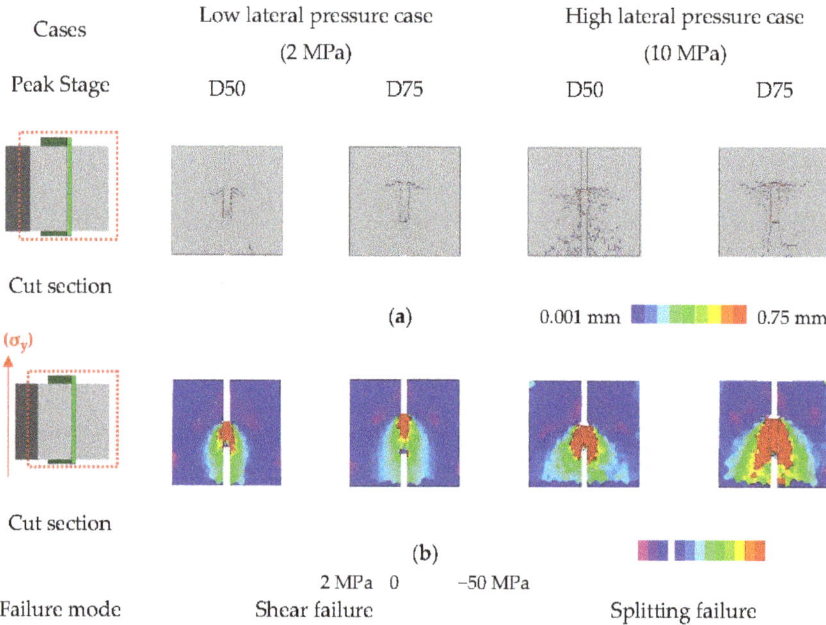


**Figure 15.** Comparison of the failure behaviors for low and high lateral pressures against 75-mm and 50-mm diameters of the holes at the peak stage: (**a**) Internal crack patterns at the cut section perpendicular to the thickness of the steel plate; (**b**) Normal stress distribution along with height (y-axis: direction corresponding to the application of push-out force) at the cut section perpendicular to the thickness of the steel plate.

Figure 15 highlights that in a low lateral pressure case (2 MPa), the internal crack patterns are almost similar for both diameters, and the cracks propagate along the shear surfaces. In a 50-mm diameter case, the large compressive stresses surround the complete dowel region, whereas in the 75-mm diameter of the hole, the large compressive stresses have been observed on the upper region of the concrete dowel. However, since the region of the large compressive stresses exists almost up to the mid-height of the concrete dowel between the vertical cracks along the shear surfaces, therefore, the increment in shear capacity is limited in a specimen subjected to a lateral pressure of 2 MPa. In high lateral pressure case (10 MPa), the internal crack propagation is less for the 75-mm diameter hole because of its greater shear resistance against the applied shear load compared with the 50-mm diameter. It is also evaluated that the region of the large compressive stresses in the 75-mm diameter case progresses more in the side concrete blocks compared with the 50-mm diameter case, and as a result, it contributes more toward the increase in shear capacity due to the wide regions of the large compressive stresses compared with the lateral pressure of 2 MPa and consequently produces the maximum shear capacity under the combined effects, as shown in Figure 14.

## 6. Conclusions

The current research involves the numerical simulation analyses of the PBL shear connector under the influence of the varying amounts of the lateral pressures applied to the side concrete blocks employing a coupled Rigid Body Spring Model (RBSM) and the nonlinear solid Finite Element Method (FEM) model, focusing on the damage progress (internal crack propagation behavior and stress distribution) of the concrete. Furthermore, the detailed internal failure mechanism of a single PBL shear connector in the simple push-out test is also revealed quantitatively. Additionally, the combined effects of the various lateral pressures and the diameters of the holes on the shear capacity of the PBL shear connector are also evaluated. Based on the outcomes of the current research, the conclusions enumerated hereafter are drawn.

The numerical evaluation of the PBL shear connector using a coupled RBSM and solid FEM model under a simple push-out test was carried out. The validation of the numerical model was confirmed through the numerical simulations for quantitative shear strength evaluation and the reproducibility of the failure modes of concrete under the influence of the varying amounts of lateral pressures.

It was confirmed numerically that the shear capacity of the PBL increased and the failure mode changed from shear failure to splitting failure with the increased amounts of lateral pressure, which was the same as the behavior shown with the test investigations.

In the low lateral pressure case, the vertical cracks propagated along with the shear surface, the large compressive stresses ($\sigma_y$ and $\sigma_z$) were concentrated only around the concrete dowel region, and the shear strength of the PBL shear connector was determined by the shear failure of two cracked surfaces around the edge of the hole in the steel plate.

While in the high lateral pressure application case, as the amount of the applied lateral pressure was increased, the shear strength of the two vertical cracked surfaces along with the shear surfaces was also enhanced, the splitting cracks propagated in the side concrete blocks perpendicular to the thickness of the perforated steel plate, and the large compressive stresses ($\sigma_y$ and $\sigma_z$) were also observed in the concrete dowel region as well as in the side concrete blocks. In the high lateral pressure application case, the shear strength of the PBL shear connector was characterized by the occurrence of the splitting cracks in the surrounding concrete blocks.

Numerical evaluation of the combined effects of the lateral pressures and the diameters of the holes revealed that there existed a clear combined effect and the increase in the shear strength was more in the large diameter case subjected to high lateral pressure because of the wide compressive regions generated around the concrete dowel region as well as in the side concrete blocks.

**Author Contributions:** Conceptualization, M.S.K., Y.Y. and H.N.; methodology, M.S.K., Y.Y. and H.N.; software, M.S.K., Y.Y. and H.N.; validation, M.S.K. and Y.Y.; formal analysis, M.S.K.; investigation, M.S.K., Y.Y. and H.N.; resources, M.S.K., Y.Y., H.N., and T.M.; data curation, M.S.K.; writing—original draft preparation, M.S.K.; writing—review and editing, M.S.K., Y.Y. and H.N.; visualization, Y.Y. and H.N.; supervision, Y.Y. and H.N.; project administration, M.S.K.; funding acquisition, Y.Y., H.N. and T.M. All authors have read and agreed to the published version of the manuscript.

**Funding:** This research received no external funding.

**Acknowledgments:** The research presented in the current study was conducted at Nagoya University, Japan.



## References

1.  Kim, S.H.; Lee, C.G.; Ahn, J.H.; Won, J.H. Experimental study on joint of spliced steel-PSC hybrid girder, Part I: Proposed parallel-perfobond-rib-type joint. *Eng. Struct.* **2011**, *33*, 2382–2397. [CrossRef]
2.  He, J.; Liu, Y.; Pei, B. Experimental study of the steel-concrete connection in hybrid cable-stayed bridges. *J. Perform. Constr. Facil.* **2013**, *28*, 559–570. [CrossRef]
3.  He, S.; Fang, Z.; Fang, Y.; Liu, M.; Liu, L.; Mosallam, A.S. Experimental study on perfobond strip connector in steel-concrete joints of hybrid bridges. *J. Constr. Steel Res.* **2016**, *118*, 169–179. [CrossRef]

4.  Liu, Y.; Xin, H.; He, J.; Xue, D.; Ma, B.; Mosallam, A.S. Experimental and analytical study on fatigue behavior of composite truss joints. *J. Constr. Steel Res.* **2013**, *83*, 21–36. [CrossRef]

5.  Li, Y.; Liu, Y.; Wang, F.; Yang, F. Load transfer mechanism of hybrid pylon joint with cells and bearing plates. *Adv. Civ. Eng.* **2018**, *2018*, 1–12. [CrossRef]

6.  Liu, Y.; Xin, H.; Liu, Y. Load transfer mechanism and fatigue performance evaluation of suspender-girder composite anchorage joints at serviceability stage. *J. Constr. Steel Res.* **2018**, *145*, 82–96. [CrossRef]

7.  Veldanda, M.R.; Hosain, M.U. Behaviour of perfobond rib shear connectors: Push-out tests. *Can. J. Civ. Eng.* **1992**, *19*, 1–10. [CrossRef]

8.  Machacek, J.; Studnicka, J. Perforated shear connectors. *Steel Compos. Struct.* **2002**, *2*, 51–66. [CrossRef]

9.  Yang, Y.; Chen, Y. Experimental study on mechanical behavior of PBL shear connectors. *J. Bridge Eng.* **2018**, *23*, 1–12. [CrossRef]

10. Zhang, J.; Hu, X.; Kou, L.; Zhang, B.; Jiang, Y.; Yu, H. Experimental study of the short-term and long-term behavior of perfobond connectors. *J. Constr. Steel Res.* **2018**, *150*, 462–474. [CrossRef]

11. Deng, W.; Xiong, Y.; Liu, D.; Zhang, J. Static and fatigue behavior of shear connectors for a steel-concrete composite girder. *J. Constr. Steel Res.* **2019**, *159*, 134–146. [CrossRef]

12. Wang, X.; Zhu, B.; Cui, S.; Liu, E.M. Experimental research on PBL connectors considering the effects of concrete stress state and other connection parameters. *J. Bridge Eng.* **2018**, *23*, 1–13. [CrossRef]

13. Taira, Y.; Asanuma, T.; Ichimiya, T.; Furuichi, K. Strength evaluation of perforated steel plate shear connector considering restrain effect. In Proceedings of the JCI Annual Convention; 2013; Volume 35, pp. 1225–1230. (In Japanese).

14. Zhao, C.; Li, Z.; Deng, K.; Wang, W. Experimental investigation on the bearing mechanism of Perfobond rib shear connectors. *Eng. Struct.* **2018**, *159*, 172–184. [CrossRef]

15. Ahn, J.E.; Lee, C.G.; Won, J.H.; Kim, S.H. Shear resistance of the perfobond-rib shear connector depending on concrete strength and rib arrangement. *J. Constr. Steel Res.* **2010**, *66*, 1295–1307. [CrossRef]

16. Vianna, J.D.C.; De Andrade, S.; Vellasco, P.C.G.D.S.; Neves, L.C. Experimental study of Perfobond shear connectors in composite construction. *J. Constr. Steel Res.* **2013**, *81*, 62–75. [CrossRef]

17. Zheng, S.; Liu, Y.; Yoda, T.; Lin, W. Parametric study on shear capacity of circular-hole and long-hole perfobond shear connector. *J. Constr. Steel Res.* **2016**, *117*, 64–80. [CrossRef]

18. Nakajima, A.; Koseki, S.; Hashimoto, M.; Suzuki, Y.; Nguyen, M.H. Evaluation of shear resistance of perfobond strip based on simple push-out test. *J. Jpn. Soc. Civ. Eng.* **2012**, *68*, 495–508. (In Japanese)

19. Jeong, Y.J.K.; Kim, H.Y.; Koo, H.B. Longitudinal shear resistance of steel-concrete composite slabs with perfobond shear connectors. *J. Constr. Steel Res.* **2009**, *65*, 81–88. [CrossRef]

20. Hosseinpour, E.; Baharom, S.; Badaruzzaman, W.H.W. Push-out test on the web opening shear connector for a slim-floor steel beam: Experimental and analytical study. *Eng. Struct.* **2018**, *163*, 137–152. [CrossRef]

21. Chen, J.J.; Wang, J.H.; Qiao, P.; Hou, Y.M.; Gu, Q.Y. Shear bearing of cross-plate joints between diaphragm wall panels—I: Model tests and shear behaviour. *Mag. Concr. Res.* **2016**, *68*, 1–14. [CrossRef]

22. Leonhardt, F.; Andrä, W.; Andrä, H.P.; Harre, W. Neues, vorteilhaftes Verbundmittel für Stahlverbund-Tragwerke mit hoher Dauerfestigkeit. *Beton-und Stahlbetonbau* **1987**, *82*, 325–331. (In German) [CrossRef]

23. Oguejiofor, E.C.; Hosain, M.U. Tests of full-size composite beams with perfobond rib connectors. *Can. J. Civ. Eng.* **1995**, *22*, 80–92. [CrossRef]

24. Kim, H.-Y.; Jeong, Y.-J. Experimental investigation on behaviour of steel-concrete composite bridge decks with perfobond ribs. *J. Constr. Steel Res.* **2006**, *62*, 463–471. [CrossRef]

25. Kim, S.H.; Ahn, J.H.; Choi, K.T.; Jung, C.Y. Experimental evaluation of the shear resistance of corrugated perfobond rib shear connections. *Adv. Struct. Eng.* **2011**, *14*, 249–264. [CrossRef]

26. Vianna, J.; Neves, L.C.; Vellasco, P.C.G.D.S.; De Andrade, S. Experimental assessment of perfobond and T-perfobond shear connectors' structural response. *J. Constr. Steel Res.* **2009**, *65*, 408–421. [CrossRef]

27. Wang, Z.; Li, Q.; Zhao, C. Ultimate shear resistance of perfobond rib connectors based on a modified push-out test. *Adv. Struct. Eng.* **2013**, *16*, 667–680. [CrossRef]

28. Su, Q.; Wang, W.; Luan, H.; Yang, G. Experimental research on bearing mechanism of perfobond rib shear connectors. *J. Constr. Steel Res.* **2014**, *95*, 22–31. [CrossRef]

29. Oguejiofor, E.C.; Hosain, M.U. A parametric study of perfobond rib shear connectors. *Can. J. Civ. Eng.* **1994**, *21*, 614–625. [CrossRef]

30. Al-Darzi, S.Y.K.; Chen, A.R.; Liu, Y.Q. Finite element simulation and parametric studies of perfobond rib connector. *Am. J. Appl. Sci.* **2007**, *4*, 122–127. [CrossRef]
31. Yu, Z.; Zhu, B.; Dou, S.; Liu, W. 3D FEM simulation analysis for PBL shear connectors. *Appl. Mech. Mater.* **2012**, *170*, 3449–3453. [CrossRef]
32. Liu, Y.; Xin, H.; Liu, Y. Experimental and analytical study on shear mechanism of rubber-ring perfobond connector. *Eng. Struct.* **2019**, *197*, 109382. [CrossRef]
33. Liu, J.; Yang, J.; Chen, B.; Zhou, Z. Mechanical performance of concrete-filled square steel tube stiffened with PBL subjected to eccentric compressive loads: Experimental study and numerical simulation. *Thin Walled Struct.* **2020**, *149*, 106617. [CrossRef]
34. Li, Z.; Zhao, C.; Shu, Y.; Deng, K.; Cui, B.; Su, Y. Full-scale test and simulation of a PBL anchorage system for suspension bridges: Experimental study and numerical simulation. *Struct. Infrastruct. Eng.* **2019**, *16*, 452–464. [CrossRef]
35. Liu, Y.; Wang, S.; Xin, H.; Liu, Y. Evaluation on out-of-plane shear stiffness and ultimate capacity of perfobond connector. *J. Constr. Steel Res.* **2019**, 105850. [CrossRef]
36. Fan, L.; Zhou, Z. Study on the shear capacity of PBH shear connector basing on PBL shear connector. *Comput. Model. New Technol.* **2014**, *18*, 174–180.
37. Munemoto, S.; Sonoda, Y. A fundamental study on the evaluation method of load bearing capacity of PBL jointing area. *J. Struct. Eng.* **2014**, *60A*, 522–530. (In Japanese)
38. Munemoto, O.; Sonoda, Y.; Koshiishi, M. A study on the strength evaluation of perforated steel plate shear connector using 3D elastic-plastic FEM. In Proceedings of the JCI Annual Convention; 2013; Volume 35, pp. 1243–1248. (In Japanese).
39. Ikuma, K.; Yamamoto, Y.; Nakamura, H.; Miura, T. Mesoscale simulation of bond behavior of deformed rebar based on coupled RBSM-FEM method. In Proceedings of the JCI Annual Convention; 2018; Volume 40, pp. 541–546. (In Japanese).
40. Karam, M.S.; Yamamoto, Y.; Nakamura, H.; Miura, T. Mesoscale analysis for the bond behavior of concrete under active confinement using coupled RBSM and solid FEM. In Proceedings of the 10th International Conference on Fracture Mechanics of Concrete and Concrete Structures (FramCoS-X), Bayonne, France, 24–26 June 2019.
41. Cusatis, G.; Nakamura, H. Discrete modeling of concrete materials and structures. *Cem. Concr. Compos.* **2011**, *33*, 865–866. [CrossRef]
42. Yamamoto, Y.; Nakamura, H.; Kuroda, I.; Furuya, N. Analysis of compression failure of concrete by three-dimensional Rigid Body Spring Model. *Doboku Gakkai Ronbunshuu* **2008**, *64*, 612–630. (In Japanese) [CrossRef]
43. Yamamoto, Y. Evaluation of Failure Behaviors under Static and Dynamic Loadings of Concrete Members with Mesoscopic Scale Modeling. Ph.D. Dissertation, Department of Civil Engineering, Nagoya University, Nagoya, Japan, 2008. (In Japanese).
44. Jr, R.M.; Yamamoto, Y.; Nakamura, H.; Miura, T. Numerical evaluation of localization and softening behavior of concrete confined by steel tubes. *Struct. Concr.* **2018**, *19*, 1956–1970.
45. Gedik, Y.H.; Nakamura, H.; Yamamoto, Y.; Kunieda, M. Evaluation of three-dimensional effects in short deep beams using a Rigid-Body-Spring-Model. *Cem. Concr. Compos.* **2011**, *33*, 978–991. [CrossRef]
46. Yamamoto, Y.; Nakamura, H.; Kuroda, I.; Furuya, N. Crack propagation analysis of reinforced concrete wall under cyclic loading using RBSM. *Eur. J. Environ. Civ. Eng.* **2014**, *18*, 780–792. [CrossRef]
47. Yang, Y.; Nakamura, H.; Yamamoto, Y.; Miura, T. Numerical simulation of bond degradation subjected to corrosion-induced crack by simplified rebar and interface model using RBSM. *Constr. Build. Mater.* **2020**, *247*, 118602. [CrossRef]
48. Yang, Y.; Nakamura, H.; Miura, T.; Yamamoto, Y. Effect of corrosion-induced crack and corroded rebar shape on bond behavior. *Struct. Concr.* **2019**, *20*, 2171–2182. [CrossRef]
49. Amalia, Z.; Nakamura, H.; Miura, T.; Yamamoto, Y. Development of simulation method of concrete cracking behavior and corrosion products movement due to rebar corrosion. *Constr. Build. Mater.* **2018**, *190*, 560–572. [CrossRef]
50. Qiao, D.; Nakamura, H.; Yamamoto, Y.; Miura, T. Crack patterns of concrete with a single rebar subjected to non-uniform and localized corrosion. *Constr. Build. Mater.* **2016**, *116*, 366–377. [CrossRef]

51. Farooq, U.; Nakamura, H.; Miura, T.; Yamamoto, Y. Proposal of bond behavior simulation model by using discretized Voronoi mesh for concrete and beam element for reinforcement. *Cem. Concr. Compos.* **2020**, *110*, 103593. [CrossRef]
52. Pantò, B.; Silva, L.; Vasconcelos, G.; Lourenço, P.B. Macro-modelling approach for assessment of out-of-plane behavior of brick masonry infill walls. *Eng. Struct.* **2019**, *181*, 529–549. [CrossRef]
53. Pantò Caliò, I.; Lourenço, P.B. A 3D discrete macro-element for modelling the out-of-plane behaviour of infilled frame structures. *Eng. Struct.* **2018**, *175*, 371–385. [CrossRef]
54. Caliò, I.; Pantò, B. A macro-element modelling approach of Infilled Frame Structures. *Comput. Struct.* **2014**, *143*, 91–107. [CrossRef]
55. Rezaei, R.; Burton, H.V.; Weinand, Y. Macroscopic model for spatial timber plate structures with integral mechanical attachments. *J. Struct. Eng. (U. S.)* **2020**, *146*, 04020200. [CrossRef]
56. Bao, Y.; Kunnath, S.K.; El-Tawil, S.; Lew, H.S. Macromodel-based simulation of progressive collapse: RC frame structures. *J. Struct. Eng.* **2008**, *134*, 1079–1091. [CrossRef]
57. Hayashi, D.; Nagai, K.; Eddy, L. Mesoscale analysis of RC anchorage performance in multidirectional reinforcement using a three-dimensional discrete model. *J. Struct. Eng.* **2017**, *143*, 04017059. [CrossRef]
58. Eddy, L.; Nagai, K. Numerical simulation of beam-column knee joints with mechanical anchorages by 3D Rigid Body Spring Model. *Eng. Struct.* **2016**, *126*, 547–558. [CrossRef]
59. Mendoza, R., Jr.; Yamamoto, Y.; Nakamura, H.; Miura, T. Modeling of composite action in concrete-filled steel tubes using coupled RBSM and Shell FEM. In Proceedings of the JCI Annual Conventionl, Sendai, Japanese, 26 June 2017; Volume 39, pp. 991–996.
60. Fuji, K.; Iwasaki, H.; Fukada, K.; Toyota, T.; Fujimura, N. Crack restraint factors in ultimate slip behavior of Perfobond strip. *Doboku Gakkai Ronbunshuu* **2008**, *64*, 502–512. (In Japanese) [CrossRef]
61. Japan Society of Civil Engineers (JSCE). *Standard Specifications for Concrete Structures-2012*; Japan Society of Civil Engineers: Tokyo, Japan, 2012.

Article

# The Prediction of Stiffness of Bamboo-Reinforced Concrete Beams Using Experiment Data and Artificial Neural Networks (ANNs)

Muhtar [1,*], Amri Gunasti [1], Suhardi [2], Nursaid [1], Irawati [1], Ilanka Cahya Dewi [1], Moh. Dasuki [1], Sofia Ariyani [1], Fitriana [1], Idris Mahmudi [1], Taufan Abadi [1], Miftahur Rahman [1], Syarif Hidayatullah [1], Agung Nilogiri [1], Senki Desta Galuh [1], Ari Eko Wardoyo [1] and Rofi Budi Hamduwibawa [1]

[1]   Faculty of Engineering, University of Jember, Jember Indonesia 68121, Indonesia; amrigunasti@unmuhjember.ac.id (A.G.); nursaid@unmuhjember.ac.id (N.); irawati@unmuhjember.ac.id (I.); ilankadewi@unmuhjember.ac.id (I.C.D.); moh.dasuki22@unmuhjember.ac.id (M.D.); sofia.ariyani@unmuhjember.ac.id (S.A.); fitriana@unmuhjember.ac.id (F.); idrismahmudi@unmuhjember.ac.id (I.M.); taufan.abadi@unmuhjember.ac.id (T.A.); miftahurrahman@unmuhjember.ac.id (M.R.); syarifhidayatullah@unmuhjember.ac.id (S.H.); agungnilogiri@unmuhjember.ac.id (A.N.); senki.desta@unmuhjember.ac.id (S.D.G.); arieko@unmuhjember.ac.id (A.E.W.); rofi.hamduwibawa@unmuhjember.ac.id (R.B.H.)
[2]   Faculty of Agricultural Technology, University of Jember, Jember Indonesia 68121, Indonesia; hardi.ftp@unej.ac.id
*   Correspondence: muhtar@unmuhjember.ac.id

Received: 3 August 2020; Accepted: 19 August 2020; Published: 27 August 2020

**Abstract:** Stiffness is the main parameter of the beam's resistance to deformation. Based on advanced research, the stiffness of bamboo-reinforced concrete beams (BRC) tends to be lower than the stiffness of steel-reinforced concrete beams (SRC). However, the advantage of bamboo-reinforced concrete beams has enough good ductility according to the fundamental properties of bamboo, which have high tensile strength and high elastic properties. This study aims to predict and validate the stiffness of bamboo-reinforced concrete beams from the experimental results data using artificial neural networks (ANNs). The number of beam test specimens were 25 pieces with a size of 75 mm × 150 mm × 1100 mm. The testing method uses the four-point method with simple support. The results of the analysis showed the similarity between the stiffness of the beam's experimental results with the artificial neural network (ANN) analysis results. The similarity rate of the two analyses is around 99% and the percentage of errors is not more than 1%, both for bamboo-reinforced concrete beams (BRC) and steel-reinforced concrete beams (SRC).

**Keywords:** bamboo-reinforced concrete (BRC); stiffness prediction; artificial neural network (ANN)

## 1. Introduction

Some of the advantages of bamboo include having high tensile strength [1], easy to split, cut, elastic fibers, optimal in bearing loads, and it is not a pollutant. At the same time, the weakness of bamboo as a construction material is easily attacked by insects, because the starch content in bamboo is quite high. Therefore, bamboo as a building material requires treatment, such as immersion in water [2,3] and the application of adhesives and waterproof layers [3]. The application of adhesive and waterproof coating has increased the load capacity and stiffness of the BRC beam [4]. Bamboo as a reinforcement of concrete structural elements has been widely used, among other things, as beam reinforcement [2,5–7], bridge frame reinforcement [8], plate or panel reinforcement [9–11], and column reinforcement [12,13].

The most important mechanical properties of bamboo-reinforced concrete beams are stress, strain, and stiffness. Some previous researchers concluded that bamboo-reinforced concrete beams have lower stiffness compared to steel reinforced concrete beams but have elastic properties and high ductility, so that they are effective in absorbing earthquake energy [14,15]. However, low rigidity will lead to reduced construction integrity and excessive structural deformation. The behavior of materials and construction elements, especially the stiffness parameters can be known through the relationship of load and deflection, as shown in Figure 1.

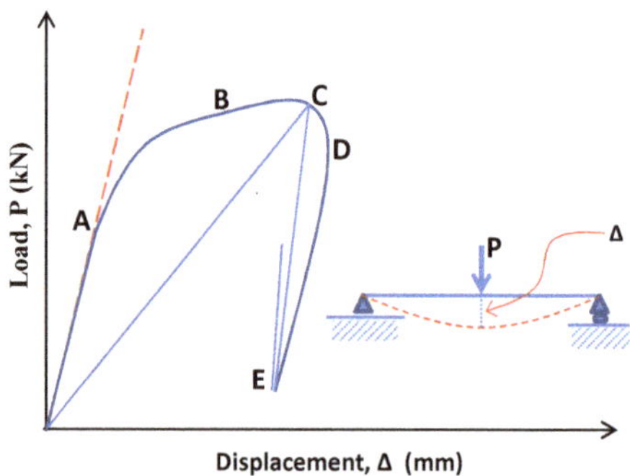


**Figure 1.** The load vs. deflection relationships of the reinforced concrete beam [15].

The stiffness of bamboo-reinforced concrete beams (*EI*) is the main factor of structural resistance to the bending deformation of BRC beams. Beam stiffness is a function of the modulus of elasticity of the material (*E*) and the moment of inertia (*I*). Moments of inertia before cracking use $I_g$, and after cracking they use $I_{cr}$. The effective inertia moment is the value between $I_g$ and $I_{cr}$. This understanding can be seen from the behavior of the load vs. deflection relationship in Figure 1. In general, the determination of beam stiffness is based on the results of the beam flexural test, while the calculation of elasticity modulus (*E*) of BRC beams for testing beams with two load points can follow Equations (1) and (2) [15].

$$E = \frac{23PL^3}{648\Delta I}(N/mm^2) \tag{1}$$

$$\Delta = \frac{23PL^3}{648EI}(mm) \tag{2}$$

where *E* is the elasticity modulus, Δ is the initial crack, *P* is the initial crack load, *L* is the span, and *I* is the inertia moment of the cross-section.

Making conclusions from the results of research on the behavior of bamboo-reinforced concrete beams (BRC) is not easy to take. Correct conclusions must go through data validation and data analysis with other methods, such as statistical analysis, the finite element method [16], or the artificial neural network (ANN) method [17]. The determination of the stiffness of bamboo-reinforced concrete beams (BRC) from the experimental results must be validated by other methods, such as the artificial neural network (ANN) method.

Artificial neural networks (ANNs) consist of many neurons. Neurons are grouped into several layers. Neurons in each layer are connected with neurons in other layers. This does not apply to the input and output layers but only to the layers in between. Information received at the input layer is continued to the layers in ANN one by one until it reaches the output layer. The layer that lies between the input and output is called the hidden layer. However, not all ANNs have a hidden layer; some are only input and output layers.

Artificial neural networks (ANNs) are a powerful tool for solving complex problems in the field of civil engineering. Many researchers have used the ANN method for many structural engineering studies, such as predicting the compressive strength of concrete [18], axial strength of composite columns [19], and determination of displacement of concrete reinforcement (RC) buildings [20]. Determination and control of BRC beam stiffness are based on load vs. deflection diagrams. Load data and deflection of experimental results are used as input data and target data in the analysis of artificial neural networks (ANNs).

Some previous researchers have concluded that artificial neural networks (ANNs) can be an alternative in calculating deflection in a reinforced concrete beam. The results of deflection calculations on reinforced concrete using ANN proved to be very effective [21]. ANN is also very well used to predict deflection in the concrete beam with a very strong correlation level of 97.27% to the test data [22]. Likewise, the use of ANN to predict deflection in cantilever beams produces very accurate outcomes [23]. In this paper, we use uniform load input data, while the target data are the deflection of laboratory test results. Distribution of ANN model data composition consists of 70% training, 15% validation, and 15% testing. The schematic of ANN architecture for rectangular beams is shown in Figure 2.

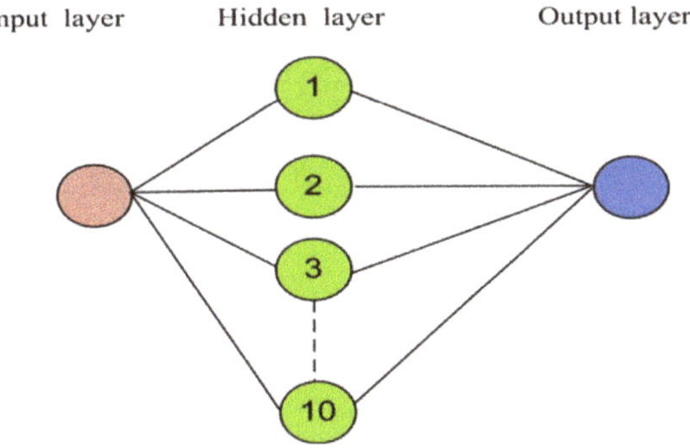


**Figure 2.** Schematic of ANN architecture for rectangular beams.

The purpose of this study is to validate the behavior and stiffness of the BRC beam experimental results with the artificial neural network (ANN) method. Errors resulting from experimental data are usually caused by some things, such as human errors, calibration of tools that have expired, test method errors, and test items that do not match. Therefore, the experimental data are evaluated and compared with the results of the artificial neural network (ANN) method. In this study, the experimental data are thought to have a large deviation from the results of the artificial neural network (ANN) method. Then, an efficient ANN-based computational technique is presented to estimate the load vs. deflection of bamboo-reinforced concrete blocks (BRC). Furthermore, stiffness observations are made at the same loading point.

## 2. Materials and Methods

Experimental data were obtained from a single reinforced BRC beam bending test with two load points based on ASTM C 78-02 [24]. The size of bamboo reinforcement is 15 mm × 15 mm, which is treated first through immersion, drying, and the waterproof coating using Sikadur®-752 [3]. As a strengthening of bamboo reinforcement used diameter hose-clamps $\frac{3}{4}''$ [8]. The number of beam test specimens were 25 pieces with a size of 75 mm × 150 mm × 1100 mm consisting of 24 BRC beams and 1 SRC beam with steel reinforcement. The detailed image of the BRC beam specimen is shown in Figure 3. The design of the concrete mixture in this study was Portland Pozzolana Cement (PPC), sand, coarse aggregate, and water with a proportion of 1:1.81:2.82:0.52. The average compressive strength of concrete at the age of 28 days is 31.31 MPa. The steel used is plain steel with $f_y = 240$ MPa.

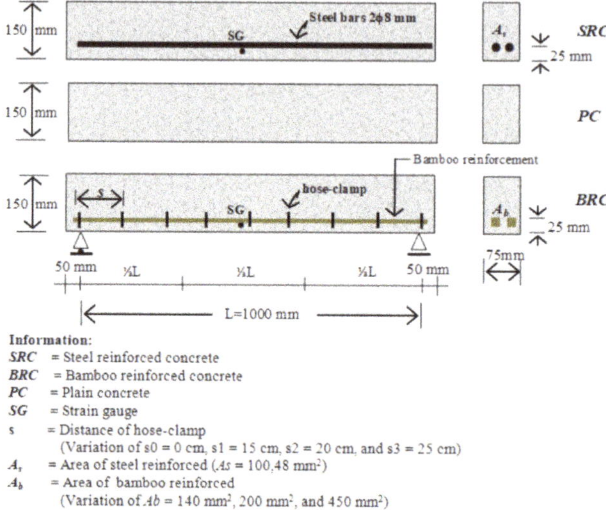



**Figure 3.** Geometry and details of bamboo-reinforced concrete beams.

The beam flexural test is carried out on two simple supports, namely joint support and roller support. Load in the form of a centralized load divided into two load points with a distance of $^1/_3$L from the support. The strain gauge is mounted on the bamboo reinforcement with a distance of $\frac{1}{2}$L from the support to determine the strain that is occurring. To detect deflection, a linear variable differential transformer is installed at a distance of $\frac{1}{2}$L from the support. To get the stages of loading from zero until the beam collapses, a hydraulic jack and load cell are used that are connected to the load indicator. Loading is carried out slowly at a speed of 8 kg/cm²–10 kg/cm². Load reading on the load indicator is used to control the hydraulic jack pump, deflection, and strain according to the planned loading stage. However, when the test specimen reaches the ultimate load, deflection readings become the control of readings of the strain and load. Hydraulic jack pumping continues to take place slowly according to the deflection reader command. The collapse pattern is observed and identified through cracks that occur, starting from the first crack until the beam collapses. The BRC beam test setting is shown in Figure 4.

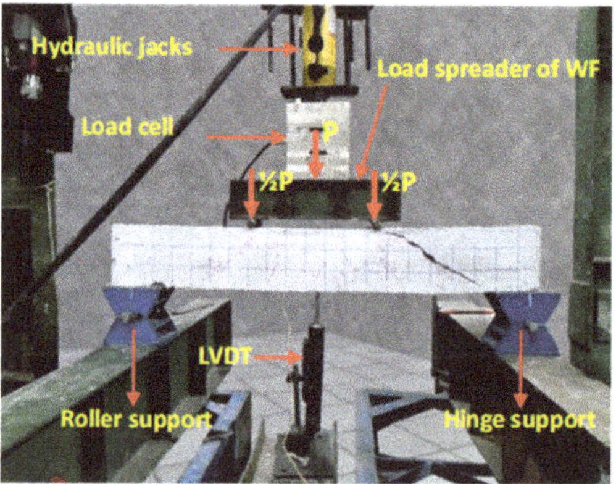



**Figure 4.** Test arrangement of bamboo-reinforced concrete beams.

## 3. Results

Mechanical properties and stress–strain characteristics of steel and bamboo materials are the dominant factors that influence the shape of the load vs. deflection relationship behavior models. The difference in the stress and strain relationship pattern of steel and bamboo is seen in the difference in melting point and fracture stress, as shown in Figures 5 and 6. Steel reinforcement shows a clear melting point, whereas bamboo reinforcement does not show a clear melting point. Both of them show a clear stress fracture point, but in bamboo reinforcement, after fracture stress occurs, the strain–stress relationship pattern tends to return to zero, as shown in Figure 5. This shows that bamboo has good elastic properties.

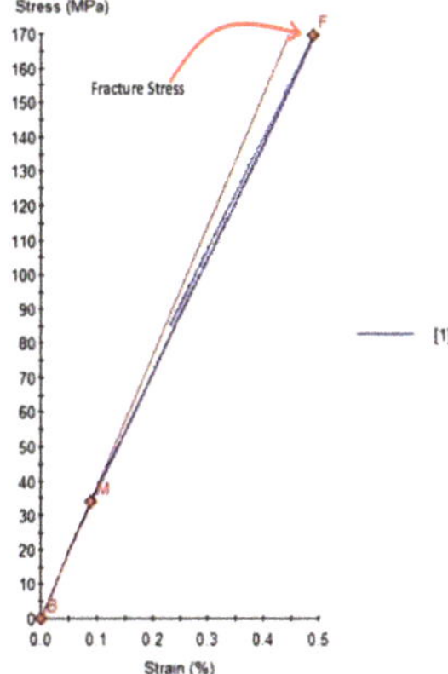


**Figure 5.** The stress–strain relationship of normal bamboo reinforcement.

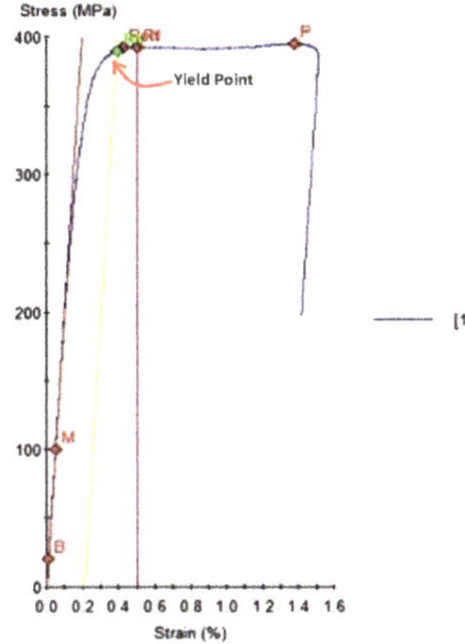


**Figure 6.** The stress–strain relationship of steel reinforcement.

Figure 7 shows the relation between load vs. deflection of the BRC beam and SRC beam from the analysis of experimental data, while Figure 8 shows the relationship between load vs. deflection of BRC beams and SRC beams resulting from the analysis of artificial neural network (ANN) methods. The BRC beam tends to have a large deflection, but when the maximum load is reached, the deflection tends to return to zero if the load is released, as shown in Figure 9. Documentation of the gradual load discharge after the ultimate load has been reached can be seen in the following link: https://goo.gl/6AVWmP [14] and the BRC beam flat back. This shows its compatibility with bamboo strain–stress behavior. The load vs. deflection relationship of the SRC beam shows the existence of an elastic limit, elasto-plastic limit, and plastic, as shown in Figure 7. While the relationship of load vs. deflection of the BRC beam shows a linear line until the maximum load limit and after the peak load, the deflection returns to zero, as shown in Figure 9.

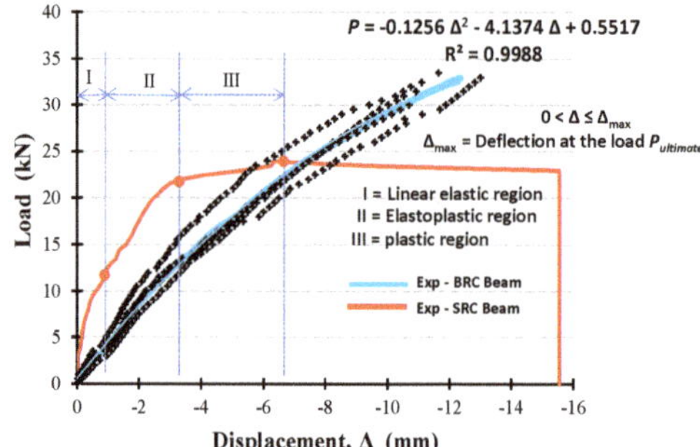


**Figure 7.** The load vs. deflection relationship of the BRC beam from experiment [14].

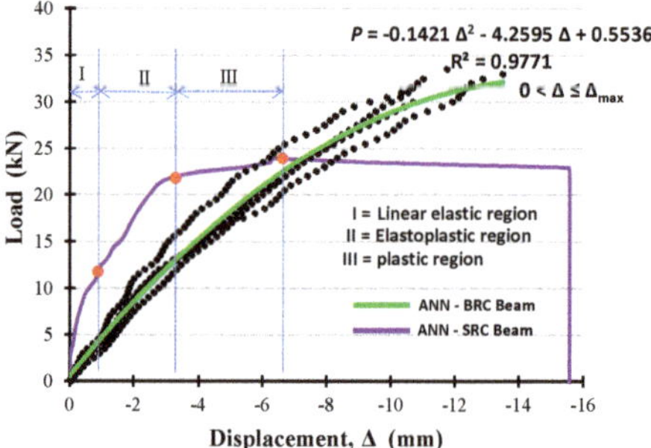


**Figure 8.** The load vs. deflection relationship of the BRC beam from the ANN method.

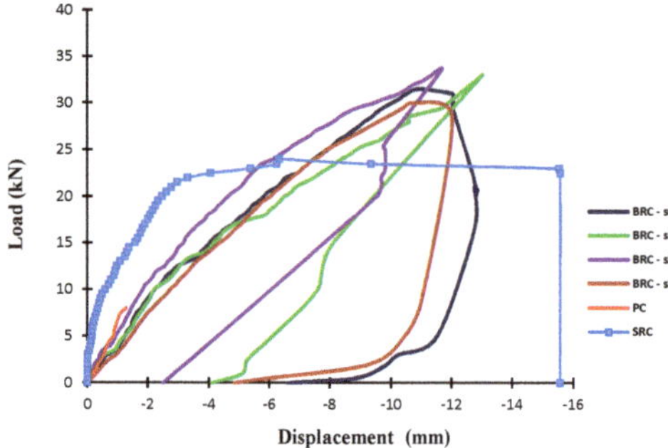


**Figure 9.** The load vs. deflection relationship of the BRC beam and the SRC beam until the gradual release of the load.

In this case, the ANN studies the network to diagnose the shape and distribution of data from the deflection of BRC beams and SRC beams with different loads. After reaching small and acceptable variations of errors, training in neural networks is stopped. Then, the neural network model is tested, and the results are validated by comparing it with the results of the analysis of experimental data. Every network created in the ANN is trained, tested, and validated for all data samples to identify the best technique. The data input for the network used is the deflection data from the experimental results of the BRC beam and the SRC beam. The deflection data file of the experimental results is saved in the form of MS Excel. Data are distributed into training (70%), testing (15%), and validation (15%).

Figures 10–13 show the prediction of the load vs. deflection relationship of the BRC beam and Figure 14 shows the prediction of the relationship of load vs. deflection of the SRC beam from the ANN method analysis. The correlation value of laboratory data by using ANN shows an average value of R Square of 0.999. The results of predictions by the ANN method show that the percentage of errors is very small, with a maximum error of 0.26%. Overall, the comparison of experimental data with the results of the ANN method predictions shows no more than a 1% error. From the data results of the two analyses and the pattern of load vs. deflection relationships, it can be concluded that the stiffness of the BRC beams is similar. Then, the stiffness prediction with the elasticity modulus parameter can be calculated based on the load vs. deflection relationship graph, as shown in Figure 15.

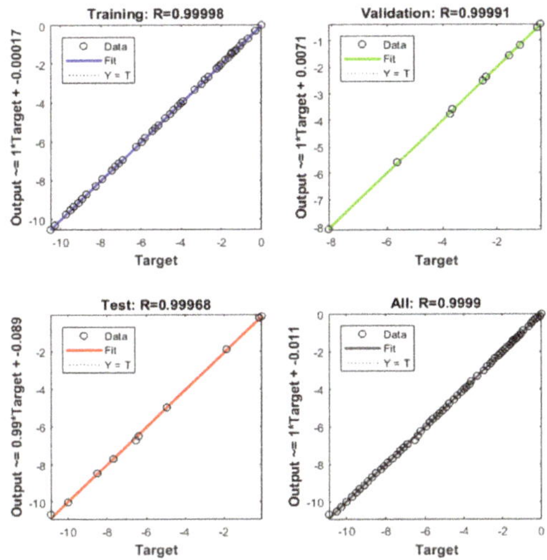

**Figure 10.** The correlation value of laboratory data and the ANN method (BRC-1 beam).

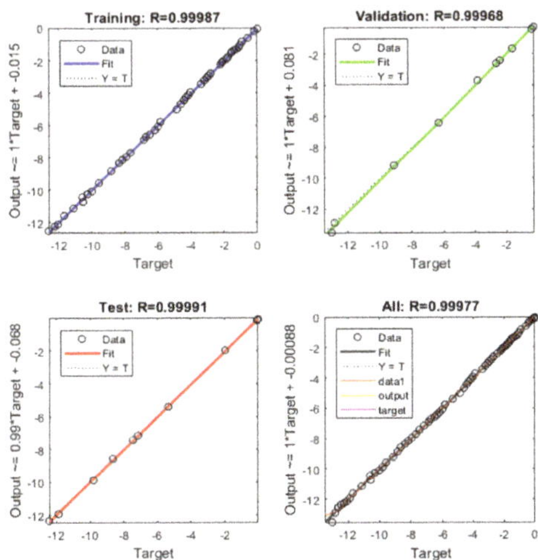

**Figure 11.** The correlation value of laboratory data and the ANN method (BRC-2 beam).

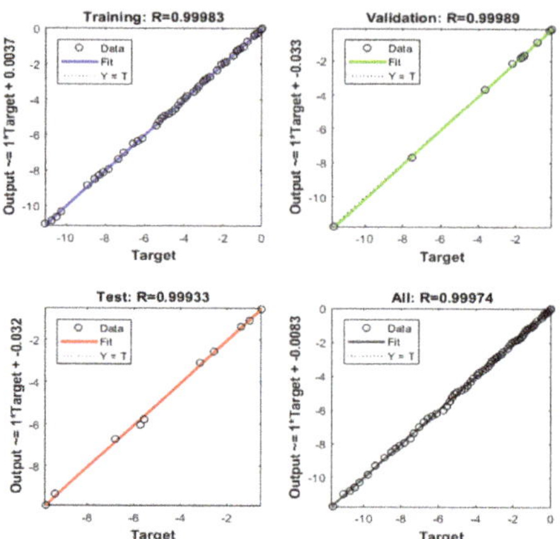

**Figure 12.** The correlation value of laboratory data and the ANN method (BRC-3 beam).

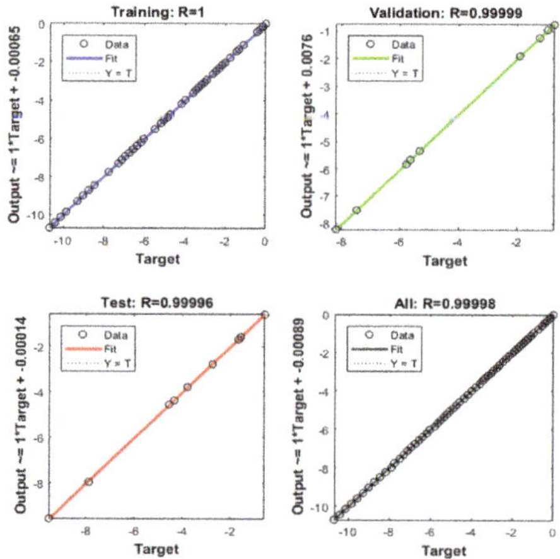

**Figure 13.** The correlation value of laboratory data and the ANN method (BRC-4 beam).

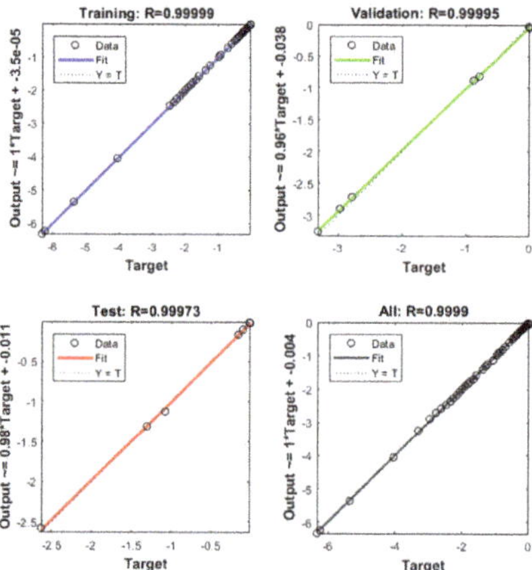

**Figure 14.** The correlation value of laboratory data and the ANN method (SRC beam).

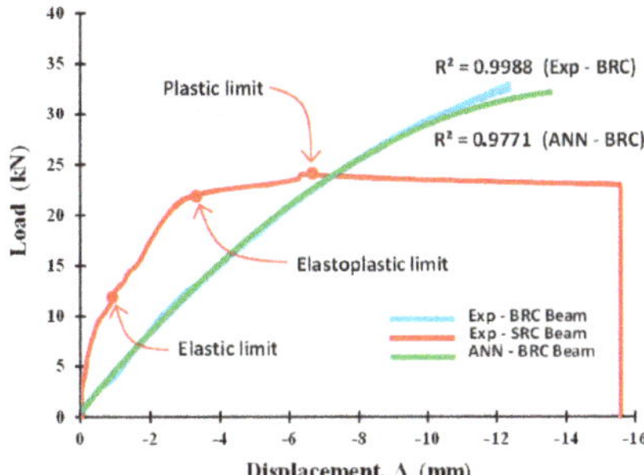

**Figure 15.** The load vs. deflection relationship the experimental results and ANN analysis.

Figure 15 shows the combined relationship of the load vs. deflection beam of the experimental BRC beam and the ANN analysis results. Figure 15 shows a graph that is coincidental with an error rate of not more than 1%, so that the combined graph of the load vs. deflection relationship can be used to determine the modulus of elasticity or the stiffness of the BRC beam.

## 4. Discussion

Figure 16 shows the results of the two methods of data analysis being a load vs. deflection pattern. From this load vs. deflection pattern, the stiffness of bamboo-reinforced concrete beams can be predicted. Prediction of stiffness with the elasticity modulus parameters can be calculated based on the load vs. deflection relationship graph. The graph of load vs. deflection relationship shows that at 40% ultimate load, the stiffness of the BRC beam has a stiffness lower to 44% than the SRC beam. Meanwhile, if viewed from the graph load vs. deflection relationship, ANN analysis results with experimental results show the same stiffness value up to 80% ultimate load. The stiffness of BRC beams at loads above 80% indicates a difference, namely the stiffness of the ANN analysis results is lower than the experimental results, as shown in Figure 16.

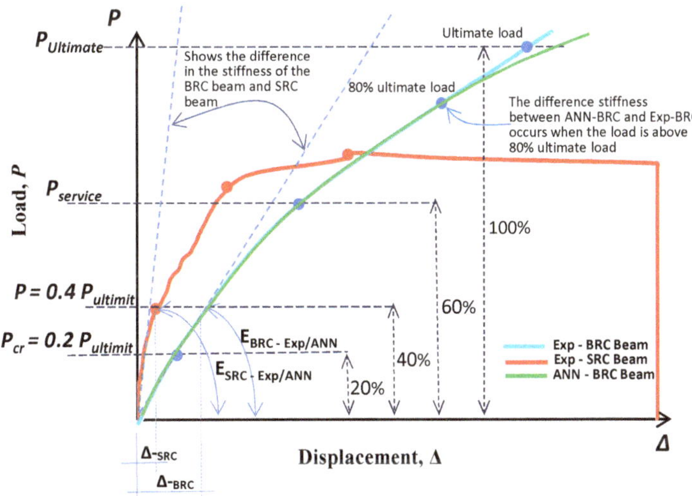

**Figure 16.** The difference in stiffness between the SRC beam, BRC beam, and BRC beam of ANN analysis result.

Table 1 shows that the initial crack (elastic region) of the BRC beam is in the range of 20% of the ultimate load and 40% of the ultimate load for the SRC beam. Whereas the effect of installing hose-clamps on bamboo reinforcement on the ultimate load of BRC beams is optimum at a distance of 20 cm (BRC-s2) and decreases at a distance of 25 cm, this indicates that installing hose-clamps that are too tight will reduce the elastic properties of bamboo reinforcement and decrease its ductility, as shown in Figure 17. Installation of hoses that are too tight does not increase the stiffness of the BRC beam but instead reduces the load capacity. The control of the load vs. deflection relationship with the ANN method is taken from the results of the regression analysis of six beam samples in each group, namely the BRC-s0, BRC-s1, BRC-s2, and BRC-s3 groups, plus one SRC beam, as shown in Figures 7 and 8. The ANN analysis results for each group are regressed back and used as the final result to determine the stiffness of the BRC beam, as shown in Figure 15. The ANN analysis results for each group are shown in Figures 10–13.

**Table 1.** The value of the average initial crack loads and ultimate loads based on theoretical calculations and experimental.

| Specimens | Theoretical Calculations | | Flexural Test Results | | |
|---|---|---|---|---|---|
| | First Crack Load (kN) | Ultimate Load (kN) | Average First Crack Load (kN) | Average Failure Load (kN) | Average Deflection at Failure (mm) |
| (a) BRC-s0 | 6.87 | 32.19 | 8.25 | 30.25 | 11.41 |
| (b) BRC-s1 | 6.87 | 32.19 | 7.25 | 32.00 | 12.60 |
| (c) BRC-s2 | 6.87 | 32.19 | 8.00 | 33.25 | 12.01 |
| (d) BRC-s3 | 6.87 | 32.19 | 7.50 | 29.75 | 9.15 |
| (e) SRC | 6.51 | 16.14 | 10.00 | 24.00 | 6.33 |

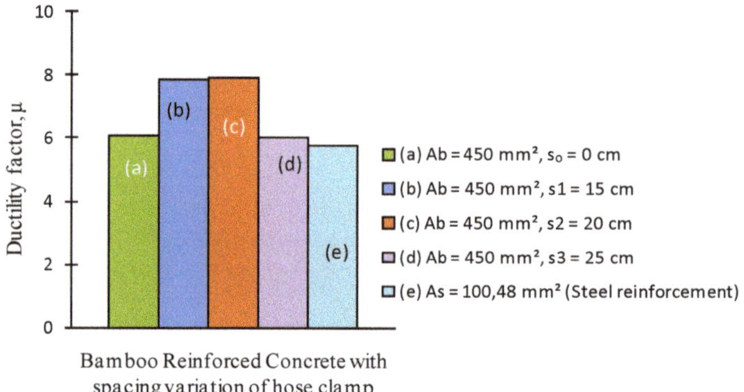


**Figure 17.** The effect of hose-clamp distance on the ductility value.

Stiffness ($EI$) is the main parameter of the resistance of structural elements to bending deformation. The basic properties and behavior of stress–strain material are the dominant factors determining the size of the rigidity of structural elements. SRC beam stiffness has a greater stiffness than the BRC beam stiffness. This is due to the steel reinforcement having an elasticity modulus greater than the elasticity modulus of bamboo. However, the BRC beam has good elastic properties, in harmony with the pattern of stress–strain relationships of bamboo. This proves that bamboo material has good earthquake energy absorption. The behavior of elastic on the BRC beam can be seen in the video at the following link: https://goo.gl/6AVWmP [14].

Integrity and rigidity in a structure are essential. Therefore, the low stiffness of the BRC beam is essential to find a solution. Solutions to solve the low stiffness of the BRC beam, such as the graph diagram in Figure 16, can be done in two ways, namely giving strength to bamboo reinforcement and applying the principle of confined concrete [7]. Strengthening of bamboo reinforcement can be achieved by using adhesive, increasing surface roughness, installing hose-clamps that function as hooks and shear connectors, and so on. An equally important solution is to increase the strength of the concrete to support increasing the stiffness of the BRC beam. Previous studies showed that the cause of the majority of BRC beam collapse is by slippage [14] and shear collapse [14]. The principle of confined concrete is fundamental to do by giving shear reinforcement to the BRC beam.

## 5. Conclusions

Predictions of bamboo-reinforced concrete beam stiffness based on experimental results and analysis results of artificial neural network (ANN) methods show very close similarities or with an error prediction of no more than 1%.

Bamboo-reinforced concrete (BRC) beams have a lower stiffness of up to 40% when compared to steel reinforced concrete (SRC) beams.

The stiffness of the BRC beam of experimental result and the artificial neural network (ANN) analysis results have in common up to 80% of the ultimate load and, afterward, show differences.

The coatings of adhesives, modification of bamboo reinforcement roughness, and the use of shear reinforcement are solutions to increase the stiffness and capacity of the BRC beam.

Installation of a hose-clamp that is too tight does not increase the stiffness of the BRC beam but reduces its elastic properties and reduces its load capacity.

**Author Contributions:** Conceptualization, M., S., and A.G.; methodology, M., S., and A.N.; software, S., S.A., A.N., and A.E.W.; validation, M., I., and I.C.D.; formal analysis, M., S., and M.D.; investigation, N., and T.A.; resources, N.; data curation, M., and S.; writing—original draft preparation, A.N. and F.; writing—review and editing, M. and S.D.G.; visualization, M.R. and S.H.; supervision, N.; project administration, R.B.H.; funding acquisition, N. and I.M. All authors have read and agreed to the published version of the manuscript.

**Funding:** APC was financed entirely by the DPRM Republic of Indonesia and LPPM of the University of Muhammadiyah, Jember, Indonesia.

**Acknowledgments:** Funding Research is fully borne by the research Program, Directorate of Research and Community Service, Directorate General of Research and Technology Strengthening and Development of the Ministry of Research, Technology, and Higher Education of the Republic of Indonesia.

**Conflicts of Interest:** The authors declare no conflict of interest.

## References

1. Ghavami, K. Bamboo as reinforcement in structural concrete elements. *Cem. Concr. Compos.* **2005**, *27*, 637–649. [CrossRef]
2. Agarwal, A.; Nanda, B.; Maity, D. Experimental investigation on chemically treated bamboo reinforced concrete beams and columns. *Constr. Build. Mater.* **2014**, *71*, 610–617. [CrossRef]
3. Muhtar, M. Experimental data from strengthening bamboo reinforcement using adhesives and hose-clamps. *Data Brief* **2019**, *27*, 104827. [CrossRef] [PubMed]
4. Muhtar, M. Cracked Pattern of Bamboo Reinforced Concrete Beams Using Double Reinforcement with the Strengthening on Tensile Reinforcement. *Int. J. Eng. Res. Technol.* **2020**, *13*, 608–612.
5. Rahman, M.M.; Rashid, M.H.; Hossain, M.A.; Hasan, M.T.; Hasan, M.K. Performance evaluation of bamboo reinforced concrete beam. *Int. J. Eng. Res. Technol. IJET-IJENS* **2011**, *11*, 113–118.
6. Dewi, S.M.; Nuralinah, D.; Munawir, A.; Wijaya, M.N. Crack behavior study of bamboo reinforced concrete beam with additional pegs in reinforcing. *Int. J. Civ. Eng. Technol.* **2018**, *9*, 1632–1640.
7. Dewi, S.M. The flexural behavior model of bamboo reinforced concrete beams using a hose clamp. *Proc. Mater. Sci. Eng. Chem.* **2019**, *276*, 1033.
8. Muhtar, A.; Gunasti Manggala, A.S.; Nusant, A.F.P.; Hanafi, A.N. Effect of reinforcement details on precast bridge frames of bamboo reinforced concrete to load capacity and crack patterns. *Int. J. Eng. Res. Technol.* **2020**, *13*, 631–636.
9. Puri, V.; Chakrabortty, P.; Anand, S.; Majumdar, S. Bamboo reinforced prefabricated wall panels for low-cost housing. *J. Build. Eng.* **2017**, *9*, 52–59. [CrossRef]
10. Maruthupandian, G.; Saravanan, R.; S Suresh, K.S.; Sivakumar, B.G. A Study on Bamboo Reinforced Concrete Slabs. *J. Chem. Pharm. Sci. A* **2016**, *9*, 978–980.

11. Daud, N.M.; Nor, N.M.; Yusof, M.A.; Yahya, M.A.; Munikanan, V. Axial and flexural load test on untreated bamboocrete multi-purpose panel. *Int. J. Integr. Eng.* **2018**, *10*, 28–31.

12. Tripura, D.D.; Singh, K.D. Mechanical behavior of rammed earth column: A comparison between unreinforced, steel and bamboo reinforced columns. *Mater. Constr.* **2018**, *68*, 174. [CrossRef]

13. Rameshwar, S.; Kale, A.; Rashmirana, P. Suitability of Bamboo as Reinforcement in Column. *Int. J. Recent Innov. Trends Comput. Commun.* **2016**, *4*, 270–272.

14. Dewi, S.M. Enhancing bamboo reinforcement using a hose-clamp to increase bond-stress and slip resistance. *J. Build. Eng.* **2019**, *26*, 100896.

15. Dewi, S.M. The Stiffness and Cracked Pattern of Bamboo Reinforced Concrete Beams Using a Hose Clamp. *Int. J. Civ. Eng. Technol.* **2018**, *9*, 273–284.

16. Muhtar, M. Numerical validation data of tensile stress zones and crack zones in bamboo reinforced concrete beams using the Fortran PowerStation 4.0 program. *Data Brief* **2020**, *29*, 105332. [CrossRef]

17. Suryanita, R.; Maizir, H.; Jingga, H. Prediction of Structural Response due to Earthquake Load using Artificial Prediction of Structural Response due to Earthquake Load using Artificial Neural Networks. In Proceedings of the International Conference on Engineering & Technology, Computer, Basic & Applied Sciences, Bangkok, Thailand, 20–21 June 2016.

18. Naderpour, H.; Kheyroddin, A.; Amiri, G.G. Prediction of FRP-confined compressive strength of concrete using artificial neural networks. *Compos. Struct.* **2010**, *92*, 2817–2829. [CrossRef]

19. Ahmadi, M.; Naderpour, H.; Kheyroddin, A. Utilization of artificial neural networks to prediction of the capacity of CCFT short columns subject to short term axial load. *Arch. Civ. Mech. Eng.* **2014**, *14*, 510–517. [CrossRef]

20. Khademi, F.; Akbari, M.; Nikoo, M. Displacement determination of concrete reinforcement building using data-driven models. *Int. J. Sustain. Built Environ.* **2017**, *6*, 400–411. [CrossRef]

21. Kaczmarek, M.; Szymanska, A. application of artificial neural networks to predict the deflections of reinforced concrete beams. *Stud. Geotech. Mech.* **2016**, *38*, 37–46. [CrossRef]

22. Abd, A.M.; Salman, W.D.; Ahmed, Q.W. Ann and statistical modelling to predict the deflection of continuous reinforced concrete. *Diyala J. Eng. Sci.* **2015**, 134–143.

23. Ya, T.T.; Alebrahim, R.; Fitri, N.; Alebrahim, M. Analysis of Cantilever Beam Deflection under Uniformly Distributed Load using Artificial Neural Networks. *MATEC Web Conf.* **2019**, *4*, 06004.

24. ASTM C78/C78M-02. *Standard Test Method for Flexural Strength of Concrete (Using Simple Beam with Third-Point Loading)*; ASTM International: Sikangshihoken, PA, USA, 2002.

*Article*

# A Coupled Modeling Simulator for Near-Field Processes in Cement Engineered Barrier Systems for Radioactive Waste Disposal


**Steven J. Benbow [1],*, Daisuke Kawama [2], Hiroyasu Takase [2], Hiroyuki Shimizu [3], Chie Oda [4], Fumio Hirano [4], Yusuke Takayama [4], Morihiro Mihara [4] and Akira Honda [4]**

1   Quintessa Limited, Henley-on-Thames RG9 1HG, UK
2   QJ Science, Yokohama 221-0052, Japan; daisuke.kawama@qjscience.co.jp (D.K.); hiroyasu.takase@qjscience.co.jp (H.T.)
3   Kajima Corporation, Minato-ku, Tokyo 107-8502, Japan; shimihir@kajima.com
4   JAEA, Tokai-Mura, Ibaraki 319-1194, Japan; oda.chie@jaea.go.jp (C.O.); hirano.fumio@jaea.go.jp (F.H.); takayama.yusuke@jaea.go.jp (Y.T.); mihara.morihiro@jaea.go.jp (M.M.); honda.akira@jaea.go.jp (A.H.)
*   Correspondence: stevenbenbow@quintessa.org


Received: 30 July 2020; Accepted: 28 August 2020; Published: 29 August 2020


**Abstract:** Details are presented of the development of a coupled modeling simulator for assessing the evolution in the near-field of a geological repository for radioactive waste disposal where concrete is used as a backfill. The simulator uses OpenMI, a standard for exchanging data between simulation software programs at run-time, to form a coupled chemical-mechanical-hydrogeological model of the system. The approach combines a tunnel scale stress analysis finite element model, a discrete element model for accurately modeling the patterns of emerging cracks in the concrete, and a finite element and finite volume model of the chemical processes and alteration in the porous matrix and cracks in the concrete, to produce a fully coupled model of the system. Combining existing detailed simulation software in this way with OpenMI has the benefit of not relying on simplifications that might be necessary to combine all of the modeled processes in a single piece of software.

**Keywords:** radioactive waste; long-term performance; degradation; modeling; finite element analysis


---

## 1. Introduction

The assessment timescales associated with geological disposal concepts for high-level radioactive waste and long-lived intermediate level and low-level wastes necessitate the long-term modeling of the near-field engineered barrier system (EBS) in order to demonstrate long-term safety of disposal (e.g., Carter et al. [1]). Interactions between engineered materials, the host formation and in situ groundwaters, together with the heat generated by the disposed waste, drive evolution within the system throughout the assessment period. To demonstrate a good understanding of the expected evolution of the disposal system and to help to understand risks associated with conceptual and parametric uncertainties, it is necessary to represent in models, at a sufficient level of detail, the couplings between the thermal, hydrogeological, mechanical and chemical (THMC) processes that control these interactions.

Traditionally, disposal system performance has been evaluated using models in which the actual design is stylized and/or simplified, usually to meet the requirements of being compatible with existing modeling tools and approaches or with existing data. This approach is not always suitable for evaluating the impact of different site conditions and engineering designs on the performance of geological disposal systems because the time-spatial representation of the model may have been simplified, or the model input parameters may have been set very pessimistically. Disposal facilities constructed in deep geological formations will take time to evolve towards a long-term stable state. Often, the only way to reflect this change in a stylized and simplified model is to set the input parameters very conservatively from the start.

In order to reflect differences in site conditions and engineering designs and to conduct a more realistic and reliable performance evaluation of geological disposal systems in the long-term it is therefore necessary to have a method for evaluating changes in geological disposal systems over a wide range of time and spatial scales, and to reflect these changes when assessing the potential nuclide migration within the site. In this paper, we attempt to address some of these issues by first focusing on the detailed representation of physical processes in models at the near-field scale.

In the performance evaluation of geological disposal, models that represent individual important phenomenon, or small subsets of relevant phenomena, are typically used to develop simulations. However, between the individual phenomena that make up the geological disposal system behavior, there are potentially positive and negative feedback loops (e.g., chemical-mechanical-mass transport feedbacks). Chemical interactions between pore waters and porous materials (e.g., cements, clays, host rocks and the waste form itself if containment is breached) can lead to the variation in time of chemical properties such as pH and Eh and concentrations of key solutes, which can affect the thermodynamic stability of materials. Disequilibrium between pore waters and porous materials will cause dissolution and/or precipitation of primary and secondary minerals, with subsequent alteration of porosity leading to modified rates of diffusive and/or advective transport of solutes, whose availability will control overall rates of chemical change. Stresses can develop in solid phases due to development of expansive secondary minerals (e.g., corrosion products) or modification of mechanical properties due to chemical and/or porosity alteration, which can contribute to deformation and cracking of materials such as concrete. When a crack occurs, it will affect the groundwater flow and therefore the mass transport and the subsequent chemical state along the new crack pathway. This may promote further change of physical properties and therefore induce further localized mechanical structural changes. Since these feedbacks are nonlinear, it is not possible to determine the behavior of the entire system by combining the results of individual phenomenon simulations of all individual barriers within the EBS. Additionally, since phenomena arising on macroscopic scales, such as cracking, will be affected by and will themselves affect phenomena occurring at the scale of the whole facility, such as groundwater flow and mechanical deformation, it is necessary to model the EBS at multiple scales.

Several reactive transport modeling tools, such as PHREEQC [2], PFLOTRAN [3] and Geochemist's Workbench [4], are in common use in the radioactive waste disposal modeling community. These are not 'single phenomena' tools, since they include several complicated coupled processes. However, these tools are not currently capable of simulating the detailed mechanical interactions that are required in order to predict the onset, location and geometry of cracking of cements alluded to above, with the associated coupled effects on groundwater flow that would arise. Similarly, standard stress analysis tools such as Abaqus [5] and COMSOL Multiphysics® [6] are not currently capable of simulating the detailed reactive transport that gives rise to alteration in material properties, and also tend to only model macroscopic development of stress and are not typically used to predict crack geometries.

There are various reasons why computer models typically only attempt to model a subset of the potential process couplings. Firstly, different numerical discretization methods are typically favoured for different processes. The finite volume (FV) framework is convenient for simulating transport processes, including reactive transport process, due to the inherent local mass balance that it provides (e.g., [7]). On the other hand, mechanical processes in continuous materials have historically been found to be well represented using finite element (FE) methods (e.g., [8]), although at smaller scales the discrete element method (DEM) has been found to be an effective approach for modeling discontinuous materials, or the cracking of continuous materials as they become discontinuous, and is capable of predicting the geometry of any potential cracks (e.g., [9]). Secondly, due to the differing complexity of solving for different process, the degree of refinement that can be afforded in grids is different for different physical processes. Crudely speaking, some processes are easier to solve than others, and so can be solved on a more refined grid in acceptable run times. As an added complication, some software will include grid refinement capabilities (e.g., [10]), allowing either fronts (regions where solution gradients are greatest), or discontinuities that develop in the underlying media to be better represented. Thirdly, focusing on a select subset of processes and couplings allows optimized numerical solution approaches to be used, whereas for general 'multi-physics' simulations, only generic solution approaches are typically applicable. Again, combining different time stepping approaches in a single computer model is not impossible but would add undesirable complexity, and some method would be needed to handle the 'out of step' transfer of information between the processes to represent couplings. Lastly, including many physical processes and their couplings in a single software application complicates the process of development and maintenance.

Much time has been invested in developing tailored computer software for solving problems of relevance to the radioactive waste disposal community involving subsets of the processes listed above. Often, they target only a single process, but because of this can offer great flexibility in terms of the parameterization of the model and the solver options that are available, so that a sufficiently complex problem can be represented and reliably solved in an acceptable time. Ideally, some benefit, other than just the intellectual knowledge gained in the development of these codes, would be obtained from the investment so far when attempting to develop new computer models that include additional processes and couplings. Future modeling simulators would benefit most obviously from these existing codes if they could be used directly without modification.

In this paper, details are presented of the development of a coupled modeling simulator, using OpenMI [11], for modeling the evolution in the near-field engineered barrier system for disposed radioactive waste. OpenMI is a software interface standard that allows pre-existing computer models to be coupled at run-time through the exchange of data to represent process couplings. The simulator has been applied to a geological repository design for transuranic (TRU) waste that is based on the Japanese disposal concept [12], and preliminary results on the development of cracks in the engineered barrier are presented. The simulator is extensible and could in future be supplemented with components to address specific modeling needs of other disposal concepts.

The example application of the coupled modeling focusses on simulating the coupled chemical-hydro-mechanical (CHM) evolution of the entire TRU Group 3/4 EBS [12]. It does not currently include coupled effects of thermal evolution processes, although these could be added in a future update, given the extensible approach that has been followed. The future inclusion of thermal processes might be expected to modify physical property values and rates of reaction, but general trends of behavior are likely to be similar. The example simulates the expected cracking of the cementitious backfill, the subsequent evolution of the cracks and reactive transport of solutes through them, their chemical interactions with the nearby backfill and the effect of these interactions and potential feedbacks.

The structure of the paper is as follows. In Section 2, an overview is given of OpenMI and of the simulation tools GARFIELD-CHEM, MACBECE and DEAFRAP that have been coupled using OpenMI in the current work. In Section 3, a description of the GARFIELD-CHEM reactive transport software is given. This is a new simulation tool that has been developed in the current work to specifically cope with the need to model the transport and reaction of solutes in domains that can undergo cracking. Section 4 provides details of the MACBECE and DEAFRAP mechanical analysis software. In Section 5, details of the application of the model to the Japanese disposal concept for TRU waste is given and results of the modeling are presented in Section 6. High-level outcomes of the approach are discussed in the conclusions in Section 7.

## 2. Modeling Software Used and Coupling with OpenMI

One possibility for allowing pre-existing computer models to be used as building blocks in the development of more complex simulation tools is to include them in a modeling framework that supports the exchange of data between models at run-time in order to represent process couplings. For example, a thermal model in such a framework could make the temperature in all locations in a modeled domain available to a second model in the framework. Provided that the second model had a capability for inputting a spatially- and time-dependent temperature then it would not need to solve for the temperature itself. If the model did not have such a capability, then it would be relatively easy to add, compared to the additional effort that would be required to add a thermal model and solver. This would provide useful added functionality to the second model, allowing temperature-dependencies of its input parameters to be represented, and the combined framework would be capable of representing a coupling that neither model could represent individually. Furthermore, if the spatial data exchange did not require each model to be using the same underlying grid to represent the domain, then different degrees of grid refinement could be used in each model. Each model in the framework could continue to use its own numerical discretization and time integration approaches, which would be optimized for its particular purposes. The framework would be required to coordinate calls to each model to solve for a particular 'global' time step and exchange data at appropriate times, at a frequency sufficient to represent the details of the coupling. A framework that possessed these capabilities would address most of the issues discussed earlier regarding the development of highly coupled models.

Frameworks such as that described above can be constructed using available software tools, such as OpenMI [11]. OpenMI is a standard that defines a set of software interfaces that allow models to exchange data at run-time. It was originally developed to simplify the linking of models used in surface hydrology simulations and was initially funded by the HarmonIT European Commission Fifth Framework Project. Computer models that implement the necessary OpenMI interfaces are OpenMI compliant and can, without any additional programming, be configured to exchange data at run-time. An advantage of this approach to coupling is that each model in the coupled system is likely to have undergone separate verification of their process modeling capabilities, so that the only additional verification that should be necessary would relate to the new code to integrate the model in the OpenMI system.

The coupled simulator that has been developed combines three computer modeling codes MACBECE (Section 4.1), DEAFRAP (Section 4.2) and GARFIELD-CHEM (Section 3). In order to describe the coupling with OpenMI that has been implemented, brief details of the three computer codes are given below.

MACBECE (Mechanical Analysis considering Chemical transition of BEntonite and CEment materials) [13–15] is a software application, written in Fortran, that has been developed to calculate the long-term mechanical behavior of the TRU tunnel and uses a finite element discretization approach. Although focused on mechanical modeling, and specifically not including any capabilities for modeling chemical evolution, the inputs to MACBECE can be parameterized in terms of the degree of chemical alteration of the repository components. For example, the mechanical properties of cement features can be characterized in terms of the porosity and the leaching of key minerals, which will vary as the cement undergoes reactions with the in situ groundwaters (e.g., due to changes in pH, carbonate concentration and temperature). MACBECE represents the effect of corrosive expansion of steel materials in the waste form and the effect of rock creep in its stress calculation, but it can only estimate the number of cracks that may arise in each modeling element when materials become overstressed and cannot directly calculate the patterns and connectivity of these cracks, which will ultimately control subsequent flow and transport through the material.

DEAFRAP (Discrete Element Analysis for FRActure Propagation) [9] is a second mechanical modeling application, written in Fortran, that is used in the coupled system and is based on the discrete element method. While MACBECE is used to calculate mechanical loadings and stresses at the TRU tunnel scale, DEAFRAP is used at discrete locations within the tunnel to predict the patterns of emerging cracks, using the displacements calculated by MACBECE as boundary conditions at the smaller scale. The discrete element method is a numerical simulation method that can be used to model the bulk behavior of granular materials. It represents media using a large number of small particles. The approach taken is similar to that used in molecular dynamics simulations, but takes into account particle shapes and contact forces, and includes rotational degrees of freedom. It is particularly suited to the purpose of crack prediction, since it allows discontinuities to develop between particles (e.g., [9]). DEAFRAP is also parameterized in terms of the degree of chemical alteration of the repository components in a similar way to MACBECE.

GARFIELD-CHEM (Grid Adaptive Refinement FInite ELement Discretization—for CHEMistry) is a C++ software application for simulating reactive geochemical transport. It has been designed to model various processes of relevance to the chemical evolution and alteration of the TRU EBS system, including thermodynamic equilibrium between aqueous solute species (basis and complex species, in the usual reactive transport modeling terminology), kinetic dissolution and precipitation of primary and secondary minerals, kinetic treatment of ideal solid solutions and coupled porosity alteration. It uses a mixed finite element discretization, allowing mineral species to be modeled discontinuously in space, in order to respect solid composition discontinuities at media interfaces within the system, while modeling solute concentrations continuously in space. In the finite element grid, diffusive transport is simulated, with transport parameters being parameterizable in terms of the evolving porosity.

The domain represented by the finite element grid in GARFIELD-CHEM is assumed to be continuous. However, time-dependent cracking patterns can be specified, and represented in GARFIELD-CHEM. For this, GARFIELD-CHEM uses a separate finite volume crack model for simulating flow, advective transport and thermodynamic equilibrium of aqueous solutes in a crack network. The crack model is fully coupled to the grid model for porous media transport and reaction. Diffusion between the crack model and the underlying finite element grid cells is represented to allow interaction between crack porewaters and matrix porewaters and solid phases. The finite element grid can also automatically adapt to track solute concentration fronts in the porous media, which allows heterogeneous (liquid-solid) interactions to be modeled more accurately.

The three computer programs, MACBECE, DEAFRAP and GARFIELD-CHEM, initially being standalone applications, were modified to make them 'OpenMI linkable components' [11]. The modification process takes the form of developing a thin 'wrapper' layer that implements OpenMI interfaces, allowing the wrapper to communicate within the OpenMI framework, and convert instructions from OpenMI into native instructions that are passed on to the models. The key instructions essentially inform the solver engines to initialize, to perform a timestep (for which the model chooses an appropriate step size, it is not imposed), to return the current time, to set input parameter values and to get output data values. The wrapping layer for each model declares the input and output data exchanges that each model either accepts or provides. More details on the wrapping of existing programs to make them OpenMI linkable components is given in [11].

The wrapped models are managed within the OpenMI framework by a controller application that is responsible for configuring the input and output data connections between the models at run-time, allowing a coupled representation of the processes that occur within the whole system to be defined. The GUI controller application, referred to as the 'OATC Configuration Editor', from the OpenMI v1.4 reference implementation [16] was used to build the coupled process model using the three OpenMI compliant DLLs, by defining the data exchanges to be performed.

OpenMI uses a 'pull mechanism' approach to data exchange [11]. Components pull (request) the data that they need (model parameters or boundary conditions) in order to perform a timestep from another component that provides it (as an output). The arguments of the pull mechanism specify the locations and times for which a component requires the data. The model receiving the request for data will perform timesteps until the data can be provided at the requested time and may request any data that it needs to perform the step from other models that can provide it. The pull mechanism therefore results in a chain of requests for data which lead to timesteps being performed in order to calculate the requested data.

Each component is responsible for determining the appropriate time step size to attempt and can base this on the state of any internal variables or can simply take a fixed time step. Internally, the model engine may take smaller timesteps, for example in order to satisfy stability criteria for the numerical time stepping methods that they implement, but it will only return output data values via OpenMI corresponding to the time step defined at the component level. The component level step therefore does not affect the accuracy of the solutions from each model, but it does determine the frequency of exchange of data with other models and therefore does control the 'system-level accuracy' with regard to the coupling between the models.

The input and output data exchanges that are accepted as input or provided as output by each model in the coupled TRU system model are as follows:

- GARFIELD-CHEM
  *provides as output:* chemically-altered porosity, $\theta_{chem}(x,t)$ (−); and the leaching ratio $C_i(x,t)/C_{i,0}$ data, where $C_i$ (mol/kgw) is the concentration of the $i$th mineral species that is simulated, for $i = 1, \ldots, N_m$, where $N_m$ is the number of mineral species that are simulated; *accepts as input:* effective diffusion data, $D_{eff}(x,t)$ (m²/s), which overrides the in-built effective diffusion model; and information regarding timings and locations of cracks that appear in the system, fluxes of water through them and details of crack growth.
- MACBECE
  *provides as output:* displacement data; and $D_{eff}(x,t)$ based on mechanical porosity change; *accepts as input:* $\theta_{chem}(x,t)$; and $C_i(x,t)/C_{i,0}$ data, which it uses to parameterize material properties.
- DEAFRAP
  *provides as output:* timing and locations of cracks that appear in the system and details of crack growth. *accepts as input:* $\theta_{chem}(x,t)$; $C_i(x,t)/C_{i,0}$ and displacement data.

These information exchanges between the three models are shown schematically in Figure 1.

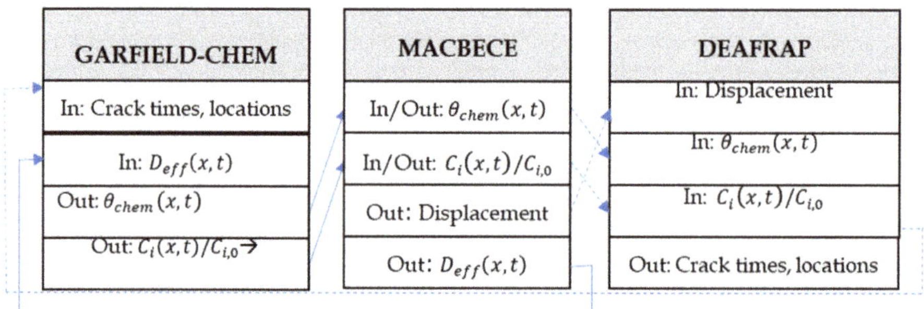


**Figure 1.** Information exchanges between the models. 'In' denotes key input data, 'Out' denotes key output data. Solid and dashed blue lines indicate the tightly and loosely coupled data exchanges respectively.

GARFIELD-CHEM and MACBECE are 'tightly' coupled using the automatic OpenMI pull mechanism to form a coupled C-M model of the TRU tunnel system. DEAFRAP calculations are computationally expensive and so are not appropriate for direct inclusion in the tightly coupled loop. Instead, DEAFRAP is included in a 'loosely' coupled loop as follows.

At each MACBECE calculation step in the GARFIELD-CHEM-MACBECE loop, stresses are monitored. MACBECE is used to perform continuum mechanical analysis using the finite element method, while DEAFRAP uses the Discrete Element Method to identify locations and patterns of crack formation. When threshold stresses are encountered that are expected to lead to crack formation or to the growth of previously identified cracks, the tightly coupled GARFIELD-CHEM-MACBECE loop is paused. The displacement computed by MACBECE are then passed into DEAFRAP in order to confirm the emergence of cracks and to determine their locations, or to compute the growth of previously identified cracks. Cracks are only of interest to GARFIELD-CHEM if they are found to connect with the boundary of the TRU tunnel (where natural host rock groundwaters will be present), since before this time they will not be advective and will not connect with groundwaters that are not already in equilibrium with the cementitious materials. If one or more cracks is found to connect with the tunnel boundary, then the location and dimensions of the cracks are passed to GARFIELD-CHEM, using OpenMI. GARFIELD-CHEM then constructs a crack network model to represent the cracks, and the porous media grid is adaptively refined to resolve the locations of the cracks to a specified resolution, so that diffusive chemical alteration in the cement matrix neighboring the cracks can be represented to required spatial accuracy. The GARFIELD-CHEM-MACBECE tightly coupled loop is then restarted and continues until threshold stresses are again encountered in the MACBECE stress monitoring, when the tight loop is paused again and the DEAFRAP analysis is repeated. The detail of the relation between MACBECE and DEAFRAP will be described in Section 4.3.

## 3. The GARFIELD-CHEM Reactive Transport Simulator

Some more detail is given here of the GARFIELD-CHEM software. GARFIELD-CHEM is used to solve for the reactive transport of solutes in the various media components comprising the TRU tunnel system.

### 3.1. Reactive Transport Equations

The set of aqueous solute species, $S$, that are included in the model is partitioned into basis and complex species. Basis species represent the fundamental 'building blocks' of the chemical system, with all other aqueous complex species and minerals in the system being expressible as reactions involving only basis species (see, e.g., [17]). The basis species and complex species concentrations in the pore water are denoted $b_i$ (mol/kg), $i = 1, \ldots, N_B$, and $c_i$ (mol/kg), $i = 1, \ldots, N_C$ respectively, where $N_B$ and $N_C$ are the number of basis and complex species that are included in the model. Minerals species concentrations per unit volume of solid are denoted $m_i$ (mol/m$^3$), $i = 1, \ldots, N_M$. The total concentration of basis species $i$, $B_i$ (mol/kg), is given by:

$$B_i = b_i + \sum_{j=1}^{N_C} \alpha_{ij} c_j \tag{1}$$

Here, $\alpha_{ij}$ is the stoichiometry of basis species $i$ in the reaction for complex species $j$, when expressed purely in terms of basis species.

The reactive transport equations are then given by:

$$\frac{\partial}{\partial t}(\rho \theta_{chem} B_i) = \nabla \cdot \left(\rho D_{eff} \nabla B_i - \rho \underline{q} B_i\right) - \sum_{k=1}^{N_M} \beta_{ik} R_k \quad i = 1, \ldots, N_B \tag{2a}$$

$$K_j = \left[c_j\right]^{-1} \prod_{i=1}^{N_B} [b_i]^{\alpha_{ij}} \qquad\qquad j = 1, \ldots, N_C \tag{2b}$$

$$\frac{dm_k}{dt} = R_k \qquad\qquad k = 1, \ldots, N_M \tag{2c}$$

Here, $\rho$ (kg/m$^3$) is the density of the pore fluid (assumed constant), $\theta_{chem}$ is the porosity available for transport, $D_{eff}$ (m$^2$/s) is the effective diffusion coefficient of the porous medium, $\underline{q}$ (m/s) is the Darcy velocity, $\beta_{ik}$ is the stoichiometry of basis species $i$ in the reaction for mineral $k$, when expressed purely in terms of basis species, $R_k$ (mol/m$^3$/s) is the rate of reaction per unit volume (positive for precipitation and negative for dissolution) of mineral $k$, and $K_j$ (−) is the thermodynamic equilibrium constant for complex species $j$. $[a_i]$ ($a = b$, or $c$) denotes the activity of aqueous species and is given by $[a_i] = \gamma_i a_i$, where $\gamma_i$ is the activity coefficient. Equation (2a) is a conservation equation for the total concentration of the basis species, incorporating transport of solutes and source/sink terms arising due to mineral dissolution/precipitation. The use of the total concentration in the equation arises as a consequence of assuming that the aqueous complex species are always in instantaneous thermodynamic equilibrium in the porewater. Equation (2b) is a mass action equation for the complex species reactions and Equation (2c) models the change in mineral abundance due to dissolution and precipitation.

The appropriate method of calculating activity coefficients depends on the application being considered. For very dilute solutions, the approximation $\gamma_i = 1$ is sometimes made. In dilute systems up to a few tenths molal, the Davies model is reasonably accurate (see e.g., [18]). For solutions up to ~3 molal the B-dot model (e.g., [19]) is applicable, and the Pitzer model [20] can be used for more concentrated solutions.

The rate of reaction per unit volume, $R_k$ of mineral species is expanded as:

$$R_k = \tilde{R}_k A_k W_k m_k \tag{3}$$

where $\tilde{R}_k$ (mol/m$^2$/s) is the specific rate of reaction per unit area of mineral $k$, $A_k$ (m$^2$/g) is its specific reactive surface area and $W_k$ (g/mol) is its molar weight. It is usually the case that $\tilde{R}_k$ is defined in models assuming transition state theory (TST), where the rate of reaction is expressed in terms of the magnitude of the porewater chemistry's departure from equilibrium between reactants and products in the mineral hydrolysis reaction. However, alternative models, including directly imposed rates of dissolution or precipitation can be applied.

Effective diffusion is modeled using Archie's law [21] to couple rates of diffusion to the porosity, which evolves due to the degree of chemical alteration of the pore space, and is given by:

$$D_{eff}(\theta) = D_{eff,0}\left(\frac{\theta}{\theta_0}\right)^m \tag{4}$$

where $D_{eff,0}$ (m$^2$/s) is the initial effective diffusivity or the intact material with porosity $\theta_0$ and $m$ (–) is a cementation factor.

Mineral precipitation and dissolution in the medium will lead to chemical alteration at a rate

$$\frac{d\theta_{chem}}{dt} = -\sum_{k=1}^{N_M} R_k V_k \tag{5}$$

where $V_k$ (m$^3$/mol) is the molar volume of mineral $k$.

GARFIELD-CHEM uses the globally implicit approach to solving the reactive transport Equations (1)–(5) [17]. The IDA solver from the SUNDIALS solver suite [22] is used to solve the coupled system of equations.

GARFIELD-CHEM has a secondary run mode, referred to as GARFIELD-RNT. In this mode, the software can reload the time-dependent outputs of a previous simulation, including porosity and pH evolution and the time-history of the grid, and can replay this history while simulating transport of radionuclide species that leach from the waste form, and which can undergo decay and ingrowth, retardation and solubility limitation, and whose transport properties can be parameterized in terms of the reloaded chemical evolution (in particular as functions of porosity and pH). The approach of reloading the results of the (computationally intensive) coupled CHM simulation allows the simpler radionuclide transport calculation to be performed repeatedly to undertake sensitivity analysis.

*3.2. Discretization*

Initially, all components in the TRU domain are assumed to be modeled as continuous porous media. At later times, cracks will develop in the cementitious materials, so that the modeled domain becomes a combination of porous media and cracks. The reactive transport Equations (2a)–(2c) are solved in both the porous media and in the cracks that emerge. Solutes are assumed to be transported by diffusion in the porous media regions and by advection in the cracks.

Mineral reactions are only modeled in the porous media. The reason for excluding mineral reactions in the cracks is that as minerals precipitate the cracks may become clogged, which would inhibit transport of solutes and potentially reduce solute transport rates to zero if the cracks were to become fully clogged. Assuming that no mineral reactions take place in the cracks therefore implies that the cracks remain open and 'fully transmissive', which is conservative from both the perspective of maximising the amount of interaction of the cementitious materials in the TRU tunnel with the host rock groundwaters that enter the cracks, which is likely to lead to maximal loss of mechanical strength (and therefore promote further cracking), and from the perspective of maximising rates of transport of radionuclides from the waste form in any subsequent radionuclide transport modeling. Mineral alteration processes within cracks in cementitious media has been considered in other studies, e.g., [23], where the precipitation and dissolution of minerals in single cracks is fully coupled to the rates of advective transport in the cracks to derive timescales for clogging. These rates can be reflected in the current model by setting time-dependent flow rates (as snapshots) in the crack model, i.e., a fully clogged crack leg would be assigned a zero flow rate.

The diffusive transport and reaction of solutes in the porous media domain is solved using mixed Q1-DG0 finite elements [8]. Q1 elements are used to model the continuous solute concentrations throughout the domain, while the DG0 discontinuous Galerkin elements are used to model the mineral concentrations, which can be discontinuous at interfaces between materials. The Deal.II finite element library [24] is used in the software for the implementation of the finite element calculations and for management of the grid.

Deal.II includes routines for adaptive local grid refinement based on error estimators that identify regions in which the solution varies most greatly or is discontinuous. These routines are used to automatically track alteration fronts in the porous media based on error estimators for the solute species. Regions in which solute concentration gradients vary most greatly are likely to coincide with regions in which minerals are undergoing the most rapid alteration. Automatically refining the grid in such regions, while coarsening it in regions where gradients are small, will lead to a more realistic shape of alteration at interfaces between materials, while maintaining a manageable number of nodes within the grid. For example, Figure 2 shows the pattern of alteration of a square quartz block centred in a domain with a fixed low silica concentration source at the left boundary. The low silica concentration source is expected to gradually dissolve the quartz block. After refinement steps at successive times, the rounding or the corners of the quartz block that would be expected in such an experiment is seen in the model. In contrast, if grid refinement had not been performed, the model would predict uniform dissolution across the whole block at a slower rate than is observed in the refined cells near the corner. Additionally plotted in the figure is the dissolved silica concentration, showing how the grid tracks gradients in the solute concentrations to identify the regions for refinement.

During refinement, neighboring cells in the grid can be refined different numbers of times. This results in nodes on the interfaces between cells that belong to a cell on only one side, which are referred to as 'hanging nodes'. This can be seen in Figure 2. Since continuity of the solute species concentration across cells must be maintained, constraint equations are set at these nodes to ensure continuity. Essentially the solution at the hanging node is set to be an average of the concentrations at the neighboring nodes on the common interface, which would match the midpoint concentration of the Q1 element solution in the larger neighboring cell. The same simple averaging approach is used for all solute species, which includes both basis and complex species, which means that the mass action Equation (2b) is not expected to be satisfied at hanging nodes. This will lead to small errors with respect to the thermodynamic system in the computed aqueous species concentrations at hanging nodes, however this is not expected to impact greatly upon the overall solution and is acceptable given the added realism in the solution that is provided by the adaptive gridding.

The timing and location of cracks that appear in the modeled domain can be specified directly in the GARFIELD-CHEM input file or can be defined at run-time via OpenMI. Regardless of the way in which the cracks are specified, they are represented in the software as collections of crack legs that are connected at junctions, as shown in Figure 3. The crack model is separate from the grid used to simulate the porous media portions of the domain. Crack legs are allowed to bifurcate in the direction of flow, but not recombine. Each crack leg is piecewise linear, and within each leg a constant volumetric flow rate is assumed over each timestep. Crack flow rates can be updated at discrete times during the evolution of the system (again either via the input file or at run-time using OpenMI). Subscript index notation $i, j$ is used to indicate crack legs $L_{i,j}$ that bifurcate from preceding leg $L_i$ (where the index $i$ can itself be a tuple). Thus, the number of values in each index reflects the number of times that the crack pathway has bifurcated to reach the leg (minus one). For example, in the crack pathway illustrated in Figure 3, the pathway bifurcates twice before the crack leg $L_{0,1,1}$ is reached.

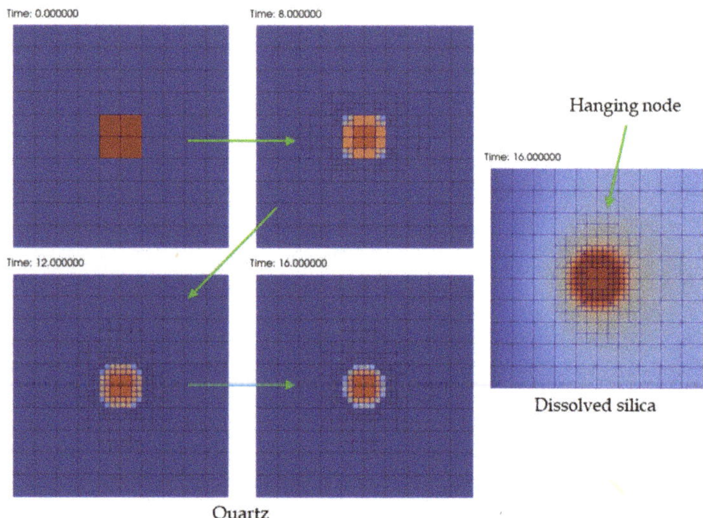


**Figure 2.** Adaptive refinement around a square quartz region in a low silica concentration domain. A fixed low concentration silica source is present at the left boundary. The left-most four plots show how refinement and coarsening at successive times leads to more plausible shapes of alteration of the quartz block than would be obtained with a fixed grid. The corresponding silica solute concentration at the final time is shown in the right-hand plot. An example hanging node is also noted.

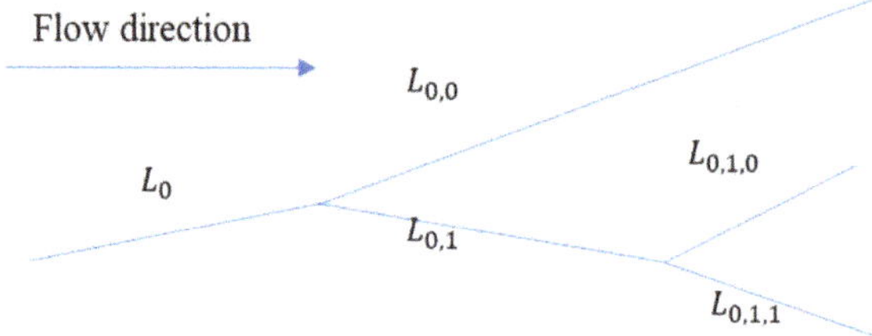


**Figure 3.** Example fracture network showing bifurcation of an initial crack leg $L_0$ into subsequent crack legs.

If the volumetric flow rate in crack leg $L_i$ is $v_i$ (m$^3$/s) and if the leg bifurcates into $n$ further legs, then the volumetric flow must be conserved across the junction, so:

$$v_i = \sum_{j=1}^{n} v_{i,j} \tag{6}$$

GARFIELD-CHEM does not calculate flow rates and only checks that the relationship (6) is satisfied.

Solute fluxes between the finite volume crack model and porous media finite element grid model are represented in the software. The resulting interactions are expected to lead to alteration of minerals adjacent to the cracks, and so GARFIELD-CHEM automatically adapts the grid when cracks are imposed to resolve the crack locations to within a specified tolerance, $x_{tol}$. The grid adaption algorithm in this case is different to the front tracking algorithm discussed earlier and proceeds as follows. The crack pathway is overlaid on the grid and cells that the pathway intersects are identified, as shown in Figure 4. The cells intersected by the pathway are repeatedly refined until the cells that the crack intersects become sufficiently small; the cells are refined until

$$\frac{d_{cell}}{w_{crack}} < x_{tol} \tag{7}$$

where $d_{cell}$ (m) is the diameter of the cell (the largest diagonal of the cell), $w_{crack}$ (m) is the width of the crack. The process results in successively refined cells closer to the crack, as shown in Figure 4.

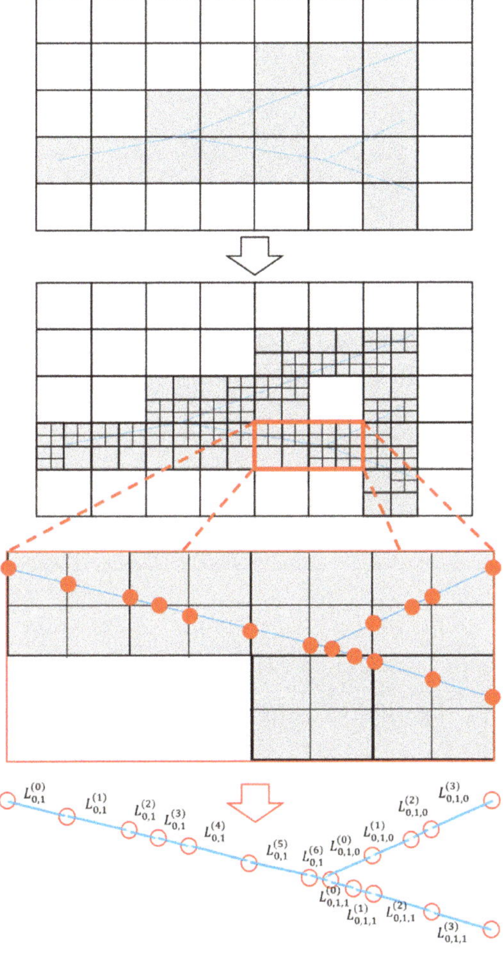

**Figure 4.** Grid refinement and discretization of each crack pathway leg into segments. The cells intersected by the pathway are repeatedly refined until the cells that the crack intersects become sufficiently small. Each segment intersects only one grid cell.

The software allows separate set of media transport properties (i.e., diffusion coefficients) to be specified for cells that are ultimately intersected by cracks, in order to represent micro-cracking damage at scale below that which can be represented directly. These properties will ultimately control rates of transport between the crack and the grid cell, as discussed below.

Once the grid is refined to resolve the crack pathways to the required resolution, the crack pathway is refined into segments such that each segment intersects only one grid cell, as shown in Figure 4. The notation $L_i^{(n)}$ is used to indicate the $n$th segment in the $i$th crack leg (where $i$ is a tuple). The crack refinement algorithm proceeds by splitting each leg at the points where they intersect the grid cell boundaries to define the segments and then computes the length of each segment.

Advection of all solute species (basis and complex species) is modeled in the crack pathway, as well as thermodynamic equilibrium of complex species. No mineral dissolution/precipitation reactions are assumed to occur in the cracks, as discussed earlier. Instead, upscaled interactions are represented by crack-grid interactions in the cracked grid cells. Thus, in the cracks the system Equations (2a)–(2c) is solved, with $R_k = 0$.

The transport in the crack pathway is modeled using a finite volume approximation to simulate transport. In a given crack pathway segment, denoted with index $k$, the transport of the $i$th reactive basis species is given by

$$
\begin{aligned}
V_k \frac{d}{dt}\left( b_i^{(k)} + \sum_{j=1}^{N_C} \alpha_{ij} c_j^{(k)} \right) &= S_i^{(k)} + Q_k^{UP}\left( b_i^{UP(k)} + \sum_{j=1}^{N_C} \alpha_{ij} c_j^{UP(k)} \right) \\
&\quad - \sum_{SEG \in DOWNSTREAM(k)} Q_k^{SEG}\left( b_i^{SEG} + \sum_{j=1}^{N_C} \alpha_{ij} c_j^{SEG} \right), \; i = 1, \ldots, N_B
\end{aligned}
\tag{8}
$$

Here, $V_k$ (m$^3$) is the volume of the crack segment with index $k$ ($k$ = crack width × segment length × domain thickness), $Q_k^{UP}$ (m$^3$/s) is the volumetric flow rate immediately upstream of the crack pathway segment with index $k$, and DOWNSTREAM(k) denotes the collection of segments that are downstream of the segment with index $k$ (there can be more than one when the segment with index $k$ is the final segment in the leg before a bifurcation point), with $Q_k^{SEG}$ denoting the downstream volumetric flow rate in each segment in DOWNSTREAM(k). The term $S_i^{(k)}$ (mol/s) is a source/sink terms representing transfers between the grid and the crack segment with index $k$, which is defined as follows.

By construction, each crack segment intersects a single grid cell, and therefore the source/sink term $S_i^{(k)}$ represents the transfer between a single grid cell and a single crack segment. The geometry of the interaction is sketched in Figure 5 for a single grid cell. The grid cell is assumed to have volume $V_G$ (m$^3$) and to be intersected by a crack with width $W_C$ (m) with corresponding crack segment length $L_S$ (m).

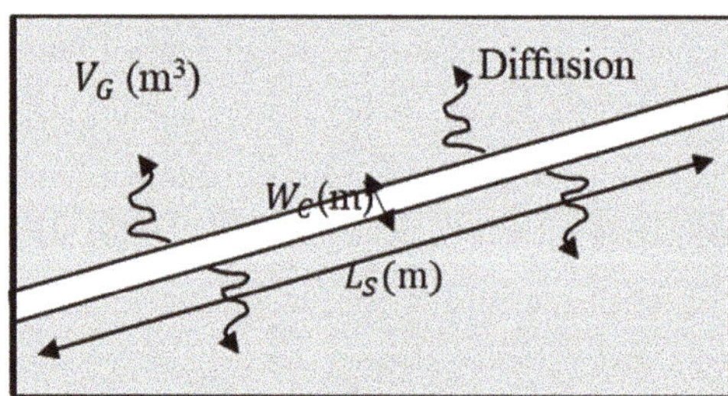


**Figure 5.** Geometry of crack-grid transfer for a single grid cell with volume $V_G$ (m$^3$) intersected by a crack of width $W_C$ (m) and segment length $L_S$ (m).

Transport between the crack and the grid is assumed to be diffusive, with diffusion coefficient controlled by the diffusive resistance of the media in the grid cell. The diffusive flux from the crack to the grid for a reactive chemical basis species is given by:

$$S_i^{(k)} = A_{CG}^{(k)} \frac{D_G(\theta_G)}{L_D} \left( b_i^{(k)} + \sum_{j=1}^{N_C} \alpha_{ij} c_j^{(k)} - b_i^{(G)} - \sum_{j=1}^{N_C} \alpha_{ij} c_j^{(G)} \right) \tag{9}$$

where $A_{CG}^{(k)}$ (m$^2$) is the crack surface area in the grid cell ($=2 \times L_S \times$ domain thickness), $D_G$ (m$^2$/s) is the diffusivity in the crack grid media, which depends on the grid cell porosity, $\theta_G$, $L_D$ (m) is the diffusion length in the grid cell, and $b_i^{(k)}$ and $c_j^{(k)}$ (mol/kg) are the concentration of the $i$th basis species and the $j$th complex species in the segment and $b_i^{(G)}$ and $c_j^{(G)}$ (mol/kg) are the corresponding quantities in the grid cell. The diffusion length in the grid cell is taken to be the half of the grid cell diameter (which is equal to the length of the largest diagonal).

## 4. Effects of Chemical Alteration on Mechanical Parameters

### 4.1. Modeling in MACBECE

MACBECE [13–15] is capable of long-term mechanical analysis of various barrier materials, such as bentonite, metals, rocks and cement. It was developed to allow the effects of the mechanical alteration of the barrier materials on the long-term transitions of the hydraulic field of the near field to be inferred in order to develop more reliable safety assessment models. Cracks in cement materials in the TRU Gr 3/4 tunnel are expected to be generated mainly by corrosion swelling of metal materials, such as the steel containers surrounding the waste, and rock creep of the host formation. The connected crack pathway would strongly affect the water flow and the alteration behavior of cement materials.

For the rock materials, the variable compliance type constitutive model proposed by Okubo [25] was adopted to evaluate the rock creep.

Metal materials are modeled using corrosion swelling elements in MACBECE. The corrosion swelling is expressed by applying the swelling pressure to the metal element as equivalent nodal force. The swelling pressure of the metal materials is calculated from the corrosion swelling rate and the rigidity of the material. Since the actual corrosion occurs on the surface of the metal material, the swelling is assumed to be anisotropic, depending on the thickness of the metal member. Therefore, the corrosion swelling rate in MACBECE is given separately in horizontal and vertical directions to consider the anisotropic corrosion swelling (see Appendix A).

Cement materials are considered to deform due to external force from corrosion swelling of metal materials and rock creep, while reducing its rigidity and strength according to the progress of alteration such as porosity change and calcium leaching. To represent this behavior like strain softening, a nonlinear elasticity model (e.g., [26]) is included in the model.

The rotating crack model [27] and an iterative calculation method for stress redistribution are applied to calculate the strain softening behavior due to reduction of strength and rigidity associated with chemical alteration. In the rotating crack model, a crack occurs when the minimum principal stress of an element reaches tensile strength. The crack generation behavior and the change of mechanical properties due to the failure are expressed in the continuum elements by reducing the rigidity of the element in the direction perpendicular to the crack.

The deformation and the stress state at a time step in MACBECE are calculated by solving the balance of forces considering the material property changes. Chemical transitions of each barrier material that cause the change of the mechanical properties and the operating forces due to additional deformation in the materials are given as input data from other calculation software (i.e., GARFIELD-CHEM).

### 4.2. Modeling in DEAFRAP

DEAFRAP is a two-dimensional mechanical modeling application based on the discrete element method (DEM) [15]. The details of the fundamental DEM algorithm can be found in [9,28,29].

In DEAFRAP, an intact cement material is modeled as a dense packing of small rigid circular particles. Neighboring particles are bonded together at their contact points. If the normal tensile stress or shear stress acting on the cross-section of the bond exceeds the tensile strength or the shear strength, the bond breaks and tensile force never act on the contact point between two particles after the breakage unless the particles become reconnected due to subsequent chemical bonding. Individual bond breakage between bonded particles is defined as a microcrack. A microcrack is generated at the contact point between two particles. In 2D, the direction of a microcrack is perpendicular to the line joining the centres of the particles, and its aperture is calculated as the distance between the two particles. Neighboring microcracks may connect with each other and form a longer crack. These connected microcracks are defined as connected crack pathways in the DEAFRAP simulation.

### 4.3. Relation between MACBECE and DEAFRAP

The rotating crack model used in MACBECE does not recognize each crack individually because it models the behavior of the element containing the cracks as a continuum. In contrast, DEAFRAP can directly represent the grain-scale microstructural features of materials, such as pre-existing flaws, pores and microcracks. These microscale discontinuities induce complex macroscopic behaviors, such as crack generation, propagation, separation and merging, that are independent of complicated constitutive laws. However, the detailed crack analysis by DEAFRAP (and all DEM tools) can be time consuming, and so it is difficult to simulate the entire TRU Gr 3/4 EBS system in acceptable times. Therefore, based on the result of mechanical analysis by MACBECE, DEAFRAP is applied only to the region where the occurrence of cracks has been identified. The displacement calculated by MACBECE is given as the forced displacement to the outer boundary of the DEFRAP model.

Although DEAFRAP itself has no function to calculate chemical alteration such as porosity change, the temporal and spatial distribution of porosity obtained from GARFIELD-CHEM can be taken as input data. From the particle arrangement of the DEAFRAP model, the continuous porosity distribution is converted to the porosity of each particle. The rigidity and strength of bonded particles are changed according to the porosity. The relationship between porosity and mechanical parameters in DEAFRAP is calibrated to be consistent with that in MACBECE.

In an actual cement material, it can be considered that generated cracks may become filled by mineral precipitation and the cracks repaired by the solidification over time. To represent this phenomenon, the porosity at the crack position is recorded at the time of bond breakage in DEAFRAP. The crack is repaired, and the particles are reconnected when the value of porosity reduction at that position reaches a specified value.

## 5. Application to TRU Gr3/4

### 5.1. GARFIELD-CHEM Model

The disposal tunnels in the example application of the TRU Gr 3/4 design feature a large cementitious backfill, whose main function in the safety case is to condition the porewater at a high pH to inhibit corrosion of the steel containers surrounding the waste, and to provide mechanical load-bearing capacity. The backfill is expected to exhibit self-sealing capabilities and to limit radionuclide leaching and transport through sorption, following late-time failure of the metal waste containers. The concrete backfill is expected to provide some capacity to limit the rate of groundwater flow, and hence the release rate of radionuclides after failure of the containers. However, there is a current lack of understanding of the processes governing its long-term degradation and so the potentially low permeability of the backfill is not explicitly accounted for in the current safety case. where it is treated pessimistically, by considering it as a porous media with a permeability similar to sand [12]. A better understanding of the degradation process may allow this pessimistic simplification to be made more realistic in future safety analyses.

The simple geological repository design for transuranic (TRU) waste based on the Japanese disposal concept [12] is shown in Figure 6. The backfill, invert, inner liner and outer liner components are all cementitious, with mineral compositions as shown in Table 1.

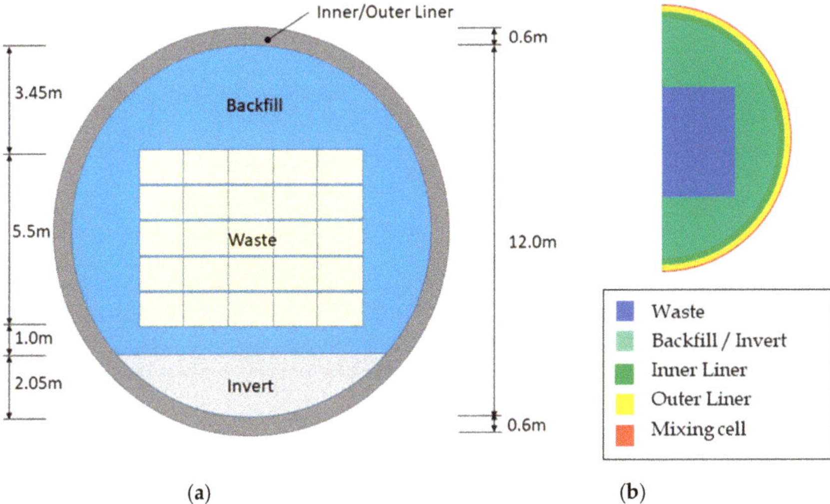


**Figure 6.** Cross-section of the TRU Gr 3/4 tunnel design (**a**) and half system that is modeled with GARFIELD-CHEM, showing the initially coarse grid at t = 0 year (**b**). The grid is constructed using quadrilateral elements and was computed using GMSH [30].

**Table 1.** Composition of cementitious features in the model. Minerals with zero initial concentration in the materials are treated as potential secondary minerals.

| Mineral | Waste (mol/m$^3$) | Backfill/Invert/Inner Liner/Outer Liner (mol/m$^3$) |
|---|---|---|
| Brucite | 0 | 0 |
| Calcite | 0 | 0 |
| CSH165 * | 1734 | 1318 |
| CSH105 * | 0 | 0 |
| Portlandite | 2142 | 1628 |
| Porosity (−) | 0.19 | 0.13 |

* CSH165 or CSH105 is a C-S-H phase with ratio Ca/Si = 1.65 or 1.05.

Thermodynamic parameters used in the modeling of the Gr 3/4 EBS materials in GARFIELD-CHEM are described in Appendix A.

The initial porewater in all tunnel materials was assumed to be identical, with composition as shown in Table 2. This composition was obtained assuming that fresh, reducing high pH (FRHP) groundwater [31] is in equilibrium with the cement material. External water compositions are also shown in Table 2 and were assumed to be present at the outer boundary, and at the upstream end of any cracks that emerge in the system and were obtained by equilibrating FRHP groundwater with calcite.

The external water is not applied directly to the tunnel boundary, since the in situ porewater at the boundary will arise as a consequence of mixing between the cementitious water diffusing from the tunnel and the host rock groundwater. Directly applying a host rock water on the tunnel boundary would therefore be too 'aggressive' in terms of the chemical alteration that can be expected in the liner regions. A mixing cell condition is therefore applied adjacent to the tunnel boundary, where the two waters are allowed to react. A turnover flow rate of host rock groundwater of $3 \times 10^{-9}$ m/s is assumed, corresponding to an assumed head gradient of $10^{-2}$ m/m perpendicular to the modeled domain, and a hydraulic conductvity of the (engineering damaged) rock region adjacent to the tunnel of $3 \times 10^{-7}$ m/s (defined as 100 times larger than that of host rock). The mixing region has a width of 50 mm.

**Table 2.** Water compositions. The external water is based on a calcite-equilibrated calculation with FRHP water [31] and the initial porewater is equilibrated with the cement material for Backfill (Table 1).

|     | Initial Porewater (mol/kg) | External Water (mol/kg) |
| --- | --- | --- |
| pH  | 13.49 | 8.46 |
| C   | $3.83 \times 10^{-4}$ | $3.54 \times 10^{-3}$ |
| Ca  | $7.09 \times 10^{-4}$ | $1.09 \times 10^{-4}$ |
| K   | $2.51 \times 10^{-1}$ | $6.15 \times 10^{-5}$ |
| Mg  | $2.15 \times 10^{-9}$ | $5.00 \times 10^{-5}$ |
| Na  | $2.19 \times 10^{-1}$ | $3.55 \times 10^{-3}$ |
| Si  | $1.18 \times 10^{-5}$ | $3.39 \times 10^{-4}$ |

The parameterization of effective diffusivity (Equation (4)) used in the model is shown in Table 3. $m$ is a cementation exponent and models how the pore network affects diffusivity. A separate parameterization of the diffusivity model is used in cells that contain cracks. This allows the effect of micro-cracks, that are below the scale of the dominant cracks that are modeled, to enhance diffusion.

**Table 3.** Transport properties.

|     | Uncracked | | Cracked | |
| --- | --- | --- | --- | --- |
|     | $D_{eff,0}$ | $m$ | $D_{eff,0}$ | $m$ |
| Waste | $1.79 \times 10^{-11}$ | 3.05 | $4.64 \times 10^{-10}$ | 1.00 |
| Backfill/Invert/Inner Liner/Outer Liner | $4.51 \times 10^{-12}$ | 3.05 | $2.95 \times 10^{-10}$ | 1.00 |

The diffusion model can be overridden at run-time via OpenMI, allowing MACBECE to specify an effective diffusivity based on a combination of the degree of chemical alteration and on the computed mechanical displacement.

The strain-porosity component model for cement materials is shown in Figure 7. $\theta_c$ [-] is the porosity value calculated by GARFIELD-CHEM (i.e., $\theta_{chem}(x,t)$), $\varepsilon_v$ [-] is a volumetric strain (positive value for compression). The model assumes that aggregate material is incompressible so that the porosity value after distortion ($\theta'_c$) can be calculated as follows,

$$\theta'_c = (\theta_c - \varepsilon_v) / (1 - \varepsilon_v) \tag{10}$$

The $\theta'_c$ value is used in GARFIELD-CHEM to recalculate the effective diffusion coefficient using Archie's law (Equation (4)).

For the analysis shown in Section 6, mechanical distortions were found to account for less than 1% of the total porosity. Since this has a negligible effect on the effective diffusion, the coupling to update $D_{eff}$ based on the mechanical evolution was omitted from the model to simplify the coupling in the current analysis. However, it can be reintroduced easily if it is found to be more significant in future applications.

The TRU Gr 3/4 tunnel geometry (Figure 6) was modeled as a half-system, assuming symmetry about the vertical axis. The initial grid that is used, which was computed using GMSH [30], is shown in Figure 6. The initial grid is intentionally coarse, with some cell sizes being tens of centimetres. As cracks appear, automatic grid refinement around the crack locations will lead to a finer grid in regions that are expected to undergo the most alteration, thus focussing computational effort in the regions of greatest relevance to potential pathways for long-term radionuclide transport.

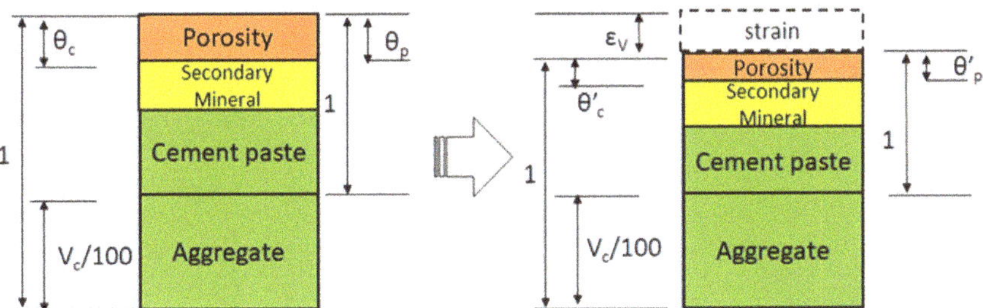



**Figure 7.** Strain-porosity component model for cement materials. $\theta_c$ is the total porosity calculated in GARFIELD-CHEM, $V_c$ is the percentage volume occupied by incompressible aggregate (non-porous) and $\theta_p$ is the corresponding porosity in the porous media fraction of the total volume. After volumetric strain $\varepsilon_V$ is applied, porous media components are distorted and total porosity and porous media porosity are reduced, while the aggregate volume remains constant.

## 5.2. MACBECE Model

The domain considered in the MACBECE stress analysis model of the TRU Gr 3/4 tunnel is shown in Figure 8. To represent the evolving stresses and displacements in the tunnel system it is necessary to include a large region of host rock surrounding the tunnel, and also to represent the internal structures in the waste form. The grid used in the tunnel region is also shown in Figure 8. A combination of quadrilateral and triangular elements is used in the finite element model.

The analysis conditions and parameters used in the MACBECE model are described in Appendix A.

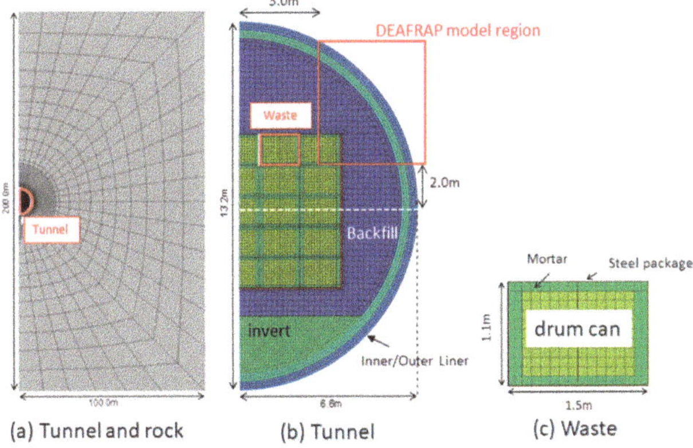



**Figure 8.** Whole model of the TRU Gr3/4 tunnel. The grid is constructed using quadrilateral and triangular elements. (**a**) Simulated tunnel and rock grid, (**b**) Zoomed view showing tunnel grid, (**c**) Zoomed view showing waste grid.

## 5.3. DEAFRAP Model

The analysis conditions of DEAFRAP are based on a case presented in [32]. That analysis aimed to simulate crack generation for the whole-tunnel (Figure 8) and did not consider any direct coupling to a chemical process model. In the modeling presented here, the domain considered by DEAFRAP is not the whole tunnel, but is instead a "corner" region of the tunnel (Figure 9). This corner region is identified by MACBECE, in the coupled loop with GARFIELD-CHEM, as being the region in which the density of cracking is expected to be greatest. Further details are given in Section 6.1.

In the DEAFRAP analysis, displacement data on the boundary must be imposed. These data are derived from the MACBECE analysis. The mechanical parameters for each material is the same as for the MACBECE analysis (Section 5.2). The treatment of chemical alteration information in DEAFRAP is also the same as for MACBECE, that is, the porosity values calculated by GARFIELD-CHEM are assigned to the corresponding DEM particles.

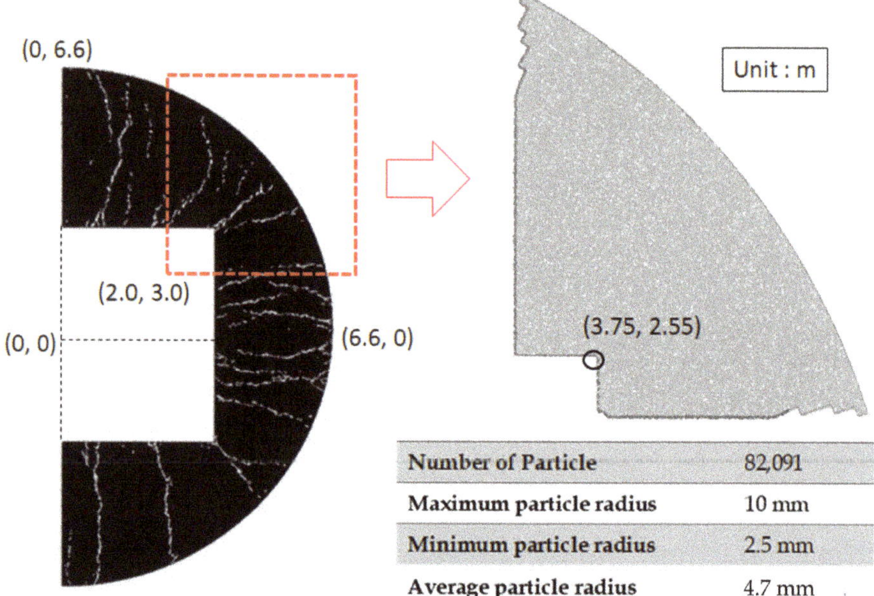


| Number of Particle | 82,091 |
|---|---|
| Maximum particle radius | 10 mm |
| Minimum particle radius | 2.5 mm |
| Average particle radius | 4.7 mm |

**Figure 9.** An example of DEAFRAP analysis for half tunnel (extraction of continuum cracks, left (from [32])) and DEAFRAP system for coupled analysis, showing DEAFRAP particle properties (right).

Figure 9 provides details of the number of DEM particles in the DEAFRAP model and their dimensions.

## 6. Results

Results of a preliminary application of the coupled modeling framework are presented. The results cover the period up to identification of the first crack that fully penetrates the backfill, and a 2000 years period thereafter.

### 6.1. Before Cracking

#### 6.1.1. Chemical Alteration

In this section, results are shown for the period up to 10,000 years. As will subsequently be explained, full penetration of cracks from the waste region to the host rock was found to occur at 10,000 years, and so the results in this section describe the evolution of the TRU Gr 3/4 system prior to cracking.

Prior to cracking, reactive transport within the system is diffusion-limited and so the most significant alteration is expected to occur at the interface between the tunnel and the host rock, where the largest porewater chemistry gradients are present. This can be seen in Figure 10, which shows profiles of calcite, CSH165 (the primary C-S-H phase with ratio Ca/Si = 1.65) and portlandite at 0 and 10,000 years. A front of calcite, which is initially assumed to not be present in the cement regions, precipitates to a depth of 0.15 m due to the supply of carbonate-rich water from the host rock. Calcium for the calcite precipitation is mostly liberated from portlandite dissolution and a smaller amount of CSH165 dissolution, due to the lower pH of the host rock pore water. Portlandite is dissolved beyond the calcite precipitation front, to a depth of 0.30 m, with corresponding precipitation of CSH165 in the region. This alteration is uniform along the boundary with the host rock, as would be expected due to the symmetry of the boundary conditions.

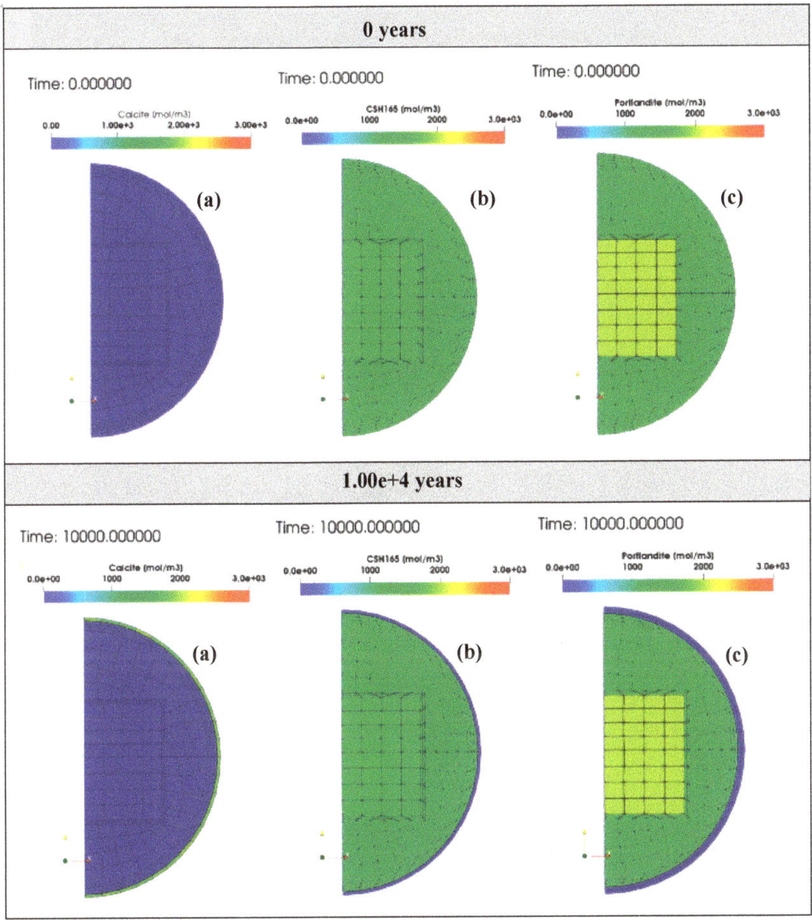

**Figure 10.** Profiles at 0 years (top) and 10,000 years (bottom) of: (**a**) calcite, (**b**) C-S-H 165 and (**c**) portlandite.

### 6.1.2. Porosity Alteration (Chemical) and Effective Diffusion Coefficients

Profiles of porosity and effective diffusion at 0 and 10,000 years are shown in Figure 11. The porosity alteration is calculated purely from chemical alteration (dissolution and precipitation of minerals) by GARFIELD-CHEM; mechanical displacements are not considered. As mentioned in Section 5.1, the coupling of the effective diffusion to mechanical displacement was not considered because the mechanical distortion effect accounts for less than 1% of porosity. Therefore, all porosity evolution in the current analysis arises purely as a consequence of chemical alteration.

By 10,000 years, the porosity value in the outermost 0.15 m is increased from 13% to ~20%, due to the portlandite dissolution. As noted above, calcite precipitation occurs in this region, but the net effect of the alteration is to increase the porosity. The porosity in the region from 0.15–0.30 m is decreased by ~4%, due to the precipitation of CSH165.

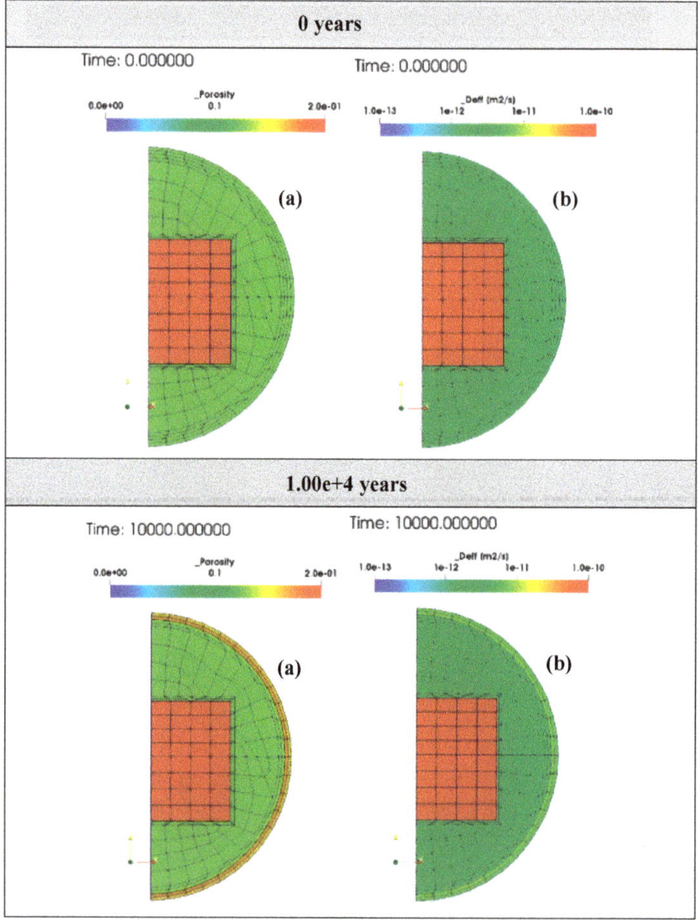

**Figure 11.** Profiles at 0 years (top) and 10,000 years (bottom) of porosity (**a**) due to chemical alteration calculated by GARFIELD-CHEM, and effective diffusion coefficients (**b**).

6.1.3. Crack Distribution

Crack aperture distributions calculated with MACBECE at 5000 and 10,000 years are shown in Figure 12. The crack aperture calculation assumes that calculated distortions result in a single crack per cell. The plots are repeated with the maximum value on the crack aperture scale set to 1 mm and 0.1 mm (i.e., cracks with apertures larger than this size are shown at the maximum plotted aperture). The cracks concentrate in the North-East corner region, with some cracks appearing to fully penetrate the backfill region by 10,000 years. Therefore, the corner region at 10,000 years is selected as a region for detailed cracking analysis with DEAFRAP.

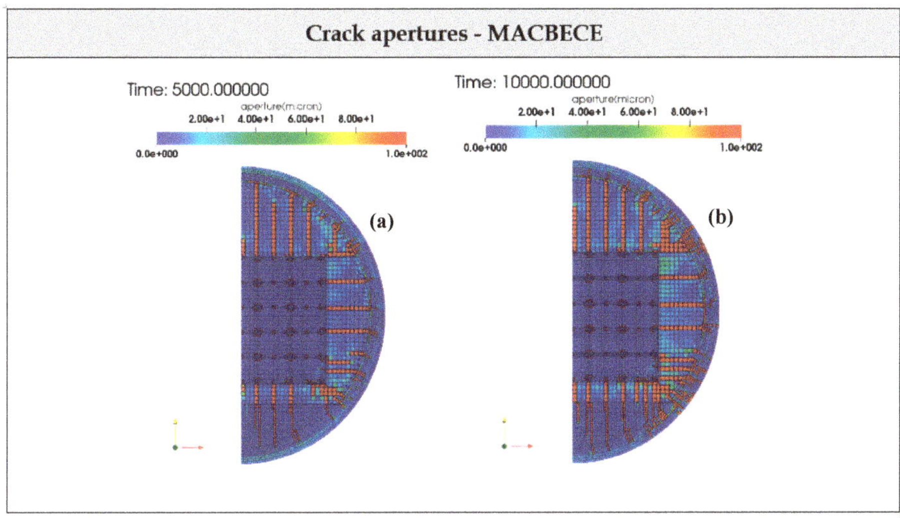

**Figure 12.** Distributions of crack apertures obtained with MACBECE with a maximum plotted aperture of 0.1 mm at 5000 years (**a**) and 10,000 years (**b**). Cracks with apertures larger than 0.1 mm appear to penetrate the backfill by 10,000 years.

The corresponding DEAFRAP analysis is shown in Figure 13. Again, plots are shown with maximum plotted crack apertures of 1 mm and 0.1 mm. The DEAFRAP results confirm that fully penetrating cracks are expected by 10,000 years and identifies potential cracking patterns and pathways. Extraction of continuous crack pathways from the model is discussed in the next section.

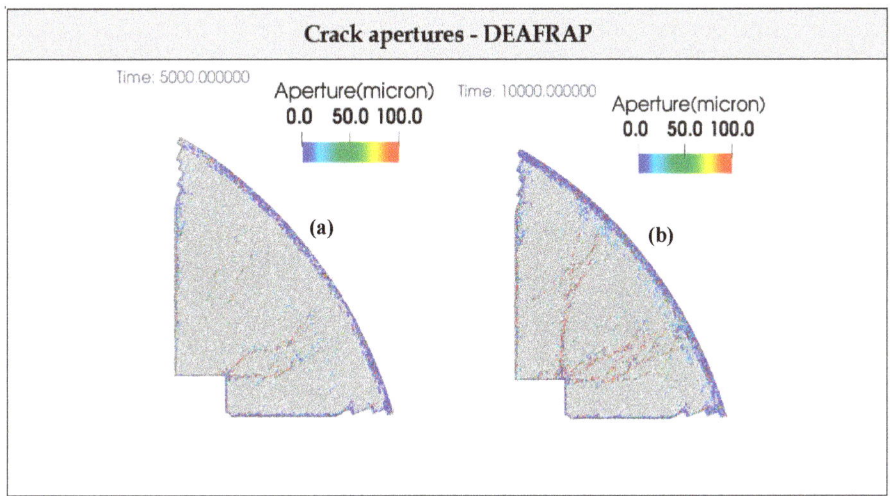

**Figure 13.** Distributions of crack apertures obtained with DEAFRAP with a maximum plotted aperture of 0.1 mm at 5000 years (**a**) and 10,000 years (**b**). Cracks with apertures larger than 0.1 mm appear to penetrate the backfill by 10,000 years.

The corresponding DEAFRAP analysis is shown in Figure 13. Again, plots are shown with maximum plotted crack apertures of 1 mm and 0.1 mm. The DEAFRAP results confirm that fully penetrating cracks are expected by 10,000 years and identifies potential cracking patterns and pathways. Extraction of continuous crack pathways from the model is discussed in the next section.

### 6.1.4. Extraction of Backfill-Spanning Cracks

Using the result of the DEAFRAP cracking analysis, a continuous crack pathway from the waste form that fully penetrates the backfill region is identified. The extracted crack is shown in Figure 14. The aperture of the crack identified in the analysis is taken to be 0.89 mm, as the 99th percentile value of the selected continuous cracks.

In the current version of the modeling system, identification of the crack pathway is performed using a separate script that identifies connected crack sections to form an overall path from the waste region to the outer boundary.

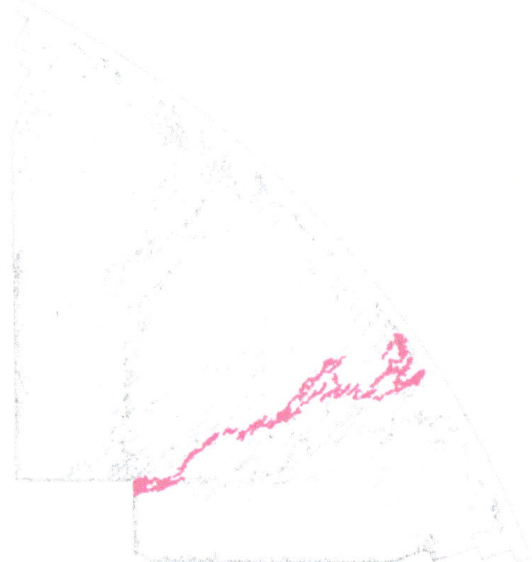

**Figure 14.** Extracted continuous crack that spans the backfill, from the waste to the host rock boundary.

### 6.2. After Cracking

### 6.2.1. Hydraulic Analysis

Since a half-system is modeled by symmetry it is assumed that once a crack pathway is identified that joins the waste region to the host rock, an analogous pathway will join the waste to host rock on the other side of the tunnel, which is assumed to be the upstream side. Advective conditions will then arise in the model, with water flowing through the cracks in the half of the system that is modeled.

Using the extracted crack geometry, a hydraulic analysis is carried out using the FEMWATER v3.0.5 software [33] to solve for the saturated flow conditions in the tunnel taking into account the separate permeabilities of the waste, cement, EDZ and rock regions in the domain. The transmissivity [m$^2$/s] of the crack is calculated from the crack aperture using the cubic law arising from assuming Poiseuille flow conditions in the crack [34],

$$T = \frac{(2b)^3 \rho g}{12\mu} \tag{11}$$

where $\rho$ (kg/m$^3$) is the fluid density, $g$ (m/s$^2$) is gravitational acceleration, $\mu$ (Pa s) is viscosity, and $2b$ (m) is the crack aperture.

### 6.2.2. Mesh Refinement after Cracking

In order to continue the coupled analysis between GARFIELD-CHEM and MACBECE following identification of the crack, the crack pathway location, aperture and volumetric flow rate are passed to GARFIELD-CHEM. The parameters are summarized in Table 4.

**Table 4.** Summary of crack parameters passed to GARFIELD-CHEM following the DEAFRAP and FEMWATER analysis.

| Parameter | Value |
| --- | --- |
| Crack aperture | $0.84 \times 10^{-3}$ (m) |
| Volumetric flux | $1.16 \times 10^{-10}$ (m$^3$/s) |
| Crack geometry | Piecewise linear pathway based on Figure 14. |

The volumetric flow rate is obtained in the separate analysis described above. The crack in the half of the domain that is modeled provides a transport pathway from the waste to the host rock. The flow rate through the pathway from the hydraulic analysis determines the volumetric flux that is applied in the GARFIELD crack model.

Figure 15 shows the GARFIELD-CHEM grid after refinement for the crack information summarized in Table 4. Note that the grid cells around the cracks are refined until the size of the "cracked cell" is smaller than 0.114 m (100 times larger than the crack aperture). This choice of scaling factor will control the accuracy of the diffusive transfers between the finite volume crack model in GARFIELD-CHEM (in which the precise crack geometry and aperture is used) and the porous media grid, and will also control the number of grid cells that arise, and therefore the numerical complexity. The actual cell size depends on the shape of the neighboring cells, and in this example application the size of the cracked cells is mostly a few centimeters.

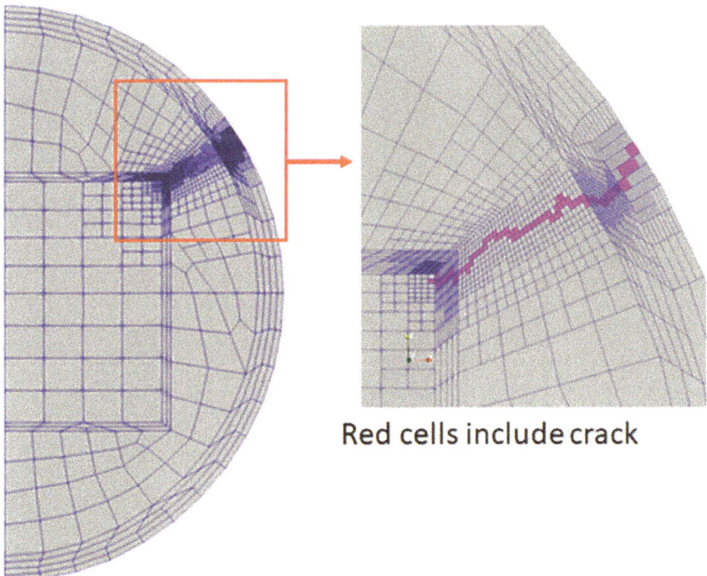



**Figure 15.** GARFIELD-CHEM refined mesh after cracking.

The water composition at entry points to the crack pathway is the same as the "External Water" in Table 2. Therefore, it is assumed that fresh water flows directly into the crack pathway, conservatively ignoring any cement buffering in the upstream half of the tunnel.

### 6.2.3. Chemical Alteration

Preliminary results of the subsequent reactive transport and alteration following the appearance of the crack are shown in Figure 16; Figure 17 at 12,000 years (2000 years after appearance of the crack). The grid outline has been removed in these figures to aid visibility. The corresponding grid can be seen in Figure 15. A front of portlandite dissolution emanates from the waste form (Figure 16b), where the fresh water enters the cementitious backfill and reaches a depth of approximately 10 cm into the matrix. A corresponding calcite precipitation pattern is seen (Figure 16d), whose extent into the cement matrix is not as extensive as the portlandite dissolution front (analogous to the portlandite dissolution front that emerged at the interface between the tunnel liner and the host rock in the early evolution, Section 6.1). CSH165 is largely unchanged during the 2000 years period (Figure 16c), in contrast to the early evolution when CSH165 precipitation in advance of the calcite front was seen. This is possibly a consequence of the reduced availability of the more rapid removal of cementitious water along the advective crack pathway compared to the diffusive mixing at the tunnel liner interface with the rock in the early evolution. pH is not significantly reduced and only falls by ~0.2 close to the waste form (Figure 16e).

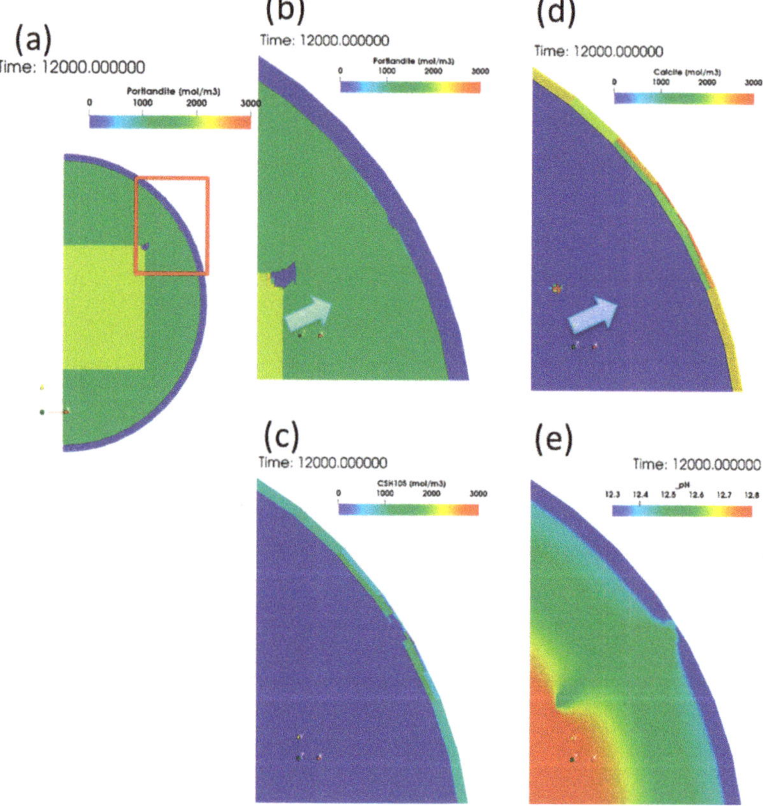


**Figure 16.** Portlandite (**a**) and its zoomed profile (**b**), at 12,000 years (2000 years after crack appearance), corresponding CSH165 (**c**) profile, calcite profile (**d**), and pH profile (**e**).

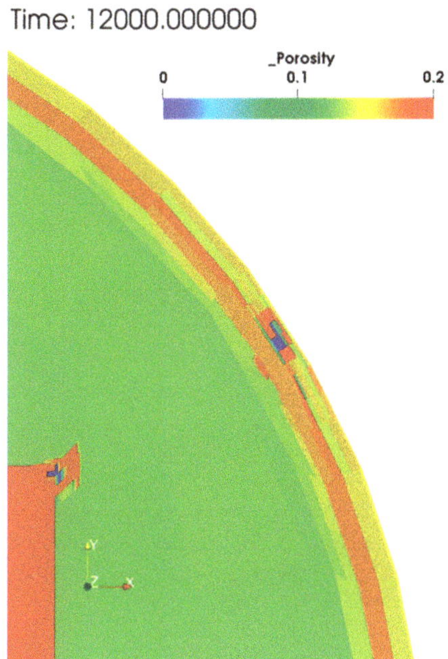


**Figure 17.** Porosity profile at 12,000 years (2000 years after crack appearance).

### 6.2.4. Porosity Alteration (Chemical)

Figure 17 shows the porosity distribution at 12,000 years (2000 years after the appearance of the crack). In the vicinity of the waste region, the porosity close to the crack pathway is only marginally reduced, since the porosity increase arising from the portlandite dissolution is mostly compensated for by precipitation of calcite. Deeper into the matrix there is a net porosity increase surrounding the crack pathway where portlandite dissolution has occurred without any compensating mineral precipitation.

### 7. Conclusions

Modeling of coupled processes associated with geological disposal concepts for radioactive waste is complicated by the fact that it is difficult to include all relevant processes in a single simulation code. There is a large body of existing simulation software that is used in radioactive waste assessment modeling that is tailored to modeling one or a small number of physical processes, but different numerical approaches to discretization, different gridding strategies and resolutions and different numerical time stepping methods are often used to simulate the processes. Ideally, some method of combining these existing tailored codes would be available to allow coupled models of systems of interest to be formed.

An approach to developing simulators for geological disposal concepts by coupling existing modeling software has been presented that uses the OpenMI standard to allow exchange of data between models at run-time. Existing mechanical models, MACBECE and DEAFRAP, have been coupled to a new reactive transport simulator, GARFIELD-CHEM, that incorporates both a finite element component for simulating reactive transport in a porous media grid and a coupled finite volume crack network model component, with grid refinement allowing cracked regions to be more finely resolved. GARFIELD-CHEM and MACBECE are coupled tightly in an OpenMI loop to allow parametric dependence of mechanical properties on the degree chemical degradation (characterized by the degree of porosity alteration and mineral leaching) and chemical properties on the degree of mechanical alteration (characterized by the dependence of effective diffusion on displacements) to be represented. DEAFRAP is used to assess cracking patterns when the stresses and displacements calculated by MACBECE are expected to be sufficient to lead to cracking, or assess growth of existing cracks, and the resultant cracking patterns are used to define crack networks that are subsequently imposed in GARFIELD-CHEM before the tightly coupled chemical-mechanical model is resumed. The resulting coupled model allows the chemical-mechanical-mass transport feedbacks that control the onset and development of cracking of the cement backfill to be simulated.

The benefits of combining existing detailed simulation software with OpenMI to construct detailed coupled system models, which do not rely on significant simplifications that might be necessary to combine all of the modeled processes in a single piece of software, has been demonstrated by simulating the cracking of the cement backfill in the Gr 3/4 concept based on the Japanese disposal concept for disposal of TRU waste [12]. Preliminary results are presented that demonstrate the effect of the coupling up to and beyond the onset of cracking. Further application will be undertaken in future to simulate the effect of the cracking over the entire assessment period to attempt to better understand the nature of the backfill degradation process. It is hoped that such analysis will allow greater realism in future assessments by better representing the potentially beneficial properties of cement barriers, as opposed to their pessimistic treatment in historical safety cases.

The coupling approach using OpenMI is generic and potentially has wider application in the radioactive waste modeling community for combining existing software to form coupled models of systems of relevance in order to help to understand the effects of couplings that would otherwise be difficult or expensive to represent in a single computer model.

**Author Contributions:** Conceptualization, S.J.B., D.K., H.T., H.S., C.O., F.H., Y.T., M.M. and A.H.; Formal analysis, S.J.B., D.K. and H.S.; Investigation, S.J.B., D.K., H.S. and C.O.; Methodology, S.J.B., D.K., H.T. and H.S.; Software, S.J.B., D.K. and H.S.; Validation, S.J.B., D.K. and H.S.; Visualization, S.J.B. and D.K.; Writing—original draft, S.J.B. and D.K.; Writing—review & editing, S.J.B., D.K., H.T., C.O., F.H., Y.T., M.M. and A.H. All authors have read and agreed to the published version of the manuscript.

**Funding:** This research was funded by the Ministry of Economy, Trade and Industry of Japan, grant name "The project for validating assessment methodology in geological disposal system (JFY2017)".

**Conflicts of Interest:** The authors declare no conflict of interest.

## Appendix A

*Appendix A.1. Thermodynamic Parameters for GARFIELD-CHEM Analysis*

Primary and secondary mineral thermodynamic data used in GARFIELD-CHEM simulations is given in Table A1. All minerals are simulated kinetically using TST reactions, parameterized as shown in Table A1. The porewaters throughout the domain were modeled as a Ca-CO3-Si-Mg system, using thermodynamic data as shown in Table A2. The modeled chemical system is intentionally a simplified one, since the primary aim of the modeling is to test the approach to coupling the various models. It is hoped that more comprehensive chemical systems may be included in future work.

**Table A1.** Thermodynamic, molar volume and kinetic rate data for primary and secondary cement minerals. All data is taken from JAEA-TDB v1.07 [35].

| Mineral | Reaction | Molar Volume (cc/mol) | Log K (25 °C) | Reaction Rate (mol m$^{-2}$ s$^{-1}$) | Specific Surface Area (m$^2$/g) |
|---|---|---|---|---|---|
| Brucite | $Mg(OH)_2 + 2H^+ = Mg^{2+} + 2H_2O$ | 24.630 | 17.1018 | $5.75 \times 10^{-17}$ | 10 |
| Calcite | $CaCO_3 + H^+ = Ca^{2+} + HCO_3^-$ | 36.934 | 1.8487 | $6.30 \times 10^{-6}$ | 0.02 |
| CSH165 | $(CaO)_{1.65}(SiO_2)(H_2O)_{2.1167} + 3.3H^+ = 1.65Ca^{2+} + H_4SiO_{4(aq)} + 1.7667H_2O$ | 77.406 | 28.8407 | $3.63 \times 10^{-9}$ | 22.43 |
| CSH105 | $(CaO)_{1.05}(SiO_2)(H_2O)_{1.5167} + 2.1H^+ = 1.05Ca^{2+} + H_4SiO_{4(aq)} + 0.5667H_2O$ | 62.131 | 15.4797 | $1.10 \times 10^{-11}$ | 62.48 |
| Portlandite | $Ca(OH)_2 + 2H^+ = Ca^{2+} + 2H_2O$ | 33.060 | 22.8002 | $1.00 \times 10^{-6}$ | 10.44 |

**Table A2.** Thermodynamic data for complex solute species. Basis species the model are $H_2O$, $H^+$, $Ca^{2+}$, $HCO_3^-$, $H_4SiO_{4(aq)}$ and $Mg^{2+}$. All thermodynamic data is taken from JAEA-TDB v1.07 [35].

| Aqueous Species | Reaction | Log K (25 °C) |
|---|---|---|
| $OH^-$ | $OH^- + H^+ = H_2O$ | 13.9951 |
| $CaCO_{3(aq)}$ | $CaCO_{3(aq)} + H^+ = Ca^{2+} + HCO_3^-$ | 7.0017 |
| $CaOH^+$ | $CaOH^+ + H^+ = Ca^{2+} + H_2O$ | 12.8501 |
| $CaHCO_3^+$ | $CaHCO_3^+ = Ca^{2+} + HCO_3^-$ | −1.0467 |
| $CO_{2(aq)}$ | $CO_{2(aq)} + H_2O = H^+ + HCO_3^-$ | −6.3813 |
| $CO_3^{2-}$ | $CO_3^{2-} + H^+ = HCO_3^-$ | 10.3288 |
| $MgHCO_3^+$ | $MgHCO_3^+ = Mg^{2+} + HCO_3^-$ | −1.0357 |
| $MgCO_{3(aq)}$ | $MgCO_{3(aq)} + H^+ = Mg^{2+} + HCO_3^-$ | 7.3499 |
| $MgOH^+$ | $MgOH^+ + H^+ = Mg^{2+} + H_2O$ | 11.6795 |
| $CaH_3SiO_4^+$ | $CaH_3SiO_4^+ + H^+ = Ca^{2+} + H_4SiO_{4(aq)}$ | 8.6998 |
| $CaH_2SiO_{4(aq)}$ | $CaH_2SiO_{4(aq)} + 2H^+ = Ca^{2+} + H_4SiO_{4(aq)}$ | 19.0646 |
| $H_3SiO_4^-$ | $H_3SiO_4^- + H^+ = H_4SiO_{4(aq)}$ | 9.8096 |
| $H_2SiO_4^{2-}$ | $H_2SiO_4^{2-} + 2H^+ = H_4SiO_{4(aq)}$ | 23.1701 |

*Appendix A.2. Parameters for MACBECE Analysis*

Appendix A.2.1. Specifications of Cement Materials

The mechanical analysis parameters are based the "Second Progress Report on Research and Development for TRU Waste Disposal in Japan" [12]. The assumed specification and detailed properties of the cement materials are summarized in Tables A3 and A4.

**Table A3.** Assumed specifications for cement components (from [12]).

| Component | Material | Aggregate Ratio | Porosity | Real Density (g/cm$^3$) | Compressive Strength (MPa) |
|---|---|---|---|---|---|
| Inner Liner, Outer Liner, Invert, Backfill | Concrete (W/C = 45%) | 67 vol% | 13% | 2.62 | 43 |
| Waste | Mortar (W/C = 55%) | 54 vol% | 19% | 2.58 | 35 |

**Table A4.** Properties for cement components.

| Property | Inner Liner/Outer Liner/ Invert/Backfill | Waste |
|---|---|---|
| Initial Young's modulus $E_0$ (MPa) | $3.04 \times 10^4$ | $1.75 \times 10^4$ |
| Poisson ratio $v(-)$ [before yield] | 0.20 | 0.20 |
| [after yield] | 0.45 | 0.45 |
| Effective overburden pressure $\sigma_{vi}'$ (Mpa) | Obtained by deadweight calculation | |
| Coefficient of earth pressure at rest $K_i$ (−) | Obtained by deadweight calculation | |
| Initial elastic shear modulus $G_0$(MPa) | $G_0 = E_0/2(1 + v)$ | |
| Initial adhesive force $c_0$ (MPa) | 21.5 | 17.5 |
| internal frictional angle $\varphi$ (deg.) | 0 | 0 |
| Initial compressive strength $\sigma_{c0}$ (MPa) | 43 | 35 |
| Minimum compressive strength $\sigma_{c\,min}$ (MPa) | 0.43 | 0.35 |
| Air dried weight of unit volume $\gamma$ (kN/m$^3$) | 22.8 | 20.9 |
| Initial tensile strength $\sigma_t$ (MPa) | 4.3 | 3.5 |
| Minimum tensile strength $\sigma_{t\,min}$ (MPa) | 0.043 | 0.035 |
| Rigidity reduction ratio after cracking (−) | 1/100 | 1/100 |
| Water viscosity (m$^2$/s) | $1.0 \times 10^{-6}$ | $1.0 \times 10^{-6}$ |
| Number of cracks per element | 1.0 | 1.0 |

Appendix A.2.2. Properties of Metal Materials

Analysis parameters for metal materials (treated as corrosion swelling elements) are summarized in Table A5.

**Table A5.** Properties for metal materials.

|  | Waste (Steel Package) | Waste (Drum Can) |
|---|---|---|
| Corrosion swelling ratio $\beta$ (−) | 3.2 | |
| Rigidity (sound region) $E_0$ (MPa) | 211,400 | |
| Rigidity (corroded region) $E_{cor}$ (MPa) | 250 | |
| Poisson ratio $\nu$ (−) | 0.333 | |

*Appendix A.3. Relation between Corrosion Swelling Rate and Distortion Increasing Rate*

The distortion increasing rate for the $x$ and $y$ directions can be described as follows (see Figure A1).

$$\Delta\varepsilon_x = \frac{100\gamma}{h_x} \; (\%/\text{y})$$
$$\Delta\varepsilon_y = \frac{100\gamma}{h_y} \; (\%/\text{y})$$

where $\gamma$ is the corrosion swelling rate (m/year) and $h_x$ ($h_y$) is the thickness of the metal part (m). The corrosion swelling rate $\gamma$ can be represented as

$$\gamma = (\beta - 1)\alpha \; (\text{m}/\text{y})$$

where $\alpha$ is the corrosion rate (m/y) and $\beta$ is the corrosion swelling ratio (treated as a magnification). The assumed corrosion rate $\alpha$ is 0.04 μm/y (considering the corrosion of both sides: [36]) and the assumed corrosion swelling ratio $\beta$ is 3.2 [36], so that $\gamma = 8.8 \times 10^{-8}$ (m/y).

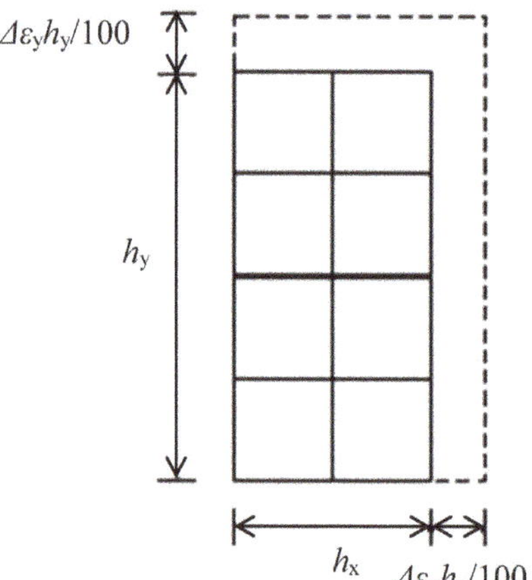


**Figure A1.** Distortion increasing rate.

*Appendix A.4. Distortion Increasing Rate for Steel Packages*

The dimensions of the steel packages and drum can are shown in Figure A2. The left figure shows a half of the cross section, and the red circled numbers correspond to the ID number shown in Table A6, which lists the distortion increasing rate at the various locations on the surface of the package.

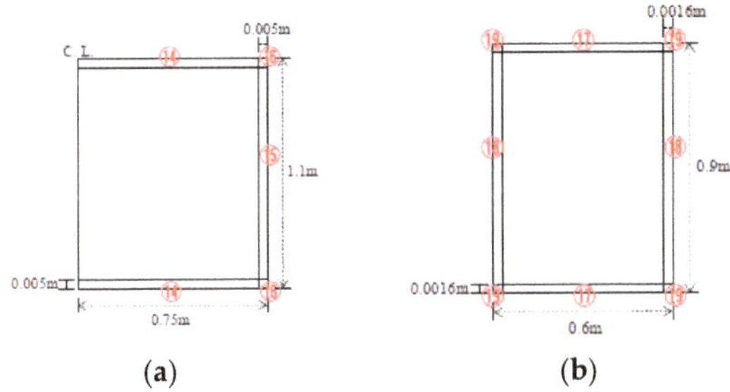

(a)           (b)

**Figure A2.** The shape of steel package (**a**) and drum can (**b**) [12].

**Table A6.** Distortion increasing rate for the steel packages. $\gamma$ represents corrosion swelling rate.

| Number | $\Delta\varepsilon_x$ (%/Year) | $\Delta\varepsilon_y$ (%/Year) |
|---|---|---|
| 14 | $100\gamma/h_x = 66.667\gamma$ <br> ($h_x = 0.75 \times 2$) | $100\gamma/h_y = 20000\gamma$ <br> ($h_y = 0.005$) |
| 15 | $100\gamma/h_x = 20000\gamma$ <br> ($h_x = 0.005$) | $100\gamma/h_y = 90.909\gamma$ <br> ($h_y = 1.1$) |
| 16 | $100\gamma/h_x = 66.667\gamma$ <br> ($h_x = 0.75 \times 2$) | $100\gamma/h_y = 90.909\gamma$ <br> ($h_y = 1.1$) |
| 17 | $100\gamma/h_x = 166.667\gamma$ <br> ($h_x = 0.6$) | $100\gamma/h_y = 62500\gamma$ <br> ($h_y = 0.0016$) |
| 18 | $100\gamma/h_x = 62500\gamma$ <br> ($h_x = 0.0016$) | $100\gamma/h_y = 111.111\gamma$ <br> ($h_y = 0.9$) |
| 19 | $100\gamma/hx = 166.667\gamma$ <br> ($hx = 0.6$) | $100\gamma/h_y = 111.111\gamma$ <br> ($h_y = 0.9$) |

Properties for the Rock Material

The mechanical analysis parameters for the rock material are based on those for the "SR-C" (soft rock) in [12]. The assumed rock property values are summarized in Table A7. The initial overburden pressure and lateral pressure coefficient assume a repository depth of 500 m. The Ohkubo model [25] was used as the stress-distortion relation model.

**Table A7.** Properties for the rock materials (nonlinear viscoelasticity).

| Property | Value |
|---|---|
| Unit weight $\gamma_t$ (kN/m$^3$) | 22.0 |
| Uniaxial compressive strength $q_u$ (MPa) | 15.0 |
| Tensile strength $\sigma_t$ (MPa) | 2.1 |
| Initial Young's modulus $E_0$ (MPa) | $3.50 \times 10^3$ |
| Initial Poisson ratio $v_0$ (−) | 0.30 |
| Initial overburden pressure $\sigma_{vi}$ (MPa) | 11.0 |
| Lateral pressure coefficient $K_i$ (−) | 1.066 |

## References


1. Carter, A.; Kelly, M.; Bailey, L. Radioactive high level waste insight modelling for geological disposal facilities. *Phys. Chem. Earth Parts A/B/C* **2013**, *64*, 1–11. [CrossRef]
2. Parkhurst, D.L.; Appelo, C. *Description of Input and Examples for PHREEQC Version 3: A Computer Program for Speciation, Batch-Reaction, One-Dimensional Transport, and Inverse Geochemical Calculations*; Techniques and Methods; U.S. Geological Survey: Denver, CO, USA, 2013.
3. Lichtner, P.C.; Hammond, G.E.; Lu, C.; Karra, S.; Bisht, G.; Andre, B.; Mills, R.T.; Kumar, J. *PFLOTRAN User Manual: A Massively Parallel Reactive Flow and Transport Model for Describing Surface and Subsurface Processes*; Office of Scientific and Technical Information (OSTI): Oak Ridge, TN, USA, 2015. [CrossRef]

4. Bethke, C.M.; Farrell, B.; Yeakel, S. *The Geochemist's Workbench, Version 12.0: GWB Essentials Guide*; Aqueous Solutions, LLC.: Champaign, IL, USA, 2018.

5. Hibbitt, D.; Karlsson, B.; Sorensen, P. *ABAQUS/CAE User's Manual, 6.11*; Simulia Corp.: Johnston, RI, USA, 2013.

6. *COMSOL Multiphysics®v. 5.4. www.comsol.com*; COMSOL AB: Stockholm, Sweden, 2018.

7. Eymard, R.; Gallouët, T.; Herbin, R. Finite volume methods. In *Handbook of Numerical Analysis*; Ciarlet, P.G., Lions, J.L., Eds.; Elsevier: Amsterdam, The Netherlands, 1997; Volume 7, pp. 713–1018. [CrossRef]

8. Johnson, C. Numerical Solution of Partial Differential Equations by the Finite Element Method. *Math. Comput.* **1989**, *52*, 247. [CrossRef]

9. Shimizu, H.; Murata, S.; Ishida, T. The distinct element analysis for hydraulic fracturing in hard rock considering fluid viscosity and particle size distribution. *Int. J. Rock Mech. Min. Sci.* **2011**, *48*, 712–727. [CrossRef]

10. Tryggvason, G.; Bunner, B.; Esmaeeli, A.; Jurič, D.; Al-Rawahi, N.; Tauber, W.; Han, J.; Nas, S.; Jan, Y.-J. A Front-Tracking Method for the Computations of Multiphase Flow. *J. Comput. Phys.* **2001**, *169*, 708–759. [CrossRef]

11. Gregersen, J.B.; Gijsbers, P.J.A.; Westen, S.J.P. OpenMI: Open modelling interface. *J. Hydroinform.* **2007**, *9*, 175–191. [CrossRef]

12. JAEA; FEPC. *Second Progress Report on Research and Development for TRU Waste Disposal in Japan—Repository Design, Safety Assessment and Means of Implementation in the Generic Phase—, JAEA-Review 2007-010*; FEPC TRU-TR2-2007-01, Technical Report; Japan Atomic Energy Agency (JAEA): Tokai, Ibaraki, Japan; The Federation of Electric Power Companies Japan (FEPC): Chiyoda, Tokyo, Japan, 2007.

13. Ohno, S.; Morikawa, S.; Mihara, M. The Rock Creep Evaluation in the Analysis System for the Long-Term Behavior of TRU Waste Disposal System. In Proceedings of the ASME 2009 12th International Conference on Environmental Remediation and Radioactive Waste Management, Liverpool, UK, 11–15 October 2009; Volume 2, pp. 291–297.

14. Mihara, M.; Hirano, F.; Takayama, Y.; Kyokawa, H.; Ohno, S. Long-term mechanical analysis code considering chemical alteration for a TRU waste geological repository. *Genshiryoku Bakkuendo Kenkyu (Internet)* **2017**, *24*, 15–25.

15. JAEA. *The Project for Validating Assessment Methodology in Geological Disposal System*; Summary Report for JFY2013-2017; Japan Atomic Energy Agency (JAEA): Tokai, Ibaraki, Japan, 2018. (In Japanese)

16. Moore, R.; Gijsbers, P.; Fortune, D.; Gergersen, J.; Blind, M. OpenMI Document Series: Part A Scope for the OpenMI (Version 1.4). IT Frameworks (HarmonIT) Contract EVK1-CT-2001-00090. Available online: https://www.openmi.org/openmi-around-the-world/development-tools (accessed on 28 August 2020).

17. Steefel, C.I.; Appelo, C.A.J.; Arora, B.; Jacques, D.; Kalbacher, T.; Kolditz, O.; Lagneau, V.; Lichtner, P.C.; Mayer, K.U.; Meeussen, J.C.L.; et al. Reactive transport codes for subsurface environmental simulation. *Comput. Geosci.* **2014**, *19*, 445–478. [CrossRef]

18. Zhu, C.; Anderson, G. *Environmental Applications of Geochemical Modeling*; Cambridge University Press: Cambridge, UK, 2002.

19. Bethke, C.M. *Geochemical and Biogeochemical Reaction Modeling*; Cambridge University Press: Cambridge, UK, 2007. [CrossRef]

20. Pitzer, K.S. *Activity Coefficients in Electrolyte Solutions*, 2nd ed.; CRC Press: Boca Raton, FL, USA, 1991.

21. Archie, G. The Electrical Resistivity Log as an Aid in Determining Some Reservoir Characteristics. *Trans. AIME* **1942**, *146*, 54–62. [CrossRef]

22. Hindmarsh, A.C.; Brown, P.N.; Grant, K.E.; Lee, S.L.; Serban, R.; Shumaker, D.E.; Woodward, C.S. SUNDIALS, suite of nonlinear and differential/algebraic equation solvers. *ACMTrans. Math. Softw.* **2005**, *31*, 363–396. [CrossRef]

23. Wilson, J.C.; Benbow, S.; Metcalfe, R. Reactive transport modelling of a cement backfill for radioactive waste disposal. *Cem. Concr. Res.* **2018**, *111*, 81–93. [CrossRef]

24. Arndt, D.; Bangerth, W.; Clevenger, T.C.; Davydov, D.; Fehling, M.; Garcia-Sanchez, D.; Harper, G.; Heister, T.; Heltai, L.; Kronbichler, M.; et al. The deal.II library, Version 9.1. *J. Numer. Math.* **2019**, *27*, 203–213. [CrossRef]

25. Okubo, S. Analytical Consideration of Constitutive Equation of Variable Compliance Type. *Shigen-to-Sozai* **1992**, *108*, 601–606. [CrossRef]

26. Motojima, M.; Kitahara, Y.; Ito, H. *The Stability of Rock Foundation with Considering Strain-Softening Material*; CRIEPI Technical Report; Central Research Institute of Electric Power Industry (CRIEPI): Tokyo, Japan, 1981. (In Japanese)
27. Rots, J. Computational Modelling of Concrete Fracture. Ph.D. Thesis, Delft University of Technology, Delft, The Netherlands, 1988.
28. Cundall, P.A.; Strack, O.D.L. A discrete numerical model for granular assemblies. *Géotechnique* **1979**, *29*, 47–65. [CrossRef]
29. Potyondy, D.; Cundall, P. A bonded-particle model for rock. *Int. J. Rock Mech. Min. Sci.* **2004**, *41*, 1329–1364. [CrossRef]
30. Geuzaine, C.; Remacle, J.-F. Gmsh: A 3-D finite element mesh generator with built-in pre- and post-processing facilities. *Int. J. Numer. Methods Eng.* **2009**, *79*, 1309–1331. [CrossRef]
31. Yui, M.; Sasamoto, H.; Arthur, R. Geostatistical and geochemical classification of groundwaters considered in safety assessment of a deep geologic repository for high-level radioactive wastes in Japan. *Geochem. J.* **2004**, *38*, 33–42. [CrossRef]
32. *JAEA Advanced Assessment Technology Development for Impact of Cement on Geological Disposal System for Long-lived Radioactive Waste*; Annual Report for JFY2014, JAEA Technical Report; Japan Atomic Energy Agency (JAEA): Tokai, Ibaraki, Japan, 2015. (In Japanese)
33. Lin, H.C.J.; Richards, D.R.; Yeh, G.T.; Cheng, J.R.; Cheng, H.P. *FEMWATER: A Three-Dimensional Finite Element Computer Model for Simulating Density-Dependent Flow and Transport in Variably Saturated Media (No. WES/TR/CHL-97-12)*; Army Engineer Waterways Experiment Station Vicksburg Ms Coastal Hydraulics Lab: Vicksburg, MS, USA, 1997.
34. Snow, D.T. Anisotropie Permeability of Fractured Media. *Water Resour. Res.* **1969**, *5*, 1273–1289. [CrossRef]
35. JAEA. *The Project for Validating Assessment Methodology in Geological Disposal System*; Annual Report for JFY2016, JAEA Technical Report; Japan Atomic Energy Agency (JAEA): Tokai, Ibaraki, Japan, 2017. (In Japanese)
36. Hirano, F.; Otani, Y.; Kyokawa, H.; Mihara, M.; Shimizu, H.; Honda, A. Crack Formation in Cementitious Materials Used for an Engineering Barrier System and Their Impact on Hydraulic Conductivity from the Viewpoint of Performance Assessment of a TRU Waste Disposal System. *Trans. At. Energy Soc. Jpn.* **2015**, *15*, 97–114. [CrossRef]

Article

# Effects of Erosion Form and Admixture on Cement Mortar Performances Exposed to Sulfate Environment

**Peng Liu [1,2,3], Ying Chen [1,2,4,*] and Zhiwu Yu [1,2]**

[1] School of Civil Engineering, Central South University, 22 Shaoshan Road, Changsha 410075, China; 216056@csu.edu.cn (P.L.); 701061@csu.edu.cn (Z.Y.)

[2] National Engineering Laboratory for High Speed Railway Construction, 22 Shaoshan Road, Changsha 410075, China

[3] School of Civil Engineering, Shenzhen University, 3688 Nanhai Road, Shenzhen 518060, China

[4] School of Civil Engineering, Central South University of Forestry and Technology, 498 Shaoshan Road, Changsha 410004, China

* Correspondence: cheny83@csu.edu.cn

Received: 10 July 2020; Accepted: 31 August 2020; Published: 1 September 2020

**Abstract:** The effects of the admixtures, erosion age, concentration of sulfate solution, and erosion form of sulfate attack on the mechanical properties of mortar were investigated. Simultaneously, the microstructure, pore characteristics, kinds and morphologies of erosion products of mortar before and after sulfate attacks were performed by Mercury Intrusion Porosimetry (MIP), Environment Scanning Electronic Microscope and Energy Dispersive Spectrometer (ESEM-EDS). In addition, the crystal form and morphology characteristics of crystallization on mortar surfaces attacked by partial immersion form were studied. The results showed that the compressive and flexural strengths of mortar attacked by sulfate for four months decreased with the increase of the replacement of cement with fly ash, and the corresponding strength of mortar containing slag first increased and then decreased. The admixtures can improve the microstructure and mechanical properties of mortar within the replacement ratio of 10%. Although the change laws of the mortar specimens containing different admixtures were similar, the mortar containing slag had an excellent sulfate resistance under the same condition. Compared with the complete immersion form, the strength variation of the mortar containing fly ash attacked by semi-immersion form was less. The porosity and average pore diameter of mortar attacked by sulfate for four months increased, and the percentage of micropore with the pore diameter less than 200 nm increased. Plenty of rod-like and plate-like erosion products were generated in mortar attacked by a sulfate solution with a high concentration. A larger number of fibrous and flocculent crystallization covered the mortar's surface containing fly ash, but it was a granular and dense crystallization formed on the mortar's surface containing slag. Much dendritic erosion product was generated in the mortar attacked by semi-immersion form, and ESEM-EDS analysis revealed that it may be scawtite, spurrite, and residue of the decomposed calcium silicate hydrate (CSH) in the inner mortar; however, the crystallization sodium sulfate was crystallized on mortar surface.

**Keywords:** degradation; mechanical properties; mechanisms; diffusion; material properties

---

## 1. Introduction

Concrete and mortar are the most commonly used construction materials in civil engineering [1–3]. Many environmental influencing factors affect the performance variation of concrete structures, and the sulfate attack is one of the most important factors in the deterioration of concrete structures [4]. The performance deterioration of cement-based materials is not only because of the expansion effect of sulfate erosion products but also due to the sulfate attack, which causes a reduction in the adhesion forces of the structure by destroying the major cementitious products of the hydrated cement, i.e., calcium aluminate hydrate (CAH), calcium hydroxide (CH), and calcium silicate hydrate (CSH) [5,6]. Therefore, the sulfate attack of cement-based materials is a complex physical and chemical process including crystallization and reaction. The concrete can be treated as the combination of mortar and coarse aggregates, so the performance variation of concrete can be indirectly characterized by the property evolution of the mortar. Sulfate attack can affect the various performance of the mortar, such as strength, length change, long-term behavior, and durability [7–11]. Therefore, it is very significant to investigate the performance variation of the mortar under sulfate environment.

Many studies on sulfate attack of mortar have been conducted which mainly focus on the concentration and kinds of sulfate solution [12,13], mechanical properties and microstructure of mortar [14–16], material composition, and exposure conditions [17]. For example, Aziez et al. [18] investigated the effect of temperature and type of sand on the magnesium sulfate attack in sulfate-resisting cement mortars, and the results indicated that high temperature improved some physical and mechanical properties. Ghafoori et al. [19] discussed the effects of the fineness and tricalcium aluminate ($C_3A$) content on the sulfate resistance of mortars, and the results showed that microsilica increased sulfate resistance more effectively than nanosilica and increasing cement fineness proved beneficial in combination with either pozzolan regardless of the cement's $C_3A$ content. Kim and Baek [20] studied the weight and strength of cement mortars under sulfate attack, and their results showed that the cement mortar using crushed sand had a good resistance against sulfate compared with river sand. Akpinar and Casanova [21] conducted a combined study of the length-change, tensile strength evolution, and durability of high- and low-$C_3A$ Portland cements, and they proposed a novel approach to assess the performance under severe and moderate sulfate attack. In general, the effective way for protecting mortar from damage of sulfate attacks is to reduce the mortar's permeability or to use mineral admixtures to reduce the amount of CH formed during the cement hydration [22]. Many studies regarding the replacement of mineral admixtures to an amount of cement have been carried out [23–25]. Simultaneously, the effects of mineral admixtures including fly ash, slag, limestone, silica fume, ground bagasse ash, micro-particle additives on the mechanical properties of mortars were investigated [1,26,27]. For example, Ryou et al. [28] represented the relationship of the replacement ratio and limestone on durability of mortar, and the results demonstrated that both the high fineness level and the replacement ratio had a negative effect in resisting sodium sulfate attacks. Chindaprasirt et al. [29] discussed the effect of fly ash fineness on strength, drying shrinkage, and sulfate resistance; they presented that the fly ashes can have a significant improvement in drying shrinkage and resistance to sulfuric acid attack. Marvila et al. [30] studied the replacement of the hydrated lime by kaolinitic clay in mortars, and their results showed that with up to 50 wt.% of the hydrated lime replacement, it was perfectly feasible to fulfil with technological parameters of standards. Markssuel et al. [31] evaluated the durability of mortars with external coating containing incorporation of 0%, 1.5%, 3.0%, and 5% of fiber, the durability results proved that the mortars with 1.5% fiber showed a better behavior than the reference mortar in all situations, including in salt spray attack. Tosun and Kim et al. [32,33] investigated the influence of the admixtures' contents on resistance to sulfate attack, and the results indicated that the low temperature and limestone replacement ratio played a little effect on the sulfate resistance of cement mortars. Lee and Kim [34,35] represented the effects of admixtures including limestone and alpha-calcium sulfate hemihydrate on variation of the setting time, strength, and strain of mortar; their results indicated that the samples incorporating higher replacement levels of limestone filler were more susceptible to sulfate attacks. However, the deterioration modes were significantly dependent on the types of sulfate solutions. Plenty of other studies also showed that nano-silica replacement in the production of cement-based materials results in outstanding performance in terms of improved mechanical properties and durability [36]. In addition, Marvila et al. [37–39] carried out many works on mortar such as preparation of gypsum plaster mortar, eco-friendly mortars, and paper mortar. Although a larger number of the existing achievements regarding influence of partial replacement of cement by admixtures on the mortars' performance were carried out [40], the coupled effects of concentration of sulfate solution, types and contents of admixtures, erosion age and form on the microstructure and mechanical property have not been discussed in detail.

The aim of this study was to evaluate the resistance to sulfate attacks of the mortars with different replacements of cement by admixtures. The effects of concentration of sulfate solution, kinds and contents of admixtures, and erosion age and form of sulfate attack on the mechanical properties of mortars were investigated. In addition, the microstructure, pore characteristic, type, and morphology of erosion products were also studied. The results of this study will be useful in the efficient selection of the admixtures that provide a positive effect on sulfate resistance in mortars and can also provide a theoretical support for the durability assessment of structural engineering under sulfate attack environment.

## 2. Experimental Procedure

### 2.1. Raw Materials

Portland cement (PC) of grade P·O 32.5 was purchased from Pingtang Cement Plant of Hunan. Class I fly ash and S95 slag were provided by Hunan Xiangtan power plant and Hunan Lianyuan Steel Group Co., Ltd., respectively. The fineness modulus of ISO standard sand is 3 (Xiamen Aisiou Standard Sand Co., Ltd., Xiamen, China). In addition, the industrial grade sodium sulfate with a purity of 99% was purchased from a local market. Table 1 lists the chemical compositions of the cementitious materials, and Table 2 lists the physical properties of cementitious materials.

**Table 1.** Chemical compositions of cementitious materials/%.

| Items | CaO | SiO$_2$ | Al$_2$O$_3$ | P$_2$O$_5$ | Fe$_2$O$_3$ | MgO | K$_2$O | Na$_2$O | Loss |
|---|---|---|---|---|---|---|---|---|---|
| Portland cement | 65.61 | 19.89 | 4.73 | 2.88 | 1.45 | 2.01 | 1.14 | 0.21 | 1.98 |
| Fly ash | 2.52 | 54.63 | 26.81 | - | 6.11 | 1.12 | 1.05 | 0.38 | 2.53 |
| Slag | 35.52 | 32.41 | 12.22 | 0.15 | - | - | - | - | 1.27 |

**Table 2.** Physical properties of cementitious materials.

| Items | Average Diameter, μm | Specific Surface Area, (m$^2$/kg) | Bulk Density, (g/cm$^3$) | Initial Setting Time, min | Final Setting Time, min | Water Requirement for Standard Consistency, % |
|---|---|---|---|---|---|---|
| Portland cement | 34.6 | 345.2 | 1.35 | 172 | 251 | 28 |
| Fly ash | 41.3 | 332.7 | 0.78 | - | - | - |
| Slag | 31.7 | 432.5 | 1.18 | - | - | - |

### 2.2. Experimental Process

According to Chinese standard (GB/T 17671-1999, Method of Testing Cements-Determination of Strength), the mortar specimens of 40 mm × 40 mm × 160 mm were prepared with water:cement:sand ratio of 0.5:1:3 by weight [41]. Fly ash and slag were used as admixture to replace the cement with replacement ratios of 5%, 10%, 15%, 20%, and 30% by weight. The specimens were cured at a temperature of 20 ± 1 °C and a relative humidity of 95% ± 3%. Subsequently, all specimens were demolded after 24 h and cured in saturated lime water for 28 d at a temperature of 20 ± 1 °C. Without special instructions, the control sample (KB) in this study was defined as the mortar specimen cured in a standard condition (i.e., temperature of 20 °C and a relative humidity of 95%) for the same curing age without sulfate attack. Simultaneously, the blank sample was the mortar specimen cured in a standard condition (i.e., temperature of 20 °C and a relative humidity of 95%) for the same curing age without admixtures, termed as KB.

The experimental process of complete immersion form of the mortar attacked by sulfate was conducted as follows. Firstly, the sulfate solutions with different concentrations (i.e., 1%, 5%, 10%, and saturated solution) were prepared. The mortar specimens were immersed in sulfate solution. The distance of mortar specimens was set to be no less than 2 cm to ensure the uniformity of sulfate solution. Secondly, the sulfate solution boxes were all covered with plastic film to prevent the water from evaporation. Moreover, in order to ensure identical concentration, the sulfate solution should be changed once a week. Then, the mortar specimens were taken out from the sulfate solution and placed in air for 3 d at room temperature, when the sulfate erosion age reached the erosion ages of two and four months. Finally, the mechanical properties of mortar were tested. The experimental process of the sulfate attack of the mortar with semi-immersion form was as follow: a half of the mortar specimen was vertically immersed into the sulfate solution, and another half part of the mortar specimen was exposed to air. In order to study the salt crystallization on specimen surface, the specimens attacked by partial immersion form were carried out, that is, the mortar specimen of 1/5 length was immersed into sulfate solution, and the other part of the mortar specimen was exposed to air.

The preparations process of mortar sample for micro properties test was as follows. Firstly, the specimens were broken into particles with sizes of 2~5 mm. Secondly, the broken particles were immersed in alcohol for 24 h to terminate reaction. Then, the broken particles were dried at 60 °C for 48 h. Finally, the processed particles were stored in the dryer. The broken particles were sprayed with gold for Environment Scanning Electronic Microscope and Energy Dispersive Spectrometer (ESEM-EDS) analysis. The samples particles of 3~5 g were weighted and put into sample cells for the Mercury Intrusion Porosimetry (MIP) analysis.

The Quanta-200 Environment Scanning Electronic Microscope and Energy Dispersive Spectrometer (ESEM-EDS) produced by FTI Company from the Czech Republic was used to measure the microstructure. The Autopore-9500 Mercury Intrusion Porosimetry was produced by Microtrac Inc. of USA. Moreover, the YAW-300D produced by Jinan Kesheng Test Equipment Co. Ltd. of China was used to measure the strength of mortar specimens.

## 3. Results and Discussion

### 3.1. Effects of Admixture and Erosion Age on Strength of Mortar Attacked by Sulfate

The effects admixtures, concentration of sulfate solution, and erosion age on the strength of the mortar attacked by sulfate were investigated. The compressive and flexural strength of the mortar attacked by the complete immersion form for two months was tested. Figure 1 shows the variation curves of mortar strength with the replacement of cement by fly ash.

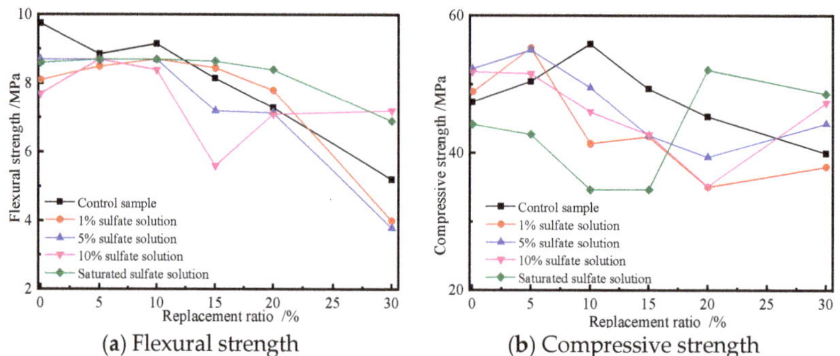


(**a**) Flexural strength      (**b**) Compressive strength

**Figure 1.** Variation curves of mortar strength with replacement ratios of cement by fly ash exposed to sulfate solution for 2 months.

As seen from Figure 1, the flexural strength of the control sample decreased with the increase of the replacement ratio of cement by fly ash. However, the flexural strength of the mortar specimens containing fly ash attacked by different concentrations of sulfate solutions first increased and then decreased with replacement ratios of cement by fly ash. The compressive strength of the control sample first increased and then decreased, and there existed a maximum value of compressive strength with the replacement ratio of 10%. Although the change laws of the flexural strength of the mortar specimens attacked by different concentrations of sulfate solution were similar, the variation of compressive strengths differed. At a low concentration of sulfate solution (i.e., less than 10%), the compressive and flexural strength of mortar specimens containing fly ash first increased and then decreased with the increase of the replacement ratio. However, the compressive strength of the mortar specimens attacked by saturated sulfate solution first decreased and then increased, and its corresponding value was less than that of the control sample. Hence, the mechanical properties of the mortar reduced with the increase of the replacement ratio. When the fly ash was used as admixture to replace a certain amount of cement in mortars, the effects of fly ash enhancing sulfate the resistance of the mortar were in terms of replacement, pozzolanic, shape, and tiny aggregate [1]. The fly ash was composed of active compositions, such as amorphous silicon oxide and aluminum oxide, which can react with calcium hydroxide (CH) and generate cementitious substances. As a result, the cement matrix became denser, and the mechanical properties of the mortar were enhanced. If there is an overly excessive replacement of cement by fly ash, the fly ash cannot completely react with CH. Thus, the amount of the hydrated products formed by fly ash is less. Therefore, a proper amount of fly ash to cement can improve the microstructure and mechanical properties of mortar specimens. However, the quantity of cementitious hydration products generated in mortar decreases when the cement in mortar was replaced by abundant of fly ash. In general, the main erosion products of sulfate attack were ettringite (AFt), monosulfoaluminate (AFm), gypsum, and salt crystal [42] which have double effects on mortar performance, i.e., positive and negative effects. On the one hand, the expansive erosion products fill in the micropores, reduce the crystallization and orientation of CH, improve the microstructure, and decrease the porosity; so, the sulfate attack had a positive effect on the performance of the mortar. On the other hand, the sulfate attack consumed the amount of CH, reduced the pH value, and decomposed the cementitious hydration products which induced the generation of coarse expansive erosion products and results in the generation of the micro damage and cracks, so it plays a negative effect on the performance of mortar. The variation of the mechanical properties of the mortar exposed to sulfate is a comprehensive result under the coupled effects of sulfate attack and admixture. If there exists a positive effect on the performance of mortar, the mechanical properties and sulfate resistance of mortar are enhanced. Conversely, the various performances of the mortar are reduced.

To discuss the effect of sulfate solution concentration on mechanical properties of the mortar containing fly ash, the strength of mortars attacked by sulfate solution for two months was measured as shown in Figure 2.

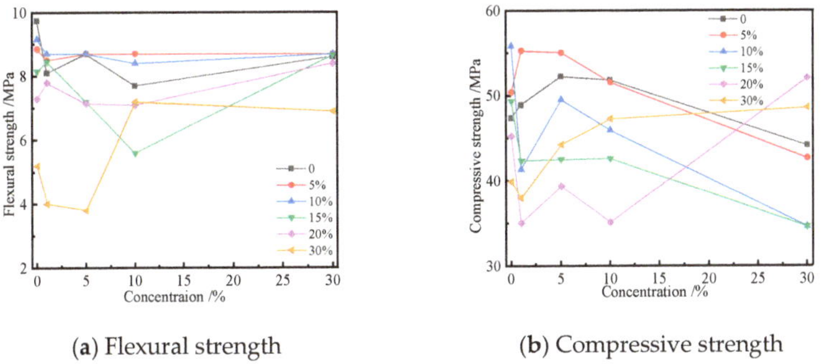


(a) Flexural strength
(b) Compressive strength

**Figure 2.** Strength variation of the mortar containing fly ash with sulfate solution concentrations.

With an increase in the concentration of the sulfate solution, the compressive and flexural strength of the mortar with the replacement ratio less than 15% decreased, but the corresponding strength of the mortar with a high replacement increased. The strength of the mortar containing more fly ash attacked by a high concentration of sulfate solution was larger than that of the control sample. This was due to the fact that the active composites of fly ash were activated by a sulfate attack and formed into plenty of expansive products which reduced the porosity and improved the microstructure of the mortar; so, the sulfate attack had a positive effect on the performance of the mortar specimens. Moreover, the filling effect of fly ash can also improve the microstructure of mortar which was proved by Reference [37]. Macroscopically, it expresses as the increase of compressive and flexural strength of mortar. Taking a mortar specimen with a replacement ratio of 30% fly ash as an example, it can be seen that the mechanical properties of the mortar first decreased and then increased with an increase of the concentration of the sulfate solution. The reasons may be as follows. On the one hand, the quantity of cementitious hydration products generated in mortar was reduced when more cement was replaced by fly ash. On the other hand, the sulfate ions of sulfate solution concentration less than 5% reacted with CH and reduced the pH value, which induced the decomposition of hydration products in mortar. Therefore, the mechanical properties of the mortar decreased. With the increase of concentration of the sulfate solution (i.e., 10%), the active compositions of fly ash are activated and generated into hydration products. Simultaneously, more expansive products, such as AFt, AFm, and gypsum are generated in mortar, which can improve the microstructure and consistency of mortar [43]. The sulfate attack and fly ash play a positive role on performance of the mortar, so the corresponding strength of the mortar specimens containing more content of fly ash attacked by a high concentration of sulfate solution increases. In addition, the mortar with a low content of fly ash attacked by low concentration of sulfate solution increases, which is due to the positive coupled effects of sulfate attack.

The erosion age also plays a significant effect on variation of the mortar strength. Taking the mortar specimens containing fly ash attacked by sulfate for four months as an example, the variation of compressive and flexural strength of the mortar with replacement ratio was investigated as shown in Figure 3.

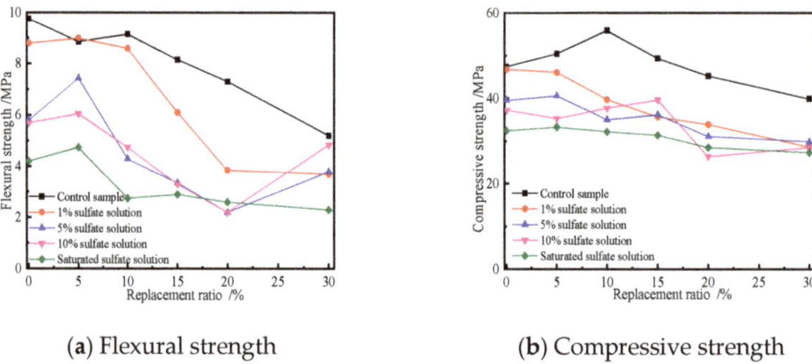


(**a**) Flexural strength      (**b**) Compressive strength

**Figure 3.** Variation of mortar strength with replacement ratios of cement by fly ash.

The compressive and flexural strength of the specimens attacked by sulfate for four months decreased with the increase of the replacement of cement by fly ash, and the corresponding values were less than that of the control sample. With the same replacement of cement by fly ash, the mechanical properties of mortar reduced with increase of concentration of the sulfate solution. Compared with Figure 1, it can be seen that the regularity of strength variation of mortar attacked by sulfate for four months becomes more significant. Although the change laws of the mechanical properties of the mortar are similar, the corresponding percentages and variations are different. Compared with Figures 1 and 3, a conclusion can be drawn that the variation of mortars containing fly ash attacked by sulfate for four months are more significant. This is due to the fact that more and more sulfate ions intrude into mortar with erosion age. Therefore, more expansive products are generated, and plenty of hydration products are decomposed, which result in the damage and micro-cracks in the mortar.

To deeply investigate the effect of sulfate solution concentration on the mechanical properties of the mortar, taking mortar containing fly ash as an example, the variation of the compressive and flexural strength of mortar attacked by sulfate solution for four months was investigated as shown in Figure 4.

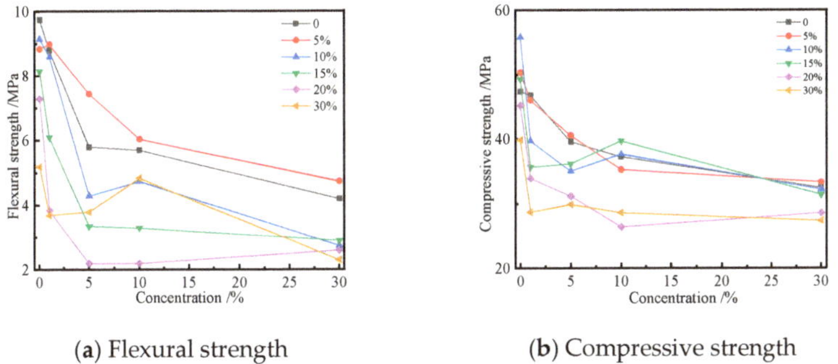

(a) Flexural strength    (b) Compressive strength

**Figure 4.** Strength variation of the mortar containing fly ash attacked by different concentrations of sulfate solution for four months.

As seen from Figure 4, the compressive and flexural strength of the mortar specimens attacked by sulfate for four months decreased with an increase in the replacement ratio of cement by fly ash. The larger the content of fly ash, the more significant the variation in the strength of the mortar. In comparison with compressive strength, the effect of the sulfate solution concentration on the flexural strength of mortar is more obvious. The compressive strength of the mortar attacked by sulfate was basically the same when the replacement of the cement by fly ash was no more than 15%. Compared with the compressive strength, the variation of the flexural strength of the mortar was more significant. Moreover, the strength of the mortar with a replacement of 5% was better than that of the blank sample. This was due to the fact that the active compositions of fly ash can be activated by CH in the mortar, resulting in the reduction of porosity and improvement of microstructure. Compared with Figure 2, the compressive and flexural strength of the mortar attacked by sulfate for four months obviously decreased, and its regularity of strength variation became more remarkable. That may be related with the coupled effects of sulfate attack and fly ash. When their coupled effects are positive, the compressive and flexural strength of the mortar specimens increase. Conversely, the corresponding strength of mortar decreases.

Moreover, the effects of solution concentration, slag content, and erosion age on the strength of mortar were also investigated. Taking the mortar attacked by sulfate for two months as an example, the variation curves of compressive and flexural strength of the mortar with the replacement ratio of cement by slag were plotted in Figure 5. Figure 6 shows the corresponding strength variation curves of the mortar with the concentration of the sulfate solution.

Figure 5 shows that the strength of the control sample decreased with the increase in the replacement of cement by slag. However, the variations of the strength of the mortar attacked by sulfate for two months differed from each other, as follows. The strength of the mortar increased with the increase of slag when the sulfate solution concentration was no more than 5%. Conversely, it decreased when the concentration of sulfate solution was more than 10%. Compared with Figure 1, the mortar had a higher strength and little variation, which revealed the mortar containing slag had an excellent sulfate resistance. The active compositions of slag were mainly calcareous and silicious ingredient, which can be easily activated by CH and generated into cementitious products. Those products can fill in the pores, reduce the porosity, and improve the microstructure of the mortar. Therefore, the strength of the mortar containing slag increased with the increase of slag content. The sulfate ions intrude into mortar and react with CH, which results in the generation of expansive products in the mortar. When the sulfate attack plays a positive effect on properties of mortar, the macro-characteristics of strength of the mortar increases. Conversely, the mechanical properties of the mortar decrease. In addition, there exist significant differences in the strength of the mortar specimens containing different replacement ratios of cement by slag or attacked by different concentrations of sulfate solution. That may be due to the fact that the sulfate attacks and slag content have the coupled effects on the performance of the mortar. The above discussion can be verified by the variation curves of strength of the mortar with different concentrations of sulfate solution for two months as shown in Figure 6.

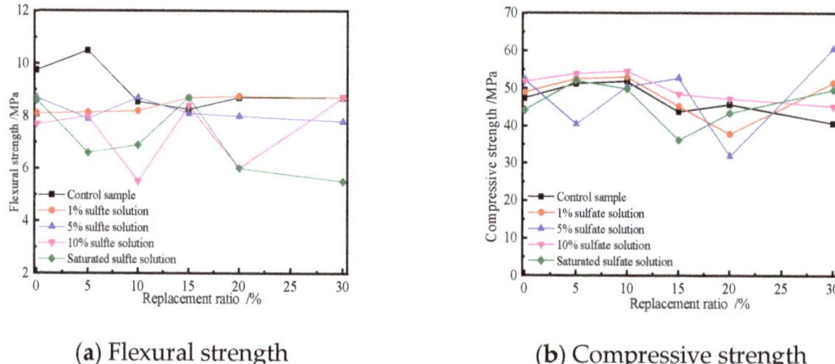

(a) Flexural strength      (b) Compressive strength

**Figure 5.** Variation curves of the strength of the mortars with replacement ratios of cement by slag.

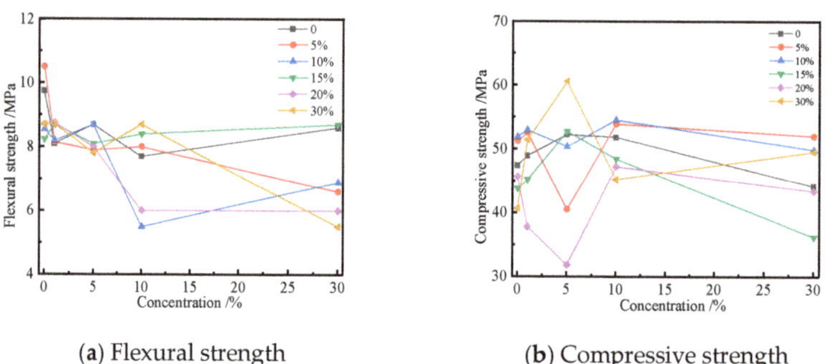

(a) Flexural strength      (b) Compressive strength

**Figure 6.** Variation curves of the strength of mortars containing slag attacked by sulfate solutions for two months.

The flexural strength of the blank sample decreased with the increase in the concentration of the sulfate solution, but the corresponding compressive strength first increased and then decreased. The flexural strength of the mortar containing slag decreased with the increase in the concentration of the sulfate solution. Moreover, Figure 6 also indicates that there exists a close relationship between compressive strength and slag content in the mortar attacked by sulfate solution. In addition, the compressive strength of the mortar specimen attacked by 5% sulfate solution differed from others. This was due to the coupled effects of sulfate attack and existence of slag. Compared with Figures 2 and 6, it can be seen that the compressive strength of the mortar containing slag attacked by sulfate for two months was higher which was due to the different active compositions and their sulfate resistance. In general, the slag contains more high-active substances than fly ash. Therefore, the change in the strength of the mortar containing slag was more significant.

To discuss the effect of erosion age on mechanical properties of the mortar containing slag well, the variations of mechanical properties of the mortar attacked by sulfate for four months were investigated as shown in Figure 7.

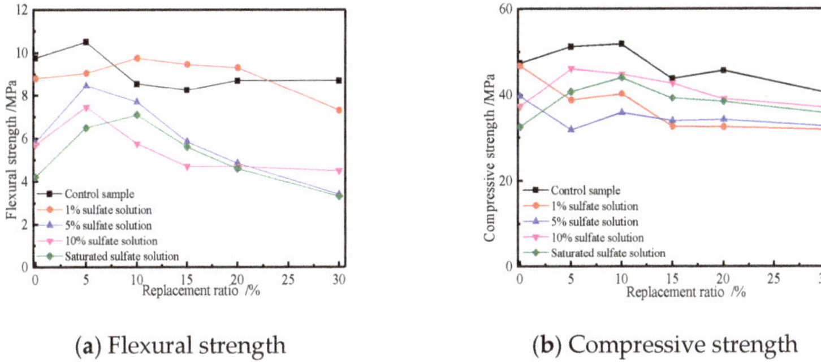

(**a**) Flexural strength        (**b**) Compressive strength

**Figure 7.** Variation curves of the strength of the mortar with slag content.

As seen from Figure 7, the compressive and flexural strength of the mortar attacked by sulfate for four months first increased and then decreased with the increase of the replacement ratio, and the corresponding strength values were less than that of the control sample. The variation of the flexural strength of the mortar attacked by sulfate was more sensitive. Although the change laws of the strength of the mortar attacked by sulfate were similar, the corresponding variations of the strength were different. There exists a maximum value of the strength of the mortar with a replacement ratio ranging from 5% to 15%. The above results indicate the sulfate solution concentration has a significant effect on the variation of the mortar's mechanical properties. In comparison with Figure 6, it can be seen that the variation of the strength of the mortar attacked by sulfate for four months was more significant and its regularity was better. This may be due to the fact that the sulfate attack plays a negative effect on the performance of the mortar. Compared with the strength variation of the mortar containing fly ash in Figure 3, the corresponding strength of the mortar containing slag was higher. Simultaneously, there exists a difference in the strength of the mortar specimens containing different replacement ratios of cement by slag. The above results indicate that the kinds and content of admixtures have a significant effect on the sulfate resistance of the mortar. Especially, the slag has better sulfate resistance performance than fly ash.

In order to reveal the effect of sulfate solution concentration on strength of the mortar, the variations of the compressive and flexural strengths of the mortar containing slag with concentrations of sulfate solution was investigated. Figure 8 shows the variation curves of the strength of mortar attacked by sulfate for four months.

The flexural strength of the mortar containing slag attacked by sulfate decreased with the increase in the concentration of the sulfate solution, but its compressive strength first decreased and then increased. The solution concentration of 5% had a significant effect on the compressive strength of the mortar specimen containing slag, but the variation of the flexural strength was not obvious. The flexural strength of the mortar specimens was more than that of the blank sample, when the replacement ratio of cement by slag was no more than 10%. The flexural strength variation of mortar became more significant under a high concentration of the sulfate solution. The corresponding compressive strength of mortar was larger than that of the blank sample when the sulfate solution concentration is larger than 10%. The variation of the compressive strength of the mortar becomes more significant with an increase of the sulfate solution concentration. Compared with two months, the strength of the mortar specimens attacked by sulfate for four months decreased obviously. In comparison with the strength variation of mortar containing fly ash in Figure 4, the strength of the mortar containing slag was higher with the same replacement ratio and sulfate solution concentration. The above results imply that the slag had an excellent sulfate resistance which was due to the active composition.

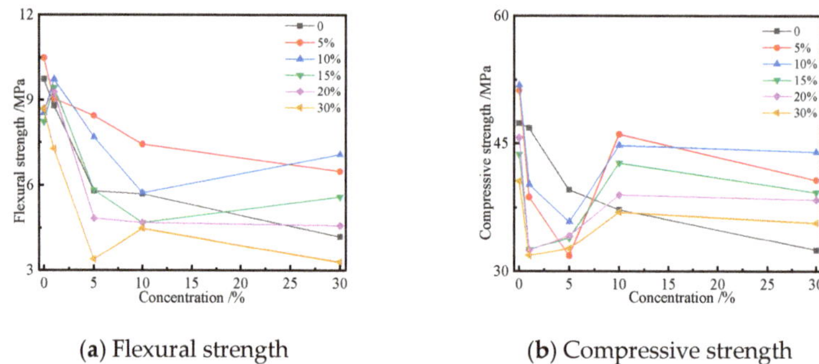

(**a**) Flexural strength

(**b**) Compressive strength

**Figure 8.** Variation curves of the strength of mortar attacked by sulfate for four months with concentrations of sulfate solution.

*3.2. Effect of Erosion Form of Sulfate Attack on the Strength of the Mortar*

The variation of the compressive and flexural strength of the mortar containing different admixtures attacked by semi-immersion form for four months was investigated, and 10% and saturated sulfate solution were used, respectively, to conduct the sulfate attack test. Figures 9 and 10 show the variation curves of the compressive and flexural strength of the mortar with the contents of fly ash and slag, respectively. The code numbers for the specimens mentioned below are explained as follow. Taken B10-F10 in Figure 9a as an example, the front part of letter and number stand for the immersion form (i.e., B is the semi-immersion form, C is the complete immersion from) and concentration of sulfate solution (i.e., the number stands for percentage, S is the saturated sulfate solution). The latter part of letter and number stand for the admixture and its substitute amount (i.e., F is the fly ash, S is the slag). Therefore, B10-F10 means the specimen containing 10% fly ash is attacked by 10% sulfate solution with semi-immersion form.

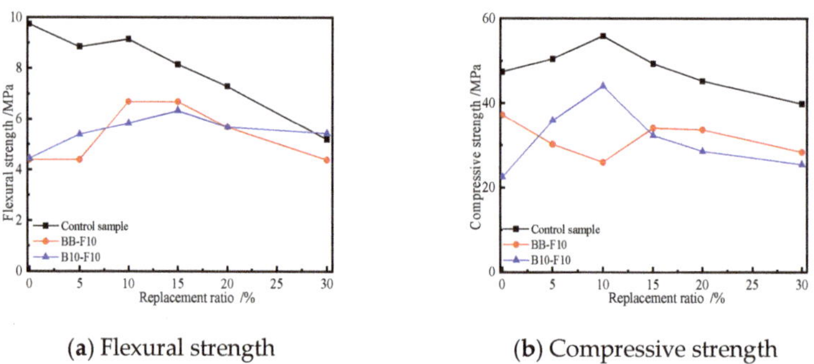

(**a**) Flexural strength

(**b**) Compressive strength

**Figure 9.** Variation curves of the compressive and flexural strengths of mortar with the replacement of cement with fly ash.

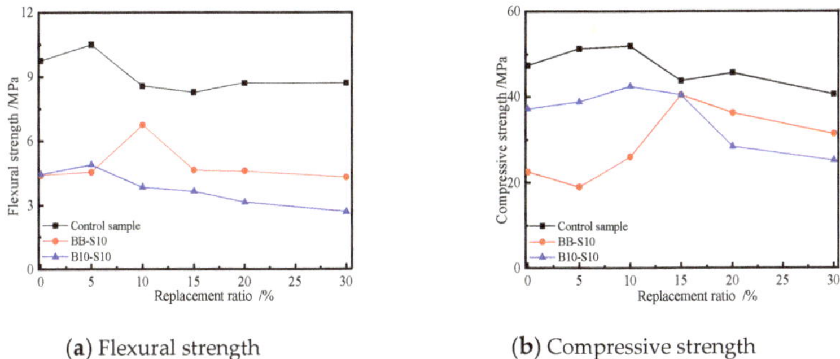

(**a**) Flexural strength    (**b**) Compressive strength

**Figure 10.** Variation curves of the compressive and flexural strength of mortar with the replacement of cement with slag.

As seen from Figure 9, the flexural strength of the control sample decreased with the increase of the replacement of cement by fly ash, but the compressive strength first increased and then decreased. However, the flexural strength of mortar attacked by semi-immersion form for four months first increased and then decreased. There exists a maximum value of strength of the mortar containing fly ash, when the replacement ratio was of 10%. The change laws of the strength of the mortar attacked by different concentrations of sulfate solution were similar, but there also existed differences in terms of variations and percentages. The compressive and flexural strength of the mortar attacked by semi-immersion form were lower than the control sample. Compared with the solution concentration of 10%, the saturated sulfate solution had a more significant effect on the strength of the mortar specimens. Figure 10 also shows that the strength of the mortar attacked by semi-immersion form for four months first increased and then decreased with the increase in the replacement of cement by slag, and the strength value were all less than the control sample. That may be due to the fact that the sulfate attack had a negative effect on the performance of the mortar attacked by sulfate for four months. The compressive strength of the mortar attacked by sulfate increased with the replacement of cement by slag no more than 10%. The concentration of sulfate solution had a different effect on the variation of the strength of the mortar containing slag, that is, the higher the concentration of sulfate solution, the more significant the decrease of the strength. Compared with the variation of the strength of the mortar containing fly ash attacked by completed immersion form (i.e., Figures 3 and 7), the variation of the flexural strength of the mortar attacked by semi-immersion form was less, but the change of the compressive strength was more significant. The above results indicate the erosion form of the sulfate attack had a significant effect on the variation of the strength of the mortar. Under the evaporated and concentrated effects, much crystallized salt occurs on the mortar surface, which can result in the micro damage and cracks in the mortar. Therefore, the mechanical properties of mortar change. The above deduction can be verified by the crystallization on the mortar attacked by semi-immersion form for four months as shown in Figure 11.

Plenty of crystallization was formed on the mortar surface. Especially, more crystallization was observed on the mortar surface attacked by saturated sulfate solution. The morphology of the crystallization on the mortar's surface containing fly ash was fibrous and flocculent, but it was granular and dense salt covering the mortar's surface containing slag. The above results show that the admixture and concentration of sulfate solution had significant effects on the type and morphology of crystal. Compared with the solution concentration of 10%, there were more sulfate ions of saturated solution, so more crystallization was generated and crystallized on the mortar's surface under the effects of evaporation and capillary. The different active compositions of admixtures resulted in the difference of the thermodynamic equilibrium constants for different crystallized salts, so the kinds and morphologies of crystallization were different. The erosion mechanism and deterioration of mortar containing different admixtures differed from each other, so the variation of the mechanical properties and microstructure of the mortar was different. This was due to the fact that the effects of the admixtures and the solution concentrations on the performance of mortar were different.

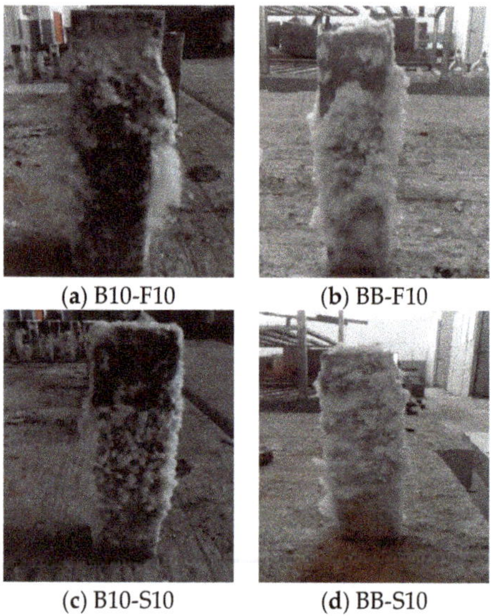

<table>
<tr><td>(a) B10-F10</td><td>(b) BB-F10</td></tr>
<tr><td>(c) B10-S10</td><td>(d) BB-S10</td></tr>
</table>

**Figure 11.** Crystallization on the surface of the mortar containing fly ash and slag attacked by semi-immersion form.

### 3.3. MIP Analysis

The effects of sulfate attack on the microstructure and variations in the pore characteristics of the mortar attacked by saturated solution for four months were studied as shown in Figure 12. Table 1 lists the characteristic parameters and distribution of pores in mortar.

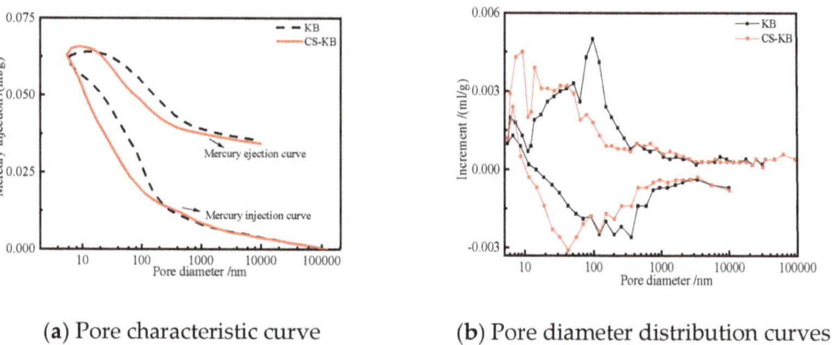



(**a**) Pore characteristic curve          (**b**) Pore diameter distribution curves

**Figure 12.** Curves of pore distribution characteristics of mortar before and after sulfate attack.

Figure 12 illustrates that the changes in the pore characteristics of the mortar before and after the sulfate attack were in terms of the porosity, average pore diameter, and pore distribution. The injection and ejection curves of mercury in Figure 12a show that the percentage of micropores with diameters less than 200 nm in the mortar attacked by sulfate for four months increased. Moreover, the difference between the mercury injection and ejection curves of the mortar attacked by sulfate decreased, which may be due to the erosion products generated by the sulfate attack refining the pores size. Compared with the blank sample, the variation of mercury curves of the mortar attacked by the saturated solution demonstrates that the number of micropores in the mortar increased. However, the variation of pores with a pore size larger than 1000 nm was not significant. The above results are in accordance with the measured data of pores listed in Table 3. The porosity and average pore diameter of mortar attacked by sulfate for four months were 13.24% and 35.1 nm, respectively. Simultaneously, the percentage of micropores in the mortar increased to 69.3%. This is due to the fact that the generated expansive products filled in some larger pores, which resulted in the decrease of the pore size. The increase in the porosity in the mortar was related to the micro damage and cracks caused by the sulfate attack.

**Table 3.** Characteristic parameters and distribution of pore in mortar.

| Items | Porosity, % | Average Pore Diameter, nm | Porosity Distribution, % | | |
|---|---|---|---|---|---|
| | | | 5–100 nm | 100 nm–10 μm | >10 μm |
| KBSJ | 12.78 | 85.3 | 54.2 | 40.1 | 5.7 |
| 4KBSJ | 13.24 | 35.1 | 69.3 | 24.7 | 6.0 |

To investigate the variation of the pore characteristics of mortar attacked by sulfate in detail, the pore distribution and characteristic curves of the mortar attacked by sulfate for four months were performed as shown in Figure 13.

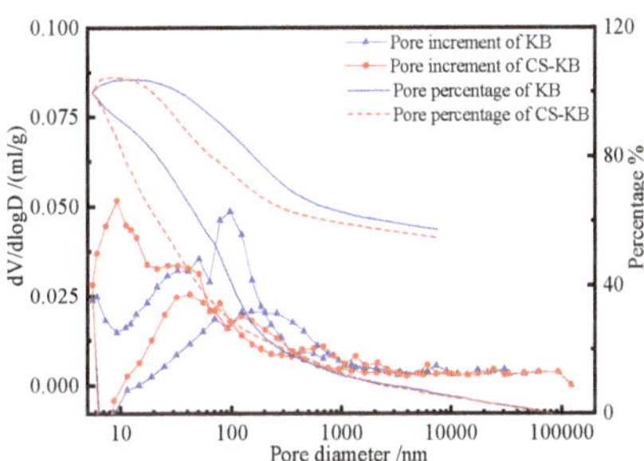


**Figure 13.** Curves of pore characteristics of mortar before and after sulfate attack.

Figure 13 shows that the porosity and the percentage of micropores with a pore diameter less than 100 nm increased when the mortar specimen was attacked by saturated sulfate solution for four months. Compared with the control sample, the percentage of macropores with a size larger than 200 nm in mortar attacked by saturated sulfate solution increased slightly, and the difference between the mercury injection and ejection curves of the mortar reduced. However, the deviation between the injection and ejection curves of the micropores with diameter less than 100 nm increased, which revealed that the micropores had a trend to translate into an ink bottle. As seen from the increment curves in Figure 13, the variations of the increment curves of the mortar specimens before and after attacked by sulfate are focused in the pore diameter less than 200 nm and 100 nm, respectively. This is due to the fact that the expansive products generated by sulfate attack in mortar fill and refine the pores, which results in the variation of porosity, compactness and crack of the mortar. Therefore, the mechanical properties of mortar specimens change before and after attacked by sulfate.

### 3.4. ESEM-EDS Analysis

The effects of admixtures and sulfate solution concentration on morphology and compositions of the mortar attacked by sulfate for four months were investigated, and the corresponding ESEM-EDS spectra are shown in Figure 14.

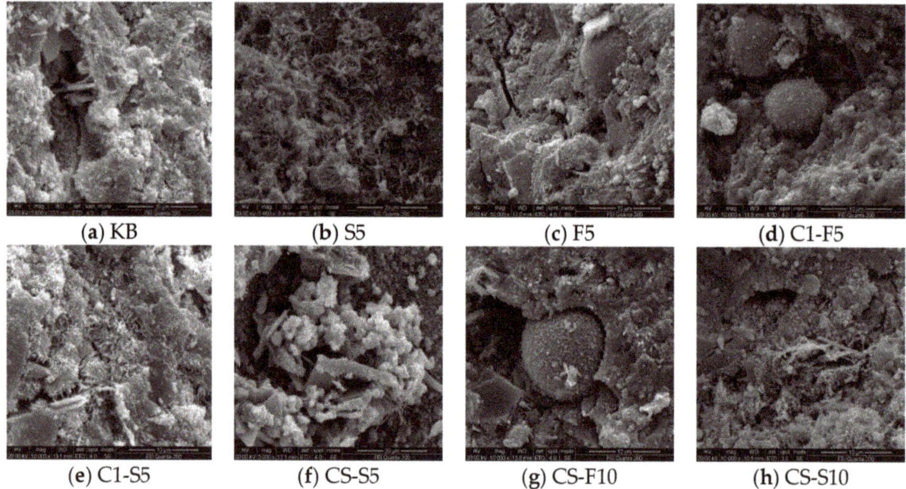

(a) KB      (b) S5      (c) F5      (d) C1-F5

(e) C1-S5      (f) CS-S5      (g) CS-F10      (h) CS-S10

**Figure 14.** ESEM-EDS spectra of mortar specimens.

The microstructure variations of the mortar containing admixtures before and after being attacked by sulfate were expressed as the compactness of microstructure, erosion products, and characteristic of pores [44] as shown in Figure 14. The major hydration products of the blank sample are a hexagonal plate CH and flocculent CAH, and some macropores can also be observed as shown in Figure 14a. The mortar containing admixtures had a more compact microstructure, less porosity, and a larger number of fibrous and flocculent hydration products as shown in Figure 14b,c. This is due to the fact that the admixtures can be activated by CH and formed into more cementitious substances. There exist differences in the microstructure of the mortar containing fly ash and slag, which is because of the difference of the active compositions and their chemical elements. Therefore, an appropriate number of admixtures added into mortar can enhance the microstructure and performances of mortar [45], which also accords well with the results in Figures 1 and 5. This is due to the fact that the actives substances of admixtures can react with CH and generate hydration products, so the microstructure and compactness of mortar are enhanced. The sulfate solution concentration plays a significant effect on microstructure, pore characteristics, erosion products. The microstructure of the mortar containing admixtures attacked by low sulfate solution concentration becomes denser, as shown in Figure 14d,e. Moreover, the fly ash particles eroded by CH can also be observed as shown in Figure 14d,g. There are plenty of rod-like and plate-like products generated by a high concentration of sulfate solution which may be considered as AFt, AFm, and gypsum based on their morphology characteristics. In addition, the replacement ratio of cement with slag also has a significant effect on erosion products as shown in Figure 14f,h.

The effect of erosion on the microstructure and morphology of mortar was studied, and the corresponding ESEM-EDS spectra of the mortar attacked by semi-immersion form for four months are shown in Figure 15.

The characteristics of microstructure and erosion products of the mortar containing admixtures were similar, and plenty of dendritic and rod-like products generated could be observed. Moreover, there were little CH and CAH. Compared with fly ash, the size of dendritic products generated by the sulfate attack in the mortar containing slag was smaller as shown in Figure 15a,b. The above morphology of erosion products in the mortar attacked by semi-immersion form differed from that of the mortar attacked by complete immersion form as shown in Figure 14. The EDS analysis showed that the major elements of dendritic products were Ca, Si, C, and O as shown in Figure 15c. Based on the morphology characteristics and composition elements, the dendritic products may be scawtite ($Ca_7(Si_6O_{18})(CO_3)\cdot 2H_2O$), spurrite ($Ca_5(SiO_4)_2(CO_3)$), and residue of decomposed CSH [46]. Simultaneously, plenty of crystallizations were formed on the mortar surface as shown in Figure 15e. The EDS analysis showed that the major elements of the crystallization were Na, S, and O as shown in Figure 15f. So. the crystallization may be sodium sulfate. Comparing with Figures 14 and 15, a conclusion can be drawn that the erosion form of sulfate attack affects the kinds and morphology of erosion products in the mortar.

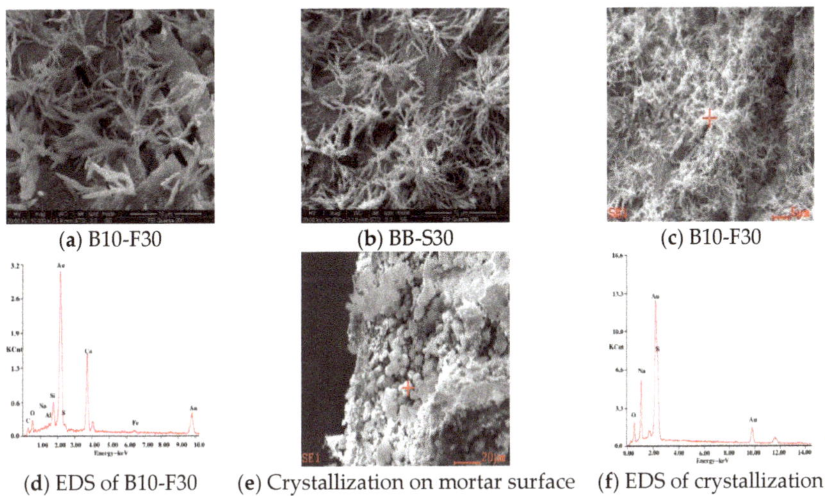

(**a**) B10-F30    (**b**) BB-S30    (**c**) B10-F30

(**d**) EDS of B10-F30    (**e**) Crystallization on mortar surface    (**f**) EDS of crystallization

**Figure 15.** ESEM-EDS spectra of mortar containing different admixtures attacked by different erosion forms.

## 4. Conclusions

(1) The effects of the admixtures, erosion age, concentration of sulfate solution, and erosion form on the mechanical properties of mortar were investigated. The results showed that the compressive and flexural strengths of the mortar specimens attacked by sulfate for four months decreased with the increase of the replacement of cement by fly ash. However, the strength of the mortar containing slag first increased and then decreased with the increase of the replacement of cement by slag. Although the change laws of the mortar specimens containing different kinds and contents of admixtures were similar, the strength variations and percentages of the mortar specimens were different. Compared with fly ash, the mortar containing slag had an excellent sulfate resistance under the same condition. The variation and regularity of the strength become more significant with the erosion age. An appropriate replacement of cement by admixtures can improve the mortar's microstructure and performances.

(2) The pore characteristics of the mortar, such as porosity and average pore diameter, were changed obviously after sulfate attack. The porosity and average pore diameter of the mortar specimen attacked by sulfate for four months increased, and the percentage of micropores with a pore diameter less than 200 nm increased. The mortar's microstructure became denser under a low concentration of sulfate solution, and there were plenty of plate-like and rod-like products in the mortar attacked by a high concentration of sulfate solution.

(3) The erosion form of sulfate attack played a significant effect on the erosion products and crystallization. The higher the concentration of the sulfate solution, the more crystallization on the mortar surface. Plenty of fibrous and flocculent crystallization were formed on the surface of the mortar containing fly ash, but they were the granular and dense crystallizations formed on the surface of mortar containing slag. Plenty of dendritic and rod-like products can be observed in the mortar containing admixtures attacked by semi-immersion form. The ESEM-EDS revealed that it may be

scawtite, spurrite, and residue of the decomposed CSH; however, the crystallization on the mortar specimen was mostly sodium sulfate.

**Author Contributions:** Writing—original draft, P.L.; investigation, Y.C.; methodology, Z.Y. All authors have read and agreed to the published version of the manuscript.

**Funding:** This study was funded by the National Natural Science Foundation of China, grant numbers 51778632 and U1934217; China Postdoctoral Science Foundation, grant numbers 2016M600675 and 2017T100647; China Railway Science and Technology Research and Development Plan Project, grant number2020-Major project-02; Basic Research on Science and Technology Program of Shenzhen, grant numbers JCYJ20170818143541342 and JCYJ20180305123935198; National Engineering Laboratory for High-Speed Railway Construction of China, grant number HSR201903; the Natural Science Foundation of Hunan Province of China, grant number 2020JJ5982.

**Conflicts of Interest:** All authors declare that they have no conflict of interest.

## References

1. Arel, H.S.; Thomas, B.S. The effects of nano- and micro-particle additives on the durability and mechanical properties of mortars exposed to internal and external sulfate attacks. *Results Phys.* **2017**, *7*, 843–851. [CrossRef]

2. Azevedo, A.R.G.; Cecchin, D.; Carmo, D.F.; Silva, F.C.; Campos, C.M.O.; Shtrucka, T.G.; Marvila, M.T.; Monteiro, S.N. Analysis of the compactness and properties of the hardened state of mortars with recycling of construction and demolition waste (CDW). *J. Mater. Res. Technol.* **2020**, *9*, 5942–5952. [CrossRef]

3. Zanelato, E.B.; Alexandre, J.; de Azevedo, A.R.G.; Marvila, M.T. Evaluation of roughcast on the adhesion mechanisms of mortars on ceramic substrates. *Mater. Struct.* **2019**, *52*, 53. [CrossRef]

4. Jiang, L.; Niu, D.; Yuan, L.; Fei, Q. Durability of concrete under sulfate attack exposed to freeze-thaw cycles. *Cold Reg. Sci. Technol.* **2015**, *112*, 112–117. [CrossRef]

5. Grengg, C.; Mittermayr, F.; Baldermann, A.; Bottcher, M.E.; Leis, A.; Koraimann, G.; Grunert, P.; Dietzel, M. Microbiologically induced concrete corrosion: A case study from a combined sewer network. *Cem. Concr. Res.* **2015**, *77*, 16–25. [CrossRef]

6. Cohen, M.D.; Mather, B. Sulfate attack on concrete: Research needs. *ACI Mater. J.* **1991**, *88*, 62–69.

7. Tanyildizi, H. Prediction of compressive strength of lightweight mortar exposed to sulfate attack. *Comput. Concr.* **2017**, *19*, 217–226. [CrossRef]

8. Kilinc, K.; Uyan, M.; Arioz, O. Effects of sulfates on strength of Portland cement mortars. In Proceedings of the 15th International Offshore and Polar Engineering Conference (ISOPE 2005), Seoul, Korea, 19–24 June 2005.

9. Kim, M.S.; Baek, D.I.; Cho, K.S. Evaluation on basic properties of crushed sand mortar in freezing-thawing and sulfate attack. *J. Ocean. Eng. Technol.* **2009**, *23*, 54–60.

10. Skaropoulou, A.; Tsivilis, S.; Kakali, G.; Sharp, J.H.; Swamy, R.N. Long term behavior of Portland limestone cement mortars exposed to magnesium sulfate attack. *Cem. Concr. Compos.* **2009**, *31*, 628–636. [CrossRef]

11. Makhloufi, Z.; Aggoun, S.; Benabed, B.; Kadri, E.H.; Bederina, M. Effect of magnesium sulfate on the durability of limestone mortars based on quaternary blended cements. *Cem. Concr. Compos.* **2016**, *65*, 186–199. [CrossRef]

12. Senhadji, Y.; Mouli, M.; Escadeillas, G.; Khelafi, A.; Bennosman, A.S.; Chihaoui, R. Behavior of limestone filler cement mortars exposed to magnesium sulfate attack. In Proceedings of the International Congress on Materials & Structural Stability, Rabat, Morocco, 26–30 November 2013.

13. Lee, S.-T. Performance of mortars exposed to different sulfate concentrations. *KSCE J. Civ. Eng.* **2012**, *16*, 601–609. [CrossRef]

14. Tanyildizi, H. The investigation of microstructure and strength properties of lightweight mortar containing mineral admixtures exposed to sulfate attack. *Measurement* **2016**, *77*, 143–154. [CrossRef]

15. Binici, H.; Arocena, J.; Kapur, S.; Aksogan, O.; Kaplan, H. Microstructure of red brick dust and ground basaltic pumice blended cement mortars exposed to magnesium sulphate solutions. *Can. J. Civ. Eng.* **2009**, *36*, 1784–1793. [CrossRef]

16. Zhang, M.H.; Jiang, M.Q.; Chen, J.K. Variation of flexural strength of cement mortar attacked by sulfate ions. *Eng. Fract. Mech.* **2008**, *75*, 4948–4957. [CrossRef]

17. Lee, S.T. Evaluation on the sulfate attack resistance of cement mortars with different exposure conditions. *J. Korean Soc. Civ. Eng. A* **2012**, *32*, 427–435.

18. Aziez, M.N.; Bezzar, A. Effect of temperature and type of sand on the magnesium sulphate attack in sulphate resisting Portland cement mortars. *J. Adhes. Sci. Technol.* **2018**, *32*, 272–290. [CrossRef]

19. Ghafoori, N.; Batilov, I.; Najimi, M. Effects of blaine and tricalcium aluminate on the sulfate resistance of nanosilica-containing mortars. *J. Mater. Civ. Eng.* **2018**, *30*, 1–13. [CrossRef]

20. Kim, M.S.; Baek, D.I. Unit weight and compressive strength of cement mortars using crushed sand under sulfate attacks. *J. Korean Soc. Civ. Eng. A* **2007**, *27*, 585–591.

21. Akpinar, P.; Casanova, I. A combined study of expansive and tensile strength evolution of mortars under sulfate attack: Implications on durability assessment. *Mater. Constr.* **2010**, *60*, 59–68. [CrossRef]

22. Idiart, A.E.; Lopez, C.M.; Carol, I. Chemo–mechanical analysis of concrete cracking and degradation due to external sulfate attack: A meso–scale model. *Cem. Concr. Compos.* **2011**, *33*, 411–423. [CrossRef]

23. Huang, Q.; Wang, C.; Luo, C.Q.; Yang, C.H.; Luo, Y.L.; Xie, H. Effect of mineral admixtures on sulfate resistance of mortars under electrical field. *Adv. Cem. Res.* **2017**, *29*, 45–53. [CrossRef]

24. Lee, S.T.; Hooton, R.D. Influence of limestone addition on the performance of cement mortars and pastes exposed to a cold sodium sulfate solution. *J. Test. Eval.* **2016**, *44*, 414–423. [CrossRef]

25. Niu, Q.L.; Feng, N.Q. Effect of different mineral admixtures on sulfate attack to mortars. *Key Eng. Mater.* **2009**, *405*, 278–285. [CrossRef]

26. Hossack, A.M.; Thomas, M.D.A. Varying fly ash and slag contents in Portland limestone cement mortars exposed to external sulfates. *Constr. Build. Mater.* **2015**, *78*, 333–341. [CrossRef]

27. Chusilp, N.C.; Chai, J.; Kraiwood, K. Effects of LOI of ground bagasse ash on the compressive strength and sulfate resistance of mortars. *Constr. Build. Mater.* **2009**, *23*, 3523–3531. [CrossRef]

28. Ryou, J.; Lee, S.; Park, D.; Kim, S.; Jung, H. Durability of cement mortars incorporating limestone filler exposed to sodium sulfate solution. *KSCE J. Civ. Eng.* **2015**, *19*, 1347–1358. [CrossRef]

29. Chindaprasirt, P.; Homwuttiwong, S.; Sirivivatnanon, V. Influence of fly ash fineness on strength, drying shrinkage and sulfate resistance of blended cement mortar. *Cem. Concr. Res.* **2004**, *34*, 1087–1092. [CrossRef]

30. Marvila, M.T.; Alexandre, J.; Azevedo, A.R.G.; Zanelato, E.B.; Xavier, G.C.; Monteiro, S.N. Study on the replacement of the hydrated lime by kaolinitic clay in mortars. *Adv. Appl. Ceram.* **2019**, *118*, 373–380. [CrossRef]

31. Markssuel, T.M.; Afonso, R.G.A.; Daiane, C.; Jonatas, M.C.; Gustavo, C.X.; Dirlanede, F.C.; Sergio, N.M. Durability of coating mortars containing acai fibers. *Case Stud. Constr Mater.* **2020**, *13*, 00406.

32. Tosun, K.; Felekoglu, B.; Baradan, B.; Altun, I.A. Effects of limestone replacement ratio on the sulfate resistance of Portland limestone cement mortars exposed to extraordinary high sulfate concentrations. *Constr Build. Mater.* **2009**, *23*, 2534–2544. [CrossRef]

33. Kim, Y.S. Effects of pozzolanic admixture replacement ratio in ternary mortar on the sulfate attack resistance. *J. Arch. Inst. Korea Struct. Constr.* **2005**, *21*, 89–96.

34. Lee, S.T.; Hooton, R.D.; Jung, H.S.; Park, D.H.; Choi, C.S. Effect of limestone filler on the deterioration of mortars and pastes exposed to sulfate solutions at ambient temperature. *Cem. Concr. Res.* **2008**, *38*, 68–76. [CrossRef]

35. Lee, B.; Kim, G.; Nam, J.; Lee, K.; Kim, G.; Lee, S.; Shin, K.; Koyama, T. Influence of alpha-calcium sulfate hemihydrate on setting, compressive strength, and shrinkage strain of cement mortar. *Materials* **2019**, *12*, 163.

36. Said, A.M.; Zeidan, M.S.; Bassuoni, M.T.; Tian, Y. Properties of concrete incorporating nano-silica. *Constr. Build. Mater.* **2012**, *36*, 838–844. [CrossRef]

37. Marvila, M.T.; Azevedo, A.R.G.; Barroso, L.S.; Barbosa, M.Z.; de Brito, J. Gypsum plaster using rock waste: A proposal to repair the renderings of historical buildings in Brazil. *Constr. Build. Mater.* **2020**, *250*, 118786. [CrossRef]

38. Azevedo, A.R.G.; Alexandre, J.; Marvila, M.T.; Xavier, G.D.; Monteiro, S.N.; Pedroti, L.G. Technological and environmental comparative of the processing of primary sludge waste from paper industry for mortar. *J. Clean. Prod.* **2020**, 119336. [CrossRef]

39. Amaral, L.F.; Delaqua, G.C.G.; Nicolite, M.; Marvila, M.T.; de Azevedo, A.R.G.; Alexandre, J.; Vieira, C.M.F.; Monteiro, S.N. Eco-friendly mortars with addition of ornamental stone waste-A mathematical model approach for granulometric optimization. *J. Clean. Prod.* **2020**, *248*, 119283. [CrossRef]

40. Ghrici, M.; Kenai, S.; Said, M.M. Mechanical properties and durability of mortar and concrete containing natural pozzolan and limestone blended cements. *Cem. Concr. Compos.* **2007**, *29*, 542–549. [CrossRef]

41. State Bureau of Quality Technical Supervision. *Method of Testing Cements-Determination of Strength, GB/T 17671–1999*; China Standard Press: Beijing, China, 1999.
42. Mehta, P.K. Mechanism of sulfate attack on Portland cement concrete—Another look. *Cem. Concr. Res.* **1983**, *13*, 401–406. [CrossRef]
43. Liu, P.; Chen, Y.; Wang, W.L.; Yu, Z.W. Effect of physical and chemical sulfate attack on performance degradation of concrete under different conditions. *Chem. Phys. Lett.* **2020**, *745*, 137254. [CrossRef]
44. Skaropoulou, A.; Tsivilis, S.; Kakali, G.; Sharp, J.H.; Swamy, R.N. Thaumasite form of sulfate attack in limestone cement mortars: A study on long term efficiency of mineral admixtures. *Constr. Build. Mater.* **2009**, *23*, 2338–2345. [CrossRef]
45. Liu, T.J.; Qin, S.S.; Zou, D.J.; Song, W. Experimental investigation on the durability performances of concrete using cathode ray tube glass as fine aggregate under chloride ion penetration of sulfate attack. *Constr Build. Mater.* **2018**, *163*, 634–642. [CrossRef]
46. Zhang, Y.Q.; Radha, A.V.; Navrotsky, A. Thermochemistry of two calcium silicate carbonate minerals: Scawtite, $Ca_7(Si_6O_{18})(CO_3) \cdot 2H_2O$, and spurrite, $Ca_5(SiO_4)_2(CO_3)$. *Geochim. Cosmochim. Acta* **2013**, *115*, 92–99. [CrossRef]

*Article*

# Thermal Performance of Alginate Concrete Reinforced with Basalt Fiber

**Seyed Esmaeil Mohammadyan-Yasouj** [1,*]**, Hossein Abbastabar Ahangar** [2]**,**
**Narges Ahevani Oskoei** [1]**, Hoofar Shokravi** [3]**, Seyed Saeid Rahimian Koloor** [4] **and Michal Petrů** [4]

[1] Department of Civil Engineering, Najafabad Branch, Islamic Azad University, Najafabad 8514143131, Iran;
narges.oskoei@yahoo.com
[2] Department of Chemistry, Najafabad Branch, Islamic Azad University, Najafabad 8514143131, Iran;
abbastabar@gmail.com
[3] School of Civil Engineering, Faculty of Engineering, Universiti Teknologi Malaysia,
Skudai 81310, Johor, Malaysia; shoofar2@graduate.utm.my
[4] Institute for Nanomaterials, Advanced Technologies and Innovation (CXI),
Technical University of Liberec (TUL), Studentska 2, 461 17 Liberec, Czech Republic;
s.s.r.koloor@gmail.com (S.S.R.K.); Michal.petru@tul.cz (M.P.)
* Correspondence: sm7093370@yahoo.com

Received: 13 July 2020; Accepted: 28 August 2020; Published: 3 September 2020

**Abstract:** The sustainability of reinforced concrete structures is of high importance for practitioners and researchers, particularly in harsh environments and under extreme operating conditions. Buildings and tunnels are of the places that most of the fire cases take place. The use of fiber in concrete composite acts as crack arrestors to resist the development of cracks and enhance the performance of reinforced concrete structures subjected to elevated temperature. Basalt fiber is a low-carbon footprint green product obtained from the raw material of basalt which is created by the solidification of lava. It is a sustainable fiber choice for reinforcing concrete composite due to the less consumed energy in the production phase and not using chemical additives in their production. On the other hand, alginate is a natural anionic polymer acquired from cell walls of brown seaweed that can enhance the properties of composites due to its advantage as a hydrophilic gelling material. This paper investigates the thermal performance of alginate concrete reinforced with basalt fiber. For that purpose, an extensive literature review was carried out then two experimental phases for mix design and to investigate the compressive strength of samples at a temperature range of 100–180 °C were conducted. The results show that the addition of basalt fiber (BF) and/or alginate may slightly decrease the compressive strength compared to the control concrete under room temperature, but it leads to control decreasing compressive strength during exposure to a high temperature range of 100–180 °C. Moreover, it can be seen that temperature raise influences the rate of strength growth in alginate basalt fiber reinforced concrete.

**Keywords:** concrete; sodium alginate; basalt fiber; compressive strength; temperature

---

## 1. Introduction

Concrete is widely used in the construction of many structures such as buildings [1,2] and bridges [3,4]. Plain concrete is weak in tension due to its brittle nature; hence fibers are added to sustain stresses and restrict the propagation of cracks [5]. There are various fibers used in concrete application that the most predominant types include synthetic, natural, carbon, glass, and steel [6]. Basalt fiber (BF) is a new kind of natural fiber manufactured from the extrusion of basalt rock in a melting temperature between 1500 and 1700 °C [7]. Basalt fibers have good resistance to fire and corrosion with advanced vibration and acoustic insulation capacity [8]. Basalt fiber is a cheaper and more eco-friendly alternative to glass, carbon, or aramid fiber for reinforcing concrete composite due to the less consumed energy in the production phase and not using chemical additives in their production [9]. Table 1 shows a summary of previous studies that investigated the behavior of basalt fiber-reinforced concrete (FRC) compared with other fibers including polypropylene, glass, and polyvinyl alcohol (PVA).

Several experiments were conducted to study the effect of using different lengths, diameters, and proportions of fibers in concrete. Jiang et al. [10] studied the effect of BF on the mechanical characteristic of the FRC. The obtained results indicated that adding BF can significantly enhance the flexural and tensile strength, as well as the toughness index of FRC where no obvious increase of compressive strength was reported. In comparison to other fiber types, higher splitting tensile and flexural strengths of BF than polypropylene [10], higher tensile strength, crack resistance, and ductility and lower flexural strength of BF than glass fiber (GF) [11], lower increase in energy absorption at high strain rate by BF than GF [12], neutral contribution of BF and significant contribution of PVA in post cracking flexural response [13] have been reported. Almost all previous research on the fiber reinforced concrete confirmed that increase in the fiber length and/or content decreases concrete workability and increases its porosity [14]. Reduction in the workability due to an increase in the fiber content requires higher water to cement ratio which leads to higher porosity in the hardened concrete. Accordingly, fiber content is also a main parameter considered by previous researchers to control the fiber effect on the concrete properties. However, selecting an appropriate amount of fiber content to reach proper workability with expected mechanical properties for concrete is necessary.

Alginate sodium is a biopolymer extracted from cell walls of brown algae, that forms swollen hydrogels in the presence of water [15,16]. It is a linear anionic polysaccharide composed of mannuronic (M) and guluronic (G) acid that is glycosidically connected via a-1→4 and ß-1→4 linkages thus hydrogel formation occurs through ionic cross-linking [17]. Sodium alginate is currently being investigated for the production of self-healing of concrete [18]. Alginate concentration influences the properties of the produced hydrogel [19]. Abbas and Mohsen [20] investigated the effect of alginate in the behavior of concrete. It is reported that the addition of alginate increased the slump and fresh density, and improved the workability of fresh concrete. Enhanced mechanical properties (compressive, splitting tensile and flexural strength) were also reported by the addition of 1% of the alginate in the concrete mixture. However, Heidari et al. [21] indicated that alginate may decrease the mechanical properties of concrete, but some observations can improve the performance of self-compacting concrete in a fresh state. There are remarks that alginate gel releases its entrained mixing water and incites the densification and more hydration of cement particles [22]. Alginates also exhibit different thermal behavior according to the type used and alginate or its derivative were stated for their self-healing effect in concrete [23]. From the same resource, it was reported that building materials containing sodium alginate and/or calcium alginate lead to significant flame-, fire-, and heat-resistance or imperviousness to the materials. Due to these reports, alginate is promising for internal curing, self-sealing, and self-healing of cracks and is also effective when a high amount of superabsorbent is required. Though, the application of appropriate amounts of alginate in concrete or PC may help improving concrete characteristics such as its crack resistance or density that are effective for long term behavior. Table 4 shows a summary of the studies using alginate in concrete mixtures.

**Table 1.** Cement concrete with fiber.

| Ref. | Conducted Tests and Sample Monitoring | Test Age (day) | Codes and Sample Size (mm) | Fiber Type | | | | Remarks |
|---|---|---|---|---|---|---|---|---|
| | | | | Amount (% of the Total Volume) | Diameter (μm) | Length (mm) | Name | |
| Jiang et al., 2014 [10] | Compression | 3-7-28-90 | Ø AS (1012.9-1999, 1012.10–2000, and 1012.11–1985) | (0.05, 0.1, 0.3, and 0.5) | 20 | 12 and 22 | Basalt (B) | Addition of fiber increased porosity and reduced workability |
| | Tension | | 100 × 200 cylinder | | | | | Small to moderate increase in compressive strength (by BF > by PP) at 7–28 day age. |
| | Flexure | | 150 × 300 cylinder | | | | | Higher toughness and splitting tensile strength (BF> PP) is achieved by increasing fiber contents. |
| | Permeability | | 75 × 75 × 305 prism | | | | | Addition to 0.3% fiber increases flexural strength; 0.5% BF content reduce flexural strength. |
| | | | | | - | 4–19 | Polypropylene (PP) | 0.3% of BF content with the length of 22 mm increases tensile and flexure strength. |
| Kabay 2014 [24] | Compression | 28 | Ø EN 12390-3 and ASTM C642 | (0.07 and 0.14) | 13–20 | 12 and 24 | B | Workability is reduced by increasing the length and/or volume of fibers. |
| | Flexure | | 71 × 71 cube | | | | | Compared to compressive strengthabrasion has a higher correlation with void content and flexural strength. |
| | Fracture energy | | 100 × 100 cube | | | | | Addition of fiber reduced compression strength. |
| | | | 70 × 70 × 285 prism (with 30 mm depth and 3 mm thickness notch) | | | | | The addition of fibers reduces abrasive wear in the range of 2–18%. |
| | | | | | | | | Most improved flexure strength was seen in mixture with 60% w/c. |
| Ayoub et al., 2014 [25] | Compression | 28-7 | Ø BS (1881-125: 1986, 1881-3:1970, 1881-102) | (1, 2, and 3) | 18 | 25 | B | Increased BF contents reduced compressive strength. |
| | Tension | | 100 × 200 cylinder | | | | | Higher tension strength is achieved for high-performance concrete with BF and 10% silica. |
| | E-value | | | | | | | BF does not correlate with E-value. |
| Kizilkanat et al., 2015 [11] | Compression | 28-7 | EN 12390-3, ASTM (C469M, C496), and JCI-S-002 | (0.25, 0.5, 0.75, and 1) | 13–20 | 12 | B | Increasing fibers caused a significant increase in fracture energy where workability and E-value were decreased |
| | Tension | | 150 × 150 × 150 cube | | | | | Slight increase in compressive strength was observed in >0.25% fiber content |

**Table 1.** *Cont.*

| Ref. | Conducted Tests and Sample Monitoring | Test Age (day) | Codes and Sample Size (mm) | Fiber Type | | | | Remarks |
|---|---|---|---|---|---|---|---|---|
| | | | | Amount (% of the Total Volume) | Diameter (μm) | Length (mm) | Name | |
| | Flexure | | 100 × 200 cylinder | | | | | 0.5% and 0.75% inclusion of BF improved compressive strength |
| | E-value | | 100 × 100 × 350 prism (with 30 mm notch) | | | | | BF has a higher impact on crack avoidancetensile strengthand ductility |
| | Fracture energy | | | | 13 | 12 | Glass (G) | GF is directly related with flexure strength |
| Fenu et al., 2016 [12] | Compression | 28 | Ø EN (197-1, 196-1) | 3 and 5 | 14 | 12 | B | Fibers' addition did not show a significant impact on the dynamic increase factor. |
| | Tension | | 40 × 40 × 160 prism (cut from bending); | | | | | A significant and slight increase in energy absorption at a high strain rate was achieved by GF and BFrespectively. |
| | Flexure | | 100 × 200 cylinder core from prism for dynamic test | | | | | Improved static flexural strength and post-peak behavior were achieved by GF and BF. |
| | SEM | | | | 14 | 12 | G | |
| Shafiqh et al., 2016 [13] | Flexure | 28 | Ø ASTM C192/C192M, C 494-86 Type G, C78, C 1609) | 1, 2, and 3 | 18 | 25 | B | Strain attaining capacity was achieved by the addition of GF and BF. |
| | SEM | | 100 × 200 × 1500 prism | | | | | Ductilitypost-cracking flexural responseand toughness were significantly improved by PVA fibers. |
| | | | | | 660 | 30 | Polyvinyl Alcohol (PVA) | No correlation between BF in post-peak flexural behavior was observed. |
| Girgin 2016 [26] | Compression | 17 and 28 | Ø EN (1170-4/5:1997, 12467) | 2 | 14–20 | 24 | B | Stress accumulation due to the cement hydration in the matrix–fiber interface influenced flexural strain capacity of BF. |
| | Flexure | | 50 × 50 cube | | | | | A 35% reduction in the average strain capacity of GF-reinforced concrete compared with the 7-day one was achieved. |
| | Heating | | 12 × 600 × 600 square sheet | | | | | |
| | SEM | | 11 × 50 × 270 square sheet | | | | | |
| | | | | | 13–20 | 24 | G | |

**Table 1.** *Cont.*

| Ref. | Conducted Tests and Sample Monitoring | Test Age (day) | Codes and Sample Size (mm) | Fiber Type | | | | Remarks |
|---|---|---|---|---|---|---|---|---|
| | | | | Amount (% of the Total Volume) | Diameter (μm) | Length (mm) | Name | |
| Afroz et al., 2017 [9] | Compression | | Ø AS (3972, 3582.1–1998, 1012.8, 1012.9, 1012.10, 1012.11) | 0.5 | 15 | 3 | B (Chinese) | Long-term compatibility of BF reinforcement fibers for chloride and sulfate solutions. |
| | Tension | | 100 × 200 cylinder | | | | | Compared to non-modified BFmodified BFs had better stability in alkaline medium. |
| | Flexure | | 150 × 300 cylinder | | | | | Coating provided good protection while exposure to extensive alkaline ions from a physiochemical analysis point of view. |
| | SEM | | 100 × 100 × 350 prism | | | | | A reduction in the mechanical properties of concrete for treated fibers was observed. |
| | EDX | | | | 15 | 25 | B (Russian) | Improved long-term flexural strengths and splitting tensile for high volume fly ash concrete with silane coated BFs even after 56 days. Difference in compressive strength was not notably improved. |
| Zhao et al., 2017 [27] | Impact | 120 | GB/T 50082-2009 | 1, 1.5, 2, and 2.5 | 15 | 18 | B | Gradual loss in performance was observed by freeze-thaw cycles. |
| | Freeze-thaw | | 40 × 40 × 160 prism | | | | | Improved anti-impact deformation characteristics were achieved using BF. |
| | | | | | | | | Anti-impact deformation characteristics were weakened by freeze-thaw cycles. |
| Katkhuda and Shatarat 2017 [28] | Compression | 28 | Ø ASTM (C33, C127, C131, C1524, C143) | 0.1, 0.3, 0.5, 1, and 1.5 | 16 | 18 | B | Flexural and splitting tensile strength of BF-reinforced concrete was significantly improved while it was minimally enhanced for the compressive strength. |
| | Tension | | 100 × 100 cube | | | | | The optimum BF content to achieve the same compressive and splitting tensile strength as the natural aggregate was 0.5% for untreated recycled coarse aggregate (RCA) and 0.3% for treated RCA while the optimum BF content for flexural strength was 0.3% for untreated RCA and 0.1% for treated RCA. |
| | Flexure | | 150 × 300 cylinder | | | | | |
| | Permeability | | 100 × 100 × 500 prism | | | | | |
| Sun et al., 2019 [29] | Compression | 28 | Ø SL352-2006, CECS13:2009 | 1, 2, 3, 4, and 5 | 17.4 | 6 and 12 | B | The compressive and splitting tensile strengths of concrete reduced with the increase of fiber content while the bending strength increased. |
| | Tension | | 150 × 150 cube | | | | | Compared to BF with the length of 12 mm 6 mm BF had better splitting tensile performance and compressive strength while variation of bending strength is slight |
| | Flexure | | 100 × 100 × 400 prism | | | | | 2% of 6 mm BF achieved the maximum strength |

Using low carbon footprint materials as a constituent of concrete is one of the ways to improve the sustainability of the construction industry. Basalt fiber and alginate are two low carbon footprint green products that are widely used for improving the performance of the produced concretes. Recently, several studies have been carried out to investigate the effect of BF-reinforced concrete exposure to elevated temperature. Few researchers have covered the effect of alginate in concrete. To the best of the authors' knowledge, no research has been directed at determining the thermal performance of alginate concrete reinforced with basalt fiber. In this study, the effect of simultaneous and separate addition of BF and alginate to improve the temperature resistance of concrete is experimentally studied. A two stage procedure was used to carry out the experimental study. The first stage was conducted to determine the appropriate water to cement ratios and also deciding for inclusion or exclusion of the aggregate smaller than 150 μm (passed through sieve No. 100). In the second stage, the thermal performance of the mortar with addition or/and elimination of ingredients (alginate and basalt fiber) in the composition mix was investigated. Basalt fiber is expected to potentially enhance concrete's resistance to crack and prevent the damages such as spalling in the concrete caused by elevated temperatures. Sodium alginate is intended to improve the microstructure of concrete to achieve higher temperature resistance capacity while exposed to elevated temperature.

## 2. Materials and Methods

### 2.1. Materials

Materials including limestone aggregates, cement, BF, and alginate were purchased from local suppliers and prepared for mixtures. Type I Portland cement was used for concretes. Type I Portland cement was used for concretes. In the primary study on the prepared materials, the mix designs were investigated with 450 and 470 kg/m$^3$ cement amounts and 0.5 and 0.55 water to cement ratios. The aggregates for the mortar were crushed limestone and sieved below through sieve number 4 to provide relatively fine aggregates of sizes less than 4.75 mm in accordance of ASTM C117 and E11. The fineness modulus of the aggregates was determined to be 2.72. The size distribution curve for the aggregates is shown in Figure 1. The curve is in accordance to the standard area of ASTM C33-03. Water absorption of the aggregates was about 1%.

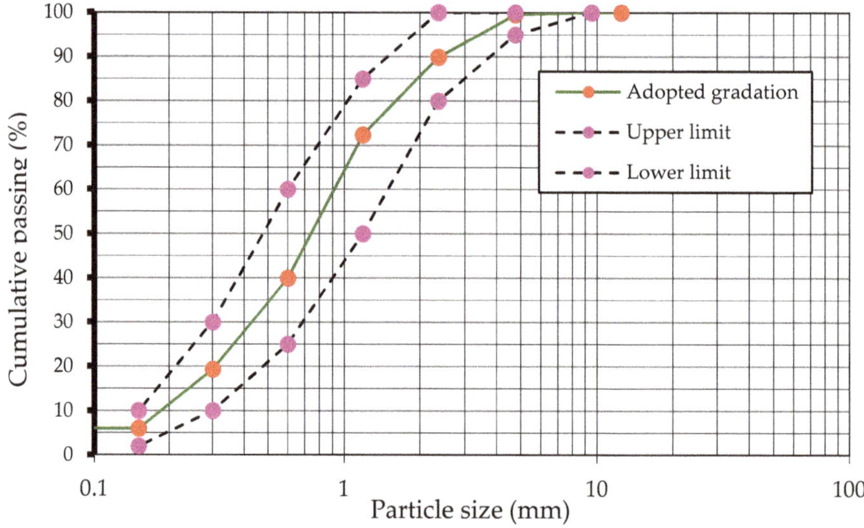


**Figure 1.** Size distribution curve for aggregate used in the present research.

Cubic mortar specimens of 50 × 50 × 50 mm were cast for testing compressive strength. The maximum size of aggregates in the mortar was limited to the aggregates smaller than 4.75 mm (passed sieve No. 4), and fineness modulus was 2.72. From Table 1, it can be seen that the length of BF used for concrete-reinforcing applications in earlier studies was varied in the range of 3–30 mm. It is evident that shorter chopped fibers can achieve more uniform distribution while longer chopped lengths are more efficient in bridging and controlling the width of cracks. As shown in Table 1, in previous studies the amount of fiber content in concrete mixes varied in the range of 0.05–5% of the total concrete volume. The findings drawn from previous studies suggest using fiber content and a chopped length of 2% and 10 mm, respectively. From Table 2, the amount of sodium alginate used in concrete mixture is within a limited range. In this study, a sodium alginate concentration of 0.1% of the total mortar volume was used for each mixture. The morphological and mechanical properties of the BF used as reinforcement of the mortar specimens in the present study are demonstrated in Table 3. Figure 2 presents the materials used in the composition of the mortar mixes of the experimental study.

**Figure 2.** The materials used in the composition of the concrete mixes of the experimental study that includes (**a**) aggregate (**b**) cement (**c**) basalt fibers (**d**) alginate powder.

**Table 2.** Alginate application in concrete.

| Ref. | Conducted Tests and Sample Monitoring | Test Age (day) | Codes and Sample Designation (mm) | Details on Alginate | | Remarks |
| | | | | Amount (% of Cement Weight) | Type | |
| --- | --- | --- | --- | --- | --- | --- |
| Ouwerx et al., 1998 [19] | E-value | 1 | Solution | - | Alginate gel bead | Alginate concentration limits the elasticity of breads. |
| | SEM | | | | | Properties of alginate gel bead are dependent on the concentration and polymer type. |
| Pathak et al., 2008 [16] | TGA | - | Solution | - | Dried algae (Sea mustard) | Alginic acid and metal alginates were prepared from fresh algae using extraction method then compared. |
| | FTIR | | | | | Conversion of alginic acid into metal alginate was confirmed by FTIR spectroscopy. |
| | DSC | | | | | Cobalt alginate became decomposed at a higher temperature in comparison to sodium and calcium alginate. |
| | SEM | | | | | Presence of metal ions impacted on pore size distribution of alginates. |
| | Porosity | | | | | Different thermal behavior was observed in different metal alginates due to the structural differences. |
| Heidari et al., 2015 [21] | Slump Flow (L-box) | 7, 28, 56, and 90 | Ø ASTM (C 462) | 0.5 and 1 | Alginic acid | Decrease in compressive, tensile, and flexural strengths by alginate. |
| | Compression | | 100 × 100 × 100 cube | | | Increase in workability, permeability, and compressive strength by silica. |
| | Tension | | 150 × 300 cylinder | | | Alginate improved performance of SCC in fresh state. |
| | Flexure | | 160 × 400 × 400 prism | | | In mixes containing alginate, nano and micro silica together, use of nano and micro-silica in upper percent of alginate increased the strength. |
| | Permeability | | | | | |
| Mignon et al., 2016 [22] | Compressive | 28 | Ø EN 196-1 | 0.5 and 1 | Alginate biopolymers (NaAlg, CaAlg) | Alginate was promising for internal curing and was effective when much of superabsorbent polymer was required. |
| | Flexure | | 40 × 40 × 160 prism | | | NaAlg was insoluble due to the divalent cations in cement filtrate and created a physically cross-linked hydrogel. |
| | Swelling | | | | | CaAlg beads of 1% reduced only 15% of the compression strength. |
| | | | | | | NaAlg of 1%, due to the higher water uptake capacity and stronger macro-pore, led to a 28% decrease in strength. |

**Table 3.** Properties of basalt fiber (BF) used in the mortar specimens.

| Cutting Length (mm) | 10 |
|---|---|
| Diameter ($\mu$m) | 17 |
| Density (gr/cm$^3$) | 2.65 |
| Elastic modulus (GPa) | 93–110 |
| Tensile strength (MPa) | 4100–4800 |
| Elongation (%) | 1.3–3.2 |
| Softening point (°C) | 1050 |
| Water absorption (%) | <0.5 |

## 2.2. Mix Design and Specimen Preparation

The experimental study was carried out in two stages. The first stage consisted of six mixes considering two different water to cement ratios of 0.5 and 0.55 and also deciding for inclusion or exclusion of the aggregate smaller than 150 $\mu$m (passed through sieve No. 100) identify water content and proportion of the mix. In the second stage, the main objective was to investigate the thermal performance of the mortar with addition and/or elimination of ingredients (alginate and basalt fiber) in the composition mix. Tables 4 and 6 present details on mix designs and the obtained compressive strength of each mix in the first stage, respectively. From the results in Table 6 that are discussed in the latter parts, $M_pW_{0.5}C_{470}E_{\#100}$ was considered as the reference mix design for concrete samples, in the second stage of the experiment.

**Table 4.** Details of mix designs in the primary study.

| Name | Water–Cement Ratio | Cement (kg/m$^3$) | Polymer (%) * | Aggregates Passed Through Sieve # 100 |
|---|---|---|---|---|
| $M_pW_{0.5}C_{450}I_{\#100}$ ** | 0.5 | 450 | - | Included |
| $M_pW_{0.5}C_{470}I_{\#100}$ | 0.5 | 470 | - | Included |
| $M_pW_{0.5}C_{450}E_{\#100}$ | 0.5 | 450 | - | Excluded |
| $M_pW_{0.5}C_{470}E_{\#100}$ | 0.5 | 470 | - | Excluded |
| $M_pW_{0.55}C_{450}E_{\#100}$ | 0.55 | 450 | - | Excluded |
| $M_pW_{0.55}C_{470}E_{\#100}$ | 0.55 | 470 | - | Excluded |

* of the total weight. ** $M_pW_xC_yI_{\#z}$ > $M_p$: mix design of the primary study, $W_x$: water to cement ratio of $x$, $C_y$: cement amount of $y$ kg/m$^3$, $I_{\#z}$: represented inclusion of aggregate passed through the sieve number $z$ in the mixture.

In the second stage, a different mix of BF and alginate was designed to be tested in various temperatures from room temperature to 180 °C that is shown in Table 5. The compositions in the mix design were obtained based on the statistical design of the experiment. Moreover, it should be reminded that in the mix designs, the water to cement ratio and cement content was kept constant equal to 0.5 and 470 kg/m$^3$, respectively. The zero temperature in subsequent tables represent the room curing temperature without applying any external source of heating that is used as the reference before heating of samples. The compressive strength of the mixes was tested at several constant temperatures including 100, 150, and 180 °C. In order to test the compressive strength, of the samples in elevated temperature, the molds were opened after 24 h, and specimens were placed in a water tub for curing (ASTM C 109). After 28 days of water curing, the specimens were placed in the oven for 3 h under the intended temperature. After 3 h of heating, to prevent cooling shock, samples were not removed rapidly and the test was conducted after 24 h.

**Table 5.** Details of concrete composition in the main study.

| Name ** | BF (%) * | Alginate (%) * | Evaluation Temperature (°C) |
|---|---|---|---|
| $MCT_{RT}$ *** | - | - | 0 |
| $MCT_{100}$ | - | - | 100 |
| $MCT_{150}$ | - | - | 150 |
| $MCT_{180}$ | - | - | 180 |
| $MCBT_{RT}$ | 0.2 | - | 0 |
| $MCBT_{100}$ | 0.2 | - | 100 |
| $MCBT_{150}$ | 0.2 | - | 150 |
| $MCBT_{180}$ | 0.2 | - | 180 |
| $MCAT_{RT}$ | - | 0.1 | 0 |
| $MCAT_{100}$ | - | 0.1 | 100 |
| $MCAT_{150}$ | - | 0.1 | 150 |
| $MCAT_{180}$ | - | 0.1 | 180 |
| $MCBAT_{RT}$ | 0.2 | 0.1 | 0 |
| $MCBAT_{100}$ | 0.2 | 0.1 | 100 |
| $MCBAT_{150}$ | 0.2 | 0.1 | 150 |
| $MCBAT_{180}$ | 0.2 | 0.1 | 180 |

* of the total weight. ** MCT, MCBT, MCAT, MCBAT are OC with no alginate and BF, with BF only, with alginate only, and with alginate and BF, respectively. *** RT represents room temperature.

## 2.3. Compressive Strength Testing

The compressive strength tests were according to ASTM C109-08. Mortar specimens were cast in triplicate cubic $50 \times 50 \times 50$ mm and tested for each mixture, after 28 days of water curing. Axial compressive load was applied to the specimen by a 3 MN capacity compressive testing machine. The compression load was applied continuously with a rate of 2.4 KN/s according to ASTM C31. The load was applied until the specimen reach its ultimate load-bearing capacity and failure occurred. Further discussion on behavior and mechanics of composites are provided in Cristescu et al. [30], Singh et al. [31], Abdi et al. [32], and Koloor and Tamin [33]. The stress and load were measured and recorded in MPa and KN, respectively. The failure patterns of the specimens under compressive load are given in Figure 3.

|     |     |     |     |
|:---:|:---:|:---:|:---:|
| (a) | (b) | (c) | (d) |

**Figure 3.** Failure pattern of samples under compression test (**a**) control sample, (**b**) BF-reinforced, (**c**) alginate mortar, (**d**) Alginate-BF mortar.

## 3. Results and Discussion

Figure 3 shows some of the specimens after failure under compression test. The investigation on the crack patterns of the specimens showed that the formation of cracks in non-fiber mortar samples was more compared to the ones in reinforced BF. It shows that the fibers act as the crack arrestors and block and divert crack formation and propagation. The failure of fiber concrete was gradual as compared to the abrupt and brittle failure of non-fiber concrete.

From Table 6, the highest compressive strength for samples of the mortars after 7 days was for mix design $M_pW_{0.5}C_{470}E_{\#100}$. Although the addition of the fine fraction of the aggregates passing through sieve No. 100 (<150 μm) may result in a richer Fuller curve gradation and achieving an optimum density and strength of the concrete mixture, the higher absorption capacity of the fine particles necessitate an increase in w/c ratio that generally leads to a fall in the compressive strength. Higher compressive strength of $M_pW_{0.5}C_{470}E_{\#100}$ can be justified by the well-established fact that lower w/c ratio of mortar mixes gave rise to compressive strength.

**Table 6.** Compressive strength of primary tests.

| Name | Age Tested | $f'_c$ (MPa) |
|---|---|---|
| $M_pW_{0.5}C_{450}I_{\#100}$ | | 29.9 |
| $M_pW_{0.5}C_{470}I_{\#100}$ | | 28.8 |
| $M_pW_{0.5}C_{450}E_{\#100}$ | 7-day | 30.6 |
| $M_pW_{0.5}C_{470}E_{\#100}$ | | 32.8 |
| $M_pW_{0.55}C_{450}E_{\#100}$ | | 23.0 |
| $M_pW_{0.55}C_{470}E_{\#100}$ | | 19.3 |

Results of compressive strength for the cement concrete samples are shown in Table 7, that are corresponding to four different temperature conditions.

**Table 7.** Compressive strength for the cement concrete.

| Name | $f'_c$ (MPa) 28-day | Variation to MCT₀ (%) | Variation after Heating (%) | | |
|---|---|---|---|---|---|
| $MCT_{RT}$ | 41.0 | - | - | | $MCT_{RT}$ Compared to |
| $MCT_{100}$ | 34.0 | −17.07 | −17.07 | | |
| $MCT_{150}$ | 31.9 | −22.19 | −22.19 | | |
| $MCT_{180}$ | 35.6 | −13.17 | −13.17 | | |
| $MCBT_{RT}$ | 29.8 | −27.32 | - | | $MCBT_0$ Compared to |
| $MCBT_{100}$ | 31.1 | −24.15 | +4.36 | | |
| $MCBT_{150}$ | 30.1 | −26.60 | +1.01 | | |
| $MCBT_{180}$ | 30.2 | −26.34 | +1.34 | | |
| $MCAT_{RT}$ | 27.6 | −32.68 | - | | $MCAT_0$ Compared to |
| $MCAT_{100}$ | 33.7 | −17.80 | +22.10 | | |
| $MCAT_{150}$ | 31.6 | −22.93 | +14.49 | | |
| $MCAT_{180}$ | 28.2 | −31.22 | +2.17 | | |
| $MCBAT_{RT}$ | 24.4 | −40.49 | - | | $MCBAT_0$ Compared to |
| $MCBAT_{100}$ | 28.2 | −31.22 | +15.57 | | |
| $MCBAT_{150}$ | 23.4 | −42.93 | −4.10 | | |
| $MCBAT_{180}$ | 22.6 | −44.88 | −7.38 | | |

The compressive strength of the mixes compared to the reference mortar specimens maintained at room temperature are shown in Figure 4. The obtained results of the samples maintained at room temperature show that the addition of BF or alginate decreased compressive strength. A smaller reduction in compressive strength was observed by the addition of 0.2% of BF in the mortar mix, compared to the specimens containing alginate. The reduction of the compressive strength by the addition of BF is in agreement with the data published by other authors [24,25,29]. This decrease is mainly due to the compactness reduction of the composite caused by the introduction of voids into the matrix during incorporation of the fibers. However, the observation during compressive tests indicated that the specimens containing FB did not exhibit brittle failure but rather a ductile, plastic failure and the mortar showed the ability to absorb a large amount of plastic energy under compressive loads.

Hydrogels in alginate concrete can be formed by a crosslinking reaction of sodium alginate. The samples with sodium alginate pose higher water absorption capacity causing a greater amount of water to be required to achieve the desired workability compared to the control mortar. Lower workability contributes to making the concrete more porous and, as a consequence, the compressive strength values would be decreased. On the other hand, releasing the encapsulated water in the mortar matrix as hydrogels leads to macropores which can reduce compactness of the composite and negatively affect the compressive strength. This result is in agreement with the findings of Heidari et al. [21] and Mignon et al. [22]. The application of both BF and alginate together decreased the compressive strength even more.

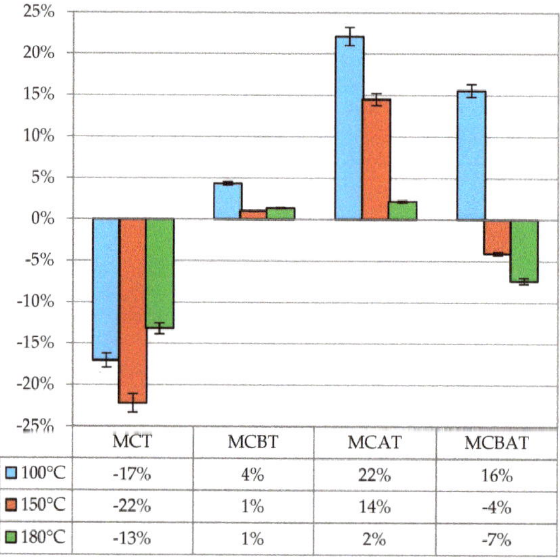

|  | MCT | MCBT | MCAT | MCBAT |
|---|---|---|---|---|
| ◻ 100°C | -17% | 4% | 22% | 16% |
| ◼ 150°C | -22% | 1% | 14% | -4% |
| ◼ 180°C | -13% | 1% | 2% | -7% |

**Figure 4.** The compressive strength of the mixes compared to the reference mortar specimens maintained at room temperature.

Variations in the compressive strength of the mix designs after exposure to elevated temperature indicated that the maximum reduction of compressive strength for the control mixture was 22.19% when exposed to 100 °C for 3 h while the mix designs containing BF or alginate, each alone, even exhibited growth in the compressive strength. The maximum growth in the compressive strength was for MCAT$_{100}$; the composition containing sodium alginate subjected to 100 °C. After the introduction of water to sodium alginate, hydrogels are formed and they tend to cross-link in the presence of mixed metal oxides in the concrete that leads to dense structures after crosslinking and drying. The addition of the sodium alginate can efficiently encapsulate the dispersed micro or macro-size water droplets within the concrete composition. The encapsulated water droplets provide enhanced fire-resistive properties for the concrete. With increasing the temperature from 0 to 100 °C the encapsulated water is gradually released to facilitate the hydration of the cementitious system and positively influence the strength growth rate of the concrete.

It was observed that after 100 °C compressive strength was decreased. The reason for the decrease in compressive strength can be attributed to the effect of stress decay due to the viscoelastic relaxation of cross-linked alginate chains. Therefore, the internal hydrostatic pressures are developed and lead to water exudation through the network, as well as fracture of the air-containing cells, or to junction zones breakdown [34]. The mix design with BF has also revealed growth in the compressive strength after heating and the maximum growth in this mix design is under 100 °C too. These samples from mixed designs including BF or alginate exhibited small growth under 150 and 180 °C temperature. However, heating the cement concrete samples, including both BF and alginate subjected to 100 °C, resulted in compressive strength growth, but the increase of temperature to higher than 100 °C led to strength reduction. This reduction may be due to the departure of bound water in the microstructure and the free water from the cement past that results in concrete porosity. It can be attributed to the bonding force between basalt fiber and the mortar matrix through the mechanical and chemical bonding. It must be mentioned that the bonding force decreases with increasing temperature [35].

## 4. Conclusions

In order to investigate the thermal performance of concrete and mortar containing alginate and BF, an experimental study was carried out. The variation in compressive strength of mortar mixes after exposure to an elevated temperature of 100, 150, and 180 °C was tested. The experimental program was performed in two stages including determination of the mix design of the control mix and afterward, investigating the thermal performance of the mortar with the addition and/or elimination of ingredients (alginate and basalt fiber) in the composition mix. The temperature variation can be divided into two ranges of 100 °C and temperatures beyond 100 °C in terms of effect on strength loss/gain. The mixes containing BF and alginate, or each alone, exhibited growth in the compressive strength when exposed to 100 °C due to enhanced hydration that exceeded the loss in strength. Basalt fiber and alginate are effective at stopping strength reduction of cement concrete under raised temperature. In view of these findings, it was concluded that the addition of BF and alginate results in an improved mortar that offers high temperature-resistive properties within the temperature range considered in the experiment.

**Author Contributions:** Conceptualization, N.A.O., S.E.M.-Y. and H.A.A.; methodology, N.A.O., S.E.M.-Y. and H.A.A.; software, N.A.O.; validation, N.A.O., S.E.M.-Y. and H.A.A.; formal analysis, N.A.O., S.E.M.-Y. and H.A.A.; investigation, N.A.O., S.E.M.-Y. and H.A.A.; resources, N.A.O., S.E.M.-Y., H.A.A, H.S., S.S.R.K. and M.P.; data curation, N.A.O.; writing—original draft preparation, N.A.O., S.E.M.-Y., H.A.A. and H.S.; writing—review and editing, N.A.O., S.E.M.-Y., H.A.A., H.S. and S.S.R.K.; visualization, N.A.O., S.E.M.-Y., H.A.A, H.S., S.S.R.K. and M.P.; supervision, S.E.M.-Y., H.A.A.; project administration, N.A.O., S.E.M.-Y., H.A.A, H.S., S.S.R.K. and M.P. funding acquisition, N.A.O., S.E.M.-Y., H.A.A, H.S., S.S.R.K. and M.P. All authors have read and agreed to the published version of the manuscript.

**Funding:** The research was funded by the Ministry of Education, Youth and Sports of the Czech Republic and the European Union (European Structural and Investment Funds—Operational Programme Research, Development and Education), Reg. No. CZ.02.1.01/0.0/0.0/16_025/0007293, as well as the financial support from internal grants in the Institute for Nanomaterials, Advanced Technologies and Innovations (CXI), Technical University of Liberec (TUL).

**Acknowledgments:** The authors would like to thank from Islamic Azad University, Najafabad Branch, for financial support, as well as facilities and services given and acknowledge the financial support by Ministry of Education, Youth and Sports of the Czech Republic and the European Union (European Structural and Investment Funds—Operational Programme Research, Development and Education), Reg. No. CZ.02.1.01/0.0/0.0/16_025/0007293, as well as the financial support from internal grants in the Institute for Nanomaterials, Advanced Technologies and Innovations (CXI), Technical University of Liberec (TUL).

**Conflicts of Interest:** The authors declare no conflict of interest.

## References

1. Mohammadyan-Yasouj, S.E.; Marsono, A.K.; Abdullah, R.; Moghadasi, M. Wide beam shear behavior with diverse types of reinforcement. *ACI Struct. J.* **2015**, *112*, 199–208. [CrossRef]
2. Moghadasi, M.; Marsono, A.K.; Mohammadyan-Yasouj, S.E. A study on rotational behaviour of a new industrialised building system connection. *Steel Compos. Struct.* **2017**, *25*, 245–255. [CrossRef]
3. Shokravi, H.; Shokravi, H.; Bakhary, N.; Koloor, S.S.R.; Petrů, M. A Comparative Study of the Data-driven Stochastic Subspace Methods for Health Monitoring of Structures: A Bridge Case Study. *Appl. Sci.* **2020**, *10*, 3132. [CrossRef]
4. Shokravi, H.; Shokravi, H.; Bakhary, N.; Koloor, S.S.R.; Petrů, M. Health Monitoring of Civil Infrastructures by Subspace System Identification Method: An Overview. *Appl. Sci.* **2020**, *10*, 2786. [CrossRef]
5. Yuan, C.; Chen, W.; Pham, T.M.; Hao, H. Effect of aggregate size on bond behaviour between basalt fibre reinforced polymer sheets and concrete. *Compos. Part B Eng.* **2019**, *158*, 459–474. [CrossRef]
6. Ahmed, A.A.M.; Jia, Y. Effect of Using Hybrid Polypropylene and Glass Fibre on the Mechanical Properties and Permeability of Concrete. *Materials* **2019**, *12*, 3786. [CrossRef]

7.   Alnahhal, W.; Aljidda, O. Effect of Fiber Volume Fraction on Behavior of Concrete Beams Made with Recycled Concrete Aggregates. In Proceedings of the MATEC Web of Conferences, 14 January 2019; EDP Sciences: Paris, France; Volume 253, p. 2004.

8.   Swathi, T.; Resmi, K.N. Experimental Studies on Fly Ash Based Basalt Fibre Reinforced Concrete. In Proceedings of the 3rd National Conference on Structural Engineering and Construction Management, 15–16 May 2019; Springer: Kerala, India, 2019; pp. 25–39.

9.   Wang, X.; He, J.; Mosallam, A.S.; Li, C.; Xin, H. The effects of fiber length and volume on material properties and crack resistance of basalt fiber reinforced concrete (BFRC). *Adv. Mater. Sci. Eng.* **2019**, *2019*. [CrossRef]

10.  Jiang, C.; Fan, K.; Wu, F.; Chen, D. Experimental study on the mechanical properties and microstructure of chopped basalt fibre reinforced concrete. *Mater. Des.* **2014**, *58*, 187–193. [CrossRef]

11.  Kizilkanat, A.B.; Kabay, N.; Akyüncü, V.; Chowdhury, S.; Akça, A.H. Mechanical properties and fracture behavior of basalt and glass fiber reinforced concrete: An experimental study. *Constr. Build. Mater.* **2015**, *100*, 218–224. [CrossRef]

12.  Fenu, L.; Forni, D.; Cadoni, E. Dynamic behaviour of cement mortars reinforced with glass and basalt fibres. *Compos. Part B Eng.* **2016**, *92*, 142–150. [CrossRef]

13.  Shafiq, N.; Ayub, T.; Khan, S.U. Investigating the performance of PVA and basalt fibre reinforced beams subjected to flexural action. *Compos. Struct.* **2016**, *153*, 30–41. [CrossRef]

14.  Sarkar, A.; Hajihosseini, M. Feasibility of Improving the Mechanical Properties of Concrete Pavement Using Basalt Fibers. *J. Test. Eval.* **2020**, *48*. [CrossRef]

15.  Mohammadyan-Yasouj, S.E.; Ghaderi, A. Experimental investigation of waste glass powder, basalt fibre, and carbon nanotube on the mechanical properties of concrete. *Constr. Build. Mater.* **2020**, *252*, 119115. [CrossRef]

16.  Pathak, T.S.; San Kim, J.; Lee, S.-J.; Baek, D.-J.; Paeng, K.-J. Preparation of alginic acid and metal alginate from algae and their comparative study. *J. Polym. Environ.* **2008**, *16*, 198–204. [CrossRef]

17.  Engbert, A.; Gruber, S.; Plank, J. The effect of alginates on the hydration of calcium aluminate cement. *Carbohydr. Polym.* **2020**, *236*, 116038. [CrossRef]

18.  Wang, J.; Mignon, A.; Snoeck, D.; Wiktor, V.; Van Vliergerghe, S.; Boon, N.; De Belie, N. Application of modified-alginate encapsulated carbonate producing bacteria in concrete: A promising strategy for crack self-healing. *Front. Microbiol.* **2015**, *6*, 1088. [CrossRef]

19.  Ouwerx, C.; Velings, N.; Mestdagh, M.M.; Axelos, M.A. V Physico-chemical properties and rheology of alginate gel beads formed with various divalent cations. *Polym. Gels Networks* **1998**, *6*, 393–408. [CrossRef]

20.  Abbas, W.A.; Mohsen, H.M. Effect of Biopolymer Alginate on some properties of concrete. *J. Eng.* **2020**, *26*, 121–131.

21.  Heidari, A.; Ghaffari, F.; Ahmadvand, H. Properties of self compacting concrete incorporating alginate and nano silica. *Asian J. Civ. Eng. BHRC* **2015**, *16*, 1–11.

22.  Mignon, A.; Snoeck, D.; D'Halluin, K.; Balcaen, L.; Vanhaecke, F.; Dubruel, P.; Van Vlierberghe, S.; De Belie, N. Alginate biopolymers: Counteracting the impact of superabsorbent polymers on mortar strength. *Constr. Build. Mater.* **2016**, *110*, 169–174. [CrossRef]

23.  DeBrouse, D.R. Alginate-Based Building Materials. U.S. Patent 8,246,733, 21 August 2012.

24.  Kabay, N. Abrasion resistance and fracture energy of concretes with basalt fiber. *Constr. Build. Mater.* **2014**, *50*, 95–101. [CrossRef]

25.  Ayub, T.; Shafiq, N.; Nuruddin, M.F. Mechanical properties of high-performance concrete reinforced with basalt fibers. *Procedia Eng.* **2014**, *77*, 131–139. [CrossRef]

26.  Girgin, Z.C.; Yıldırım, M.T. Usability of basalt fibres in fibre reinforced cement composites. *Mater. Struct.* **2016**, *49*, 3309–3319. [CrossRef]

27.  Zhao, Y.-R.; Wang, L.; Lei, Z.-K.; Han, X.-F.; Xing, Y.-M. Experimental study on dynamic mechanical properties of the basalt fiber reinforced concrete after the freeze-thaw based on the digital image correlation method. *Constr. Build. Mater.* **2017**, *147*, 194–202. [CrossRef]

28.  Katkhuda, H.; Shatarat, N. Improving the mechanical properties of recycled concrete aggregate using chopped basalt fibers and acid treatment. *Constr. Build. Mater.* **2017**, *140*, 328–335. [CrossRef]

29.  Sun, X.; Gao, Z.; Cao, P.; Zhou, C. Mechanical properties tests and multiscale numerical simulations for basalt fiber reinforced concrete. *Constr. Build. Mater.* **2019**, *202*, 58–72. [CrossRef]

30. Cristescu, N.D.; Craciun, E.-M.; Soós, E. *Mechanics of Elastic Composites*; CRC Press: New York, NY, USA, 2003; Volume 1, ISBN 0203502817.
31. Singh, A.; Das, S.; Craciun, E.-M. Effect of thermomechanical loading on an edge crack of finite length in an infinite orthotropic strip. *Mech. Compos. Mater.* **2019**, *55*, 285–296. [CrossRef]
32. Abdi, B.; Koloor, S.S.R.; Abdullah, M.R.; Amran, A.; Bin Yahya, M.Y. Effect of strain-rate on flexural behavior of composite sandwich panel. *Appl. Mech. Mater. Trans Tech. Publ.* **2012**, *229*, 766–770.
33. Koloor, S.S.R.; Tamin, M.N. Mode-II interlaminar fracture and crack-jump phenomenon in CFRP composite laminate materials. *Compos. Struct.* **2018**, *204*, 594–606. [CrossRef]
34. Mancini, M.; Moresi, M.; Rancini, R. Mechanical properties of alginate gels: Empirical characterisation. *J. Food Eng.* **1999**, *39*, 369–378. [CrossRef]
35. Wang, J.; Zhou, S.; Huang, J.; Zhao, G.; Liu, Y. Interfacial modification of basalt fiber filling composites with graphene oxide and polydopamine for enhanced mechanical and tribological properties. *RSC Adv.* **2018**, *8*, 12222–12231. [CrossRef]

*Article*

# The Impact Resistance and Deformation Performance of Novel Pre-Packed Aggregate Concrete Reinforced with Waste Polypropylene Fibres

Fahed Alrshoudi [1,*], Hossein Mohammadhosseini [2,*], Rayed Alyousef [3,*], Mahmood Md. Tahir [2], Hisham Alabduljabbar [3] and Abdeliazim Mustafa Mohamed [3]

[1]  Department of Civil Engineering, College of Engineering, King Saud University, Riyadh 12372, Saudi Arabia
[2]  Institute for Smart Infrastructure and Innovative Construction (ISIIC), School of Civil Engineering, Faculty of Engineering, Universiti Teknologi Malaysia, Skudai 81310, Johor, Malaysia; mahmoodtahir@utm.my
[3]  Department of Civil Engineering, College of Engineering, Prince Sattam bin Abdulaziz University, Alkharj 11942, Saudi Arabia; h.alabduljabbar@psau.edu.sa (H.A.); a.bilal@psau.edu.sa (A.M.M.)
*   Correspondence: Falrshoudi@ksu.edu.sa (F.A.); mhossein@utm.my (H.M.); r.alyousef@psau.edu.sa (R.A.)

Received: 23 July 2020; Accepted: 31 August 2020; Published: 6 September 2020

**Abstract:** Pre-packed aggregate fibre-reinforced concrete (PAFRC) is an innovative type of concrete composite using a mixture of coarse aggregates and fibres which are pre-mixed and pre-placed in the formwork. A flowable grout is then injected into the cavities between the aggregate mass. This study develops the concept of a new PAFRC, which is reinforced with polypropylene (PP) waste carpet fibres, investigating its mechanical properties and impact resistance under drop weight impact load. Palm oil fuel ash (POFA) is used as a partial cement replacement, with a replacement level of 20%. The compressive strength, impact resistance, energy absorption, long-term drying shrinkage, and microstructural analysis of PAFRC are explored. Two methods of grout injection are used—namely, gravity and pumping methods. For each method, six PAFRC batches containing 0–1.25% fibres (with a length of 30 mm) were cast. The findings of the study reveal that, by adding waste PP fibre, the compressive strength of PAFRC specimens decreased. However, with longer curing periods, the compressive strength enhanced due to the pozzolanic activity of POFA. The combination of fibres and POFA in PAFRC mixtures leads to the higher impact strength energy absorption and improved ductility of the concrete. Furthermore, drying shrinkage was reduced by about 28.6% for the pumping method PAFRC mix containing 0.75% fibres. Due to the unique production method of PAFRC and high impact resistance and energy absorption, it can be used in many pioneering applications.

**Keywords:** impact resistance; pre-packed aggregate fibre-reinforced concrete; strength; long-term shrinkage; microstructure; waste polypropylene fibres

---

## 1. Introduction

In general, conventional concrete is prepared by pre-mixing all the components after which the mixture of fresh concrete is poured into designed formwork. Concrete components may also be manufactured by first packing the coarse aggregate into the designed formwork, followed by injecting a special type of mortar (in the form of grout) in between the aggregates. Pre-packed aggregate concrete (PAC), also called two-stage concrete, is a specific sort of concrete which was designated originally in the 1930s [1,2]. This specific method of concreting is produced by initially placing coarse aggregates of various sizes and shapes, depending on the application, in pre-planned fitments after which a specially prepared mortar (in the form of grout) is poured into the gaps amongst the aggregate particles [3]. This specific technique of concrete construction can be used in components with complex reinforcement. It can also be used in specific arrangements where the conventional type of concrete may interrupt the pre-planned formworks, such as conduits, pipes, and openings [4]. Another application of the PAC method is mass concreting, such as in the cases of piers and bridge abutments. Moreover, the PAC technique is one of the most preferred methods for concreting underwater in which aggregate particles with different sizes are pre-placed in water, and then, a mixture of grout is injected into the gaps between the aggregate, thus replacing the water [5].

According to the literature [1–3], for pre-packed aggregate concrete (PAC), the grouting process can be attained by gravity or pumping methods. In the PAC technique of concreting, the grouting method may vary depending on the minimum size of aggregate used, as well as the depth of formworks. Generally, PAC grouting is done by either gravity or pumping methods [6]. In the gravity method, the grout mixture is poured onto the aggregates packed into the formwork and allowed to penetrate slowly downwards, under its own weight. The gravity method is usually used for thin concrete components with a depth of up to 300 mm and larger size of aggregates; for instance, road pavement slabs and floor slabs. However, with an increase in the depth of sections, the grout cannot completely cross through the entire thickness of the formwork. Therefore, in components with smaller-sized aggregates and deeper sections, the pumping method of grout injection is preferred. In this method, a pump with a pressure measuring device is used. The grout is pumped in between the pre-placed aggregates from the bottom of the formworks. Pipes with perforations, which are extracted upwards, inject and uniformly distribute the grout mixture into the sections. In this method, the entire depth of the formworks is poured with grout [7]. PAC can be used in massive concrete where placement by conventional practices is very difficult [8,9]. As the coarse aggregates are pre-packed, the manufacturing process makes PAC one of the most desirable methods in the construction of nuclear power plants, where heavy minerals can be used together with coarse aggregates in the production of concrete without any segregation [10,11].

Concrete, in many different forms, is the most extensively used construction material worldwide. Plain concrete, however, is considered a brittle material due to its low tensile strength. Therefore, higher impact strength and energy absorption capacity are vital in different usages for PAC and conventional concrete. Other components are needed to develop these properties of concrete, where these features are vital [12,13]. Consequently, new materials which can develop the performance of concrete with higher energy absorption capability and superior ductility are in high demand. A related promising solution to achieve these properties in concrete is the addition of fibres at various dosages into the concrete mixture [14]. Pre-packed aggregate fibre-reinforced concrete (PAFRC) is a novel material made of pre-placed coarse aggregates of different sizes, which are mixed with short fibres and cementing materials mixed with sand in the form of grout, which is injected into the mixture of pre-packed aggregates and fibres. The utilisation of various types of fibres, such as polymeric and metallic fibres, in conventional concrete at volume fractions of 0.1% to 2.0%, has been recognised by many researchers [15,16]. Numerous kinds of research works have been conducted to investigate the ductility performance of fibre-reinforced concrete (FRC) with short fibres through impact resistance tests. In this regard, Mohammadhosseini et al. [17], as well as Awal and Mohammadhosseini [18], reported that the inclusion of waste polypropylene carpet fibres remarkably developed the impact strength and energy absorption of conventional concrete. However, PAFRC is a new sort of concrete reinforced with short fibres. The utilisation of recycled CFRP fibres at different lengths and dosages has been studied by Mastali et al. [19], who reported that these waste fibres significantly improved the impact resistance of conventional fibre-reinforced concrete. Furthermore, Alyousef et al. [15] reported that the inclusion of plastic fibres at different dosages could significantly enhance the performance of concrete. Ong et al. [20] found that the use of polymeric base fibres in combination with steel fibres remarkably enhanced the impact strength and energy absorption capacity of concrete slabs.

The production of polymeric fibres, as employed in the carpet and textiles industries, has been steadily rising, presently about 70 million tons annually [21,22]. However, a massive amount of these fibres is sent to landfill as waste during the manufacturing process. Like other countries, Malaysia produces carpets, and about 50 tons of polymeric fibres are sent to landfill as waste every year [23]. Indeed, this option of waste disposal is gradually becoming unviable, owing to the increasing cost of landfilling in addition to the limited accessibility of discarding sites [24]. Consequently, the demand to reutilise waste materials is growing, due to a lack of landfill places globally and the preservation of natural resources [25]. In Malaysia, approximately 5 million tons of waste palm oil ashes (POFA) are produced per year as a waste material [26,27]. However, according to past studies, with proper treatment, POFA can be utilised as partial cement replacement in concrete, with adequate performance [28,29].

So far, there has been no study on the strength and impact resistant of PAC comprising waste carpet fibres and POFA. Given that POFA is categorised as a pozzolanic material and considering the local availability of industrial waste carpet fibre, these materials were used in the manufacturing of PAC. Therefore, the purpose of this research work was to study the combined influences of waste polypropylene carpet fibres and POFA on the mechanical properties and impact resistance of novel pre-packed aggregate concrete, in addition to understanding the way that waste carpet fibres contribute to the improvement in energy absorption, in comparison to plain PAC without any fibres.

## 2. Materials and Methods

### 2.1. Materials

In this study, a type I cement complying with the specifications of ASTM C 150-07 [30] was used. Ordinary Portland cement (OPC) was substituted by POFA at substitution level of 20%. Initially, the raw palm oil fuel ash particles were collected as waste from the local mill industry. The ash particles were dried at a temperature of $100 \pm 5$ °C and then sieved to remove the larger particles (i.e., over 150 μm). Subsequently, the small-sized ash particles were kept in a crushing machine and the grinding process was continued for about two hours per each 4 kg of ash. The ground POFA particles were then tested following the specifications of ASTM C618-15 [31] and BS 3892: Part 1–97 [32], in order to achieve the desired properties. The ashes which passed the standard requirements with the desired chemical compositions and physical properties (as given in Table 1) were then used as cementing materials. Moreover, sets of trial mixes were carried out using different percentages of POFA content. The results revealed that inclusion of POFA at more than 20% led to a reduction in the compressive strength of concrete, particularly at the early ages. However, according to the trial mix results and to reduce the consumption of OPC, 20% POFA content was selected as an optimum level to be used in the main experimental work.

**Table 1.** Characteristics of used palm oil fuel ash (POFA) and OPC.

| Composition | OPC (%) | POFA (%) |
|---|---|---|
| $SiO_2$ | 20.40 | 62.60 |
| $Al_2O_3$ | 5.20 | 4.65 |
| $Fe_2O_3$ | 4.19 | 8.12 |
| CaO | 62.39 | 5.70 |
| MgO | 1.55 | 3.52 |
| $K_2O$ | 0.005 | 9.05 |
| $SO_3$ | 2.11 | 1.16 |
| LOI | 2.36 | 6.25 |
| *Physical properties* | | |
| Specific gravity | 3.15 | 2.42 |
| Blaine fineness ($cm^2/g$) | 3990 | 4930 |
| Soundness (mm) | 1.0 | 2.0 |

Fine and coarse aggregates are the main constituents in the production of PAC. Therefore, in this study, natural river sand with a maximum size of 4.75 mm was used to produce a grout mixture. The fineness modulus, specific gravity, and water absorption of sand particles were found as 2.3,

2.6 g/cm$^3$, and 0.7%, respectively. Moreover, the coarse aggregate particles—which are the main skeleton of PAC—were selected following the specifications of ACI 304.1R-05 [10]. Crushed granite coarse aggregates of 20–38 mm in size, 2.7 g/cm$^3$ specific gravity, and 0.5% water absorption were employed. The aggregates were cleaned and washed before placing in formworks, in order to eliminate any impurities. Additionally, multi-filament polypropylene waste fibres collected from the local carpet industry were used in this study to reinforce the PAC specimens. Initially, the fibres were collected in the form of waste yarn. Then, in order to use this yarn as a fibrous material in the concrete, several tests were carried out. After they satisfied the standard requirements (as given in Table 2), the yarns were cut and fabricated to the desired length of 30 mm with an aspect ratio (l/d) of 67, as illustrated in Figure 1.

**Table 2.** Typical properties of fabricated polypropylene (PP) waste fibres.

| Waste PP Fibre | Length (mm) | Diameter (mm) | Density (kg/m$^3$) | Melting Point (°C) | Tensile Strength (MPa) | Reaction with Water |
|---|---|---|---|---|---|---|
| Multi-filament polypropylene | 30 | 0.45 | 910 | 170 | 400 | Hydrophobic |

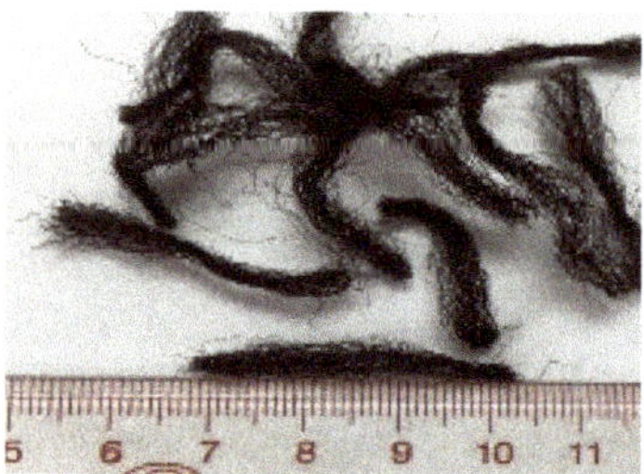

**Figure 1.** Waste PP fibres with 30 mm length.

## 2.2. Mix Proportions

In this study, two groups of pre-packed aggregate concrete mixes, based on the method of grouting, were made: gravity (G) and pumping (P) mixes. For each group, five mixes were cast, where one was a control mix without any fibres (G0, P0) and another four mixes were reinforced with polypropylene (PP) fibre volume fractions of 0.25%, 0.50%, 0.75%, and 1.0%. In all mixes, OPC was replaced by 20% POFA, and the water/binder (w/b) ratio of 0.5 and cement/sand (c/s) ration of 1/1.15 was kept constant. As the effect of fibres at different volume fractions was the main parameter to be investigated, the w/b and c/s ratios were kept constant to attain similar conditions for all PAFRC mixes. Table 3 lists the quantities of the various components used in the manufacture of PAFRC mixes.

**Table 3.** Proportions of various components used in the pre-packed aggregate fibre-reinforced concrete (PAFRC) mixes.

| | Mix | Water (kg/m$^3$) | Cement (kg/m$^3$) | POFA (kg/m$^3$) | Fine Aggregate (kg/m$^3$) | Coarse Aggregate (kg/m$^3$) | $V_f$ (%) |
|---|---|---|---|---|---|---|---|
| Gravity | G0 | 186 | 304 | 76 | 545 | 1320 | 0.0 |
| | G1 | 186 | 304 | 76 | 545 | 1320 | 0.25 |
| | G2 | 186 | 304 | 76 | 545 | 1320 | 0.50 |
| | G3 | 186 | 304 | 76 | 545 | 1320 | 0.75 |
| | G4 | 186 | 304 | 76 | 545 | 1320 | 1.00 |
| | G5 | 186 | 304 | 76 | 545 | 1320 | 1.25 |

**Table 3.** *Cont.*

|  | Mix | Water (kg/m$^3$) | Cement (kg/m$^3$) | POFA (kg/m$^3$) | Fine Aggregate (kg/m$^3$) | Coarse Aggregate (kg/m$^3$) | V$_f$ (%) |
|---|---|---|---|---|---|---|---|
| | P0 | 186 | 304 | 76 | 545 | 1320 | 0.0 |
| Pump | P1 | 186 | 304 | 76 | 545 | 1320 | 0.25 |
| | P2 | 186 | 304 | 76 | 545 | 1320 | 0.50 |
| | P3 | 186 | 304 | 76 | 545 | 1320 | 0.75 |
| | P4 | 186 | 304 | 76 | 545 | 1320 | 1.00 |
| | P5 | 186 | 304 | 76 | 545 | 1320 | 1.25 |

## 2.3. Sample Preparation

The manufacture of PAFRC specimens was carried out in two phases. First, the dry mixture of coarse aggregate particles and PP fibres was placed and packed into the design formworks and moulds; then, the pre-mixed mortar made of blended cement and river sands, with adequate flowability and in the form of the grout, was injected in between the gaps amongst aggregates and fibres, either by the gravity or pumping methods. In the grouting process by the gravity method, a polyvinyl chloride (PVC) pipe with a diameter of 5 mm was placed at the centre of the cylindrical moulds of size 100 × 200 mm and 150 × 300 mm, and then, grout was injected under the force of gravity, as illustrated in Figure 2b. The grouting process in the pumping method was more complicated. In this method, as illustrated in Figure 2a, unplasticized polyvinyl chloride (UPVC) pipes of 100 and 150 mm diameter were used as moulds. The lengths of the pipes varied between 1 and 2 m, based on the number of required samples. The pipes were then arranged in a formwork made of plywood to prevent movement during the grouting process. A pump with a pressure control device was used, which was attached to the hopper for grout injection purposes. To avoid the overflow of grout, as well as to avoid the uplifting of the aggregate particles, a cap made of plywood was fixed to the top surface of the UPVC pipes. The entire process of grouting was monitored, in order to avoid any leakage of grout from the moulds as well as to prevent overflowing. After the casting process was carried out, the PAFRC specimens were cured for 24 h at an ambient temperature. The samples were then removed from the moulds and kept in a water tank at room temperature until further testing.

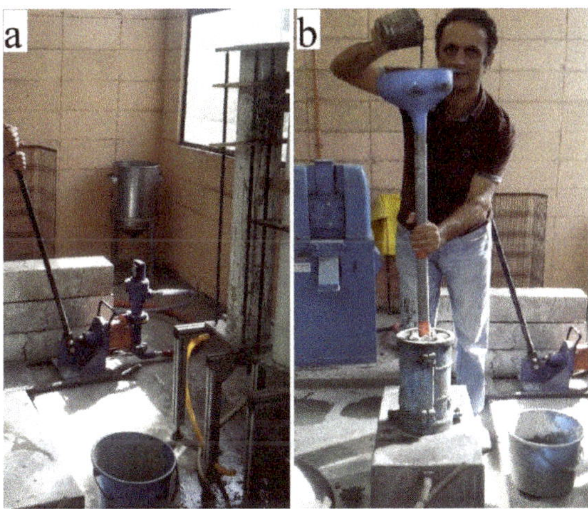


**Figure 2.** Methods of pre-packed aggregate concrete (PAC) grouting: (**a**) pumping; (**b**) gravity.

### 2.4. Testing Methods

In order to explore the influence of POFA as a partial cement replacement on the consistency of the grout mixture, the flow property of the grout was assessed following the ASTM C939-16 [33]. The compressive strength test of PAFRC specimens was carried out following the specifications of ASTM C39M-18 [34], using cylindrical moulds of size 100 mm × 200 mm and 150 mm × 300 mm for gravity and pumping techniques. Further, scanning electron microscopy (SEM) was used to examine the microstructure of the grout paste. In addition, following the specifications of ASTM C512-10 [35], the long-term drying shrinkage test was also carried out on cylindrical specimens of size 100 mm × 200 mm, as shown in Figure 3.

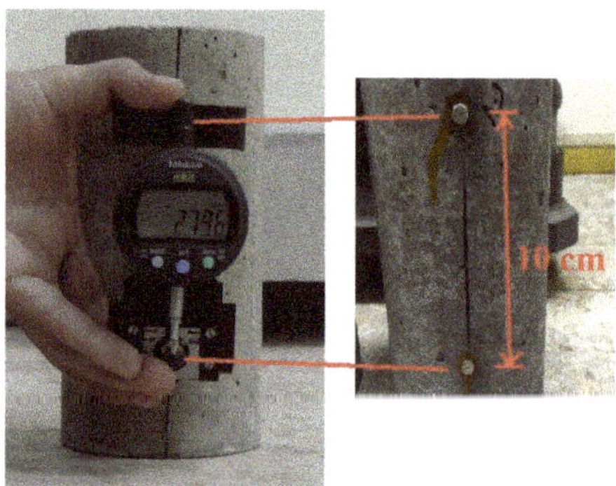


**Figure 3.** Drying shrinkage test setup.

Impact resistance and energy absorption tests were carried out on the PAFRC specimens in accordance with ACI 544.2R-99 [36]. For each batch of concrete, three disk samples with 64 mm thickness and 150 mm diameter were cut from the cylinders of size 150 × 300 mm and subjected to an impact load induced by a steel hammer with mass 4.45 kg, frequently released from a height of 457 mm on a stainless steel ball 63.5 mm in diameter positioned on the top surface of the centre of the concrete specimens. Figure 4 illustrates the drop weight impact resistance test device and the concrete disk samples. The number of drops for the first crack and failure were recorded as the first crack impact resistance (N1) and ultimate impact resistance (N2), correspondingly. Moreover, the impact energy induced by the hammer for blows at first crack and failure was evaluated according to Equations (1)–(4) [18]:

$$U = \frac{mV^2}{2} \tag{1}$$

$$m = \frac{W}{g} \tag{2}$$

$$V = gt \tag{3}$$

$$H = \frac{gt^2}{2} \tag{4}$$

where $U$ is the impact energy of each drop (kN mm); $W$ is the weight of the hammer (4.45 kg); $m$ is the mass of the hammer; $g$ is the acceleration of gravity (9.81 m/s$^2$); $V$ is the velocity of the hammer; $t$ is the required time for the hammer to drop (0.3053 s), and $H$ is the falling height of the hammer (457 mm).

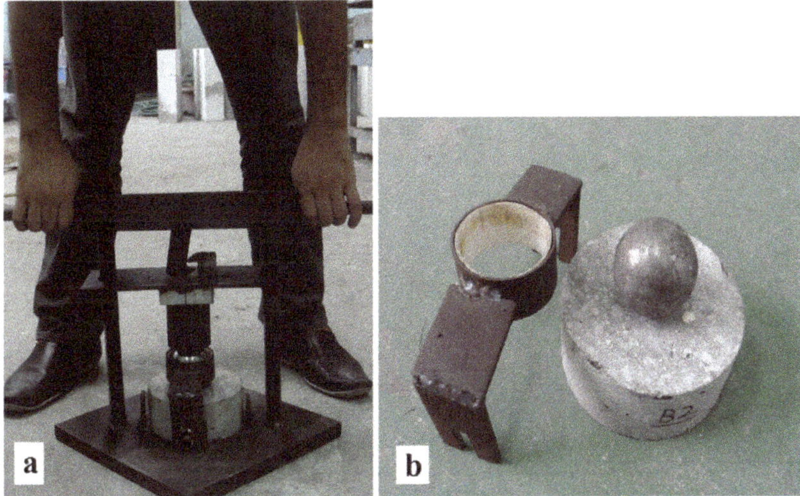



**Figure 4.** (**a**) Impact resistance test device and (**b**) steel ball and disk sample.

## 3. Results and Discussion

### 3.1. Grout Fluidity

In PAC, flowability and uniformity of the grout are essential, as they have a significant influence on its pumpability and permeability. The fluidity of grout was assessed through the flow cone test. Figure 5 shows the outcomes of the fluidity test of the 100% OPC grout mixture and the mixture containing 20% POFA. It was observed that the fluidity of the grout mixture (with constant w/c and c/s) increased when the OPC was replaced by 20% POFA. For the POFA-based grout mixture, the fluidity was noted as 13.1 s, which was comparatively lower than that (15.2 s) found for the grout mixture with 100% OPC. The higher flowability of the POFA-based grout could be due to the finer size of POFA ashes, as compared to OPC particles. The smaller particle size of POFA could, therefore, ease in the movement of the grout mixture and result in higher fluidity [37,38].

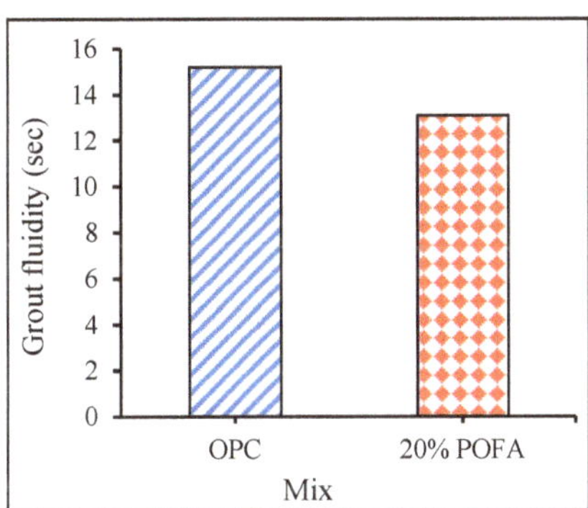



**Figure 5.** Grout fluidity of the OPC and POFA grout mixtures.

### 3.2. Compressive Strength

The compressive strength values of PAFRC specimens at 28, 90, and 180 days were measured and are presented in Figure 6. The compressive strength of PAFRC specimens shows that reinforcement of plain PAC with PP fibres resulted in a slight reduction in the strength values. The decrease in the compressive strength of PAFRC specimens might be due to the presence of cavities in the matrix, which would be increased with the inclusion of fibres at high dosage. Higher fibre content induces the balling effect, void formation, and clustering, consequently making the matrix weaker and more prone to cracking, therefore reducing the amount of grout injected between the mixture of coarse aggregates and fibres, leading to the reduction in compressive strength of the concrete [39,40]. It can be observed that the compressive strength of gravity method PAFRC specimens was reduced by 2.3%, 6.5%, 8.8%, 12.2%, and 16.1% when adding 0.25%, 0.5%, 0.75%, 1%, and 1.25% fibres, respectively, as compared to the plain PAC mix. Likewise, the pumping method specimens showed a minor reduction in compressive strength values, as compared to that of the plain PAC mix. The compressive strength reduced by about 3.4%, 6.4%, 8.5%, 11.4%, and 18%, respectively, for the same fibre dosages as above. However, a remarkable rise in compressive strength values was noted at the curing period of 90 days, as compared to those recorded at 28 days. This could be attributed to the pozzolanic nature of POFA for which the rate of hydration is slow during the early curing periods [41]. Additionally, the reduction in compressive strength values was considered for the delay in the hydration process and lower development of the C–S–H gel, which negated the increase in compressive strength. To improve the hydration process, Chandara et al. [29] suggested the use of finer ash particles in the cement composite. In a detailed study, they demonstrated that ground POFA is highly pozzolanic and can be used as supplementary cementing material, for up to 40% by weight of OPC.

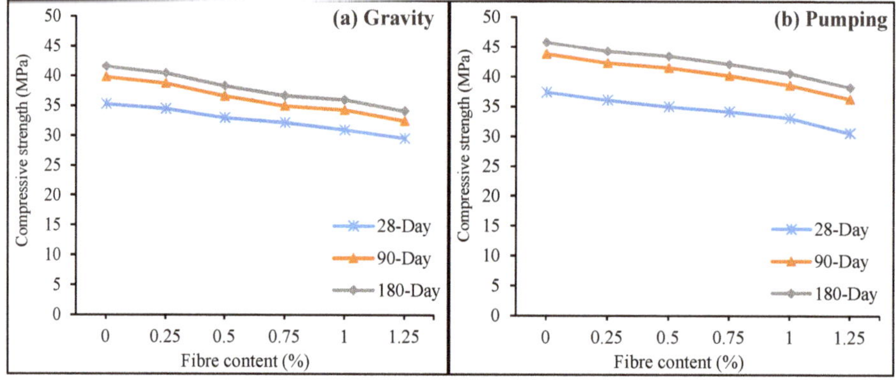


**Figure 6.** Variation in the compressive strength of (**a**) gravity and (**b**) pumping method PAFRC specimens.

Moreover, beyond 28 days, the compressive strength of the PAFRC mixes tended to increase with the curing age for all fibre volume fractions and gave higher compressive strength than those at early ages. This can be explained by the fact that the higher-fineness POFA develops pozzolanic properties and particle packing density [42]. It was also detected that the pumping method samples obtained a higher strength value, in comparison with the specimens using the gravity technique. The higher strength values of the pumping method specimens were due to the uniform distribution of grout mixture under pressure across the formwork, which provided a dense matrix. As the volume and quality of cement paste directly influence the strength of concrete, consistent distribution of grout under the control pressure results in a denser microstructure and higher strength values [5].

### 3.3. Impact Resistance and Energy Absorption

In this study, the number of drops for the first crack and ultimate failure were recorded, and based on these numbers, the impact resistance and energy absorption were determined, the results of which are shown in Figures 7 and 8. The increase in the first crack and ultimate crack impact resistance and impact energy, which indicated how many times it increased in comparison to the reference mixes (G0 and P0), are also shown. As shown in Figure 7, in the gravity method specimens, the number of drops for the first crack was recorded as 16, 28, 45, 63, 84, and 102 for the G0, G1, G2, G3, G4, and G5 mixes, correspondingly. It can be seen that, by adding fibres and with an increase in the fibre dosages, the number of drops remarkably improved. Likewise, the number of drops to failure in the specimens were noted as 21, 41, 74, 82, 108, and 131 for G0, G1, G2, G3, G4, and G5 mixes, correspondingly. These outcomes indicate that the mixes with higher fibre volume fractions performed better under impact loads. A similar tendency was noted for the pumping method of PAFRC specimens. As revealed in Figure 7, the impact resistance of PAFRC samples at the first crack was enhanced by about 90%, 215%, 320%, 405%, and 555% with the addition of 0.25%, 0.5%, 0.75%, 1%, and 1.25% PP fibres, respectively, as compared to the plain mix (P0). Additionally, for the same fibre dosages, the impact resistance at failure was improved by 104%, 248%, 296%, 404%, and 544%, respectively. As the test was carried out after 90 days of curing, owing to the pozzolanic nature of POFA and the creation of additional hydration products, such as C–S–H gels, the strength of specimens increased, therefore resulting in the higher impact resistance of PAFRC specimens [43]. Khankhaje et al. [42] reported that hydration products (e.g., C–S–H gels) are developed at longer curing periods, resulting in a denser matrix and, therefore, higher strength values.

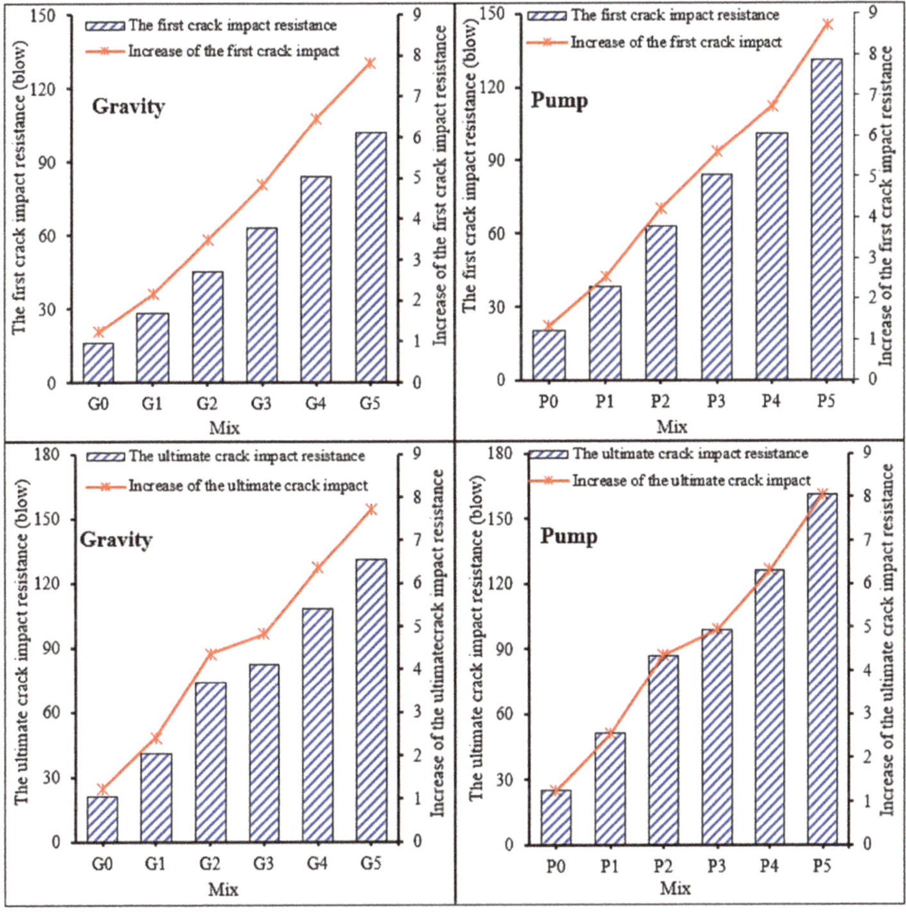

**Figure 7.** Impact resistance of PAFRC specimens for gravity and pumping techniques.

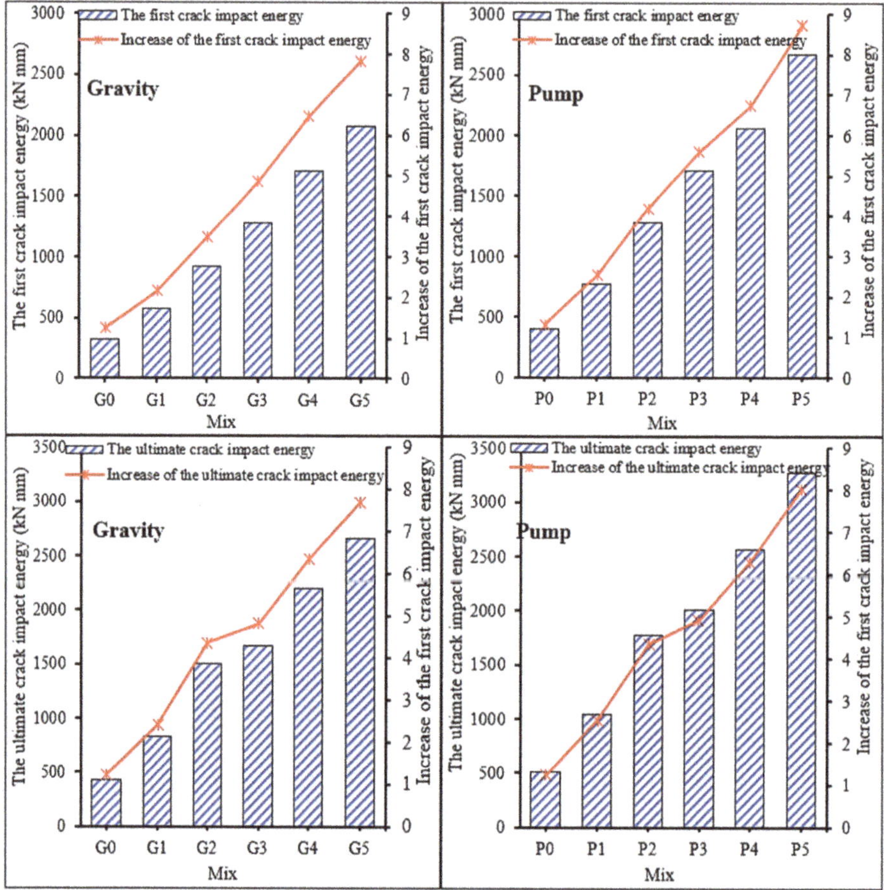

**Figure 8.** Impact energy of PAFRC specimens for gravity and pumping methods.

Furthermore, the impact energy levels of the PAFRC samples at first crack and failure were calculated following Equation (1). It is clear that the inclusion of fibre into the PAFRC specimens noticeably increased the number of repeated drops to the first crack and failure, with reference to the plain PAC specimens. The obtained impact energy values of the PAFRC specimens are demonstrated in Figure 8. In the gravity method group, the highest impact energies were 2076 and 2667 kN mm, recorded for the PAFRC mix reinforced with 1.25% PP carpet fibres (G5) at initial crack and failure, respectively.

A similar tendency to that of the gravity method was noticed for the impact energy of the pumping method group. Higher fibre content resulted in more considerable impact energy in the PAFRC specimens. From Figure 8, it can be seen that the highest impact energies of 2666 and 3277 kN mm were noted for the PAFRC specimens reinforced with 1.25% fibres at first crack and ultimate failure, respectively, in comparison with that of the plain PAC (P0) mix. By comparing the obtained results of the impact energy, it was revealed that the initial crack and failure impact energies of the PAFRC specimens (for all fibre volume fractions), which were grouted by the pumping technique, were considerably higher than those of the gravity samples. The results obtained in this study confirmed the findings of Abirami et al. [14] for layered concrete slabs reinforced with steel fibres and Mastali et al. [44] with the inclusion of recycled CFRP fibres into the concrete. They both reported that the addition of short fibres resulted in higher impact resistance and energy absorption under impact loads.

The PAFRC samples used for impact strength test and the failure modes after the impact test are illustrated in Figure 9. It was detected that the plain specimens without any fibres absorbed less energy and were broken into a few pieces with a small number of drops. This could be due to the brittle nature and lower energy absorption capacity of plain concrete under impact loads. However, by adding PP fibres and with an increase in the fibre dosages, the ductility and energy absorption capacity of PAFRC specimens were significantly enhanced. As shown in Figure 9, the failure modes of the reinforced specimens changed, the specimens became more ductile without sudden failure, and cracks only occurred on the surface of specimens. The enhancement in the ductility and energy absorption of the PAFRC specimens may have been due to the bridging action of PP fibres, which arrests cracks and prevents sudden failure in the specimens. In addition, the uniform distribution of fibres resulted in the propagation of cracks on the top surface of disks and, therefore, absorbed a higher amount of energy before failure [8].

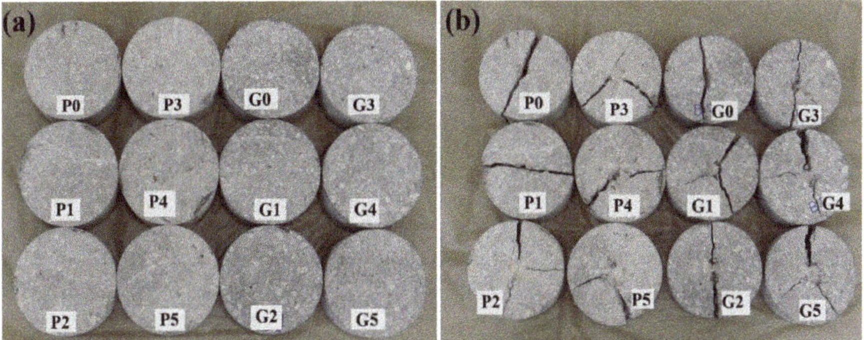


**Figure 9.** PAFRC specimens (**a**) before and (**b**) after impact load.

### 3.4. Long-Term Drying Shrinkage

Drying shrinkage is one of the most common causes of the cracking in concrete structures, which directly affects the durability and strength properties of concrete. Drying shrinkage of concrete cannot be recovered by rewetting. Therefore, the addition of PP provides an alternative solution to reduce drying shrinkage, which is bound to occur over time [21]. As PAC is a preferred type of concrete in massive concreting, thermal cracks are one of the critical problems, which affect the long-term performance of the structure. Consequently, the addition of short PP fibres may also provide an alternative solution to minimise such cracks [1]. Therefore, in this study, the long-term drying shrinkage of PAFRC specimens containing PP fibres and POFA was measured. The main purpose of this experimental work was to investigate the influence of fibres and POFA on the shrinkage behaviour of pre-packed aggregate concrete over a one-year testing period. The experiments performed indicated that the inclusion of PP fibres (up to a specific fibre volume fraction) and POFA has a dominating effect on the drying shrinkage of PAFRC. As shown in Figure 10, the presence of fibres in the PAFRC mixes reduced the drying shrinkage values in which the maximum shrinkage value was detected for the plain mix of gravity method (G0). It can be seen that, up to 28 days of testing, the rate of shrinkage was high for all mixes. However, over time, the rate of variation in the shrinkage of the specimens reduced. It was also detected that the pumping method PAFRC samples attained smaller drying shrinkage values than gravity technique samples. This could have been due to the better distribution of the grout mixture amongst the mixture of aggregates and fibres.

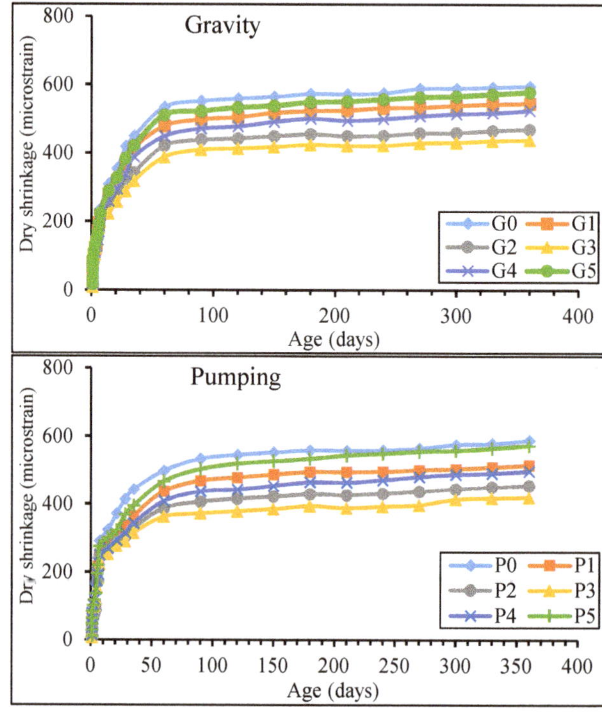

**Figure 10.** Long-term drying shrinkage of PAFRC specimens reinforced with PP fibres.

It is well-known that the shrinkage of concrete depends on the quality of the cement paste, the volume of voids, and (to some extent) on the strength level of the concrete [38]. Therefore, the proper injection of grout leads to a denser matrix, which causes lower drying shrinkage values. Moreover, at the end of the testing period of 365 days, the addition of 0.25%, 0.5%, 0.75%, 1%, and 1.25% PP fibres into the gravity method PAFRC specimens resulted in the reduction in drying shrinkage of 8.5%, 21.2%, 26.3%, 11.9%, and 2.3%, correspondingly, as compared to that of the plain mix (G0) without any fibre. Likewise, in the pumping method PAFRC samples with the same fibre dosages, the shrinkage values were reduced by 12.4%, 22.5%, 28.6%, 15.3%, and 2.9%, correspondingly, as compared to that of the plain mix (P0). The results show that the pumping method specimens obtained lower drying shrinkage values than the gravity method specimens, which may be due to the dense microstructure of the pumping method specimens, a consequence of the uniform distribution of grout into the formworks filling up the voids and the consequent lower volume of pores [3,6].

It can be observed that all mixtures reinforced with PP fibres obtained lower drying shrinkage values after the one-year testing period. However, a further rise in fibre dosages beyond 0.75% caused higher shrinkage values, particularly in the gravity method specimens. This could be due to the existence of the larger amount of fibres in the mixture preventing the uniform distribution of grout amongst the aggregates and, therefore, increasing the volume of voids and causing higher shrinkage values. These outcomes are in agreement with the results reported by Karahan and Atis [45]. They reported that the application of PP fibres in concrete composites was able to considerably reduce drying shrinkage by improving the tensile performance and bridging action along the forming cracks. Medina et al. [46] also demonstrated that the inclusion of PP fibres into concrete composite reduced its drying shrinkage, while also observing a decrease in drying shrinkage related to the increment of fibre content after a certain amount. They stated that the utilisation of PP fibres controlled the movements of the fine cracks in cement paste through a bridging action.

### 3.5. Microstructural Analysis

Microstructural analysis of the grout paste was carried out in terms of SEM analysis, in order to explore the effects of POFA replacement on the hydration product. In addition, the role of PP fibres in enhancing the interfacial transition zone in the matrix was assessed. It was observed that the substitution of OPC with 20% POFA in PAFRC mixes resulted in higher strength values at the ultimate ages. This can be attributed to the pozzolanic nature of POFA, which contains a large volume of reactive $SiO_2$. With the existence of moisture, over time, this reactive $SiO_2$ chemically reacts with the released CH—which is the hydration product of OPC—and forms extra C–S–H gels in the matrix. Consequently, these supplementary hydration products, which are due to the pozzolanic action of POFA, fill up the cavities and decrease the porosity in the matrix, leading to a dense microstructure. This denser microstructure then results in the enhanced strength and durability of concrete [43,47]. Figure 11 displays the SEM image of POFA-based paste in PAFRC samples at the ages of 28 and 90 days. The uniform spreading of C–S–H gels at the age of 90 days is revealed, indicating the denser matrix at the ultimate ages, as compared with the 28-day paste. Besides, the homogeneous distribution of hydration products, such as C–S–H gels is shown in the 90 days SEM image, which filled up the cavities in the matrix and provided a solid microstructure.

**Figure 11.** SEM images indicating the hydration products at the ages of (**a**) 28 and (**b**) 90 days.

Figure 12 illustrates the uniform distribution of PP fibres in the matrix and the linking action of fibres, in addition to the interfacial transition zone among fibres and cement paste in the concrete matrix afterwards. The SEM image reveals a robust bond among fibres and the blended cement paste. This strong bonding between fibres and paste provides a reliable solid microstructure in the matrix, which results in a reduction in crack formation at the interface zone [48]. Consequently, the lower volume of cracks leads to strength development and high impact resistance, as well as the better ductility and energy absorption of PAFRC specimens. The utilisation of waste PP fibres in conventional fibre reinforced concrete has also been studied by Mohammadhosseini et al. [17], who revealed that waste PP fibres and cement paste provide a strong interfacial transition zone; therefore, the consequent concrete specimens could absorb higher energy, resulting in better ductility performance under impact loads.

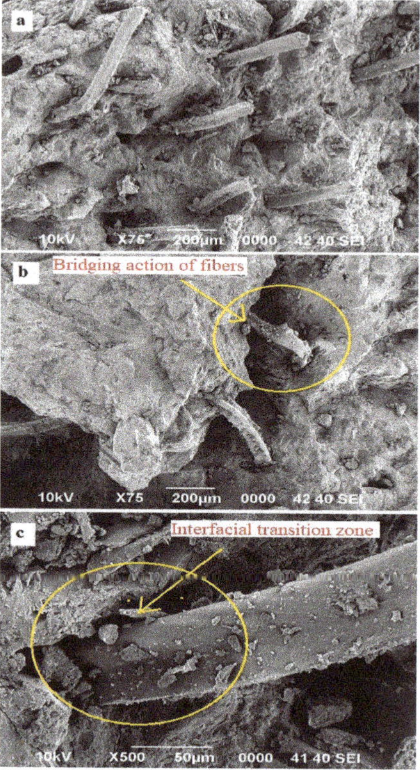



**Figure 12.** (**a**) Distribution of PP fibres; (**b**) bridging action of fibres; (**c**) the interfacial transition zone between paste and fibre.

## 4. Statistical and Analytical Analyses

In this section, the probability distribution of tensile (Ft) and compressive (Fc) strengths, first crack (FC), and ultimate crack (UC) impact resistance of PAFRC specimens containing PP carpet fibres are discussed. In this study, to better understand the performance of PAFRC specimens under impact loads, statistical analyses of the obtained data were deemed necessary. The effects of fibre dosage on distribution and statistical parameters of compressive and tensile strengths and impact resistance of PAC were also considered significant. Empirical relationships were developed for the compressive strength of PAFRC specimens by gravity and pumping methods. To carry out the statistical calculations, the Statistical Package for the Social Sciences (SPSS) software was used. Thus, the Kolmogorov–Smirnov (K–S test) technique, as proposed by Mastali and Dalvand [19], was applied to test the normality of the obtained experimental data. The K–S test method is based on the maximum deviation of the detected cumulative histogram from the hypothesised cumulative distribution function. The normal distribution can well-characterise the strength variability of most ductile materials, such as FRC. Furthermore, most recent studies on FRC have used a normal distribution [19,44], and as such, this study also used a normal distribution to analyse the strength properties of the PAFRC specimens using carpet fibres. From Figure 13, it can be seen that the histogram of the strength and impact resistance of PAFRC specimens reinforced with PP carpet fibres followed a normal distribution. According to the statistical analysis results, p-values of above 0.05 were obtained for all tests, which quantified the strength of evidence against the null hypothesis. The hypotheses for the normal distribution were as follows:

**H1.** The compressive strength of pumping and gravity method PAFRC specimens reinforced with PP fibres follows the normal distribution.

**H2.** The tensile strength of pumping and gravity method PAFRC specimens reinforced with PP fibres follows the normal distribution.

**H3.** The first crack impact resistance of pumping and gravity method PAFRC specimens reinforced with PP fibres follows the normal distribution.

**H4.** The ultimate crack impact resistance of pumping and gravity method PAFRC specimens reinforced with PP fibres follows the normal distribution.

The obtained p-values for all tests were found to be above 0.05. This finding corroborates the null hypothesis at the 0.05 significance level. Consequently, the compressive strength, tensile strength, and first and ultimate crack impact resistance of PAFRC reinforced with waste PP fibres followed normal distributions. The attained outcomes of the current study were similar to the results described by Song et al. [49] and Mohammadhosseini et al. [50], who reported that the mechanical properties of conventional concrete reinforced with PP fibres followed a normal distribution, while the first and the ultimate crack impact resistance hardly followed a normal distribution.

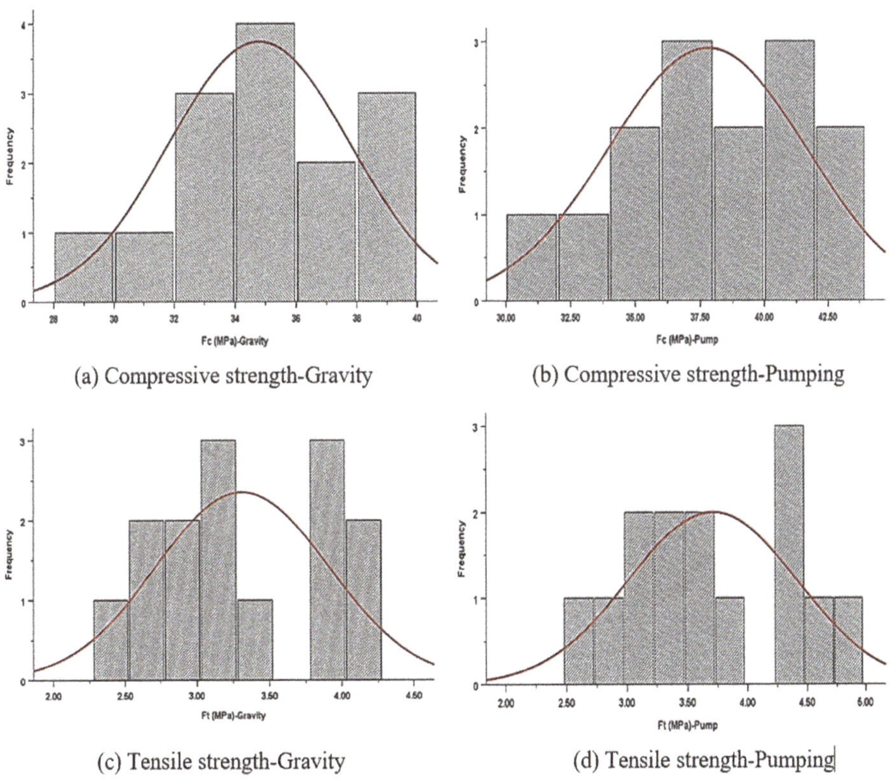


(a) Compressive strength-Gravity

(b) Compressive strength-Pumping

(c) Tensile strength-Gravity

(d) Tensile strength-Pumping

**Figure 13.** *Cont.*

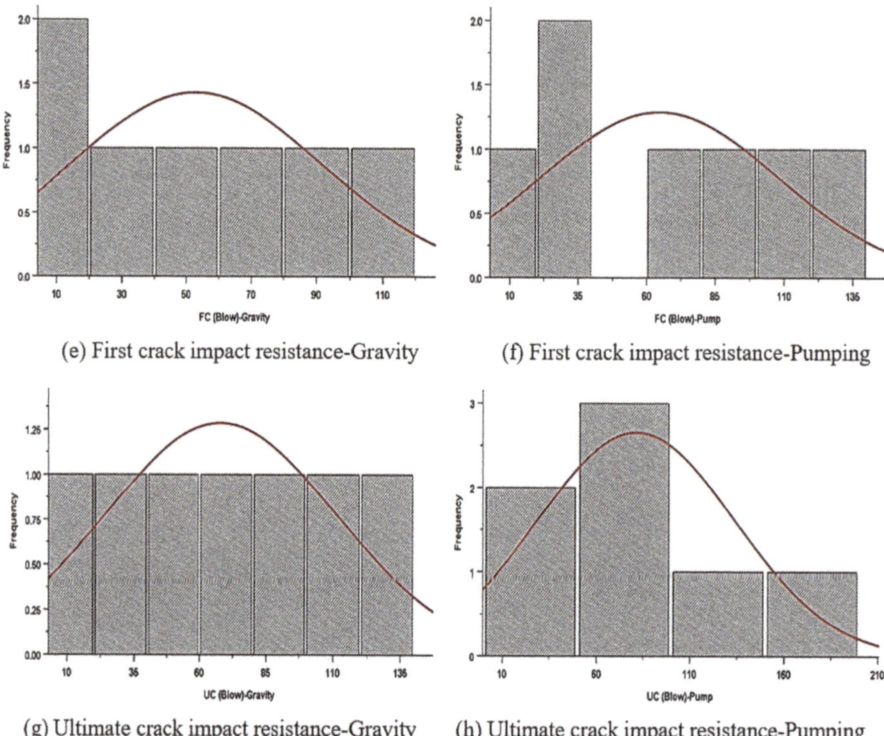

(e) First crack impact resistance-Gravity

(f) First crack impact resistance-Pumping

(g) Ultimate crack impact resistance-Gravity

(h) Ultimate crack impact resistance-Pumping

**Figure 13.** Histograms of strength properties of PAFRC specimens.

According to the collected experimental data, the mechanical properties and impact resistance of the PAFRC specimens could be correlated through empirical relationships with a high coefficient of determination ($R^2$). The correlations were developed by a regression analysis of the given data. Linear regression analyses, as revealed in Figure 14a,b were used to correlate the compressive strength and tensile strength values of PAFRC for the pumping and gravity methods specimens in which a high $R^2$ value was found, thus showing a good relationship between the variables. Concerning the developed Equations (5) and (6), the obtained $R^2$ values were 0.9156 and 0.9843, respectively, which signified a great confidence level for the relationships:

$$f_{cG} = 0.7467f_{cP} + 6.5691, \qquad R^2 = 0.9156 \tag{5}$$

$$f_{tG} = 0.851f_{tP} + 0.1743, \qquad R^2 = 0.9843 \tag{6}$$

where $f_{cG}$ and $f_{cP}$ are the compressive strength (MPa) of the PAFRC specimens by gravity and pumping methods, respectively; $f_{tG}$ and $f_{tP}$ are the tensile strength (MPa) of the PAFRC specimens by pumping and gravity methods, respectively.

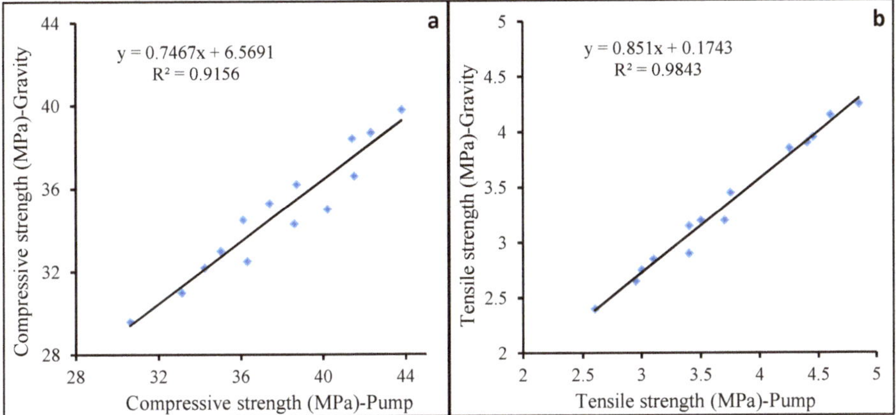



**Figure 14.** The correlations amongst (**a**) compressive strength and (**b**) tensile strength values of PAFRC specimens by gravity and pumping techniques.

Furthermore, the relationship between the first crack impact resistance versus the ultimate crack impact resistance of gravity and pumping methods PAFRC specimens is illustrated in Figure 15a,b. Relating to the obtained Equations (7)–(10), a linear regression analysis was carried out amongst the impact resistance values for the first and ultimate cracks and fibre contents of the gravity and pumping groups PAFRC specimens with $R^2$ values of more than 0.98. It can be seen that the first and ultimate crack impact resistance values of the P5 mix had the highest slope. This indicated that the reinforcement of PAFRC specimens with 1.25% PP carpet fibres had the maximum rate of first crack and failure impact strength improvement.

$$P: FC = 88.625V_f + 17.094, \qquad R^2 = 0.9956 \tag{7}$$

$$G: FC = 70.15V_f + 12.563, \qquad R^2 = 0.9954 \tag{8}$$

$$P: UC = 107.05V_f + 23.938, \qquad R^2 = 0.9897 \tag{9}$$

$$G: UC = 88.6V_f + 20.25, \qquad R^2 = 0.989 \tag{10}$$

Here, FC and UC are the first crack and ultimate crack impact strength of PAFRC specimens, respectively, and $V_f$ is the fibre dosage.

Finally, to explore the impact energy reliability, the relationships between the impact energies at first and ultimate cracks and fibre content of the gravity and pumping groups PAFRC mixtures were correlated; the results are demonstrated in Figure 16a,b. The results demonstrate that there exists a linear relationship between the first and ultimate crack impact energies of PAFRC specimens. Following the developed Equations (11)–(14), the $R^2$ values were found to be higher than 0.9. According to the obtained results, the best performance in increasing the impact energy was detected in the pumping method for PAFRC specimens. The results of this study confirmed the findings by Song et al. [49], with regards to the existence of linear relationships among the first and ultimate crack impact energy of concrete reinforced with PP fibres.

$$P: E_{FC} = 1803.9V_f + 347.92, \qquad R^2 = 0.9956 \tag{11}$$

$$G: E_{FC} = 1427.8V_f + 255.7, \qquad R^2 = 0.9954 \tag{12}$$

$$P: E_{UC} = 2178.9V_f + 487.22, \qquad R^2 = 0.9897 \tag{13}$$

$$G: E_{UC} = 1803.4V_f + 412.17, \qquad R^2 = 0.989 \tag{14}$$

Here, $E_{FC}$ and $E_{UC}$ are the first crack and ultimate crack impact energies of PAFRC specimens.

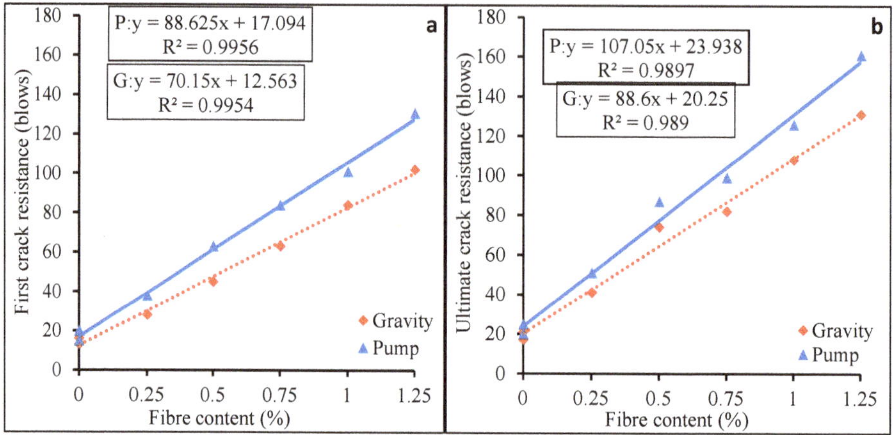

**Figure 15.** Impact resistance at (**a**) first and (**b**) ultimate crack vs. fibre content of PAFRC specimens.

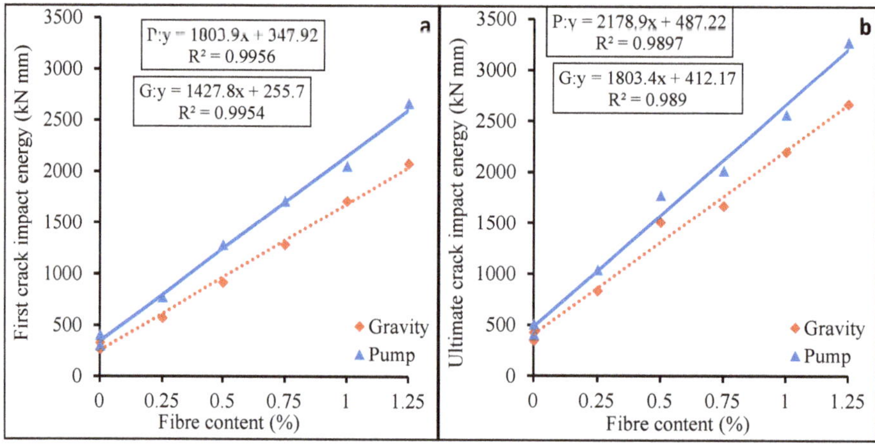

**Figure 16.** (**a**) The first and (**b**) ultimate crack impact energy vs. fibre content of PAFRC specimens.

## 5. Conclusions

In this paper, the strength and impact resistance of a new pre-packed aggregate concrete reinforced with waste polypropylene fibres were investigated, both experimentally and statistically. The following conclusions could be drawn:

With the replacement of OPC by POFA, the fluidity of the fresh grout mixture was increased.

With the addition of fibres into the PAC mixtures, a minor reduction in the compressive strength was detected. At the early age of 28 days, owing to the slow rate of pozzolanic activity in POFA particles, the improvement in strength was marginally low. However, at the ultimate age of 180 days, the obtained strength values were improved, and the strength values were in the acceptable range for structural applications using gravity and pumping techniques.

In spite of lesser developments in compressive strength, noteworthy improvements were detected in the impact strength and energy absorption of PAFRC specimens. Furthermore, the maximum impact strength and energy absorption values were noted for the PAFRC mix containing 1.25% fibres. Overall, the specimens using the pumping technique performed better than those of the gravity method.

Waste PP fibres had a great influence on the long-term drying shrinkage of both gravity and pumping method PAFRC specimens. The lowest shrinkage values were recorded for the pumping method PAFRC mix comprising 0.75% fibres.

The existence of POFA resulted in providing a dense and amorphous microstructure in PAFRC specimens with less porosity and microcracks, as a result of the superior pozzolanic activity of POFA at the ultimate age.

Statistical analyses showed that the strength properties of all PAFRC mixes were normally distributed. Furthermore, the strength properties of PAFRC mixtures were correlated linearly, together with a high coefficient of determination ($R^2$) values.

**Author Contributions:** Conceptualisation, M.M.T. and H.M.; methodology, F.A., R.A., and H.A.; validation, A.M.M. and H.A.; formal analysis, H.M. and F.A.; investigation, H.M. and R.A.; resources, F.A. and R.A.; data curation, H.A. and A.M.M.; writing, original draft preparation, H.M.; writing, review, and editing, H.M., F.A., and R.A.; visualisation, A.M.M.; supervision, M.M.T.; project administration, H.A. All authors have read and agreed to the published version of the manuscript.

**Funding:** The project was supported by the Deanship of Scientific Research at Prince Sattam bin Abdulaziz University (Saudi Arabia) under the research project No. 2020/01/16810.

**Acknowledgments:** The authors gratefully acknowledge the technical and financial support received from the Universiti Teknologi Malaysia (UTM).

**Conflicts of Interest:** The authors declare no conflict of interest.

## References

1. Najjar, M.F.; Soliman, A.M.; Nehdi, M.L. Critical overview of two-stage concrete: Properties and applications. *Constr. Build. Mater.* **2014**, *62*, 47–58. [CrossRef]
2. Awal, A. Creep Recovery of Prepacked Aggregate Concrete. *J. Mater. Civ. Eng.* **1992**, *4*, 320–325. [CrossRef]
3. Abdelgader, H.S. How to design concrete produced by a two-stage concreting method. *Cem. Concr. Res.* **1999**, *29*, 331–337. [CrossRef]
4. Awal, A.S.M.A. Manufacture and Properties of Pre-Packed Aggregate Concrete. Master's Thesis, University of Melbourne, Melbourne, Australia, 1984.
5. Swaddiwudhipong, S.; Zhang, J.; Lee, S.L. Prepacked Grouting Process in Concrete Construction. *J. Mater. Civ. Eng.* **2003**, *15*, 567–576. [CrossRef]
6. Mohammadhosseini, H.; Awal, A.S.M.A.; Sam, A.R.M. Mechanical and thermal properties of prepacked aggregate concrete incorporating palm oil fuel ash. *Sādhanā* **2016**, *41*, 1235–1244. [CrossRef]
7. Najjar, M.F.; Nehdi, M.; Soliman, A.; Azabi, T. Damage mechanisms of two-stage concrete exposed to chemical and physical sulfate attack. *Constr. Build. Mater.* **2017**, *137*, 141–152. [CrossRef]
8. Murali, G.; Ramprasad, K. A feasibility of enhancing the impact strength of novel layered two stage fibrous concrete slabs. *Eng. Struct.* **2018**, *175*, 41–49. [CrossRef]
9. Abdelgader, H.S.; Elgalhud, A.A. Effect of grout proportions on strength of two-stage concrete. *Struct. Concr.* **2008**, *9*, 163–170. [CrossRef]
10. American Concrete Institute. *ACI 304.1. Guide for the Use of Preplaced Aggregate Concrete for Structural and Mass Concrete Applications*; ACI Committee: Farmington Hills, MI, USA, 2005.
11. Mohammadhosseini, H.; Tahir, M.M.; Alaskar, A.; Alabduljabbar, H.; Alyousef, R. Enhancement of strength and transport properties of a novel preplaced aggregate fibre reinforced concrete by adding waste polypropylene carpet fibres. *J. Build. Eng.* **2020**, *27*, 101003. [CrossRef]
12. Mohammadhosseini, H.; Awal, A.S.M.A. Physical and mechanical properties of concrete containing fibres from industrial carpet waste. *Int. J. Res. Eng. Technol.* **2013**, *2*, 464–468.
13. Song, P.S.; Hwang, S. Mechanical properties of high-strength steel fibre-reinforced concrete. *Constr. Build. Mater.* **2004**, *18*, 669–673. [CrossRef]
14. Abirami, T.; Loganaganandan, M.; Murali, G.; Fediuk, R.; Sreekrishna, R.V.; Vignesh, T.; Januppriya, G.; Karthikeyan, K. Experimental research on impact response of novel steel fibrous concretes under falling mass impact. *Constr. Build. Mater.* **2019**, *222*, 447–457. [CrossRef]

15. Alyousef, R.; Mohammadhosseini, H.; Alrshoudi, F.; Alabduljabbar, H.; Mohamed, A.M. Enhanced Performance of Concrete Composites Comprising Waste Metalised Polypropylene Fibres Exposed to Aggressive Environments. *Crystals* **2020**, *10*, 696. [CrossRef]

16. Ramkumar, V.R.; Murali, G.; Asrani, N.P.; Karthikeyan, K. Development of a novel low carbon cementitious two stage layered fibrous concrete with superior impact strength. *J. Build. Eng.* **2019**, *25*, 100841. [CrossRef]

17. Mohammadhosseini, H.; Awal, A.A.; Yatim, J.M. The impact resistance and mechanical properties of concrete reinforced with waste polypropylene carpet fibres. *Constr. Build. Mater.* **2017**, *143*, 147–157. [CrossRef]

18. Awal, A.S.M.A.; Mohammadhosseini, H. Green concrete production incorporating waste carpet fibre and palm oil fuel ash. *J. Clean. Prod.* **2016**, *137*, 157–166. [CrossRef]

19. Mastali, M.; Dalvand, A.; Sattarifard, A. The impact resistance and mechanical properties of the reinforced self-compacting concrete incorporating recycled CFRP fibre with different lengths and dosages. *Compos. Part B Eng.* **2017**, *112*, 74–92. [CrossRef]

20. Ong, K.; Basheerkhan, M.; Paramasivam, P. Resistance of fibre concrete slabs to low velocity projectile impact. *Cem. Concr. Compos.* **1999**, *21*, 391–401. [CrossRef]

21. Mohammadhosseini, H.; Yatim, J.M.; Sam, A.R.M.; Awal, A.A. Durability performance of green concrete composites containing waste carpet fibers and palm oil fuel ash. *J. Clean. Prod.* **2017**, *144*, 448–458. [CrossRef]

22. Wang, Y. Fibre and textile waste utilisation. *Waste Biomass Valoris.* **2010**, *1*, 135–143. [CrossRef]

23. Mohammadhosseini, H.; Tahir, M.M.; Sam, A.R.M.; Lim, N.H.A.S.; Samadi, M. Enhanced performance for aggressive environments of green concrete composites reinforced with waste carpet fibres and palm oil fuel ash. *J. Clean. Prod.* **2018**, *185*, 252–265. [CrossRef]

24. Mohammadhosseini, H.; Alyousef, R.; Lim, N.H.A.S.; Tahir, M.M.; Alabduljabbar, H.; Mohamed. A.M.; Samadi, M. Waste metalised film food packaging as low cost and ecofriendly fibrous materials in the production of sustainable and green concrete composites. *J. Clean. Prod.* **2020**, *258*, 120726. [CrossRef]

25. Alrshoudi, F.; Mohammadhosseini, H.; Tahir, M.M.; Alyousef, R.; Alghamdi, H.; Alharbi, Y.R.; AlSaif, A. Sustainable Use of Waste Polypropylene Fibers and Palm Oil Fuel Ash in the Production of Novel Prepacked Aggregate Fiber-Reinforced Concrete. *Sustainability* **2020**, *12*, 4871. [CrossRef]

26. Lim, S.K.; Tan, C.S.; Lim, O.Y.; Lee, Y.L. Fresh and hardened properties of lightweight foamed concrete with palm oil fuel ash as filler. *Constr. Build. Mater.* **2013**, *46*, 39–47. [CrossRef]

27. Mohammadhosseini, H.; Lim, N.H.A.S.; Tahir, M.M.; Alyousef, R.; Samadi, M.; Alabduljabbar, H.; Mohamed, A.M. Effects of Waste Ceramic as Cement and Fine Aggregate on Durability Performance of Sustainable Mortar. *Arab. J. Sci. Eng.* **2019**, *45*, 3623–3634. [CrossRef]

28. Tangchirapat, W.; Khamklai, S.; Jaturapitakkul, C. Use of ground palm oil fuel ash to improve strength, sulfate resistance, and water permeability of concrete containing high amount of recycled concrete aggregates. *Mater. Des.* **2012**, *41*, 150–157. [CrossRef]

29. Chandara, C.; Azizli, K.A.M.; Ahmad, Z.A.; Hashim, S.F.S.; Sakai, E. Heat of hydration of blended cement containing treated ground palm oil fuel ash. *Constr. Build. Mater.* **2012**, *27*, 78–81. [CrossRef]

30. Wilsdon, B.H. Discrimination by Specification Statistically Considered and Illustrated by the Standard Specification for Portland Cement. *Suppl. J. R. Stat. Soc.* **1934**, *1*, 152. [CrossRef]

31. American Society for Testing and Materials. *ASTM C618. Standard Specification for Coal Fly Ash and Raw or Calcined Natural Pozzolan for Use in Concrete*; American Society for Testing and Materials: Philadelphia, PA, USA, 2015.

32. British Standard. *BS 3892-1. Pulverised-Fuel Ash Part 1. Specification for Pulverised-Fuel Ash for Use with Portland Cement*; British Standard: Herndon, VA, USA, 1997.

33. American Society for Testing and Materials. *ASTM C939. Standard Test Method for Flow of Grout for Preplaced-Aggregate Concrete (Flow Cone Method)*; American Society for Testing and Materials: Philadelphia, PA, USA, 2016.

34. American Society for Testing and Materials. *ASTM C39. Standard Test Method for Compressive Strength of Cylindrical Concrete Specimens*; American Society for Testing and Materials: Philadelphia, PA, USA, 2018.

35. American Society for Testing and Materials. *ASTM C512. Standard Test Method for Creep of Concrete in Compression*; American Society for Testing and Materials: Philadelphia, PA, USA, 2010.

36. American Concrete Institute. *ACI 544.2R. Measurement of Properties of Fiber Reinforced Concrete*; ACI Committee: Farmington Hills, MI, USA, 1999.

37. Mohammadhosseini, H.; Alyousef, R.; Lim, N.H.A.S.; Tahir, M.M.; Alabduljabbar, H.; Mohamed, A.M. Creep and drying shrinkage performance of concrete composite comprising waste polypropylene carpet fibres and palm oil fuel ash. *J. Build. Eng.* **2020**, *30*, 101250. [CrossRef]

38. Alrshoudi, F.; Mohammadhosseini, H.; Tahir, M.M.; Alyousef, R.; Alghamdi, H.; Alharbi, Y.; Alsaif, A. Drying shrinkage and creep properties of prepacked aggregate concrete reinforced with waste polypropylene fibres. *J. Build. Eng.* **2020**, *32*, 101522. [CrossRef]

39. Lim, N.H.A.S.; Mohammadhosseini, H.; Tahir, M.M.; Samadi, M.; Sam, A.R.M. Microstructure and Strength Properties of Mortar Containing Waste Ceramic Nanoparticles. *Arab. J. Sci. Eng.* **2018**, *43*, 5305–5313. [CrossRef]

40. Alsubari, B.; Shafigh, P.; Jumaat, M.Z. Utilisation of high-volume treated palm oil fuel ash to produce sustainable self-compacting concrete. *J. Clean. Prod.* **2017**, *137*, 982–996. [CrossRef]

41. Mastali, M.; Naghibdehi, M.G.; Naghipour, M.; Rabiee, S. Experimental assessment of functionally graded reinforced concrete (FGRC) slabs under drop weight and projectile impacts. *Constr. Build. Mater.* **2015**, *95*, 296–311. [CrossRef]

42. Khankhaje, E.; Hussin, M.W.; Mirza, J.; Raieizonooz, M.; Salim, M.R.; Siong, H.C.; Warid, M.N.M. On blended cement and geopolymer concretes containing palm oil fuel ash. *Mater. Des.* **2016**, *89*, 385–398. [CrossRef]

43. Muthusamy, K.; Mirza, J.; Zamri, N.A.; Hussin, M.W.; Majeed, A.P.A.; Kusbiantoro, A.; Budiea, A.M.A. Properties of high strength palm oil clinker lightweight concrete containing palm oil fuel ash in tropical climate. *Constr. Build. Mater.* **2019**, *199*, 163–177. [CrossRef]

44. Mastali, M.; Dalvand, A. The impact resistance and mechanical properties of self-compacting concrete reinforced with recycled CFRP pieces. *Compos. Part B Eng.* **2016**, *92*, 360–376. [CrossRef]

45. Karahan, O.; Atiş, C.D. The durability properties of polypropylene fibre reinforced fly ash concrete. *Mater. Des.* **2011**, *32*, 1044–1049. [CrossRef]

46. Medina, N.F.; Barluenga, G.; Olivares, F.H. Enhancement of durability of concrete composites containing natural pozzolans blended cement through the use of Polypropylene fibers. *Compos. Part B Eng.* **2014**, *61*, 214–221. [CrossRef]

47. Huseien, G.F.; Tahir, M.M.; Mirza, J.; Ismail, M.; Shah, K.W.; Asaad, M.A. Effects of POFA replaced with FA on durability properties of GBFS included alkali activated mortars. *Constr. Build. Mater.* **2018**, *175*, 174–186. [CrossRef]

48. Alyousef, R.; Alabduljabbar, H.; Mohammadhosseini, H.; Mohamed, A.M.; Siddika, A.; Alrshoudi, F.; Alaskar, A. Utilization of sheep wool as potential fibrous materials in the production of concrete composites. *J. Build. Eng.* **2020**, *30*, 101216. [CrossRef]

49. Song, P.S.; Wu, J.C.; Hwang, S.; Sheu, B.C. Statistical analysis of impact strength and strength reliability of steel–polypropylene hybrid fibre-reinforced concrete. *Constr. Build. Mater.* **2005**, *19*, 1–9. [CrossRef]

50. Mohammadhosseini, H.; Alrshoudi, F.; Tahir, M.M.; Alyousef, R.; Alghamdi, H.; Alharbi, Y.R.; Alsaif, A. Performance evaluation of novel prepacked aggregate concrete reinforced with waste polypropylene fibers at elevated temperatures. *Constr. Build. Mater.* **2020**, *259*, 120418. [CrossRef]

Article

# Determination of Mohr-Coulomb Parameters for Modelling of Concrete

**Selimir Lelovic** [1,*] and **Dejan Vasovic** [2,*]

1   Faculty of Civil Engineering, University of Belgrade, 11000 Belgrade, Serbia
2   Faculty of Architecture, University of Belgrade, 11000 Belgrade, Serbia
*   Correspondence: lelovic@grf.bg.ac.rs (S.L.); d.vasovic@arh.bg.ac.rs (D.V.)

Received: 23 July 2020; Accepted: 7 September 2020; Published: 13 September 2020

**Abstract:** Cohesion is defined as the shear strength of material when compressive stress is zero. This article presents a new method for the experimental determination of cohesion at pre-set angles of shear deformation. Specially designed moulds are created to force deformation (close to $\tau$-axis) at fixed pre-set values of angle with respect to normal stress $\sigma$. Testing is performed on series of concrete blocks of different strengths. From the compressive side, cohesion is determined from the extrapolation of the linear Mohr–Coulomb (MC) model, as the intercept on the shear stress axis. From the tensile stress side (from the left), cohesion is obtained using the Brazilian test results: first, indirect tensile strength of material $\sigma_t{}^{BT}$ is measured, then Mohr circle diagram values are calculated and cohesion is determined as the value of shear stress $\tau^{BT}$ on the Mohr circle where normal stress $(\sigma)t = 0$. A hypothesis is made that cohesion is the common point between two tests. In the numerical part, a theory of ultimate load is applied to model Brazilian test using the angle of shear friction from the MC model. Matching experimental and numerical results confirm that the proposed procedure is applicable in numerical analysis.

**Keywords:** cohesion; angle of shear deformation; Mohr–Coulomb model; induced tensile strength; concrete samples; Brazilian test; finite element method (FEM)

---

## 1. Introduction

In order to describe the deformation behaviour of brittle material, complex issues are discussed for the Brazilian test and forced shear test. The control of many variables connected with various samples and testing equipment is a difficult and expensive task aimed to obtain representative stress strain curves for real material. Concrete, as a brittle material, is inhomogeneous, its mechanical properties vary even within the same batch, and therefore, the number of measurements must follow any definition of the material properties where the statistical distribution of the experimental results is performed. Many authors obtained close mathematical description for the complex deformation behaviour of a real material in some cases for a specific test [1]. However, the same mathematical model may not be valid in describing the deformation behaviour under a different set of experimental conditions.

It is widely known that ordinary concrete is a non-homogeneous mixture of Portland cement, aggregate (gravel and sand), water, and different admixtures. The hardening of concrete is a continuous and long-lasting process and the compressive strength of the concrete, being the most used and most important characteristic of the concrete, is standardised as compressive strength at the age of 28 days. Mechanical properties of concrete may be described as those of brittle ceramic material. It is widely known that concrete is an anisotropic material and has high compressive and low tensile strength [2]. Therefore, in structural applications, concrete is used as a material mostly subjected to compression [3]. With the increase of the load, concrete exhibits the following behaviour: elastic deformation and plastic flow in early stages [4], followed by the appearance and propagation of microcracks [5].

In the Brazilian test (split cylinder test), compression induces tension. Therefore, in modelling such tests, compression may be described with the mechanics of plastic deformation, while induced reactive tension may be described with fracture mechanics [6]. The plastic behaviour of concrete was modelled either as elastic-perfectly plastic material [7] or as elastic-plastic material with strain hardening [8]. Various authors used a large number of different techniques to investigate the microstructure of concrete, in order to explain plain concrete behaviour, such as SEM [9], various synchrotron radiation techniques to characterise cement-based material [10], non-destructive techniques (such as X-ray tomography and ultrasonic imaging) to visualise inclusions in concrete samples [11] and CT scanning for density profiles' air pockets distribution [12].

Bazant et al. [13] attempted to estimate the cumulative probability of the concrete sample fail under a given load. To describe the size effect of cracks in a brittle material, they used the Weibull statistical model. The paper presented an attempt to combine the reliability theory with fracture mechanics. There were two phases perceived synergistically: first, crack initiation phase, statistically calculated (according to the reliability theory), and second, crack propagation phase, calculated as the energy part (described as fracture nonlinearity).

Some authors tried to understand where the initial microcracks form, by applying moderate external load to induce microcracks, which formed at the interface between the cement paste and coarse gravel particles [14]. It is noted that increased load induced the coalescence of the microcracks into the large macro-cracks.

Kansan and Jirsa [15] indicated that, for the elements subjected to uni-axial compression, as a function of axial deformation, plastic hardening occurs first, followed by plastic softening. Chen [16] justified the assumption that concrete can be considered as perfectly plastic material applicable for certain classes of problems in mechanics. He described the argument as an "idealisation which contains the essential features of certain material behaviour: the tangent modulus when loading in the plastic range is small compared with the elastic modulus, and the unloading response is elastic. Further, the strain level of interest in a problem determines the choice of flow stress. Thus, in a sense, this perfectly plastic flow stress somewhat represents an averaging work-hardening or work-softening of the material over the field of flow". Further illustrations of the above can be found in the literature [17,18].

In a classic paper, the relationship between compressive and tensile strength [19] is reported on rocks. Linear fit correlation indicated that one single factor might be of a prime importance in the fracture process. A suggested factor is the presence of microcracks, which are inherent in most rock types. In another report, a two-parameters parabolic Mohr criterion model is developed to analyse the failure modes of rock materials using Brazilian test results [20]. The first parameter is called materials strength expressed as the ratio $m = \sigma_t/\sigma_c$ and the second parameter is $N = (m + 2 - 2\sqrt{m + a}) \cdot \sigma_t$. Four failure modes are defined as combination of tensile and shear stress: pure tensile failure, shear damage, pure shear failure and shear failure under compression. Failure modes, as presented in the paper by Ma et al. [21], are classified into five categories: tensile failure across and along the weakness planes, shear failure across and along the weakness planes, and mixed failure.

In a similar paper, cohesion and angle of friction are determined from tensile strength and uniaxial compression [22]. Presented are empirical equations which correlate compressive strength ($\sigma_c$), tensile strength ($\sigma_t$), and cohesion (c). For example, $1/m$ parameter looked as the ratio of $\sigma_t/\sigma_c$ is expected to lie in the range from $1/5$ to $1/20$, with an average value of about $1/10$. Test results of compressive strength, and tensile strength for lightly cemented sand (54% of cement) showed $1/m$ to be about $1/10$. This paper is based on the referenced article by Piratheepan et al. [23], where cohesion and friction angle are derived from first principles to be as follows:

$$c = \frac{\sigma_c \cdot \sigma_t}{2\sqrt{\sigma_t \cdot (\sigma_c - 2\sigma_t)}}, \quad \phi = \sin^{-1}\left(\frac{\sigma_c - 4\sigma_t}{\sigma_c - 2\sigma_t}\right) \tag{1}$$

Results obtained in the experimental testing of 35 sets of rock specimens, for the predicted value of cohesion, appear to be more realistic than for the predicted friction angle values. Using linear fit approximation, the following equation is obtained for cohesion: $c = 1.82 \cdot \sigma_t$ with correlation factor R2 = 0.92. Fitting scatter is attributed to heterogeneity in the rock specimens and anisotropy.

Some authors noticed the importance of determining the value of cohesion in order to apply the Mohr–Coulomb model to different types of concrete and used cohesion parameters to describe the concrete softening/hardening and accumulative plastic deformations [24–27].

### 1.1. Brazilian Test

The Brazilian test is described in terms of the contact angle between specimen and loading plates [28]. The contact angle may affect the shear deformation in a region just under the loading plates, while tensile stress plays a major role in the centre of the specimen. According to the standard ASTM D 3967-08, the tensile strength equation is [29]

$$\sigma_t^{BT} = \frac{2 \cdot P_{max}}{\pi \cdot D \cdot B} \tag{2}$$

where Pmax is the ultimate load reached during the test, D is the diameter and B is the sample thickness. As pointed out in the literature [30], it is possible that shear stresses in the contact region reach critical magnitude and create initial cracks long before the tensile stresses reach their limit in the centre. In such cases, the most likely place of splitting a disk in the Brazilian test is not diametrical but is on surfaces indicated with border lines of the loading area.

The Brazilian test is looked at as the tensile stress, but in reality, it is a more complex stress state. In this paper, it is proposed to look at the Brazilian complex stress state as if it were composed of pure shear and uniaxial compression. The proposed model is based on illustration reported by Hobbs for the Brazilian test, in which shear stress state is in the sample region just under the loading plates. Figure 1 shows block diagrams with the proposed stress state for the Brazilian test for massive rocks, as composed of pure shear ($\tau = \sigma_t$) and added uniaxial compression of $\sigma c = 2\sigma_t$. The radius for the Mohr circle for pure shear segment is R1 = $\tau$ = $\sigma_t$ and radius for the Mohr circle of the added compression is R2 = $\sigma_c$ = $\sigma_t$, making the total radius for the Mohr circle for the Brazilian test for rocks to be R = R1 + R2 = $2\sigma_t$, as shown in Figure 1b.

The strain distribution of the Brazilian specimen and transition between shear and tensile failure modes is reported in the paper by Li et al. [31]. A numerical analysis described in this paper showed that the maximum tensile stress and strain occurs about 5 mm away from the loading points along the central loading diameter of the 50-mm diameter Brazilian disc. This location corresponds to a failure mode transition between shear and tensile failure, as shown in Figure 1c; illustration reported by Hobbs.

A recently published paper showed a new interesting practical test method for determination of cohesion values of rock materials using double shear jaws [32]. Steel jaws are designed and manufactured for testing rock materials. Effects of various testing parameters are reported with respect to induced shear stress failure mechanisms.

Previous standard method of testing the shear strength for intact rock mass [33] is carried out in special moulds, which enable shear at the desired angle $\alpha$. Angles at which testing is performed are 30°, 45° and 70° with respect to normal stress. However, from the practical point of view, it is extremely difficult to properly measure samples at 30°. We made several attempts to test rock samples, but attempts were not successful: many samples crushed before being tested, either on the edges or any rough surfaces or planes, at very high values of applied loads. It was decided to avoid the region of testing over 45° with respect to normal stress, and to stay close to the shear stress axis.

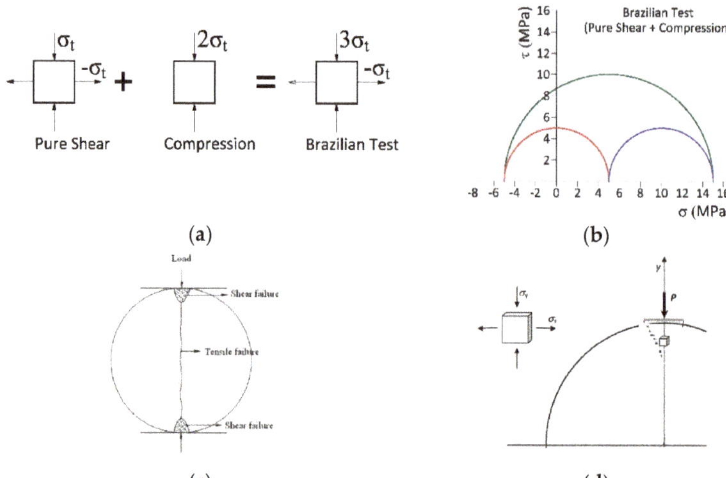

**(a)**

**(b)**

**(c)**

**(d)**

**Figure 1.** (**a**) Block diagram showing Brazilian test as composed of pure shear with the radius of $\tau = \sigma_t$ and uniaxial compression of $\sigma_c = 2\sigma_t$; (**b**) Mohr circles showing the Brazilian test (blue circle), pure shear with the radius of $\tau = \sigma_t$ (red circle) and uniaxial compression of $\sigma_c = 2\sigma_t$ (green circle). (**c**) Transition between shear and tensile failure modes in a solid disk in the Brazilian test (as illustrated by Hobbs). (**d**) Block diagram showing the state of shear stress under loading plates.

## 1.2. Shear Test at Pre-Set Angles

The novelty of this proposed new method is to stay close to $\tau$-axis, to force shear deformation at 80° and 70° with respect to normal stress $\sigma$-axis. It is a simple, efficient procedure for the application of large size samples of inhomogeneous materials, such as rocks and concrete blocks. The relation between experimental measurements and parameters such as cohesion and the angle of friction is straightforward. As indicated in Figure 2, cohesion is obtained by extrapolation of the linear fit points to obtain intercept with shear stress axis, and the angle of friction is obtained from the linear fit slope.

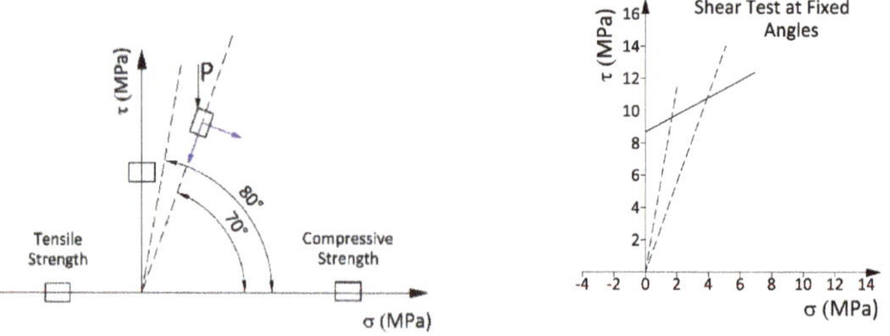

**Figure 2.** Schematic drawing showing shear test at fixed angles with respect to normal stresses ($\sigma$), in order to obtain points for linear Mohr–Coulomb (MC) model.

A presented method for determining shear strength limit in rock mechanics is based on Yugoslavian standard JUS B7.130 reference [34], with the idea that, for brittle material, it is possible to hold the angle of shear deformation constant while applying load. New testing moulds were created with angles at which testing is to be carried out close to the shear stress axis, chosen to be at fixed angles of 80° and 70° with respect to normal stress, as indicated in Figure 3.

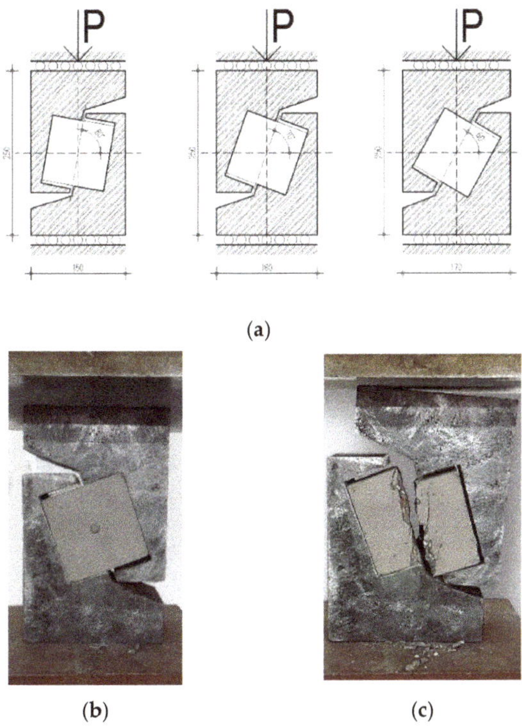

(a)

(b)                              (c)

**Figure 3.** Experimental setup for shear strength testing: (**a**) schematic drawings of mould; (**b**) photo of experimental setup in the laboratory; (**c**) photo of failed sample.

The following equations are used to describe normal and shear stress

$$\sigma = P \cdot \cos \alpha / A \tag{3}$$

$$\tau = P \cdot \sin \alpha / A \tag{4}$$

where P is the failure force, A is the surface and $\alpha$ is the angle wrt to $\sigma$.

## 2. Materials and Testing

In order to define parameters of different types of concrete, several test batches with different mixture proportions were performed according to EN 206-1. As a binder, cement type CEM I 42,5 R (Lafarge BFC, Serbia, according to EN 197-1:2000) was used. In order to test the influence of aggregate type on concrete parameters, two different types of aggregate were used: concrete type A with crushed andesite aggregate from "Šumnik" quarry in Raška, Serbia, and concrete type M with natural river aggregate from river Morava, Serbia. Third concrete type mixture (Mohr–Coulomb (MC)) was prepared with natural river aggregate and addition of styrene-butadiene latex. Polycarboxylate superplasticising admixture (PCA) were added to some mixtures in order to achieve predicted strength and designed Abrams' cone slump of 18–20 cm. Table 1 shows mixture proportions of different concrete batches.

Concrete was casted in 15 cm cubic moulds. All the mixtures were casted in series of 6 samples: 3 for compressive strength testing (EN 12390-2), and 3 for indirect tensile strength (Brazilian test—EN 12390-6). Testing was performed in the Laboratory for Material Testing at the University of Belgrade, Faculty of Civil Engineering. Testing equipment used for this experiment is AMSLER hydraulic load frame, max. capacity of 2500 kN, equipped with two inductive sensors (LVDT) calibrated to

1 micron resolution (Schaffhausen, Swicerland). After 28 days, samples were tested on the hydraulic compression frame of the 200-tons (2000 kN) capacity, synchronised with two inductive sensors (LVDT) calibrated to 1-micron resolution. Those sensors provide automatic acquisition of data with capacity of 20 to 40 data points per second. The results of the concrete samples are shown in Table 2.

**Table 1.** Mixture proportion of different concrete batches.

| Concrete Mixture (kg per m$^3$) | Type A | | | Type M | | | Type MC | |
|---|---|---|---|---|---|---|---|---|
| Aggregate | crushed andesite | | | natural river | | | natural river | |
| | A1 | A2 | A3 | M1 | M2 | M3 | MC1 | MC2 |
| 0–4 mm (45%) | 860 | 845 | 825 | 870 | 845 | 830 | 835 | 820 |
| 4–8 mm (16%) | 305 | 300 | 290 | 310 | 300 | 295 | 300 | 290 |
| 8–16 mm (39%) | 750 | 735 | 715 | 755 | 730 | 720 | 725 | 710 |
| Cement: CEM I 42,5 R | 320 | 370 | 420 | 300 | 350 | 400 | 360 | 400 |
| Water | 160 | 148 | 147 | 165 | 175 | 152 | 180 | 160 |
| PCA Superplasticizer | 1.92 | 2.96 | 4.2 | - | - | 3.2 | - | 3.2 |
| SB Latex (50% dry solid content) | - | - | - | - | - | - | 18 | 20 |

**Table 2.** Compressive and indirect tensile strength for concrete types A, M and MC after 28 days.

| Concrete Type | Compressive Strength (MPa) | | | | Indirect Tensile Strength (MPa) | | | |
|---|---|---|---|---|---|---|---|---|
| sample | 1 | 2 | 3 | Ave. | 4 | 5 | 6 | Ave. |
| A1 | 43.70 | 44.38 | 39.62 | 42.57 | 3.71 | 3.16 | 3.48 | 3.45 |
| A2 | 71.11 | 77.24 | 72.93 | 73.76 | 4.24 | 4.21 | 4.73 | 4.39 |
| A3 | 86.36 | 87.73 | 82.00 | 85.36 | 5.10 | 5.89 | 5.69 | 5.56 |
| M1 | 31.25 | 30.92 | 32.52 | 31.56 | 3.11 | 2.92 | 3.30 | 3.11 |
| M2 | 48.67 | 50.06 | 45.20 | 47.98 | 2.77 | 2.97 | 3.30 | 3.01 |
| M3 | 68.44 | 66.89 | 68.70 | 68.01 | 5.12 | 4.90 | 5.36 | 5.13 |
| MC1 | 37.87 | 41.24 | 41.33 | 40.15 | 3.05 | 2.78 | 3.28 | 3.04 |
| MC2 | 64.49 | 59.60 | 64.76 | 62.95 | 4.65 | 4.38 | 4.54 | 4.52 |

Two additional series of concrete cube samples were designed and casted, in order to provide input data for determination of MC parameters for concrete modelling. Concrete mixtures were designed to represent the concrete of nominal compressive strength of around 30 MPa (labelled MB 30) and concrete of nominal compressive strength of around 50 MPa (labelled MB 50). Additional concrete mixture as a representative for high strength concrete was designed, casted and labelled MB100. Concrete mixture proportions are shown in Table 3.

**Table 3.** Mixture proportion of concrete batches labelled as MB30 and MB 50.

| Concrete Mixture (kg per m$^3$) | MB 30 | MB 50 | MB 100 |
|---|---|---|---|
| Aggregate | natural river | natural river | crushed andesite |
| 0–4 mm (44%) | 840 | 810 | 0–2 mm: 660 |
| 4–8 mm (15%) | 290 | 290 | 2–4 mm: 220 |
| 8–16 mm (41%) | 780 | 750 | 4–8 mm: 620 |
| Cement: CEM I 42,5 R | 320 | 370 | 700 |
| Water | 170 | 180 | 210 |
| PCA Superplasticizer | - | 2.2 | 21 |

Eighteen 100 mm concrete cubes were casted of both MB30 and MB 50 mixture: three were used to determine compressive strength (shown in Table 4).

The remaining fifteen concrete cubes were used for tensile and shear stress data, presented in Table 5. Due to omission in casting, two samples of MB30 batch were not appropriate for testing, so thirteen samples were tested. Six samples of MB 100 were casted: two for compressive strength, and four for shear test (70° and 80°).

**Table 4.** Compressive and indirect tensile strength for concrete types MB 30 and MB 50 after 28 days.

| Concrete Type | Compressive Strength (MPa) | | | |
|---|---|---|---|---|
| sample | 1 | 2 | 3 | Ave. |
| MB 30 | 35.22 | 36.18 | 34.62 | 35.34 |
| MB 50 | 58.26 | 61.12 | 58.48 | 59.29 |
| MB 100 | 100.2 | 97.7 | - | 99.0 |

**Table 5.** Shear test data at fixed angles of 80°, 70° and 60° with respect to $\sigma$- axis, and tensile test data for concrete cube samples MB 30 and MB 50.

| | | Concrete Cube Samples Series MB 30 | | | | | Concrete Cube Samples Series MB 50 | | | | |
|---|---|---|---|---|---|---|---|---|---|---|---|
| Sample | Angle | Load (kN) | $\sigma$ (MPa) | $\tau$ (MPa) | Ave $\sigma$ (MPa) | Ave $\tau$ (MPa) | Load (kN) | $\sigma$ (MPa) | $\tau$ (MPa) | Ave $\sigma$ (MPa) | Ave $\tau$ (MPa) |
| 1 | 80 | 54 | 0.94 | 5.32 | | | 47 | 0.82 | 4.63 | | |
| 2 | 80 | 35 | 0.61 | 3.45 | 0.792 | 4.46 | 38 | 0.66 | 3.74 | 0.76 | 4.33 |
| 3 | 80 | 47 | 0.82 | 4.63 | | | 45 | 0.78 | 4.43 | | |
| 4 | 80 | | | | | | 46 | 0.80 | 4.53 | | |
| 5 | 70 | 67 | 2.29 | 6.30 | | | 96 | 3.28 | 9.02 | | |
| 6 | 70 | 84 | 2.87 | 7.89 | 2.82 | 7.75 | 99 | 3.39 | 9.30 | 3.34 | 9.19 |
| 7 | 70 | 94 | 3.21 | 8.83 | | | 88 | 3.01 | 8.27 | | |
| 8 | 70 | 85 | 2.91 | 7.99 | | | 108 | 3.69 | 10.15 | | |
| 9 | 60 | 144 | 7.20 | 12.47 | | | 183 | 9.15 | 15.85 | | |
| 10 | 60 | 101 | 5.05 | 8.75 | 6.60 | 11.43 | 107 | 5.35 | 9.27 | 7.51 | 13.01 |
| 11 | 60 | 151 | 7.55 | 13.08 | | | 157 | 7.85 | 13.60 | | |
| 12 | 60 | | | | | | 154 | 7.70 | 13.34 | | |
| 13 | BT | 52 | 3.31 | | | | 36 | 2.29 | | | |
| 14 | BT | 43 | 2.74 | | 2.78 | | 48 | 3.06 | | 2.97 | |
| 15 | BT | 36 | 2.29 | | | | 56 | 3.57 | | | |

## 3. Experimental Data

Tests of shear strength through mass were carried out on numerous cubic samples. Dimensions of the samples were 100 × 100 × 100 mm. With values determined for $\sigma$ and $\tau$, a graph of strength limit state is made, approximated to straight line, whose inclination determines the value of the internal friction angle, and the intersection with ordinate defines the value of cohesion c.

The new shear test method is applied for testing concrete blocks as an example of brittle material. Concrete is a material with complex function of many input variables associated with its components (such as type of cement and aggregate, chemical composition, particle size distribution, etc.) [35]. The tensile strength of most concrete types generally does not exceed 10% of the corresponding compressive strength [36]. As indicated in the referenced paper by O'Neil, compressive strength for conventional concrete is generally less than 40 MPa. Concretes designed with compression strength of 41.4 MPa or greater are known as "high-strength".

In this paper, an attempt is made to create concrete series with designed values for compressive strength, and to determine the dependence of tensile strength, and of cohesion. A series of indirect tensile strength (Brazilian test) was performed on concrete series type A, M, MC, MB30, MB50, MB100. Sets of these tests are shown in the Figure 4.

Table 6 presents the summary of experimental data for concrete series made using different input parameters such as aggregate material. For example, concrete series M is created using grain aggregate from the Morava river in Serbia. Looking at the data presented for concrete series M1 and M2, compressive strength increased from 31.5 MPa to 48 MPa with no significant change in tensile strength at around 3 MPa. Figure 5 shows functional dependence for all samples from Table 6, tensile strength being a function of designed compressive strength.

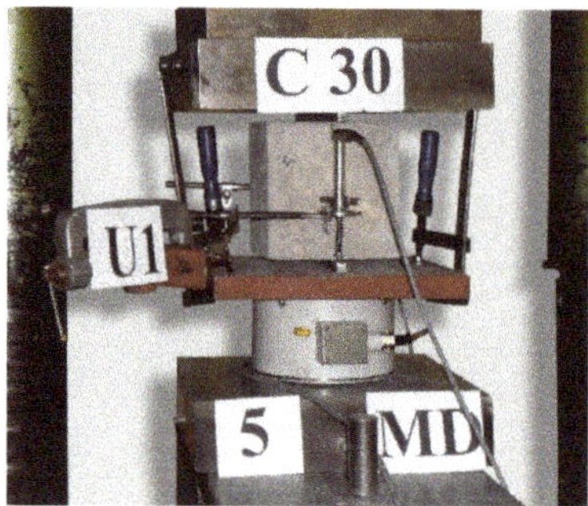

**Figure 4.** Experimental setup for induced tensile load capacity in a concrete sample.

**Table 6.** Average compressive and tensile strengths for different concrete series.

| Concrete Series | Compressive Strength (MPa) | Tensile Strength (MPa) |
| --- | --- | --- |
| M1 | 31.56 | 3.11 |
| MC1 | 40.15 | 3.04 |
| A1 | 42.60 | 3.45 |
| M2 | 47.97 | 3.01 |
| MC2 | 62.95 | 4.52 |
| M3 | 68.01 | 5.13 |
| A2 | 73.76 | 4.39 |
| A3 | 85.36 | 5.56 |

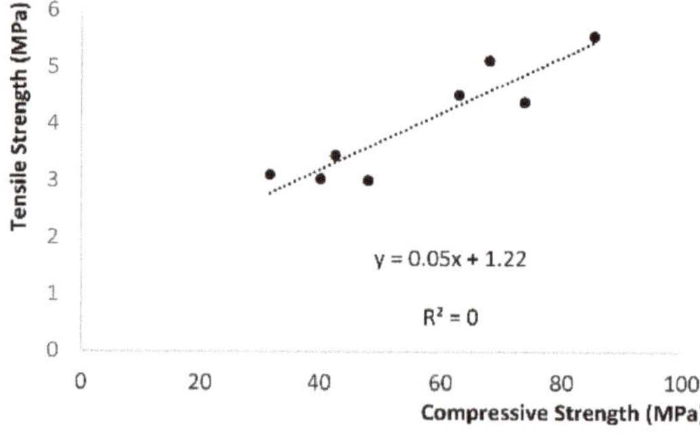


**Figure 5.** Relationship between compressive and tensile strengths for concrete series presented in Table 1.

The presented linear fit is showing an intercept at 1.22 MPa and a small gain of only 0.05 with a large spread of experimental data indicated with correlation function of only 0.85. In fact, it is a difficult assignment to create a recipe for concrete series with designed values for both compressive and tensile strengths. As mentioned previously, Hobbs showed linear dependence between compressive and tensile strengths for massive rocks: compressive strength increased four times going from limestone to sandstone (6000 lb/in2/41.37 MPa to 24,000 lb/in2/165.48 MPa), tensile strength increased 3.7 times (2600 lb/in2/17.93 MPa to 9500 lb/in2/65.50 MPa).

Nevertheless, concrete is a more heterogeneous material in comparison to rocks, with a complex structure, and therefore, the increase in compressive strength does not provoke the increase in tensile strength. Attempts to make a correlation between compressive and tensile strengths for concrete are reported [37,38]. Empirical equations for tensile strength as function of compressive strength ranged from $\sigma_t = 0.6 \cdot \sqrt{\sigma_c}$ to $\sigma_t = 0.2 \cdot (\sigma_c)^{0.7}$ while American Concrete Institute Committee reported [39] $\sigma_t = 0.59 \cdot \sqrt{\sigma_c}$ in the range for compressive strength between 21 and 83 MPa. Further attempts at making systematic correlation between compressive strength and tensile strength of ultra-high strength concrete are made on 284 splitting test specimens and 265 flexural test specimens [40]. Test results showed, and a regression analysis carried out on all those samples suggested, new empirical equation for tensile strength of ultra-high strength concrete as function of compressive strength $\sigma_t = 0.8 \cdot \sqrt{\sigma_c}$.

Two additional series of concrete cube samples are designed and produced, one with a mixture for nominal compressive strength of 30 MPa (designated MB 30) and a second series of concrete cubes with nominal compressive strength of 50 MPa (designated MB 50). Tensile test data and shear stress data are presented in Table 5. As shown in Table 5 and Figure 6, concrete blocks designed with significant increase in compressive strengths from 30 MPa to 50 MPa (67% increase) showed no significant change in tensile strengths at 3 MPa, with cohesion values following trend of tensile strength and staying flat at 4 MPa. This means that process parameters, which showed significant effect on compressive strength, did not have an effect on tensile strength and cohesion. The presented results showed a strong argument towards the importance of measuring cohesion directly, to avoid making empirical functional dependence between $\sigma_c$ and $\sigma_t$ and then making a prediction about cohesion.

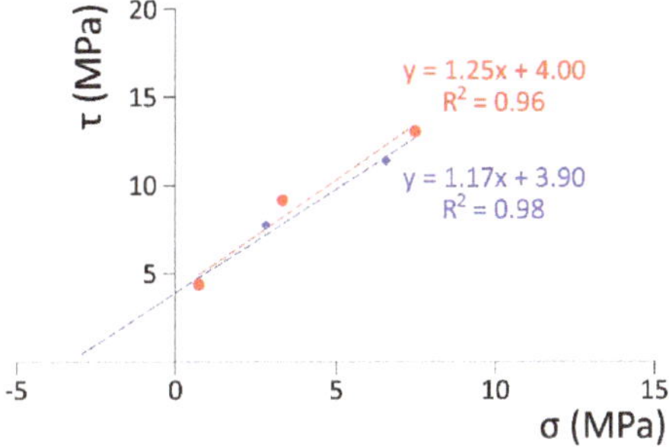


**Figure 6.** Experimental data points from Table 5 presented in σ–τ space with linear fit to determine cohesion from the Mohr–Coulomb model.

The Mohr circle approach is used to look at functional dependence between tensile strength and cohesion for concrete. Since cohesion is inherent in materials property, it is a common point regardless of whether it is determined from the right side as intercept of the linear fit at $\sigma_c = 0$, or from the left side as the value on the Mohr circle at $\sigma_t = 0$. In this report, proposed is the stress diagram for the Brazilian test for concrete samples, as if it were composed of pure shear with radius R1 (R1 = $\tau$ = $\sigma_t$) and additional compressive stress with radius R2 (R2 = $\sigma_c$ = $\sigma_t/3$), as presented in Figure 7. Therefore, overall radius for the Mohr circle for the Brazilian test for concrete samples is R = R1 + R2 = $\sigma_t$ + $\sigma_t/3$.

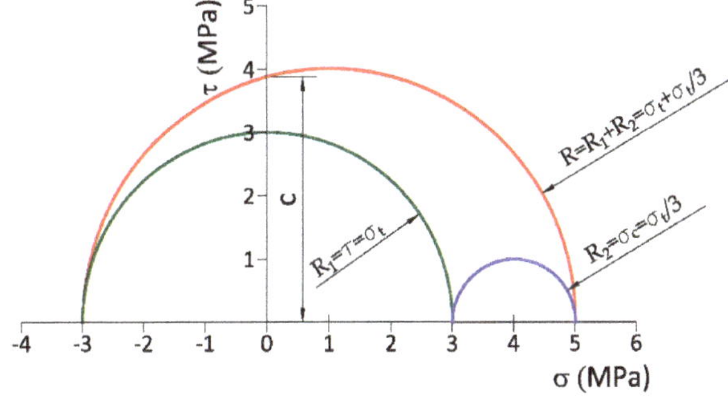


**Figure 7.** Mohr circle for the Brazilian test of concrete: $\sigma_t = 3$ MPa and c = 3.9 MPa.

The calculated Mohr circle showed remarkably good matching with experimental data for concrete series MB-30 and MB-50 ($\sigma_t$ = 3MPa and c = 3.9 MPa). Furthermore, in another example for the very high strength concrete with 100 MPa compressive strength (and in accordance with linear fit presented in Figure 8b), for the tensile strength of $\sigma_t$ = 6 MPa, the proposed Mohr circle model makes cohesion c = 7.7 MPa. This value is again a very good match with experimental measurements of cohesion at 7.6 MPa, as shown in Figure 8b. Therefore, preliminary observations showed that functional dependence between tensile strength and cohesion for concrete samples as proposed with Mohr circle with radius R = R1 + R2 = $\sigma_t$ + $\sigma_t$/3 showed remarkably good match with experimental data for concrete samples, with compressive strengths of 30 MPa, 50 MPa and 100 MPa. Again, it is important to point to the fact that, in this paper, only preliminary observations have been presented, while a more systematic study with increased sample size is needed in order to confirm the proposed correlation between cohesion and tensile strength in concrete.

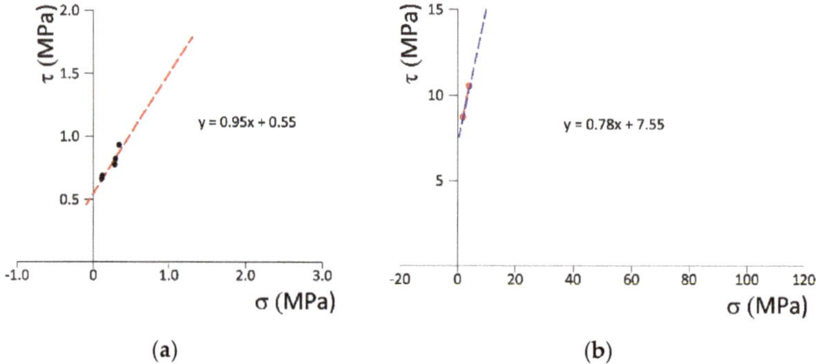


**Figure 8.** Experimental data points for (**a**) Ytong blocks with compression strength of 3 MPa showing cohesion of 0.55 MPa and (**b**) very high strength concrete blocks with compression strength of 100 MPa, showing cohesion of 7.55 MPa.

Figure 8 shows that new method is applicable for testing the wide range of materials, such as light concrete blocks, "Ytong" with compressive strength of only 3 MPa, to the very high compressive strength concrete of 100 MPa. Cohesion value measured for Ytong blocks is 0.55 MPa, while cohesion for very high strength concrete blocks is 7.55 MPa.

## 4. Numerical Modelling

For homogeneous materials like rocks, angle of internal friction is its inherent materials property. However, angle of internal friction for inhomogeneous materials like plain concrete is not its inherent property, but varies with process parameters. Even within one series, it may vary from batch to batch. As shearing force is greater than cohesive force, microcracks are formed, then angle of internal friction is first related to the resistance to crack propagation between grains and second to shear frictional (SF) resistance of cracked concrete to sliding [41,42]. From the process point of view, friction angle is a complex function of many variables associated with grain particle size, its material structure, chemical composition, anisotropy, porosity, etc.

For the numerical modelling, angle of internal friction is obtained from the bearing capacity model. From the theory of limit analysis (Citovich [43] and Nielsen and Hoang [44]), bearing capacity may be expressed as:

$$q = \frac{\pi \cdot c}{1 + (\phi - \pi/2) \cdot \tan\phi} \tag{5}$$

Starting from the known experimental values for bearing capacity q and cohesion c, it is possible to calculate the angle of internal friction. From concrete mixtures used in this paper, the friction angle obtained is of 36°. As reported in Reference [39], measured values for friction coefficient varied between 0.6 and 0.7 (angles are 31° and 35°). The report is based on Eurocode 2 [45], for concrete series with interface roughness from smooth to rough. Therefore, the value for angle of friction of 36° is reasonable.

*Analysis of Concrete Splitting Tests*

Non-linear analysis of the problem was performed using the incremental finite element method (FEM). The basic equations of the incrementally iterative procedure with the applied algorithm for the integration of constituent equations, used in this paper, are given in Reference [1]. Integration of elastoplastic constitutive relations is done with return mapping algorithm, as described by Simo and Ortiz [46].

The constitutive equations are implemented in the discrete finite element (FE) code, especially developed for the PhD thesis [47]. This code was used for the calculation of the squared or cylindrical sample bearing capacity, exposed to the Indirect Tensile Test. Simulation is done by displacement-controlled method. As presented in the Figure 9a, thanks to double symmetry of the sample, it was possible to perform the numerical analysis for the one quarter, with constant strain triangular elements under plane strain conditions. The dimension of the loading plate was b/2 = 6 mm, and the dimensions of the model are D/2 = 50 mm. The number of triangular finite elements is 5000 (50 × 50 × 2) for modelling one quarter of the concrete block. Boundary conditions are shown on Figure 9b (at the left side and at the bottom).

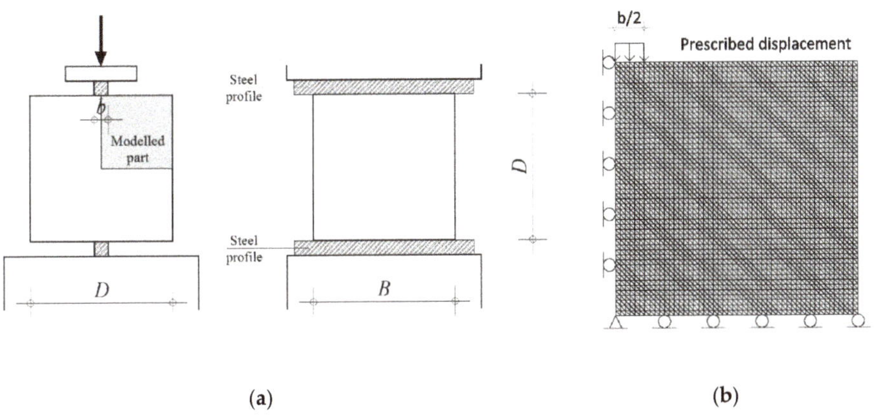


(a)                                                                                          (b)

**Figure 9.** (a) Schematic drawing showing square test arrangement and (b) model for Finite Element numerical analysis, where the width and height D/2 = 50 mm, with b/2 = 6 mm.

Compressive load from the bearing plate was applied to the concrete sample. Compression induced reactive tensile force is perpendicular to the direction of the compressive force. Curves for reactive load versus displacement for concrete series are shown in Figure 10 for selected concrete series. As presented, all curves showed a similar trend: phase 1—initial elastic behaviour, phase 2—plastic deformation followed by sharp bend, phase 3—change in slope until the final failure. Parameters used for simulation are $v = 1/9$ and $\varphi = 36°$.

The comparison between bearing capacities, i.e., measured maximum indirect tensile load and load calculated using numerical data from MC model, is shown in the Table 7. Bearing capacities showed a remarkably good correlation with a small difference from 2 to 4%.

Figure 11 shows yielding contours development for concrete series MB-30. The top left figure shows profile under displacement of 0.015 mm. Initially, the plastic zone developed right next to the edge of the bearing plate, as shown in the first profile. For the bearing plate displacement of 0.017 mm, plastic domains and the failure zone increased and became arc like shape, to reach the reflected boundary on the left. At the displacement of 0.036 mm peak load is reached, plastic zone went deeper down. With further displacement, the plastic zone spreads in the elastic region under the bearing plate, load decreased and the end of testing occurred. Observation is similar to schematic Figure 11 c, where shear fracture occurs under the bearing plate. Load bearing capacity and fracture of the sample are reached at the same point.

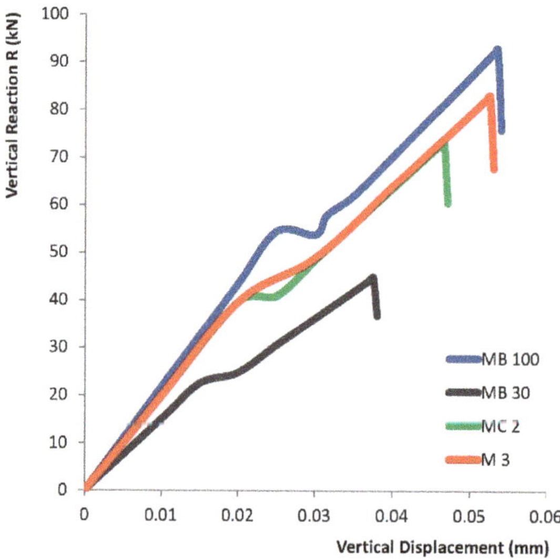


**Figure 10.** Load-displacement curves for concrete series.

**Table 7.** Summary of experimental and numerical results for ultimate load for concrete series.

| Sample Series | Experimental | | | Numerical | | |
|---|---|---|---|---|---|---|
| | $P_{max}$ (kN) | $\sigma_t$ (MPa) | E (GPa) | c (MPa) | $P^*_{max}$ (kN) | $\Delta$ (%) |
| MB 30 | 44.0 | 2.8 | 31 | 3.9 | 44.85 | 1.9 |
| MC 2 | 70.7 | 4.5 | 41 | 6.0 | 73.50 | 4.0 |
| M 3 | 80.1 | 5.1 | 43 | 6.8 | 83.03 | 3.7 |
| MB 100 | 89.5 | 5.7 | 45 | 7.6 | 92.85 | 3.8 |

Pmax—Measured value; P*max—Calculated value.

For the performed FEM analysis of concrete split cylinder test, Chen [2] used Drucker–Prager plasticity model. It was observed that initial crack started to form at the edge of the loading plate. Following the development of the initial crack, the bearing plate was pushed down into the concrete sample. Consequently, changes in distribution and ratio between horizontal and vertical stresses in their profiles appeared. When the crack reached the reflected boundary, the fracture point was reached. Diagram of the deformations presented in Figure 10 shows similarities with the modes of failure calculated using linear dynamic analysis presented in Reference [48]. According to the linear theory equations, this express the relationship between vertical and horizontal stresses and vertical diameter of the cylinder—the value of the compressive stress is high in the vicinity of the loading point, while the value of the tensile stress is high along the vertical centreline of the cylinder. As a result of the linear dynamic analysis, crack initiation started to develop at the cylinder centre. However, results of the non-linear dynamic analysis, indicated first crack propagation at the point of approximately $0.2 \cdot D$ (D = diameter of the cylinder). A further development of the crack appeared along the vertical centreline, in both directions, toward the top and bottom surfaces.

The main advantage of the presented FEM model is that it uses only four material parameters in nonlinear material analysis: Young's modulus E, Poisson's ratio $v$, cohesion c and angle of internal friction, $\varphi$. Load deformation curves obtained with FEM model showed remarkable matching with test results. It confirms that the proposed method is applicable in numerical analysis.

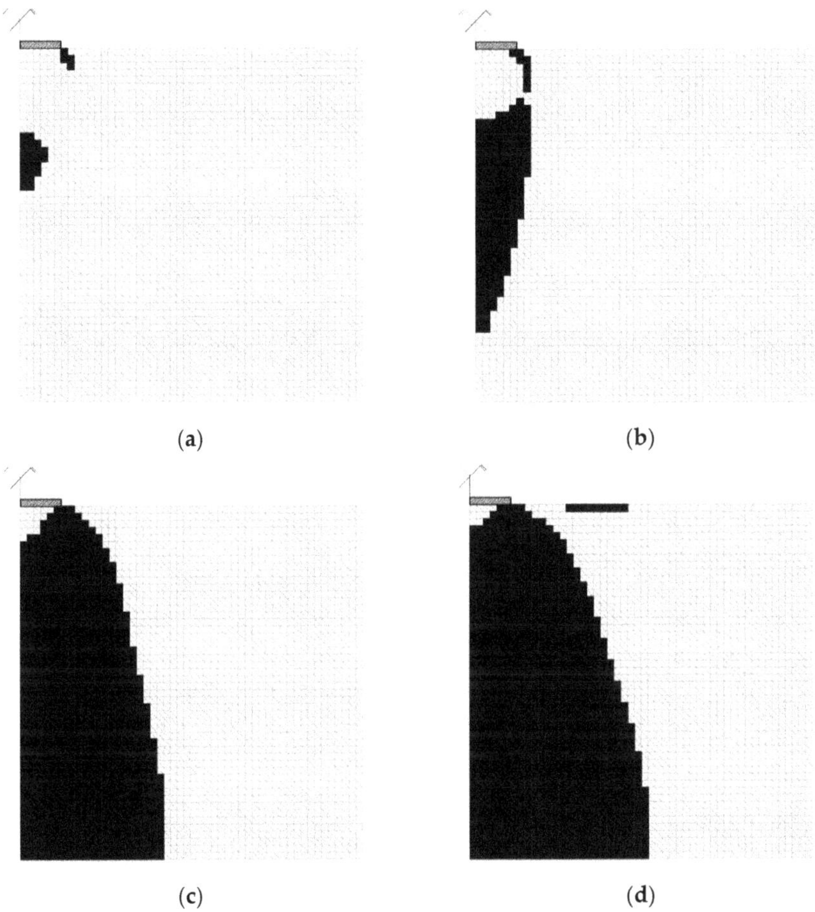

(a)      (b)

(c)      (d)

**Figure 11.** Spread of yielded zone series MB30. (**a**) Vertical displacement v = 0.015 mm. (**b**) Vertical displacement v = 0.017 mm. (**c**) Vertical displacement v = 0.036 mm. (**d**) Vertical displacement v = 0.040 mm.

## 5. Conclusions

This paper presents a new method of experimental approach to the cohesion determination from the shear stress axis with tests at pre-set values of the deformation angle. Specially designed moulds are used for testing deformation of material at pre-set low values of the shear angle. Measurements were performed on a series of samples and statistical analysis were applied on experimental results in attempt to average material's deformational behaviour. Experimental cohesion is the common value between indirect tensile test and forced shear test. Starting with the measured values for bearing capacity and cohesion, friction angle is determined. An application of the Mohr–Coulomb model for numerical analysis was performed. Data obtained from the calculations were compared to representative experimental measurements. There was a good fit between experimental data and numerical modelling. These results confirm that the proposed procedure, using only four material parameters, is applicable in numerical analysis and modelling of concrete.

**Author Contributions:** S.L. designed the model, the computational framework and performed the numerical analysis. D.V. designed the concrete mixtures and molds for the experiment. S.L. took the lead in writing the manuscript. Both authors have read and agreed to the published version of the manuscript.

**Funding:** This research received no external funding.

**Acknowledgments:** The authors wish to thank Aleksandar Radević from the University of Belgrade for his help with shear test measurements.

**Conflicts of Interest:** The authors declare no conflict of interest.

## References

1. Potts, D.M.; Zdravkovic, L. *Finite Elements Analysis in Geotechnical Engineering—Theory*; Thomas Telford Publishing: London, UK, 1999.
2. Chen, W.F. *Plasticity in Reinforced Concrete*; McGraw Hill: New York, NY, USA, 1982.
3. Chen, A.C.T.; Chen, W.F. Constitutive Relations for Concrete. *ASCE J. Eng. Mech.* **1975**, *101*, 465–481.
4. Lubliner, J.; Oliver, J.; Oller, S.; Oñate, E. A plastic-damage model for concrete. *Int. J. Solids Struct.* **1989**, *25*, 299–326. [CrossRef]
5. Krajcinovic, D. Continuous Damage Mechanics Revisited: Basic Concepts and Definitions. *J. Appl. Mech.* **1985**, *52*, 829–834. [CrossRef]
6. Lubliner, J. *Plasticity Theory*; Mac Millan: New York, NY, USA, 1990.
7. Vukelic, S. Analysis of Mechanical Deformation of Concrete. Ph.D. Thesis, University of Belgrade, Faculty of Civil Engineering, Belgrade, Serbia, 1981.
8. Hu, H.-T.; Schnobrich, W.C. Constitutive Modeling of Concrete by Using Nonassociated Plasticity. *J. Mater. Civ. Eng.* **1989**, *1*, 199–216. [CrossRef]
9. Mehta, P.K.; Monteiro, P.J. *Concrete: Structure, Properties and Materials*; Prentice Hall: Upper Saddle River, NJ, USA, 1993.
10. Chae, S.R.; Moon, J.; Yoon, S.; Bae, S.; Levitz, P.; Winarski, R.; Monteiro, P.J.M. Advanced Nanoscale Characterization of Cement Based Materials Using X-Ray Synchrotron Radiation: A Review. *Int. J. Concr. Struct. Mater.* **2013**, *7*, 95–110. [CrossRef]
11. Daigle, M.; Fratta, D.; Wang, L.B. Ultrasonic and X-ray Tomographic Imaging of Highly Contrasting Inclusions in Concrete Specimens. In Proceedings of the GeoFrontier Conference, Austin, TX, USA, 24–26 January 2005; pp. 1–12.
12. Caliskan, S. Examining Concrete Cores by Non-destructive Techniques. In Proceedings of the 4th Middle East NDT Conference and Exhibition, Manama, Bahrain, 2–5 December 2007.
13. Bazant, Z.P.; Pang, S.D.; Verchovsky, M.; Novak, D.; Pukl, R. Statistical size effect in quasibrittle materials computation and extreme value theory. In *Fracture Mechanics of Concrete Structures*; Li, V.C., William, K.J., Billington, S.L., Eds.; IA-FraMCoS: Vail, CO, USA, 2004; pp. 189–196.
14. Lowes, L. Finite Element Modeling of Reinforced Concrete Beam Column Bridge Connections. Ph.D. Thesis, UC Berkeley, Berkeley, CA, USA, 1999.
15. Karsan, I.D.; Jirsa, J.O.; Chen, W.F. Behavior of concrete under compressive loadings. *J. Struct. Div. ASCE* **1969**, *95*, 2535–2563.
16. Chen, W.F. *Limit Analysis and Soil Plasticity*; J. Ross Publishing: Fort Lauderdale, FL, USA, 2007; p. 543.
17. Gopalaratnam, V.S.; Shah, S.P. Softening response of plain concrete in direct tension. *ACI J.* **1985**, *82*, 311–323.
18. Taquieddin, Z.N. Elasto-plastic and Damage Modeling of Reinforced Concrete. Ph.D. Thesis, Louisiana State University, Baton Rouge, LA, USA, 2008.
19. Hobbs, D.W. The tensile strength of rocks. *Int. J. Rock Mech. Min. Sci.* **1964**, *1*, 385–396. [CrossRef]
20. Fan, Q.; Gu, S.C.; Wang, B.N.; Huang, R.B. Two Parameter Parabolic Mohr Strength Criterion Applied to Analyze The Results of the Brazilian Test. *Appl. Mech. Mater.* **2014**, *624*, 630–634. [CrossRef]
21. Ma, T.; Peng, N.; Zhu, Z.; Zhang, Q.; Yang, C.; Zhao, J. Brazilian Tensile Strength of Anisotropic Rocks: Review and New Insights. *Energies* **2018**, *11*, 304. [CrossRef]

22. Sivakugan, N.; Das, B.M.; Lovisa, J.; Patra, C.R. Determination of c and w of rocks from indirect tensile strength and uniaxial compression tests. *Int. J. Geotech. Eng.* **2014**, *8*, 59–65. [CrossRef]

23. Piratheepan, J.; Gnanendran, C.T.; Arulrajah, A. Determination of c and w from an IDT and unconfined compression testing and numerical analysis. *J. Mater. Civ. Eng.* **2012**, *24*, 1153–1163. [CrossRef]

24. Meschke, G. Consideration of aging of shotcrete in the context of a 3D viscoplastic material model. *Int. J. Numer. Methods Eng.* **1996**, *39*, 3123–3143. [CrossRef]

25. Altoubat, S.A.; Lange, D.A. Early Age Shrinkage and Creep of Fiber Reinforced Concrete for Airfield Pavement. In Proceedings of the 1997 Airfield Pavement Conference, Seattle, WA, USA, 17–20 August 1997.

26. Galic, M.; Marovic, P.; Nikolic, Ž. Modified Mohr-Coulomb—Rankine material model for concrete. *Eng. Comput.* **2011**, *28*, 853–887. [CrossRef]

27. Wu, D.; Wang, Y.; Qiu, Y.; Zhang, J.; Wan, Y.-K. Determination of Mohr–Coulomb Parameters from Nonlinear Strength Criteria for 3D Slopes. *Math. Probl. Eng.* **2019**, 6927654. [CrossRef]

28. Japaridze, L. Stress-deformed state of cylindrical specimens during indirect tensile strength testing. *J. Rock Mech. Geotech. Eng.* **2015**, *7*, 509–518. [CrossRef]

29. ASTM. *Standard Test Method for Splitting Tensile Strength of Intact Rock Core Specimens*; D 3967-08; ASTM International: West Conshohocken, PA, USA, 2008.

30. Japaridze, L. Shear Stresses in the Indirect Test of Tensile Strength of Rocks and other Hard Materials. *Bull. Georgian Natl. Acad. Sci.* **2016**, *10*, 45–54.

31. Li, D.; Wong, L.N.Y. The Brazilian Disc Test for Rock Mechanics Applications: Review and New Insights. *Rock Mech. Rock Eng.* **2012**, *46*, 269–287. [CrossRef]

32. Kömürlü, E.; Demir, A.D. Determination of Cohesion values of Rock Materials using Double Shear Jaws. *Period. Polytech. Civ. Eng.* **2018**, *62*, 881–892. [CrossRef]

33. Majstorovic, J.; Andelkovic, V.; Dimitrijevic, B.; Savkovic, S. Laboratory testing of shear parameters through rock mass and along discontinuities. *Undergr. Min. Eng.* **2012**, *21*, 135–142.

34. Institute for Standardization of Yugoslavia, JUS B.B7.130/1988, Rock Mechanics. Testing of physical and mechanical properties. In *Method for Determining Limiting Shear Strength*; Collection of Yugoslavian Standards for Building Constructions, Book 6-1 Geotechnic and Foundations; Faculty of Civil Engineering: Belgrade, Serbia, 1995; pp. 173–175. ISBN 86-80049-20-4.

35. Lelovic, S.; Vasovic, D.; Stojic, D. Determination of the Mohr-Coulomb Material Parameters for Concrete under Indirect Tensile Test. *Tech. Gaz.* **2019**, *26*, 412–419.

36. O'Neil, E.; Neeley, B.; Cargile, J.D. Tensile Properties of Very-High-Strength Concrete for Penetration-Resistant Structures. *Shock Vib.* **1999**, *6*, 237–245. [CrossRef]

37. Carrasquillo, R.L.; Nilson, A.H.; Slate, F.O. Properties of high strength concrete subjected to short-term load. *ACI J.* **1981**, *78*, 171–178.

38. Oluokun, F. Prediction of Concrete Tensile Strength from its Compressive Strength: An Evaluation of Existing Relations for Normal Weight Concrete. *ACI Mater. J.* **1991**, *88*, 302–309.

39. Building code requirements for structural plain concrete (ACI 318.1-83) and commentary. *Int. J. Cem. Compos. Lightweight Concr.* **1985**, *7*, 60. [CrossRef]

40. Bae, B.-I.; Choi, H.-K.; Choi, C.-S. Correlation Between Tensile Strength and Compressive Strength of Ultra High Strength Concrete Reinforced with Steel Fiber. *J. Korea Concr. Inst.* **2015**, *27*, 253–263. [CrossRef]

41. Mohamad, M.; Ibrahim, I.; Abdullah, R.; Rahman, A.A.; Kueh, A.; Usman, J. Friction and cohesion coefficients of composite concrete-to-concrete bond. *Cem. Concr. Compos.* **2015**, *56*, 1–14. [CrossRef]

42. Chen, Y.; Visintin, P.; Oehlers, D.J. Concrete Shear-Friction Material Properties: Derivation from Actively Confined Compression Cylinder Tests. *Adv. Struct. Eng.* **2015**, *18*, 1173–1185. [CrossRef]

43. Citovich, N.A. *Soil Mechanics*, 4th ed.; Student Press: Moscow, Russia, 1979; p. 341.

44. Nielsen, M.P.; Hoang, L.C. *Limit Analysis and Concrete Plasticity*, 3rd ed.; CRC Press: Boca Raton, FL, USA, 2010; p. 176.

45. *EN 1992-1-1 (2004) (English): Eurocode 2: Design of concrete structures—Part 1-1: General Rules and Rules for Buildings*; European Committee for Standardization, December 2004; Available online: https://www.phd.eng.br/wp-content/uploads/2015/12/en.1992.1.1.2004.pdf (accessed on 9 September 2020).

46. Ortiz, M.; Simo, J.D. An analysis of a new class of integration algorithms for elastoplastic constitutive relations. *Int. J. Numer. Methods Eng.* **1986**, *23*, 353–366. [CrossRef]
47. Lelovic, S. Constitutive Equations for Sand and Their Application in Numerical Analysis of Strip Foundation Behavior. Ph.D. Thesis, University of Belgrade, Beograd, Serbia, 2013.
48. Tedesco, J.W. *Numerical Analysis of Dynamic Splitting Tensile and Direct Tension Tests*; Air Force Report (ESL-TR-89-45) Air Force Engineering, Tyndall AFB FL 32403; National Technical Information Service: Springfield, VA, USA, 1990.

*Article*

# Prediction of Properties of FRP-Confined Concrete Cylinders Based on Artificial Neural Networks

**Afaq Ahmad [1], Vagelis Plevris [2,*] and Qaiser-uz-Zaman Khan [1]**

[1]  Department of Civil Engineering, University of Engineering and Technology, Taxila 47080, Pakistan; afaq.ahmad@uettaxila.edu.pk (A.A.); dr.qaiser@uettaxila.edu.pk (Q.-u.-Z.K.)

[2]  Department of Civil Engineering and Energy Technology, OsloMet—Oslo Metropolitan University, 0166 Oslo, Norway

*  Correspondence: vageli@oslomet.no

Received: 20 July 2020; Accepted: 7 September 2020; Published: 14 September 2020

**Abstract:** Recently, the use of fiber-reinforced polymers (FRP)-confinement has increased due to its various favorable effects on concrete structures, such as an increase in strength and ductility. Therefore, researchers have been attracted to exploring the behavior and efficiency of FRP-confinement for concrete structural elements further. The current study investigates improved strength and strain models for FRP confined concrete cylindrical elements. Two new physical methods are proposed for use on a large preliminary evaluated database of 708 specimens for strength and 572 specimens for strain from previous experiments. The first approach is employing artificial neural networks (ANNs), and the second is using the general regression analysis technique for both axial strength and strain of FRP-confined concrete. The accuracy of the newly proposed strain models is quite satisfactory in comparison with previous experimental results. Moreover, the predictions of the proposed ANN models are better than the predictions of previously proposed models based on various statistical indices, such as the correlation coefficient (R) and mean square error (MSE), and can be used to assess the members at the ultimate limit state.

**Keywords:** artificial neural networks; confined concrete; strength model; FRP; strain model; RMSE

---

## 1. Introduction

The lateral confinement due to fiber-reinforced polymers (FRPs) increases the efficiency of concrete compression members by enhancing their axial load carrying capacity and ductility. The increase in axial strength and strain of reinforced concrete (RC) columns is the main reason for using FRP confinement techniques. Over the last decade, the use of FRP composites for the rehabilitation, retrofitting, strengthening, and ductility enhancement of RC compression members has been one of the most popular techniques due to its advantages [1]. The advantages of the FRP composites include not only the corrosion resistance and low weight, but also provide a high strength to weight ratio. The critical parameters affecting FRP are the thickness of FRP ($nt$), angle of orientation of FRP wraps, the elastic modulus of FRP ($E_f$), total number of FRP sheets, and the unconfined strength of concrete material ($f'_{co}$) [2]. FRPs offer many advantages, such as easy handling, low disturbance of concrete members, easy installation, and little time for application [3]. Due to these advantages, FRPs are preferred over the use of conventional steel jacketing techniques [4].

The technique of strengthening RC structures using FRPs is very vital for increasing the capacity of RC structural members at the ultimate limit state (ULS) [5,6]. Poor and low strength concrete or concrete structures damaged due to lateral seismic loading usually lose strength, and they need retrofitting and rehabilitation to regain their strength at ULS [7]. The use of fiber and plastic makes the FRP high corrosion resistance and suitable for use in undersea structures [8]. The applications of

FPRs are not limited to RC building members but are also extended to RC bridge piers, girders, and slabs to enhance their strength at ULS [9–16].

Many researchers have done work for estimating the axial strength and strain of externally confined concrete members [17–24] based on empirical techniques [25–41], i.e., Karbhari et al., Miyauchi et al. [42,43], Lam and Teng [44], and Mander et al. [45], as described in Table 1. These proposed empirical equations predict the strength and strain capacity of confined concrete members with low accuracy [46]. The use of artificial neural networks (ANNs) for modeling various complex problems in structural engineering is increasing [47–52]. One can determine and capture the interactions between various variables of a complex system by using ANNs, despite the unknown and usually unpredictable nature of these interactions. ANN models can predict the compressive strength of confined plastic concrete accurately [53–56]. Khademi et al. [57] and Reddy [58] used ANN techniques for the prediction of the axial strength of confined concrete. Moreover, various researchers proposed analytical models for the prediction of the ultimate conditions of confined concrete members using different techniques such as ANN, regression analysis, soft computing, and genetic programming [25–35]. An attempt has been made [59,60] to determine the confinement effects of the FRP on RC members at the ULS.

The present work aims to propose a prediction model for the axial strength and axial strain models for FRP-confined concrete cylinders. Similar approaches have already been proposed in the past, mainly for strength models [18], using only a limited database. In the present work, previous experimental results of concrete cylinders externally confined with FRPs, in particular 708 specimens for strength and 512 for strain, were collected and evaluated based on some statistical indices, such as the correlation coefficient (R) and mean square error (MSE). This was done using the previously proposed models for the ultimate conditions of FRP-confined concrete members. In addition, ANN and empirical models are proposed for the prediction of strength and strain of concrete members with FRPs. The evaluation of previous and currently proposed models for the ultimate conditions of FRP-confined concrete members show that the proposed models provide superior performance. The significance of the present study is that the proposed models can also accurately capture the axial strain behavior of confined concrete, which is useful for the analysis and design of confined concrete members.

## 2. Mechanics of FRP Confinement

It was observed that the models for the ultimate conditions of FRP-confinement contain some common variables, such as confinement stiffness ratio ($\rho_k$), strain ratio ($\rho_\varepsilon$), hoop rupture strain of fibers ($\varepsilon_{h,rup}$), and maximum confinement stress ($f_l$). The relations for $\rho_k$ and $\rho_\varepsilon$ have been presented by Teng et al. [61], as described in Equation (1) and Equation (2).

$$\rho_\varepsilon = \frac{\varepsilon_{h,rup}}{\varepsilon_{co}} \tag{1}$$

$$\rho_k = \frac{2E_f \cdot t}{\left(\frac{f'_{co}}{\varepsilon_{co}}\right) \cdot D} \tag{2}$$

where $\varepsilon_{co}$ is the axial strain of unconfined concrete, $t$ is the thickness of FRP wraps, $f'_{co}$ is the unconfined strength of concrete, and $E_f$ is the tensile modulus of confinement sheets in the transverse direction of the specimen. Figure 1 represents the confinement stresses due to FRP-wraps with a hoop diameter $D$.

The relationship between confinement stress and some critical variables is illustrated in Figure 1 and Equation (3). The hoop rupture strain can be represented by Equation (4) [62].

$$f_l = \rho_\varepsilon \rho_\kappa f'_{co} = \frac{2E_f \varepsilon_{h,rup} \cdot t}{D} \tag{3}$$

$$\varepsilon_{h,rup} = \frac{\varepsilon_f}{f'^{0.125}_{co}} \tag{4}$$

where $\varepsilon_f$ is the maximum tensile strain of fibers, and $f'_{co}$ is expressed as in MPa units. The initial strength and strain model for the FRP confinement was given by Fardis and Khalili [63] based on the concept that the maximum strain of FRP confined concrete is increased with the enhancement of the circumferential stiffness. Mander et al. [45] proposed a model for the predictions of the axial compressive strain of FRP that was initially implemented by ACI 440-2R-02 [64]. Samaan et al. [59] proved this concept to be wrong by developing a strain model with the concept that the actual hoop rupture strain is less than the maximum tensile strength of FRPs at the failure stage. Karbhari and Gao [42] proposed a model for strain by assuming that the FRP wraps reach their ultimate tensile strength during the application of load, which is not possible for the actual conditions. By considering the concept of the actual rupture strain of FRPs, Toutanji [65] suggested a new strain model. Lam and Teng [44] proposed a new strain model by considering the actual failure strain of FRPs. Later on, Teng et al. [66] refined this model by performing tests on 18 FRP confined concrete members.

*Constructed Database for Modeling*

A large database was prepared in a previous experimental work, which contains 708 specimens for strength. Most of the works also contain data for strain, but not all of them. Therefore, the specimens for strain are fewer (572), and these strain specimens are essentially a subset of the full set of specimens used for stress. All essential parameters associated with the strength and strain enhancement of concrete cylinders confined with FRPs are included in the databases. Tables 2 and 3 provide the statistical information about the variation of certain parameters associated with the specimens considered for the strength and the strain model data, respectively. The field Range means the difference between the maximum and the minimum value for each quantity (*Range = Max–Min*). *StDev* denotes the standard variation, while *CV* is the coefficient of variation, i.e., the standard deviation divided by the mean value.

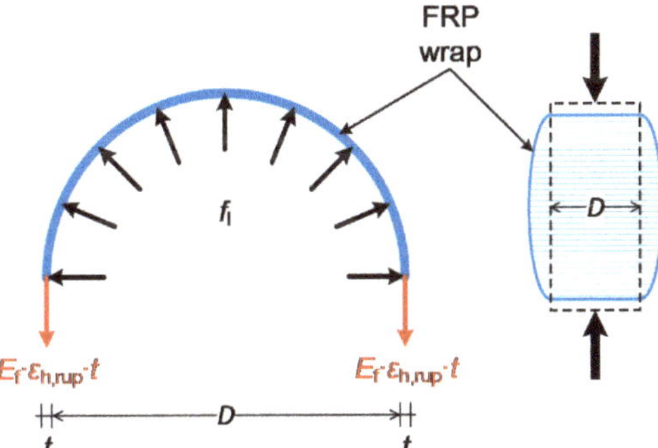


**Figure 1.** Mechanism of confinement in fiber-reinforced polymers (FRP)-confined concrete.

**Table 1.** Previously proposed strength and strain models of fiber-reinforced polymers (FRP) confined concrete specimens.

| Model | Relationship for Strength | Relationship for Strain |
|---|---|---|
| Karbhari and Gao [42] | $\dfrac{f'_{cc}}{f'_{co}} = 1 + 2.1\left(\dfrac{f_l}{f'_{co}}\right)^{0.87}$ | $\varepsilon_{cc} = \varepsilon_{co} + 0.01\dfrac{f_l}{f'_{co}}$ |
| Saafi et al. [67] | $\dfrac{f'_{cc}}{f'_{co}} = 1 + 2.2\left(\dfrac{f_l}{f'_{co}}\right)^{0.84}$ | - |
| Matthys et al. [60] | $\dfrac{f'_{cc}}{f'_{co}} = 1 + 2.3\left(\dfrac{f_l}{f'_{co}}\right)^{0.85}$ | - |
| Fardis and Khalili [63] | $\dfrac{f'_{cc}}{f'_{co}} = 1 + 3.7\left(\dfrac{f_l}{f'_{co}}\right)^{0.86}$ | $\varepsilon_{cc} = 0.002 + 0.001\dfrac{E_f t}{D f'_{co}}$ |
| Richart et al. [68] | $\dfrac{f'_{cc}}{f'_{co}} = 1 + 4.1\dfrac{f_l}{f'_{co}}$ | - |
| Newman and Newman [69] | $\dfrac{f'_{cc}}{f'_{co}} = 1 + 3.7\left(\dfrac{f_l}{f'_{co}}\right)^{2}$ | - |
| Mander et al. [45] | $\dfrac{f'_{cc}}{f'_{co}} = 2.254\sqrt{1 + 7.94\dfrac{f_l}{f'_{co}}} - 2\dfrac{f_l}{f'_{co}} - 1.254$ | $\dfrac{\varepsilon_{cc}}{\varepsilon_{co}} = 1 + 5\left(\dfrac{f'_{cc}}{f'_{co}} - 1\right)$ |
| Lam and Teng [44] | $\dfrac{f'_{cc}}{f'_{co}} = 1 + 3.3\dfrac{f_l}{f'_{co}}$ | $\dfrac{\varepsilon_{cc}}{\varepsilon_{co}} = 1.75 + 12\rho_k\rho_\varepsilon^{1.45}$ |
| Toutanji [65] | $\dfrac{f'_{cc}}{f'_{co}} = 1 + 3.3\dfrac{f_l}{f'_{co}}$ | $\dfrac{\varepsilon_{cc}}{\varepsilon_{co}} = 1 + \left(310.57\varepsilon_{h,rup} + 1.90\right)\left(\dfrac{f'_{cc}}{f'_{co}} - 1\right)$ |
| Samaan et al. [59] | $f'_{cc} = f'_{co} + 6.0 f_l^{0.70}$ | $\varepsilon_{cc} = \dfrac{f'_{cc} - f_o}{E_2}$ where $f_o = 0.872 f'_{co} + 0.371 f_l + 6.258$ and $E_2 = 245.61 f_{co}'^{0.2} + 1.3456\dfrac{E_f t}{D}$ |
| Teng et al. [61] | $\dfrac{f'_{cc}}{f'_{co}} = 1 + 3.5(\rho_K - 0.01)\rho_\varepsilon$ | $\dfrac{\varepsilon_{cc}}{\varepsilon_{co}} = 1.75 + 6.5\rho_K^{0.8}\rho_\varepsilon^{1.45}$ |
| Miyauchi et al. [70] | $\dfrac{f'_{cc}}{f'_{co}} = 1 + 3.485\dfrac{f_l}{f'_{co}}$ | - |

**Table 2.** Statistic properties of the parameters included in the constructed database of the strength model (708 samples).

| | Min | Max | Range | Mean | Median | StDev | CV |
|---|---|---|---|---|---|---|---|
| $D$ (mm) | 51 | 406 | 355 | 153.33 | 152.00 | 43.15 | 0.281 |
| $H$ (mm) | 102 | 812 | 710 | 306.86 | 304.00 | 86.32 | 0.281 |
| $nt$ (mm) | 0.09 | 5.9 | 5.81 | 0.88 | 0.47 | 1.04 | 1.182 |
| $E_s$ (GPa) | 10 | 663 | 653 | 174.68 | 219.00 | 118.80 | 0.680 |
| $f'_{co}$ (MPa) | 12.41 | 188.2 | 175.79 | 42.48 | 37.70 | 22.38 | 0.527 |
| $f'_{cc}$ (MPa) | 18.5 | 302.2 | 283.7 | 76.25 | 67.91 | 35.00 | 0.459 |

**Table 3.** Statistic properties of the parameters included in the constructed database of the *strain* model (572 samples).

| | Min | Max | Range | Mean | Median | StDev | CV |
|---|---|---|---|---|---|---|---|
| $D$ (mm) | 51 | 406 | 355 | 154.16 | 152.00 | 41.97 | 0.272 |
| $H$ (mm) | 102 | 812 | 710 | 307.99 | 304.00 | 84.45 | 0.274 |
| $nt$ (mm) | 0.09 | 5.9 | 5.81 | 0.88 | 0.46 | 1.05 | 1.194 |
| $E_s$ (GPa) | 10 | 663 | 653 | 163.08 | 118.34 | 121.32 | 0.744 |
| $\varepsilon_{co}$ (%) | 0.168 | 1.53 | 1.362 | 0.268 | 0.24 | 0.146 | 0.543 |
| $\varepsilon_{cc}$ (%) | 0.33 | 4.62 | 4.29 | 1.394 | 1.25 | 0.645 | 0.462 |

## 3. Artificial Neural Networks (ANN)

In neuroscience, a (biological) neural network describes a population of physically interconnected neurons or a group of disparate neurons whose inputs or signaling targets define a circuit. It was found that mammalian brains learn as connections between neurons are strengthened. ANNs mimic the biological neural networks and, like human beings, learn by example. ANNs can evaluate the complex functions of numerous variables whose interrelationships are not clear. Learning, categorization, and generalization are processes that are adopted by the ANNs to predict the outputs.

The ANNs perform these tasks because of their ability to save the calculation data in memory during the training process. They contain a number of hidden layers consisting of a complex interrelationship of hidden "neurons". There is a link between every two neurons of consecutive layers, having a specific weight. The predictions of neurons are then multiplied with these weights. During the latter process, the predictions of neurons are transferred through the links and added to the bias, as presented in Figure 2 [71].

Based on experience [72,73], it is considered that the multilayer feed-forward ANNs (MLFNNs) are the most suitable for this type of problem. In MLFNNs, there is an input layer, an output layer, and one or several hidden layers. In this study, we particularly use back-propagation neural networks (BPNNs). A back-propagation neural network is a feed-forward, multilayer network of standard structure, i.e., neurons are not interconnected within a layer, but they are connected with all the neurons of the previous layer and the subsequent layer. In this type of ANN, the output values are cross-checked with the target answer to get the error value. To reduce the error value, different techniques are used [72,73] by changing the weights of each connection during a large number of training cycles. In this case, one would say that the network has learned a certain target function. As the algorithm's name implies, the errors propagate backwards from the output nodes to the input nodes. The selected architecture of the NN defines the number of hidden layers and the number of neurons in each layer. Figure 2 depicts an example of a BPNN composed of one input layer with four neurons, two hidden layers with three neurons each, and an output layer with two neurons, i.e., a 4-3-3-2 BPNN with two hidden layers.

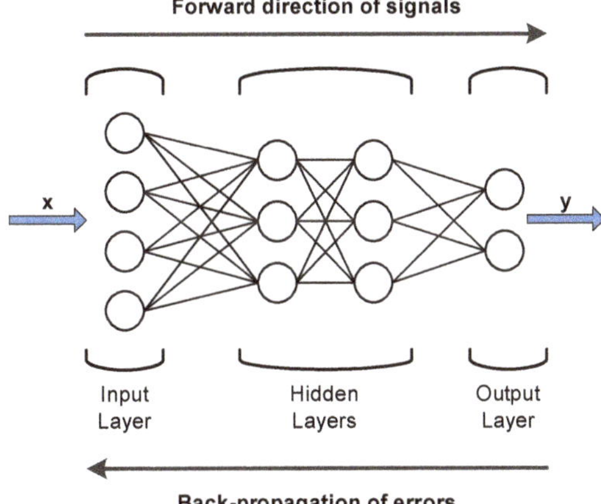


**Figure 2.** A three-layer 4-3-3-2 back-propagation neural network (BPNN) (input typically not counted as a layer).

As shown in Figure 2, a BPNN consists of several layers, and each layer contains a complex system of interconnected "neurons". There is a link between neurons of subsequent layers, having a specific weight. The values generated by the neurons are then multiplied with these weights. During the latter process, the values generated by the neurons are shifted via links and then summed with bias, as shown in Figure 3, which depicts a single node (neuron) of a hidden layer, with a single R-element input vector.

This summation is then presented into a predefined activation function, as shown in Equation (5).

$$a = f\left(\sum w_i p_i + b\right) \tag{5}$$

where $a$ is the output of the neuron, $w_i$ is the weight coefficients, $p_i$ the input values, and $b$ is the bias value. In matrix form, the above equation can be written as Equation (6),

$$a = f(\boldsymbol{wp} + b)\tag{6}$$

where the sum $\boldsymbol{wp}$, is the dot product of the (single row) matrix $\boldsymbol{w} = \{w_1, \dots, w_R\}$ and the vector $\boldsymbol{p} = [p_1, \dots, p_R]^T$.

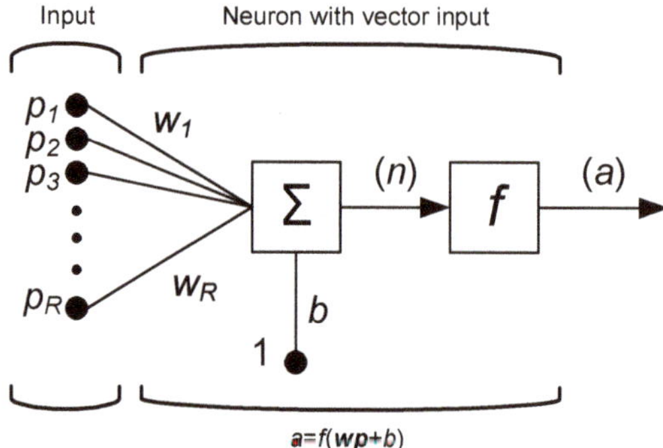


**Figure 3.** A neuron with a single correlation coefficient (R)-element input vector.

The outputs of the activation function generate the values for the input process of neurons of the next layer. The final values of the weights are obtained from the learning process based on available data after assigning random initial weights in the beginning. In the present study, ANNs are utilized to predict the axial compressive strength and the axial compressive strain of confined concrete cylindrical elements. The error resulting from this process can be calculated by Equation (7),

$$E(w) = \frac{1}{2}\sum_{i}(T_i - O_i)^2\tag{7}$$

where $T$ is the target defined in the database, and $O$ is the output value predicted by the ANN. For the strength model, we use a BPNN model, where the five input parameters are the diameter of the cylinder $D$, the height of the cylinder $H$, the total thickness $nt$, the elastic modulus of FRP $E_f$, and the compressive strength of unconfined concrete $f'_{co}$. The single output parameter is the compressive strength of confined concrete $f_{cc}$. For the strain model, we similarly used a BPNN model, where the five input parameters are $D$, $H$, $nt$, $E_f$, and the compressive strain $\varepsilon_{co}$. In this model, the single output parameter is the compressive strain $\varepsilon_{cc}$ corresponding to the compressive strength of confined concrete $f_{cc}$. The parameters for both NN models are shown in Table 4.

**Table 4.** The five inputs and the single output of the neural network (NN) strength and strain models examined.

| NN Model | Inputs (5) | Outputs (1) |
| --- | --- | --- |
| Strength model | $D, H, nt, E_f, f'_{co}$ | $f_{cc}$ |
| Strain model | $D, H, nt, E_f, \varepsilon_{co}$ | $\varepsilon_{cc}$ |

The back-propagation technique or Delta rule [74] was used to minimize the error of Equation (7). This iterated search process was carried out in the opposite direction, i.e., from right to left of the NN, as shown in Figure 2. This process aimed to adjust the values of the various weights, which were randomly selected in the beginning so that the output values realistically agreed with the required target values of the database. This process was repeated until there was no further improvement in the mean square error value, and the errors became acceptable. In the end, the trained ANN can be used to predict results even for data that have not been presented to it before, with small error.

### 3.1. ANN Architecture

A number of parameters need to be set up for the ANN, for example, the activation functions between consequent layers, the total number of neurons in each layer, the activation functions between the hidden layers and the output layer, and others. The choice of these parameters depends on the given problem and can be determined by a trial and error process. For this investigation, we took six ANN architectures into account, as shown in Table 5. The 6 + 6 Models (6 for strength and 6 for strain) had different numbers of hidden layers and nodes in every hidden layer. The logistic and hyperbolic activation functions were utilized between the input and the middle layer, based on the authors' experience [71], while the hyperbolic activation function was utilized between the output layer and its previous layer.

### 3.2. Normalization of the Databases

The ANNs work best with normalized data for inputs and outputs. This process helps avoid low learning rate issues [48]. In the present study, all the parameters associated with confined concrete cylinders were normalized using Equation (8). This process made all the parameters unitless despite the different units that were used to describe the various inputs or outputs.

$$X = \frac{\Delta X}{\Delta x} x + \left( X_{\max} - \frac{\Delta X}{\Delta x} x_{\max} \right) \tag{8}$$

**Table 5.** Architectures of the examined artificial neural network (ANN) strength (F) and strain (E) models.

| NN Model | Number of Hidden Layers | Number of Neurons in the Hidden Layer(s) |
|---|---|---|
| FANN-1 and EANN-1 | 1 | 5 |
| FANN-2 and EANN-2 | 1 | 10 |
| **FANN-3** and EANN-3 | 2 | 5-5 |
| FANN-4 and **EANN-4** | 2 | 10-10 |
| FANN-5 and EANN-5 | 3 | 5-5-5 |
| FANN-6 and EANN-6 | 3 | 10-10-10 |

As will be shown in detail in the next sections, the best performance was the one of the third models for strength and the fourth model for the strain (FANN-3 and EANN-4), corresponding to 5 and 10 neurons in each of the two hidden layers, respectively. These have been highlighted in Table 5 with bold.

In Equation (8), $x$ is the actual value, $X$ is the new normalized value, $\Delta x$ is the difference between the maximum and minimum $x$ value, $x_{\max}$ is the maximum $x$ value, and $X_{\max}$ and $\Delta X$ are the new desired maximum value of $X$ and the new desired difference between the maximum and minimum $X$ value, respectively. In our case, we used the parameters $X_{\max} = 0.9$ and $\Delta X = 0.8$ to end up with normalized values that were in the range [0.1, 0.9]. The same equation, Equation (8), in a rearranged form, was used for the denormalization of the strength and strain values, i.e., to end up to the real values of strength and strain from the normalized ones provided by the NN.

### 3.3. Performance of the Various ANN Models

It is very important to calibrate the proposed ANN models based on the experimental results to investigate the accuracy of the generated model. The multilayer free forward back-propagation (MLFFBP) process [48] and the testing results were used for the calibration of the current ANN model. The training process of ANN was adopted as described in the authors' work [75].

Overtraining (or over-fitting) can be a problem as it causes the ANN to lose its generalization capability. To avoid overtraining, the size of the network and the number of training epochs should be in correspondence with the complexity of the problem, especially if the training data contains a significant amount of noise. An excessive number of training epochs, or an excessive size of the neural network, can cause network overtraining. In our case, to enhance the performance of ANN and to avoid overtraining, the data were divided into three sets: one for training, one for validation, and one for testing purposes. Sixty percent of sample points were used for training, 20% were used validation, and 20% were used for testing purposes. The ANN models were developed using MATLAB [49].

There were 100 epochs for the training of each ANN model that included the MLFFBP process. The MLFFBP process continued until any of the following conditions was completed: (i) 20 failures occurred in the validation process during the iteration, to avoid overtraining, (ii) the performance goal (the difference between the target and the output values) became equal to a preset value of 0.0001, (iii) the minimum performance gradient became equal to $10^{-10}$. The rate at which the MLFFBP process adjusts the weights of the sample points is known as the minimum performance gradient. These conditions were used, as suggested by the Levenberg–Marquardt back-propagation technique [49]. This technique is a second-order algorithm that works with second derivatives.

Tables 6 and 7 present some statistical information about the ANN model predictions (target values) for the strength and the strain models, respectively. The tables contain the minimum values (*min*), the maximum values (*max*), the difference between the two (*Range = Max–Min*), the average values (*Mean*), the median values (*Median*), the standard deviation (*StDev*) and the coefficient of variation (*CV = StDev/Mean*). Some negative values can be found for the *Min* values of $f_{cc}/f'_{co}$ and $\varepsilon_{cc}/\varepsilon_{co}$ in the table, which make no sense and are only artifacts of the method that do not affect the overall results.

**Table 6.** Performance of the different ANN models for strength (values of $f_{cc}/f'_{co}$) on the training data set.

|  | Min | Max | Range | Mean | Median | StDev | CV |
|---|---|---|---|---|---|---|---|
| Experimental (target values) | 1.02 | 3.90 | 2.89 | 1.92 | 1.76 | 0.66 | 0.345 |
| FANN-1 | 0.70 | 4.26 | 3.56 | 1.93 | 1.79 | 0.65 | 0.335 |
| FANN-2 | 0.88 | 3.85 | 2.97 | 1.92 | 1.79 | 0.60 | 0.315 |
| FANN-3 | −0.06 | 3.99 | 4.05 | 1.91 | 1.76 | 0.63 | 0.328 |
| FANN-4 | −0.79 | 3.95 | 4.74 | 1.93 | 1.80 | 0.65 | 0.334 |
| FANN-5 | 1.08 | 4.04 | 2.96 | 1.92 | 1.76 | 0.61 | 0.315 |
| FANN-6 | 0.72 | 3.89 | 3.17 | 1.91 | 1.77 | 0.61 | 0.321 |

**Table 7.** Performance of the different ANN models for strain (values of $\varepsilon_{cc}/\varepsilon_{co}$) on the training data set.

|  | Min | Max | Range | Mean | Median | StDev | CV |
|---|---|---|---|---|---|---|---|
| Experimental (target values) | 1.38 | 17.27 | 15.90 | 5.56 | 4.88 | 2.73 | 0.492 |
| EANN-1 | 0.64 | 11.65 | 11.01 | 5.60 | 5.44 | 2.07 | 0.369 |
| EANN-2 | −5.11 | 14.41 | 19.52 | 5.48 | 5.23 | 2.23 | 0.406 |
| EANN-3 | 0.63 | 18.96 | 18.33 | 5.59 | 5.52 | 1.97 | 0.352 |
| EANN-4 | −2.04 | 19.00 | 21.04 | 5.43 | 5.18 | 2.18 | 0.402 |
| EANN-5 | −1.36 | 19.26 | 20.62 | 5.51 | 5.38 | 2.01 | 0.365 |
| EANN-6 | −3.19 | 10.71 | 13.90 | 5.76 | 5.71 | 1.84 | 0.320 |

Equations (9)–(11) show the formulas used for the calculation of the correlation factor R, the mean squared error (MSE), and the mean absolute error (MAE), respectively. These three metrics were used to evaluate the different models. When the ANN outputs $O$ match the targets $T$ perfectly, then $R \rightarrow 1$, $MSE \rightarrow 0$, $MAE \rightarrow 0$ [76–78].

$$R = \frac{\sum\limits_{i=1}^{n}\left[(T_i - \overline{T})(O_i - \overline{O})\right]}{\sqrt{\sum\limits_{i=1}^{n}(T_i - \overline{T})^2 \cdot \sum\limits_{i=1}^{n}(O_i - \overline{O})^2}} \tag{9}$$

$$MSE = \frac{\sum\limits_{i=1}^{n}(T_i - O_i)^2}{n} \tag{10}$$

$$MAE = \frac{\sum\limits_{i=1}^{n}|T_i - O_i|}{n} \tag{11}$$

In the above formulas, $T_i$ and $O_i$ are the values established experimentally (target values) and with the use of the ANN, respectively. The number of data points is denoted by $n$, while $\overline{T}$ and $\overline{O}$ are the average values of the experimental and the predicted values, respectively.

Table 8 shows the prediction performance of the six NN models for strength and strain. It has to be noted that these values correspond to the normalized values of $f_{cc}/f'_{co}$ and $\varepsilon_{cc}/\varepsilon_{co}$, for the strength and strain models, respectively.

Figure 4 shows the values of MSE, MAE, and R for the six different NN models for strength and strain. The models corresponding to the least values of MAE and MSE and the highest value of R can be considered as the best models which can accurately predict the axial strength or strain of confined concrete. It can be concluded that the ANN strength model with the highest value of R (91.2%), together with the lowest values of MSE (5.68%) and MAE (5.45%), was the third model, i.e., FANN-3.

**Table 8.** Prediction performance of the different ANN models for strength and strain (for the normalized values of $f_{cc}/f'_{co}$ and $\varepsilon_{cc}/\varepsilon_{co}$, respectively.

| Strength Models | | | | Strain Models | | | |
|---|---|---|---|---|---|---|---|
| Model | R (%) | MSE (‰) | MAE (%) | Model | R (%) | MSE (‰) | MAE (%) |
| FANN-1 | 89.7 | 6.81 | 6.02 | EANN-1 | 73.8 | 8.61 | 6.89 |
| FANN-2 | 90.5 | 6.10 | 5.76 | EANN-2 | 72.1 | 9.25 | 6.99 |
| FANN-3 | 91.2 | 5.68 | 5.45 | EANN-3 | 67.8 | 10.24 | 7.00 |
| FANN-4 | 89.8 | 6.74 | 5.38 | EANN-4 | 71.9 | 9.29 | 6.80 |
| FANN-5 | 88.6 | 7.28 | 6.10 | EANN-5 | 66.4 | 10.68 | 7.05 |
| FANN-6 | 90.7 | 6.00 | 5.64 | EANN-6 | 68.9 | 10.03 | 7.59 |

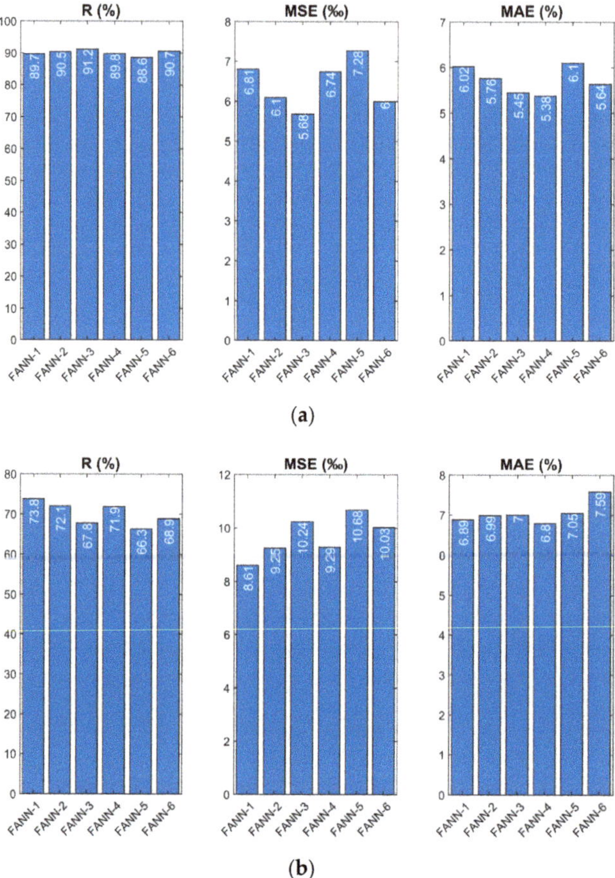

**Figure 4.** Performance of the various ANN models for the case of cylinders: (**a**) strength models (FANN); (**b**) strain models (EANN).

Similarly, the ANN strain model with the highest value of R (71.9%) along with the lowest values of MSE (9.29‰) MAE (6.80%) was the fourth model, i.e., EANN-4. Thus, FANN-3 (5-5-5-1 architecture) and EANN-4 (5-10-10-1 architecture) were selected as the final models for the prediction of the axial strength and axial strain of cylinders. Figures 5 and 6 depict the output values of the different NN models vs. the target values for strength and strain graphically, respectively.

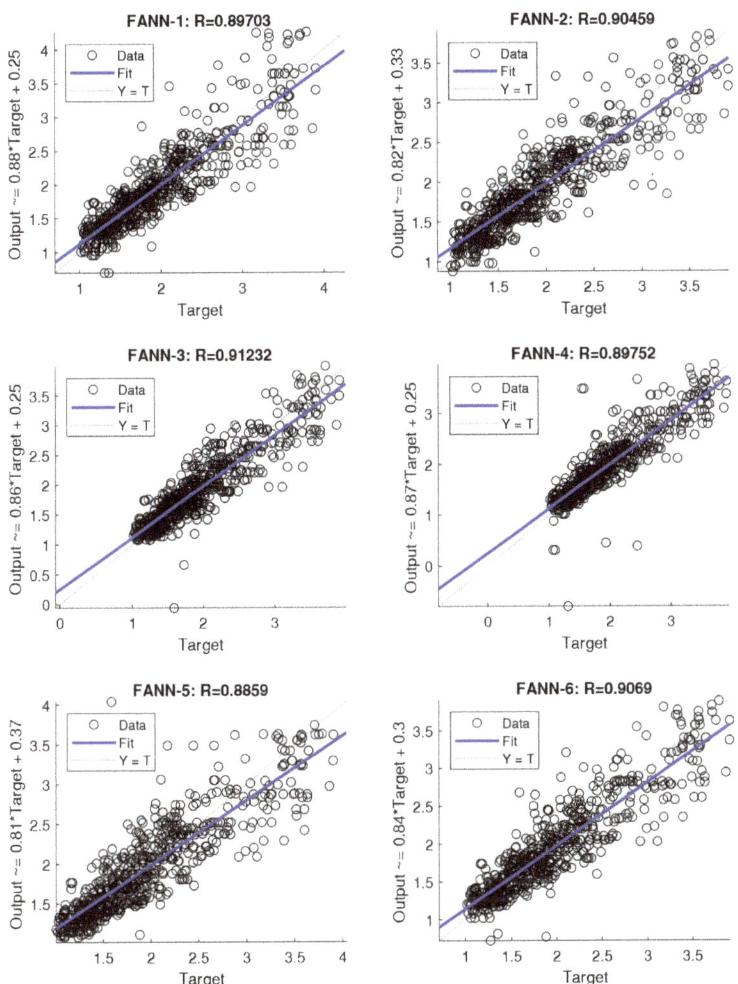

**Figure 5.** Predictions of the six examined ANN models for strength.

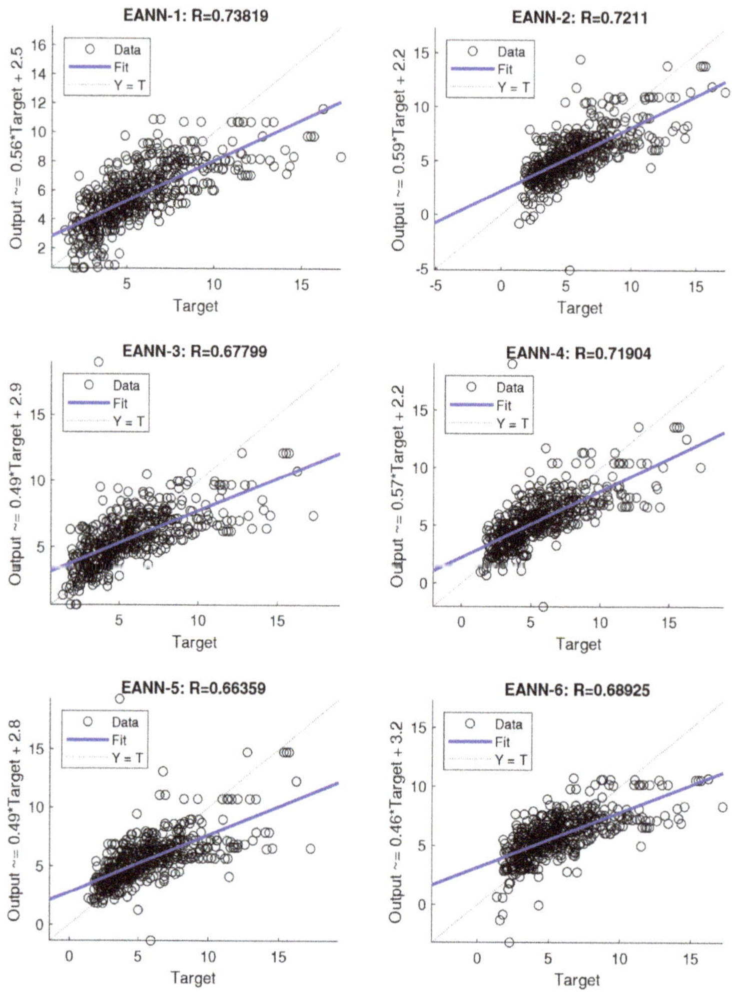

**Figure 6.** Predictions of the six examined ANN models for strain.

## 4. Analytical Models

### 4.1. Analytical Models from the Literature

In the present work, design-oriented strength and strain models were proposed based on large experimental databases of 708 confined concrete cylinders for strength and 572 for strain. The whole process was performed using ANN techniques giving the best fit after minimizing certain statistical parameters, such as R, MSE, and MAE, between the predictions of models and the experimental values.

Next, we evaluated the performance of other strength and strain models proposed in the literature using the collected database. Namely, for *strength*, we examined the following twelve models: 1. Richart et al. [68], 2. Fardis and Khalili [63], 3. Newman and Newman [69], 4. Mander et al. [45], 5. Karbhari and Gao [42], 6. Samaan et al. [59], 7. Toutanji [65], 8. Lam and Teng [44], 9. Teng et al. [61], 10. Saafi et al. [67], 11. Miyauchi et al. [70], 12. Matthys et al. [60]. Figure 7 shows the performance of these twelve strength models.

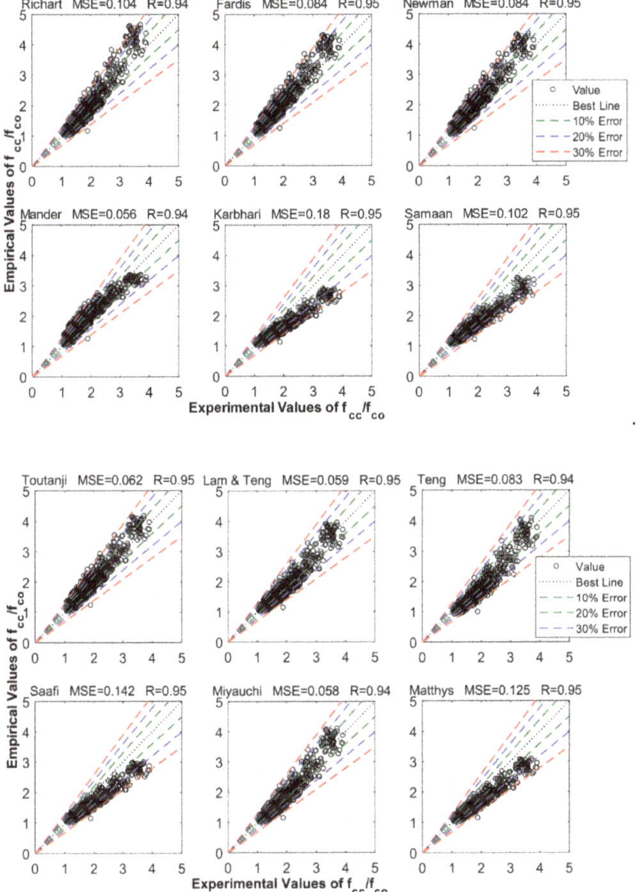

**Figure 7.** Performance of previously proposed strength models of confined concrete on the developed database.

For *strain*, we examined the following seven models: 1. Fardis and Khalili [63], 2. Mander et al. [45], 3. Karbhari and Gao [42], 4. Samaan et al. [59], 5. Toutanji [65], 6. Lam and Teng [44], 7. Teng et al. [66]. Figure 8 shows the performance of these seven strain models. The values of R and MSE are given in the titles of the sub-figures, for each examined model, for both strength (Figure 7) and strain (Figure 8). Looking at the figures, it is obvious that the strength models were more successful than the strain models, in general. The spread of the strain models was larger, and the error of these models was above the 30% threshold (red dashed line) in many cases.

## 4.2. Proposed Analytical Model

In this section, we propose two new analytical models, one for strength and one for strain, based on regression analysis. *Regression Analysis* is a statistical tool used to determine the probable change in one variable for a given amount of change in another. This means the value of the unknown variable can be estimated from the known value of another variable.

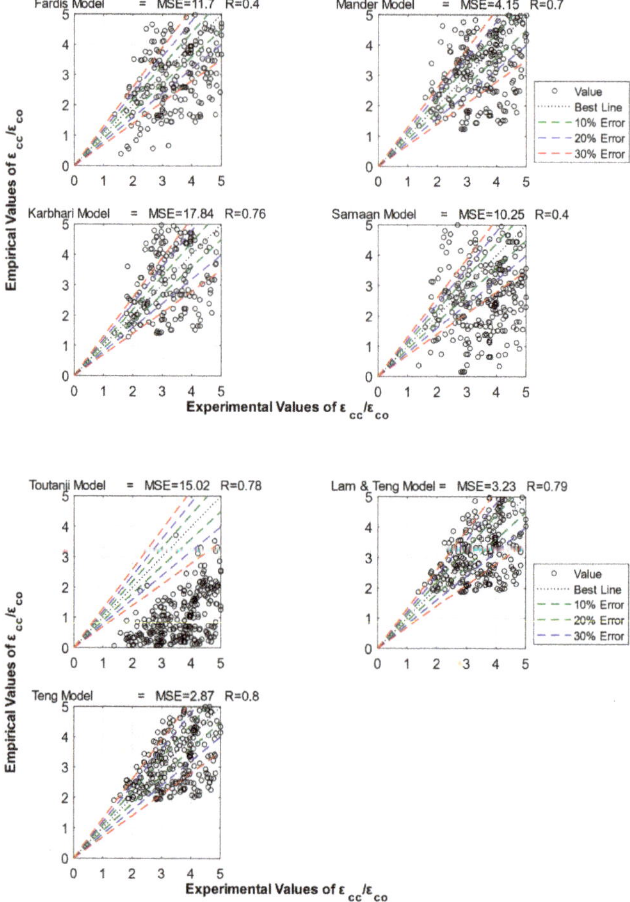

**Figure 8.** Performance of previously proposed strain models of confined concrete on the developed database.

*4.3. Proposed Equation for Strength*

The form of the new proposed analytical strength model was based on the mechanics of published models, as represented by Equation (12),

$$\frac{f_{cc}}{f_{co}} = 1 + k \cdot \left(\frac{f_l}{f_{co}}\right)^n \tag{12}$$

where $k$ and $n$ are constants. Using general regression analysis, we end up with the values $k = 3.1$ and $n = 0.83$ for the two variables of Equation (13) and the final equation becomes

$$\frac{f_{cc}}{f_{co}} = 1 + 3.1 \cdot \left(\frac{f_l}{f_{co}}\right)^{0.83} \tag{13}$$

Thus, the FRP confined concrete axial compressive strength can be finally expressed by Equation (14)

$$f_{cc} = f_{co} + 3.1 f_{co} \cdot \left( \frac{f_l}{f_{co}} \right)^{0.83} \tag{14}$$

Figure 9 shows the performance of the newly proposed analytical model for strength (on the left) and the corresponding performance of the best ANN model for strength (on the right). It has to be noted that the R-value was better for the analytical model compared to the ANN model (0.95 > 0.91), and the same trend was found for the MSE value of the analytical model compared to the one of the ANN model (0.041 < 0.074). The performance of the proposed analytical model for strength was superior to any of the examined models in the literature, exhibiting better values for both R and MSE.

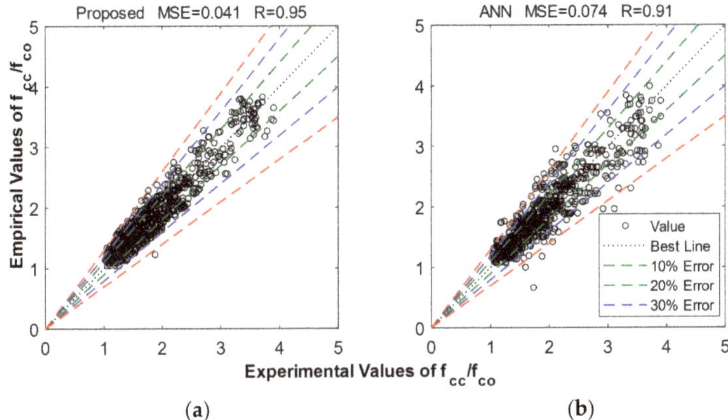


**Figure 9.** Performance for strength, for: (**a**) the proposed analytical model; (**b**) the corresponding ANN model.

### 4.4. Proposed Equation for Strain

As shown in Figure 8, the published strain model proposed by Fardis and Khalili [63] exhibited a poor performance with R = 0.4 and MSE = 11.7 for predicting the strain of confined concrete cylinders at ULS. Furthermore, the Mander et al. [45] model showed a relatively good performance with R = 0.7 and MSE = 4.15. This is the model that has been recommended by the American Concrete Institute ACI 440-2R-02 [64]. Therefore, to propose a new analytical equation for strain, the general form of this model was selected as given by Equation (15), with a total of four parameters to adjust ($\alpha$, $\beta$, $\lambda_1$, $\lambda_2$).

$$\frac{\varepsilon_{cc}}{\varepsilon_{co}} = a + \beta \cdot \rho_\kappa^{\lambda_1} \cdot \rho_\varepsilon^{\lambda_2} \tag{15}$$

Using general regression analysis, we ended up with the values $\alpha = 1.85$, $\beta = 7.46$, $\lambda_1 = 0.71$, and $\lambda_2 = 1171$ and the final equation becomes

$$\frac{\varepsilon_{cc}}{\varepsilon_{co}} = 1.85 + 7.46 \rho_\kappa^{0.71} \cdot \rho_\varepsilon^{1.171} \tag{16}$$

Finally, the relationship for the axial compressive strain of FRP confined concrete compression members can be written as

$$\varepsilon_{cc} = \left( 1.85 + 7.46 \rho_\kappa^{0.71} \cdot \rho_\varepsilon^{1.171} \right) \cdot \varepsilon_{co} \tag{17}$$

Figure 10 shows the performance of the proposed analytical model for strain (on the left) and the corresponding performance of the ANN model for strain (on the right). Again, the R-value was better for the analytical model compared to the ANN model (0.8 > 0.72), and the same trend was found for the MSE value of the analytical model, compared to the one of the ANN model (2669 < 3.66). Again, the performance of the proposed analytical model for strain was superior to any of the examined models in the literature, exhibiting better values for both R and MSE.

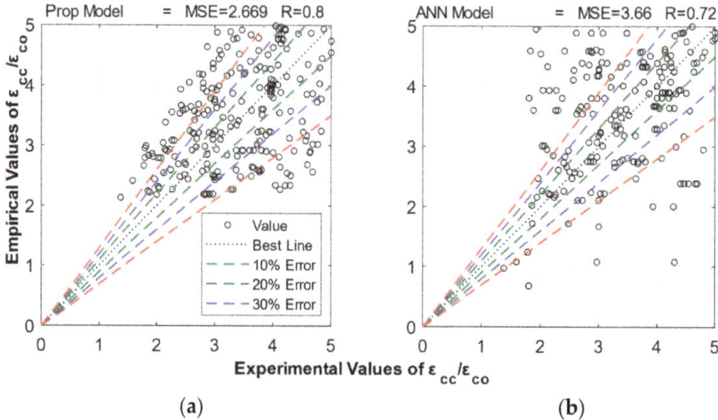


**Figure 10.** Performance for strain, for: (**a**) the proposed analytical model; (**b**) the corresponding ANN model).

## 5. Comparative Study

For a comparative study of the proposed ANN model, the proposed analytical model, and other models from the literature, we drew the probability density functions (PDFs) of the results of the various methods as normal distributions based on the mean value and the standard deviation of the experimental ratio ($f_{cc}/f'_{co}$ and $\varepsilon_{cc}/\varepsilon_{co}$) divided by the corresponding predicted ratio for each method. A good prediction would mean an average value close to 1 and the PDF curve being tall and as narrow as possible (standard deviation as low as possible). Wider curves denote a higher standard deviation from the mean value.

Figure 11 shows the results for the strength (a) and the strain (b) models. For illustration purposes, for the strength model, the comparison was made with the results from Richart [68], Fardis [63], Newman [69], and Mander [45], while for the strain model, the results were compared with the ones from Fardis [63], Mander [45], Lam and Teng [44], and Teng [61].

For the *strength* model, the proposed analytical model and the proposed ANN model showed the best results with mean values close to 1 (1.00 and 1.01, respectively), while from the other models, the one of Richart had the best mean value (0.99) with a standard deviation of 0.132. The standard deviation of the proposed analytical model and the proposed ANN model were 0.109 and 0.146, respectively.

For the *strain* model, again, the two proposed models showed very good results with a mean value of 1.05 (for both) and standard deviations 0.306 for the proposed equation and 0.360 for the proposed ANN. For comparison, the Teng and the Lam and Teng models had mean values of 1.13 and 1.12, respectively, with standard deviations 0.352 and 0.370.

It should be noted that the results were better for the strength models in general (better mean values and smaller deviation), compared to the strain model. The proposed models for confined concrete are capable of providing accurate results for confined concrete strength and strain, in comparison to other models in the literature.

Figure 12 shows the distributions of the ratios $f_{cc}/f'_{co}$ and $\varepsilon_{cc}/\varepsilon_{co}$ provided by each method. The black bar denotes the experimental (target) values. Again, the proposed equation and the proposed ANN model gave the best results with distributions that were closer to the targets.

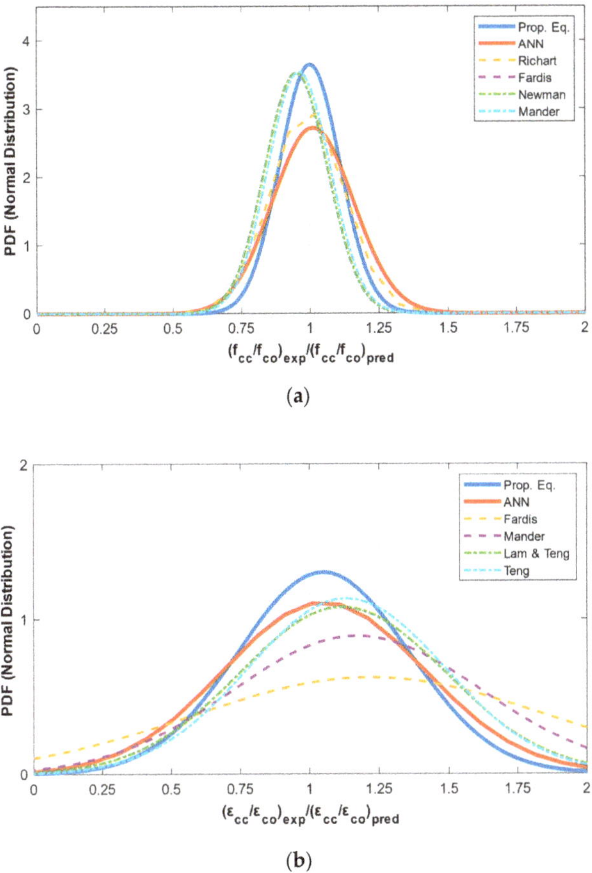

(a)

(b)

**Figure 11.** Probability density functions (as normal distributions) for the various prediction models: (**a**) strength; (**b**) strain.

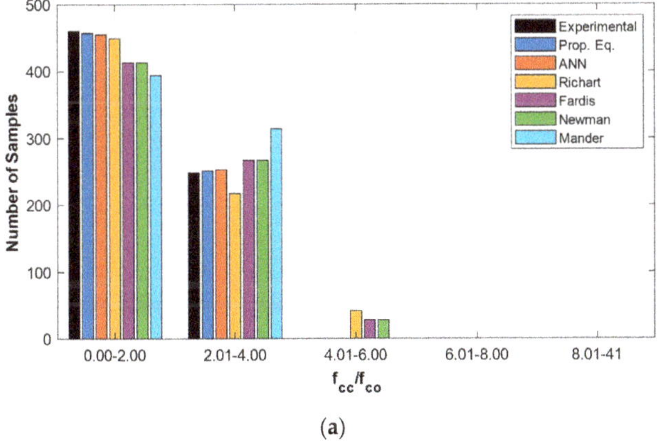

(a)

**Figure 12.** *Cont.*

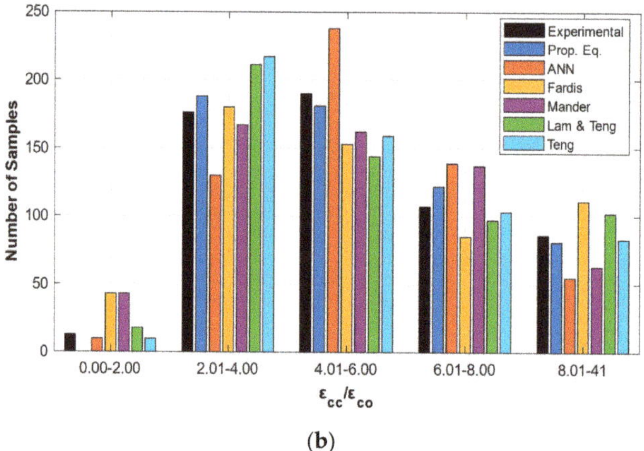


(**b**)

**Figure 12.** Distribution of values for FRP-wrapped cylinders: (**a**) strength ($f_{cc}/f'_{co}$); (**b**) strain ($\varepsilon_{cc}/\varepsilon_{co}$).

## 6. Conclusions

In the current study, a large database for FRP confined concrete compression members was constructed from data available in the literature. The database contained 708 specimens for strength and 572 specimens for strain. A preliminary evaluation of previously proposed models for ultimate conditions of FRP-confined concrete compression members was performed to select more general forms of the proposed empirical models by using statistical metrics, such as R, MSE, and MAE. The outcome of the present work is summarized as follows.

1.  Six different ANN models were generated for the strength (FANNs) and the strain (EANNs) of FRP-confined concrete compression members, separately. The ANN strength model with the highest value of R, together with the lowest values of MAE and MSE for the normalized values, was FANN-3, which is a double layer ANN model with five neurons in each hidden layer (5-5-5-1 network architecture).

2.  Similarly, the ANN strain model with the highest value of R along with the lowest values of MAE and MSE for the normalized values was EANN-4, which is again a double layer ANN model with 10 neurons in each hidden layer (5-10-10-1 network architecture).

3.  Two analytical models were also proposed for strength and strain, based on general formulas and regression analysis procedures. The performance of the proposed analytical models was superior to any one of the examined models in the literature, exhibiting better values for both R and MSE when applied to the databases. The analytical models also performed slightly better than the ANN models in terms of the R and MSE values achieved.

4.  The proposed ANN and the proposed analytical models can predict the axial behavior of FRP-confined concrete with high accuracy and can be helpful tools for designers in analyzing and designing confined concrete compression members. Both the ANN and the analytical proposed models outperformed other empirical models in their predictions. Once trained, the ANN models need only trivial computational time to provide their outputs.

**Author Contributions:** Conceptualization, A.A. and Q.-u.-Z.K.; methodology, V.P.; software, A.A. and V.P.; validation, A.A. and V.P.; formal analysis, A.A. and V.P.; investigation, Q.-u.-Z.K.; resources, V.P.; data curation, A.A.; writing—original draft preparation, A.A. and Q.-u.-Z.K.; writing—review and editing, V.P.; visualization, V.P.; supervision, Q.-u.-Z.K.; project administration, A.A. and V.P.; funding acquisition, V.P. All authors have read and agreed to the published version of the manuscript.

**Funding:** This research received no external funding. The Article Processing Charge (APC) was funded by the Oslo Metropolitan University (OsloMet).

**Acknowledgments:** The authors acknowledge the help of OsloMet and the University of Engineering and Technology, Taxila, in providing the necessary facilities. The support of OsloMet in covering the APC is also greatly acknowledged and appreciated.

**Conflicts of Interest:** The authors declare no conflict of interest.

## References


1. Hollaway, L.C.; Chryssanthopoulos, M.; Moy, S.S. *Advanced Polymer Composites for Structural Applications in Construction: ACIC 2004*; Woodhead Publishing: Cambridge, UK, 2004.
2. Parvin, A.; Jamwal, A.S. Effects of wrap thickness and ply configuration on composite-confined concrete cylinders. *Compos. Struct.* **2005**, *67*, 437–442. [CrossRef]
3. Li, G.; Kidane, S.; Pang, S.-S.; Helms, J.E.; Stubblefield, M.A. Investigation into FRP repaired RC columns. *Compos. Struct.* **2003**, *62*, 83–89. [CrossRef]
4. Demers, M.; Neale, K.W. Confinement of reinforced concrete columns with fibre-reinforced composite sheets-an experimental study. *Can. J. Civ. Eng.* **1999**, *26*, 226–241. [CrossRef]
5. Prota, A.; Manfredi, G.; Cosenza, E. Ultimate behavior of axially loaded RC wall-like columns confined with GFRP. *Compos. Part B Eng.* **2006**, *37*, 670–678. [CrossRef]
6. Li, L.; Guo, Y.; Liu, F.; Bungey, J. An experimental and numerical study of the effect of thickness and length of CFRP on performance of repaired reinforced concrete beams. *Constr. Build. Mater.* **2006**, *20*, 901–909. [CrossRef]
7. Țăranu, N.; Oprișan, G.; Ențuc, I.; Munteanu, V.; Cozmanciuc, C. The Efficiency of Fiber Reinforced Polymer Composites Solutions in the Construction Industry. In Proceedings of the 6th International Conference on Management of Technological Changes, MTC, Alexandroupolis, Greece, 3–5 September 2009; pp. 733–736.
8. De Lorenzis, L.; Tepfers, R. Comparative study of models on confinement of concrete cylinders with fiber-reinforced polymer composites. *J. Compos. Constr.* **2003**, *7*, 219–237. [CrossRef]
9. Takemura, H.K.K. Effect of loading hysteresis on ductility capacity of reinforced concrete bridge piers. *J. Struct. Eng.* **1997**, *43*, 849–858.
10. Hoshikuma, J.; Kawashima, K.; Nagaya, K.; Taylor, A.W. Stress-strain model for confined reinforced concrete in bridge piers. *J. Struct. Eng.* **1997**, *123*, 624–633. [CrossRef]
11. Kawashima, K.; Shoji, G.; Sakakibara, Y. A cyclic loading test for clarifying the plastic hinge length of reinforced concrete piers. *J. Struct. Eng.* **2000**, *46*, 767–776.
12. Saeed, H.Z.; Ahmed, A.; Ali, S.M.; Iqbal, M. Experimental and finite element investigation of strengthened LSC bridge piers under Quasi-Static Cyclic Load Test. *Compos. Struct.* **2015**, *131*, 556–564. [CrossRef]
13. Tahir, M.F.; Ahmad, A.; Iqbal, M. Seismic Evaluation of Repaired and Retrofitted Circular Bridge Piers of Low-Strength Concrete. *Arab. J. Sci. Eng.* **2015**, *40*, 3057–3066.
14. Khan, Q.U.; Ahmad, A.; Mehboob, S.; Iqbal, M. Energy dissipation characteristics of retrofitted damaged low strength concrete bridge pier. In *Proceedings of the Institution of Civil Engineers—Bridge Engineering*; Thomas Telford Services Ltd.: London, UK, 2020; pp. 1–29.
15. Ahmad, A.; Mehboob, S.; Nouman, M. Experimental and numerical investigation of T-joint enhanced confinement using flat steel strips. *Asian J. Civ. Eng.* **2020**, *21*, 1–10.
16. Raza, A.; Khan, Q.U.; Ahmad, A. Prediction of Axial Compressive Strength for FRP-Confined Concrete Compression Members. *KSCE J. Civ. Eng.* **2020**. [CrossRef]
17. Ann, K.; Cho, C.-G. Constitutive behavior and finite element analysis of FRP composite and concrete members. *Materials* **2013**, *6*, 3978–3988. [CrossRef]
18. Cascardi, A.; Micelli, F.; Aiello, M.A. An Artificial Neural Networks model for the prediction of the compressive strength of FRP-confined concrete circular columns. *Eng. Struct.* **2017**, *140*, 199–208. [CrossRef]
19. Naderpour, H.; Nagai, K.; Fakharian, P.; Haji, M. Innovative models for prediction of compressive strength of FRP-confined circular reinforced concrete columns using soft computing methods. *Compos. Struct.* **2019**, *215*, 69–84. [CrossRef]
20. Stamopoulos, A.; Tserpes, K.; Dentsoras, A. Quality assessment of porous CFRP specimens using X-ray Computed Tomography data and Artificial Neural Networks. *Compos. Struct.* **2018**, *192*, 327–335. [CrossRef]

21. Micelli, F.; Cascardi, A.; Aiello, M.A. A Study on FRP-Confined Concrete in Presence of Different Preload Levels. In Proceedings of the 9th International Conference on Fibre-Reinforced Polymer (FRP) Composites in Civil Engineering—CICE, Paris, France, 17–19 July 2018; pp. 17–19.

22. Ghiassi, M.; Saidane, H.; Zimbra, D. A dynamic artificial neural network model for forecasting time series events. *Int. J. Forecast.* **2005**, *21*, 341–362. [CrossRef]

23. Cascardi, A.; Micelli, F.; Aiello, M. Analytical model based on artificial neural network for masonry shear walls strengthened with FRM systems. *Compos. Part B Eng.* **2016**, *95*, 252–263. [CrossRef]

24. Liu, H.-J.; Xue, X.-H. Artificial neural network model for prediction of seismic liquefaction of sand soil. *Rock Soil Mech.* **2004**, *12*. Available online: http://en.cnki.com.cn/Article_en/CJFDTotal-YTLX20041200G.htm (accessed on 13 September 2020).

25. Cevik, A.; Cabalar, A.F. A genetic-programming-based formulation for the strength enhancement of fiber-reinforced-polymer-confined concrete cylinders. *J. Appl. Polym. Sci.* **2008**, *110*, 3087–3095. [CrossRef]

26. Cevik, A.; Guzelbey, I.H. Neural network modeling of strength enhancement for CFRP confined concrete cylinders. *Build. Environ.* **2008**, *43*, 751–763. [CrossRef]

27. Cevik, A.; Göğüş, M.T.; Güzelbey, İ.H.; Filiz, H. Soft computing based formulation for strength enhancement of CFRP confined concrete cylinders. *Adv. Eng. Softw.* **2010**, *41*, 527–536. [CrossRef]

28. Naderpour, H.; Kheyroddin, A.; Amiri, G.G. Prediction of FRP-confined compressive strength of concrete using artificial neural networks. *Compos. Struct.* **2010**, *92*, 2817–2829. [CrossRef]

29. Wu, Y.-B.; Jin, G.-F.; Ding, T.; Meng, D. Modeling confinement efficiency of FRP-confined concrete column using radial basis function neural network. In Proceedings of the 2nd International Workshop on Intelligent Systems and Applications (ISA), Wuhan, China, 22–23 May 2010; pp. 1–6.

30. Cevik, A. Modeling strength enhancement of FRP confined concrete cylinders using soft computing. *Expert Syst. Appl.* **2011**, *38*, 5662–5673. [CrossRef]

31. Oreta, A.W.; Ongpeng, J. Modeling the confined compressive strength of hybrid circular concrete columns using neural networks. *Comput. Concr.* **2011**, *8*, 597–616. [CrossRef]

32. Elsanadedy, H.; Al-Salloum, Y.; Abbas, H.; Alsayed, S.H. Prediction of strength parameters of FRP-confined concrete. *Compos. Part B Eng.* **2012**, *43*, 228–239. [CrossRef]

33. Jalal, M.; Ramezanianpour, A.A. Strength enhancement modeling of concrete cylinders confined with CFRP composites using artificial neural networks. *Compos. Part B Eng.* **2012**, *43*, 2990–3000. [CrossRef]

34. Mashrei, M.A.; Seracino, R.; Rahman, M.J.C.; Materials, B. Application of artificial neural networks to predict the bond strength of FRP-to-concrete joints. *Constr. Build. Mater.* **2013**, *40*, 812–821. [CrossRef]

35. Pham, T.M.; Hadi, M.N. Predicting stress and strain of FRP-confined square/rectangular columns using artificial neural networks. *J. Compos. Constr.* **2014**, *18*, 04014019. [CrossRef]

36. Di Trapani, F.; Malavisi, M.; Marano, G.C.; Sberna, A.P.; Greco, R. Optimal seismic retrofitting of reinforced concrete buildings by steel-jacketing using a genetic algorithm-based framework. *Eng. Struct.* **2020**, *219*, 110864. [CrossRef]

37. Papavasileiou, G.S.; Charmpis, D.C.; Lagaros, N.D. Optimized seismic retrofit of steel-concrete composite buildings. *Eng. Struct.* **2020**, *213*, 110573. [CrossRef]

38. Falcone, R.; Carrabs, F.; Cerulli, R.; Lima, C.; Martinelli, E. Seismic retrofitting of existing RC buildings: A rational selection procedure based on Genetic Algorithms. In *Structures*; Elsevier: Amsterdam, The Netherlands; pp. 310–326.

39. Cavaleri, L.; Trapani, F.; Ferrotto, M.; Davi, L. Stress-strain models for normal and high strength confined concrete: Test and comparison of literature models reliability in reproducing experimental results. *Ing. Sismica* **2017**, *34*, 114–137.

40. Ahmad, A.; Raza, A. Reliability analysis of strength models for CFRP-confined concrete cylinders. *Compos. Struct.* **2020**, *244*, 112312. Available online: https://www.sciencedirect.com/science/article/pii/S0263822319327503 (accessed on 13 September 2020). [CrossRef]

41. Raza, A.; Ahmad, A. Reliability analysis of proposed capacity equation for predicting the behavior of steel-tube concrete columns confined with CFRP sheets. *Comput. Concr.* **2020**, *25*, 383–400.

42. Karbhari, V.M.; Gao, Y. Composite jacketed concrete under uniaxial compression—Verification of simple design equations. *J. Mater. Civ. Eng.* **1997**, *9*, 185–193. [CrossRef]

43. Miyauchi, K.; Inoue, S.; Kuroda, T.; Kobayashi, A. Strengthening effects of concrete column with carbon fiber sheet. *Trans. Jpn. Concr. Inst.* **2000**, *21*, 143–150.

44. Lam, L.; Teng, J. Design-oriented stress-strain model for FRP-confined concrete. *Constr. Build. Mater.* **2003**, *17*, 471–489. [CrossRef]

45. Mander, J.B.; Priestley, M.J.; Park, R. Theoretical stress-strain model for confined concrete. *J. Struct. Eng.* **1988**, *114*, 1804–1826. [CrossRef]

46. Bisby, L.A.; Dent, A.J.S.; Green, M.F. Comparison of confinement models for fiber-reinforced polymer-wrapped concrete. *ACI Struct. J.* **2005**, *102*, 62–72.

47. Ashrafi, H.R.; Jalal, M.; Garmsiri, K. Prediction of load–displacement curve of concrete reinforced by composite fibers (steel and polymeric) using artificial neural network. *Expert Syst. Appl.* **2010**, *37*, 7663–7668. [CrossRef]

48. Asteris, P.G.; Plevris, V. Anisotropic masonry failure criterion using artificial neural networks. *Neural Comput. Appl.* **2017**, *28*, 2207–2229. [CrossRef]

49. Lagaros, N.D.; Plevris, V.; Papadrakakis, M. Neurocomputing strategies for solving reliability-robust design optimization problems. *Eng. Comput.* **2010**, *27*, 819–840. [CrossRef]

50. Plevris, V.; Asteris, P.G. Modeling of masonry failure surface under biaxial compressive stress using Neural Networks. *J. Constr. Build. Mater.* **2014**, *55*, 447–461. [CrossRef]

51. Shah, V.S.; Shah, H.R.; Samui, P.; Murthy, A.R.; Merono, P.; Gomez, F.; Marin, F.; Zhang, W.; Ni, P.; Liu, B. Prediction of fracture parameters of high strength and ultra-high strength concrete beams using minimax probability machine regression and extreme learning machine. *J. Comput. Mater. Contin.* **2014**, *44*, 73–84.

52. Zhuang, X.; Zhou, S. The Prediction of Self-Healing Capacity of Bacteria-Based Concrete Using Machine Learning Approaches. *J. Comput. Mater. Contin.* **2019**, *1*, 57–77. [CrossRef]

53. Ghanizadeh, A.R.; Abbaslou, H.; Amlashi, A.T.; Alidoust, P. Modeling of bentonite/sepiolite plastic concrete compressive strength using artificial neural network and support vector machine. *Front. Struct. Civ. Eng.* **2019**, *13*, 215–239. [CrossRef]

54. Yuvaraj, P.; Murthy, A.R.; Iyer, N.R.; Sekar, S.; Samui, P. ANN model to predict fracture characteristics of high strength and ultra high strength concrete beams. *J. Comput. Mater. Contin.* **2014**, *41*, 193–213.

55. Yıldızel, S.; Öztürk, A. A study on the estimation of prefabricated glass fiber reinforced concrete panel strength values with an artificial neural network model. *J. Comput. Mater. Contin.* **2016**, *52*, 42–51.

56. Yang, Q.; Du, S. Prediction of Concrete Cubic Compressive Strength Using ANN Based Size Effect Model. *J. Comput. Mater. Contin.* **2015**, *47*, 217–236.

57. Khademi, F.; Akbari, M.; Jamal, S.M.; Nikoo, M. Multiple linear regression, artificial neural network, and fuzzy logic prediction of 28 days compressive strength of concrete. *Front. Struct. Civ. Eng.* **2017**, *11*, 90–99. [CrossRef]

58. Reddy, T.C. Predicting the strength properties of slurry infiltrated fibrous concrete using artificial neural network. *Front. Struct. Civ. Eng.* **2018**, *12*, 490–503. [CrossRef]

59. Samaan, M.; Mirmiran, A.; Shahawy, M. Model of concrete confined by fiber composites. *J. Struct. Eng.* **1998**, *124*, 1025–1031. [CrossRef]

60. Matthys, S.; Toutanji, H.; Audenaert, K.; Taerwe, L. Axial load behavior of large-scale columns confined with fiber-reinforced polymer composites. *ACI Struct. J.* **2005**, *102*, 258.

61. Teng, J.; Jiang, T.; Lam, L.; Luo, Y.Z. Refinement of a design-oriented stress-strain model for FRP-confined concrete. *J. Compos. Constr.* **2009**, *13*, 269–278. [CrossRef]

62. Lim, J.C.; Karakus, M.; Ozbakkaloglu, T. Evaluation of ultimate conditions of FRP-confined concrete columns using genetic programming. *Comput. Struct.* **2016**, *162*, 28–37. [CrossRef]

63. Fardis, M.N.; Khalili, H.H. FRP-encased concrete as a structural material. *Mag. Concr. Res.* **1982**, *34*, 191–202. [CrossRef]

64. Bakis, C.E.; Ganjehlou, A.; Kachlakev, D.I.; Schupack, M.; Balaguru, P.; Gee, D.J.; Karbhari, V.M.; Scott, D.W.; Ballinger, C.A.; Gentry, T.R.; et al. Guide for the design and construction of externally bonded FRP systems for strengthening concrete structures (ACI-440.2R-02). *Rep. ACI Comm.* **2002**. Available online: https://citeseerx.ist.psu.edu/viewdoc/download?doi=10.1.1.455.4053&rep=rep1&type=pdf (accessed on 13 September 2020).

65. Toutanji, H. Stress-strain characteristics of concrete columns externally confined with advanced fiber composite sheets. *Mater. J.* **1999**, *96*, 397–404.

66. Teng, J.G.; Yu, T.; Wong, Y.; Dong, S. Hybrid FRP-concrete-steel tubular columns: Concept and behavior. *Constr. Build. Mater.* **2007**, *21*, 846–854. [CrossRef]

67. Saafi, M.; Toutanji, H.A.; Li, Z. Behavior of concrete columns confined with fiber reinforced polymer tubes. *Mater. J.* **1999**, *96*, 500–509.

68. Richart, F.E.; Brandtzæg, A.; Brown, R.L. *Failure of Plain and Spirally Reinforced Concrete in Compression*; College of Engineering, University of Illinois at Urbana Champaign: Champaign, IL, USA, 1929.

69. Newman, K.; Newman, J.B. Failure theories and design criteria for plain concrete. *Struct. Solid Mech. Eng. Des.* **1971**, 963–995. Available online: https://www.abebooks.com/Structure-Solid-Mechanics-Engineering-Design-Proceedings/22384738675/bd (accessed on 14 September 2020).

70. Miyauchi, K. Estimation of strengthening effects with carbon fiber sheet for concrete column. In Proceedings of the 3rd International Symposium on Non-Metallic (FRP) Reinforcement for Concrete Structures, Sapporo, Japan, 14–16 October 1997; pp. 217–224.

71. Ahmad, A.; Kotsovou, G.; Cotsovos, D.M.; Lagaros, N.D. Assessing the accuracy of RC design code predictions through the use of artificial neural networks. *Int. J. Adv. Struct. Eng.* **2018**. [CrossRef]

72. Cladera, A.; Marí, A. Shear design procedure for reinforced normal and high-strength concrete beams using artificial neural networks. Part I: Beams without stirrups. *Eng. Struct.* **2004**, *26*, 917–926. [CrossRef]

73. Cladera, A.; Mari, A. Shear design procedure for reinforced normal and high-strength concrete beams using artificial neural networks. Part II: Beams with stirrups. *Eng. Struct.* **2004**, *26*, 927–936. [CrossRef]

74. LeCun, Y.A.; Bottou, L.; Orr, G.B.; Müller, K.-R. Efficient backprop. In *Neural Networks: Tricks of the Trade*; Springer: Berlin/Heidelberg, Germany, 2012; pp. 9–48.

75. Ahmad, A.; Cotsovos, D.M.; Lagaros, N.D. Framework for the development of artificial neural networks for predicting the load carrying capacity of RC members. *SN Appl. Sci.* **2020**, *2*, 1–21. [CrossRef]

76. Krogh, A.; Vedelsby, J. Neural Network Ensembles, Cross Validation, and Active Learning. *Adv. Neyrul Inf. Process. Syst.* **1995**, *7*, 21–238.

77. Utans, J.; Moody, J.; Rehfuss, S.; Siegelmann, H. Input Variable Selection for Neural Networks: Application to Predicting the U.S. Business Cycle. In Proceedings of the Conference on Computational Intelligence for Financial Engineering, New York, NY, USA, 9–11 April 1995; pp. 118–122.

78. Castellano, G.; Fanelli, A.M. Variable Selection Using Neural-Network Models. *Neurocomputing* **2000**, *31*, 1–13. [CrossRef]

*Article*

# Constitutive Modeling of New Synthetic Hybrid Fibers Reinforced Concrete from Experimental Testing in Uniaxial Compression and Tension

**S. M. Iqbal S. Zainal [1,2], Farzad Hejazi [1,*], Farah N. A. Abd. Aziz [1] and Mohd Saleh Jaafar [1]**

[1] Department of Civil Engineering, Universiti Putra Malaysia, Selangor 43400, Malaysia; iqbal.zainal@ums.edu.my (S.M.I.S.Z.); farah@upm.edu.my (F.N.A.A.A.); msj@upm.my (M.S.J.)

[2] Department of Civil Engineering, Universiti Malaysia Sabah, Sabah 88400, Malaysia

* Correspondence: farzad@fhejazi.com

Received: 28 July 2020; Accepted: 3 September 2020; Published: 1 October 2020

**Abstract:** Hybridization of fibers in concrete yields a variety of applications due to its benefits compared to conventional concrete or concrete with single type-fiber. However, the Finite Element (FE) modeling of these new materials for numerical analyses are very challenging due to the lack of analytical data for these specific materials. Therefore, an attempt has been made to develop Hybrid Fiber Reinforced Concrete (HyFRC) materials with High Range Water-Reducing Admixture (HRWRA) during the concrete mixing process and conduct experimental study to evaluate the behavior of the proposed materials. Constitutive models for each of the materials are formulated to be used as analytical models in numerical analyses. The acquired data are then used to formulate mathematical equations, governing the stress–strain behavior of the proposed HyFRC materials to measure the accuracy of the proposed models. The experimental testing indicated that the Ferro with Ferro mix-combination improved the performance of concrete in the elastic stage while the Ferro with Ultra-Net combination has the highest compressive strain surplus in the plastic stage. In tension, the Ferro with Ferro mix displayed the highest elastic behavior improvement while the Ferro with Ultra-Net designs proved superior in the plastic range, providing additional toughness to conventional concrete.

**Keywords:** forta fibers; synthetic fibers; hybrid fiber reinforced concrete; constitutive modeling; uniaxial test; slump test

## 1. Introduction

Concrete is commonly used in construction because it is economical, easy to procure in the market, and has a wide range of applications. The disadvantage of concrete however is that it is very brittle, which results in poor resistance to crack initiation and propagation as well as low tensile, strain, and low energy absorption capacity. The unreinforced matrices deform elastically under tension until fracture because of the development of micro cracks and localized macro cracking. Fibrous reinforced concrete improves the post-cracking behavior of brittle concrete beyond the elastic stage depending on several factors such as matrix strength, fiber type, fiber orientation, fiber strength, fiber modulus, surface treatment of fibers, fiber aspect ratio, fiber content, and aggregate size [1].

The numerous benefits of using single-fiber reinforced concrete prompted further investigations into combining multiple type of fibers into cementitious composites for improved performance in mechanical properties. The hybridization of polypropylene, natural, glass, asbestos, and carbon fibers in concrete yield positive findings whereby positive synergistic effects were observed between organic and inorganic fibers in improving mechanical properties of concrete [2]. The combination of carbon and steel fibers also showed that steel fibers were effective in increasing strength while carbon fibers improved toughness [3]. Synthetic fibers such as polyethylene and polypropylene were combined and the result proved effective in increasing the performance of concrete in impact loading, flexural strength, as well as toughness. The combination of aluminum, carbon, and polypropylene fibers were studied and has been shown to improve the performance of composites in peak load capacity as compared to polypropylene-only composites [4]. Polyvinyl alcohol fibers and steel fibers were hybridized and the findings showed performance increase of cementitious composites in post-crack behavior [5].

It was also observed that the various length of steel fibers impacted the shrinkage strains of steel, polypropylene, and polyvinyl alcohol hybrid fibers in concrete more profoundly [6]. The combination of steel fiber with polypropylene synthetic fibers were observed to be effective in further enhancing the performance of steel fiber reinforced concrete and ultimate load [7]. The hybridization of steel, Carbon Mesophase Pitch-based (CMP), Carbon Isotropic Pitch-based (CIP), and polypropylene fibers were studied and the findings showed that CIP HyFRC (Hybrid Fiber Reinforced Concrete) produced higher performance than CMP hybrid counterpart [8]. The various forms of polypropylene fibers were examined in concrete and the findings discovered were that fibrillated polypropylene fibers were more effective compared to its monofilament counterpart in reducing plastic shrinkage cracks [9]. Additionally, it has been studied that the use of synthetic polyvinyl alcohol and recycled polyethylene terephthalate (PET) fibers significantly improved the mechanical properties of Strain-Hardening Cementitious Composite (SHCC) even when 50% of the polyvinyl alcohol fibers were replaced by the recycled PET fibers [10]

The combination between different types of synthetics were studied from a polypropylene-polyethylene fiber hybrid and the test results modified the failure behavior of concrete by effectively distributing the loads [11]. The various forms of steel fibers and polypropylene fiber hybrids were investigated and the combination showed that the anchorage and length of fibers do not provide significant improvement from each other but materials do, as shown in the effectiveness of using steel fibers [12]. It was discovered that the shear capabilities of concrete were improved with the addition of polypropylene and polyethylene fibers and changed its mode of failure without stirrups from brittle to ductile [13]. The structural application of steel and polypropylene fibers were tested and it was observed that polypropylene fibers produced better damage mitigation performance due to its higher deformation capacity compared to steel fibers [14]. The use of specially selected polyethylene fibers improved the strain hardening of cementitious composites with the tensile capacity averaging up to 8% [15].

In addition, further tests were conducted using polypropylene fibers and the findings showed that fibrillated polypropylene fibers minimized more damage and permits serviceability to the structure tested [16]. The use of recycled PET synthetic fibers were experimentally tested and the results highlighted significant improvement in compression and flexural enhancement for the case of low-strength concretes [17]. Furthermore, the bridging effect of HyFRC steel and polypropylene fibers enhanced the bond stress of rebar [18]. The use of polypropylene, polyethylene, and polyvinyl alcohol synthetic fibers have been shown to improve the mechanical properties of cementitious composites and the key fiber parameters have been identified [19]. The investigation on the tensile behavior of high-strength strain-hardening cement-based composites (HS-SHCC) using high density polypropylene, aramid, poly($p$-phenylene-2,6-benzobisoxazole) (PBO), and high density-PBO fibers were conducted and the results exhibited promising improvements [20].

It can be deduced that the use of fibrous concrete improved the overall mechanical properties of concrete which then correlates to the increased lifecycle of a structure. However, the rheological properties of the HyFRC in the reviews were seldom disclosed and fibers were known to deteriorate fresh concrete workability. Hence a parametric study using HRWRA (High Range Water-Reducing Admixture) was conducted in this research to observe and assess the effect of the admixture on the behavior of the HyFRC so that a standardized mixing guideline can be established for future applications on the proposed materials.

Conjointly, conducting numerical analyses for unconventional materials proved challenging as most commercial FE (Finite Element) software material libraries are constrained to conventional concrete. Most of the studies that have been conducted in the reviews performed small-scale experimental testing in compression, tension, and flexure while large-scale tests were narrowed to using steel fibers hybrid combinations as it has higher chances of obtaining favorable results, thus lowering the cost of experimental testing. This poses a problem as combination of other types of fibers in large-scale tests were limited to steel fiber hybrids due to the possible risk of unfavorable or mediocre results. Hence this paper attempts to develop non-steel HyFRC synthetic fiber combinations and formulate constitutive modelling of these materials for FE modelling and numerical analyses. This would lower costs of experimental testing while broadening the opportunity of these materials to be numerically tested in various structural applications, unconstrained by overhead costs.

## 2. Proposed Synthetic Hybrid Fiber Reinforced Concrete

A study is conducted in this paper to develop different types of synthetic fiber combinations in cementitious composites and reduce the reliance of steel fibers as primary fibers in any hybridization mixes. This is because synthetic fibers are more economical than steel fiber and provide similar performance to a certain degree [21]. Additionally, the non-corrosive nature of synthetics proved advantageous as corroding steel fibers deteriorate the performance of concrete. It was also proven that the production of synthetics reduce the carbon footprint more than steel [22]. The workability problems of fresh concrete associated with fibrous concrete would be addressed in the admixture parametric study.

The proposed Hybrid Fiber Reinforced Concrete is made up of concrete and synthetic fibers from FORTA Corporation (Grove City, PA, USA), namely the Ferro macrofiber as the primary fiber and the Ultra-Net, Super-Net, Econo-Net, and Nylo-Mono microfibers as the secondary fiber in the combination mix-design. All of the fibers have varying mechanical properties, bonding power, manufactured form, materials, and fiber volume fraction used as shown in Table 1. A total of five hybrid mix-designs were proposed with one plain concrete as the controlled specimen. The volume fraction of fibers used and designations for the proposed materials are shown in Table 2. The range of the fiber volume fractions were defined based on a previously conducted study on fiber hybridization [23]. Normal-mix concrete was used to design the concrete as follows; cement = 409 kg/m$^3$; water = 225 kg/m$^3$; sand = 836 kg/m$^3$; 10 mm coarse aggregates = 302 kg/m$^3$; 20 mm coarse aggregates = 604 kg/m$^3$. Portland cement Type II was used with the following properties, specific gravity = 3.15; chemical composition = tricalcium silicate (3CaO·SiO$_2$), dicalcium silicate (2CaO·SiO$_2$), tetra-calcium aluminoferrite (4CaO·Al$_2$O$_3$Fe$_2$O$_3$); particle size distribution = 1.2 μm ($D_{5\%}$), 18 μm ($D_{50\%}$), 67 μm ($D_{95\%}$). A pan type concrete mixer was used during concrete mixing with a rotational speed of 15 rpm. Subsequently, 24 cylindrical and dog-bone specimens were casted and demolded after 24 h. All of the specimens were cured for 28-days in a water tank prior to the uniaxial testing in compression and tension.

**Table 1.** Specifications for FORTA fibers. (U: Ultra-Net; S: Super-Net, E: Econo-Net; N: Nylo-Mono; FF: Ferro).

| Type | Length (mm) | Form | Bonding Power | Class | Material | Tensile Strength (MPa) |
|------|------|------|------|------|------|------|
| U | 54 | Fibrillated Twisted bundle | Extra heavy-duty | Micro | Polypropylene and additives | 570–660 |
| S | 38 | Fibrillated | Heavy-duty | | | 570–660 |
| E | 38 | Fibrillated | Medium-duty | | | 570–660 |
| N | 19 | Monofilament | Light-duty | | Virgin nylon | 966 |
| FF1 | 38 | Fibrillated Twisted bundle | Heavy-duty | Macro | Polyethylene, polypropylene, and additives | 1100 |
| FF2 | 54 | | Heavy-duty | | | 570–660 |

**Table 2.** Selected Hybrid Fiber Reinforced Concrete (HyFRC) for constitutive modeling. (C: Control; U: Ultra-Net; S: Super-Net; E: Econo-Net; F: Ferro).

| Specimens | Designation | Type of Fibers (Vol. of Fraction, %) | | | | | | Total Vol. Fraction, (%) |
|---|---|---|---|---|---|---|---|---|
| | | MacroFibers | | MicroFibers | | | | |
| | | FF1 | FF2 | UN | SN | EN | NM | |
| 1 | Control | - | - | - | - | - | - | - |
| 2 | FFC | 0.6 | 0.6 | - | - | - | - | 1.20 |
| 3 | F6U3 | 0.6 | 0.6 | 0.30 | - | - | - | 1.50 |
| 4 | F6S3 | 0.6 | 0.6 | - | 0.30 | - | - | 1.50 |
| 5 | F6E3 | 0.6 | 0.6 | - | - | 0.30 | - | 1.50 |
| 6 | F6N3 | 0.6 | 0.6 | - | - | - | 0.30 | 1.50 |

## 3. Experimental Program

The experimental program is divided into three phases. The first phase is the admixture parametric study, with the aim to investigate the effect of using a HRWRA on the slump workability, compressive, and tensile strength of the proposed HyFRC. After standardized mix-guidelines were established from this study, cylindrical and dog-bone specimens were casted from the standard mix guide to be tested under uniaxial compressive and tensile tests for constitutive modeling.

### 3.1. Admixture Parametric Study

A study on the effect of HRWRA on the selected five HyFRC designs were conducted using ADVA Cast 512 polymer-based HRWRA from GCP Applied Technologies (Cambridge, MA, USA). The HRWRA complies with the Standards Specification of Chemical Admixtures for Concrete [24] and is categorized under Type F, with the purpose of water-reducing, high-range, and accelerating admixtures. The Guide for the Use of High-Range Water-Reducing Admixtures (HRWRAs) in Concrete [25] was followed during the tests. Trial mixes were conducted to obtain a HRWRA optimum dosage for each HyFRC based on the workability test of fresh concrete as well as compressive and tensile tests of hardened concrete. Table 3 lists all the HRWRA dosages used for the trial mixes. The dosage rate was determined from the mass of cement.

**Table 3.** Admixture tests on developed HyFRC (X: Tested).

| Specimen | Designation | Admixture Dosage (%) | | | | | | | | |
|---|---|---|---|---|---|---|---|---|---|---|
| | | 0.20 | 0.30 | 0.40 | 0.50 | 0.60 | 0.70 | 0.80 | 0.90 | 1.00 |
| 1 | C (Plain) | | | | | | | | | |
| 2 | FFC | X | | X | X | X | | | | |
| 3 | F6U3 | | | | | X | X | X | | X |
| 4 | F6S3 | | | | | X | X | X | | |
| 5 | F6E3 | | | | X | X | X | | | |
| 6 | F6N3 | | | | | X | X | X | | |

The workability tests were carried out according to the Standard Test Method for Slump of Hydraulic-Cement Concrete [26]. Wet HyFRC was poured into the mold at one third of its volume for three times, with each layer stroked 25 times. The mold was then removed, followed by the measurement of the vertical difference between the top of the mold and the displaced wet concrete.

The compressive strength of the HyFRC was evaluated in adherence to the Complementary British Standard to BE EN 206: Specification for constituent materials and concrete [27]. A total of thirty-six

100 mm cubes were casted, water-cured, and tested at 28-days using a Universal Testing Machine (UTM). The UTM has a maximum capacity of 5000 kN and the loading rate was defined at 6 kN/s. The equation used to determine the compressive strength is shown below:

$$\text{Compressive Strength (MPa)} = \frac{P}{A} \tag{1}$$

whereby P denotes the maximum load obtained from the UTM (N) while A is the cross-sectional area of the cube (mm$^2$).

The tensile strength of the HyFRC was determined using the Standard Test Method for Splitting Tensile Strength of Cylindrical Concrete Specimens [28]. A total of thirty-six 100 × 200 mm cylinders were cured for 28-days and split-tested using a 5000 kN Universal Testing Machine (UTM) at a configured loading rate of 1.57 kN/s. The splitting tensile strength is calculated as follows:

$$T = \frac{2P}{\pi ld}. \tag{2}$$

whereby T is the splitting tensile strength (MPa) and P is the maximum load from the UTM (N). l is the length and d is the diameter of cylinder (mm).

### 3.2. Uniaxial Compression Test

The uniaxial compressive behavior of the HyFRC was tested using a modified set up from the Standard Test Method for Static Modulus of Elasticity and Poisson's Ratio of Concrete in Compression [29]. The improvised arrangement was taken from a previous study [30] and shown in Figure 1. A total of twelve 150 × 300 mm cylinders were prepared and tested for compression using a 2000 kN UTM machine, with a loading rate of 0.02 mm/s. Two 50 mm Linear Variable Displacement Transducers (LVDT) were clamped on a circular jig around the cylinder to measure the compressive load in the elastic stage while two other 100 mm LVDTs were placed parallel to the cross-head movement of the test machine to record the load during the plastic stage.

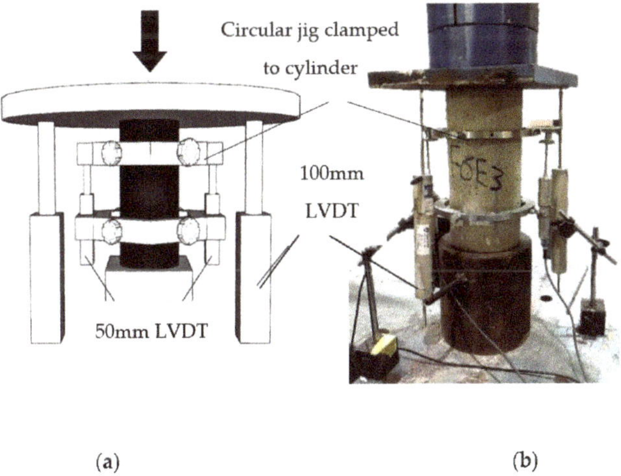


**Figure 1.** The uniaxial compression test: (**a**) conceptual, (**b**) actual.

The Concrete Damaged Plasticity (CDP) models were determined using the relationship between the damage parameters and the compressive strength of respective HyFRC via the plastic hardening strain $\varepsilon_c^{pl,h}$. The equations governing the compressive characteristics are shown as follows:

$$\sigma_c = (1 - d_c)E_0 \left(\varepsilon_c - \varepsilon_c^{pl,h}\right) \tag{3}$$

$$\begin{cases} \varepsilon_c^{in,h} = \varepsilon_c - \frac{\sigma_c}{E_0} \\ \varepsilon_c^{pl,h} = \varepsilon_c - \frac{\sigma_c}{E_0}\left(\frac{1}{1-d_c}\right) \end{cases} \tag{4}$$

$$\varepsilon_c^{pl,h} = \varepsilon_c^{in,h} - \frac{d_c}{1 - d_c}\frac{\sigma_c}{E_0} \tag{5}$$

In addition, this study uses the parabolic constitutive model developed by Kent and Park [31] for unconfined concrete, which is generally expressed by the equation:

$$\sigma_c = \sigma_{cu}\left[2\left(\frac{\varepsilon_c}{\acute{\varepsilon}_c}\right) - \left(\frac{\varepsilon_c}{\acute{\varepsilon}_c}\right)^2\right] \tag{6}$$

whereby nominal compressive stress and strain are shown as $\sigma_c$ and $\varepsilon_c$ while the ultimate compressive strength and strain are represented as $\sigma_{cu}$ and $\acute{\varepsilon}_c$. The value of $\acute{\varepsilon}_c$ was defined at 0.002 in this study as has been previously reported by Park [32].

The illustration of the model is shown in Figure 2a. The curve exhibited from point A to B shows the hardening stage of concrete while the linear behavior from point B to C displays the strain-softening state for both confined and unconfined concrete. The softening phase continued until 20% of the unconfined cylinder compressive strength at point C—achieving a plastic behavior from point C to D onwards. For simplicity, the constitutive model was considered to be a parabolic curve. Equation (6) assumed a nonlinear behavior such that the constitutive model came into effect when the compressive strength was 60% of the total strength. The elastic modulus was defined up to 40% of the concrete's strength in the elastic phase. Subsequently, the inelastic strain hardening in compression was derived as shown below:

$$\varepsilon_c^{in,h} = \varepsilon_c - \frac{\sigma_c}{E_0} \tag{7}$$

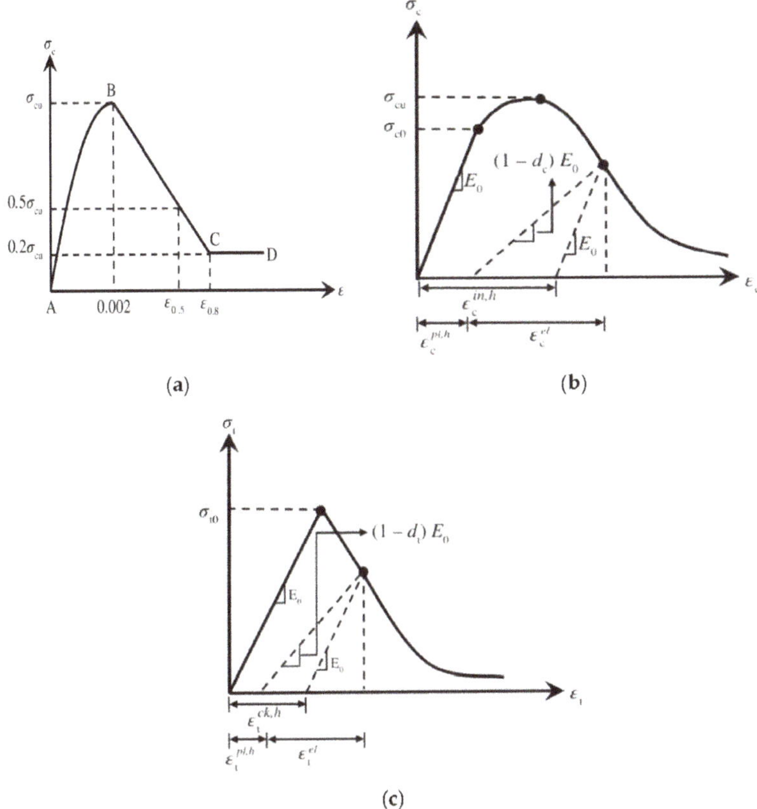

**Figure 2.** The (**a**) model for confined and unconfined concrete [31] as well as the (**b**) compressive and (**c**) tensile behavior of concrete in uniaxial conditions [33].

The inelastic strain hardening in compression $\varepsilon_c^{in,h}$ controls the unloading curve of the concrete in compression, which is an effective parameter in defining damage in compression $d_c$. Given that the $\varepsilon_c^{in,h}$ has a direct correlation with $d_c$, the following equation can be expressed:

$$d_c = 1 - \frac{\sigma_c}{\sigma_{cu}} \tag{8}$$

From Figure 2b, it can be observed that the tangent of the curve declined with respect to the modulus of elasticity $E_0$. This is as a result of the damage from an inclining plasting strain of brittle materials. The damage parameters $d_c$ was 0 at the maximum point and decreases up to 0.8—which was 20% of the remaining strength in large strains.

### 3.3. Uniaxial Direct Tensile Test

The uniaxial tensile tests were conducted using the proposed set up by a previous study [30] using a 250 kN UTM with added support jigs welded to the plates for the dog-bone specimens, as shown in Figure 3. The loading rate was configured at 0.07 mm/s and a total of 12 dog-bone specimens were prepared with a length of 500 mm and a 120 × 80 mm base which changes to a prismatic shape of 80 × 80 mm. A 10 mm-deep notch was sawed across the middle-section of the prism to define the cracking plane and two 25 mm LVDTs were clamped on the specimen, near the notch-area, to record the tensile deflections. The calculations to extract the required tensile data were referred from the Recommendations for Design and Construction of High Performance Fiber Reinforced Cement Composites with Multiple Fine Cracks (HPFRCC) [34].

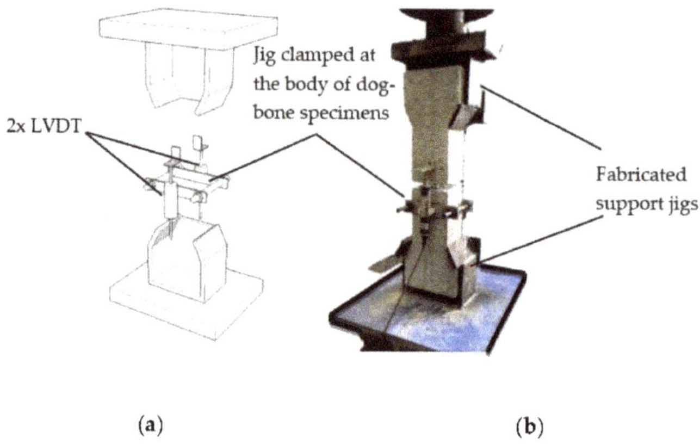


**Figure 3.** The uniaxial tensile test: (**a**) conceptual, (**b**) actual.

There are numerous constitutive models developed for concrete in the tension, however, most research deemed it insignificant due to the minimal differences between each model. Concrete in general is brittle in tension and is mostly supported with steel reinforcements. Hence in numerical modelling, the tensile behavior is simplified with more focus directed on the interaction between concrete and the embedded steel reinforcements. Hence, for simplicity, the plasticity hardening strain in tension $\varepsilon_t^{pl,h}$ was derived from the equations below:

$$\sigma_t = (1 - d_t)E_0 \left( \varepsilon_t - \varepsilon_t^{pl,h} \right) \tag{9}$$

$$\begin{cases} \varepsilon_c^{ck,h} = \varepsilon_t - \frac{\sigma_t}{E_0} \\ \varepsilon_c^{pl,h} = \varepsilon_t - \frac{\sigma_t}{E_0}\left(\frac{1}{1-d_t}\right) \end{cases} \tag{10}$$

$$\varepsilon_c^{pl,h} = \varepsilon_c^{ck,h} - \frac{d_t}{1-d_t}\frac{\sigma_t}{E_0} \tag{11}$$

The models used 7–10% of maximum compressive strength $\sigma_{cu}$ as tensile strength $\sigma_{t0}$ such that the maximum value can be represented as $\sigma_{t0} = 0.1\sigma_{cu}$. This study applies 1% of the tensile strength during analysis regardless of the actual condition to avoid instability during the numerical analyses. In addition, the corresponding strain value was taken as 10 times the percentage of the strain whereby stress was equal to ultimate tensile strength. From Figure 2c, a direct correlation can be established with the hardening cracking strain and the tensile damage—as the cracking strain increased, so does the tensile damage. This could be expressed as below:

$$d_t = 1 - \frac{\sigma_t}{\sigma_{t0}} \tag{12}$$

## 4. Results and Discussion

### 4.1. Admixture Rheological Impact

The effect of using ADVA Cast 512 polymer-based High Range Water-Reducing Admixture (HRWRA) on the developed HyFRC was studied to improve the workability as well as the corresponding compressive and tensile strengths. Each of the different type of fibers used in this study were selected mainly the FFC, F6U3, F6S3, F6E3, and F6N3 to test and observe the fiber combination efficacy.

As indicated in Table 4, the slump parametric study for the FFC was conducted at 0.2% dosage, however, the results showed poor workability up until the 0.4% dosage rate where the first slump was recorded. Subsequently, both the F6U3 and F6S3 were tested with the 0.4% baseline but recorded zero slump, even at a 0.6% dosage rate. Good workability behavior was only observed at 0.7–1.0% dosage rate—however concrete bleeding effect was observed at this maximum range. Therefore, a 0.8% threshold limit was imposed to avoid this effect, which may influence the corresponding compressive and tensile strengths of the HyFRC. In the case for the F6E3 and F6N3, both the hybrids only indicated slump values when more than 0.6% HRWRA dosage was applied.

**Table 4.** Average slump for developed HyFRC ($\sigma$ = standard deviation).

| HRWRA (%) | Average Slump (mm) | | | | | |
|---|---|---|---|---|---|---|
| | FFC | F6U3 | F6S3 | F6E3 | F6N3 | C |
| 0.2 | 0 | - | - | - | - | |
| 0.3 | - | - | - | - | - | |
| 0.4 | 132.5 ($\sigma$: 8.22) | 0 ($\sigma$: 0) | - | - | - | 90 * ($\sigma$: 5.48) |
| 0.5 | 90 ($\sigma$: 6.23) | - | - | 0 ($\sigma$: 0) | - | |
| 0.6 | 128 ($\sigma$: 1.10) | 0 ($\sigma$: 0) | 0 ($\sigma$: 0) | 130 ($\sigma$: 10.45) | 57.50 ($\sigma$: 10.98) | |
| 0.7 | - | 85 ($\sigma$: 10.85) | 106.5 ($\sigma$: 10.07) | 119 ($\sigma$: 6.57) | 50 ($\sigma$: 10.84) | |
| 0.8 | - | 130 ($\sigma$: 10.91) | 115 ($\sigma$: 4.38) | - | 102.5 ($\sigma$: 8.22) | |
| 0.9 | - | - | - | - | - | |
| 1.0 | - | 110 ($\sigma$: 10.93) | - | - | - | |

* HRWRAs were not added to plain concrete.

The average slump results observed in Figure 4a showed that the FFC and F6E3 design require 0.6% admixture dosage or less to obtain slump while the F6U3, F6S3, and F6N3 require more than 0.6% to achieve a workable state. For the purpose of simplicity, the admixture dosage rates in this study are classified into three levels—low, moderate, and high. Low levels require less than 0.6% dosage (<0.6%), moderate at 0.6% (=0.6%), and high, which demands HRWRA dosage more than 0.6% (>0.6%). These classifications are further discussed as below:

1. Low HRWRA application—the FFC can be classified into this tier because of the minimum 0.4% dosage to achieve a workable state and obtain a slump value. It is the only fiber combination without the use of microfibers and consists of only macro-sized blend of polypropylene and polyethylene in a fibrillated twisted form.

2. Moderate HRWRA application—this level consists of the F6E3 and F6N3 hybrids, whereby the microfibers are composed of fibrillated polypropylene and monofilament nylon. Both recorded slump values at a minimum 0.6% dosage rate and exhibit a reduction in slump when the HRWRA dosage was increased to 0.7%. The addition of microfibers results in an increase of available surface area of fibers that needs to be coated by mortar. Insufficient amount of mortar in fresh concrete that is available to bind the aggregates may cause the developed wet HyFRC to lose its workability, which explains why more HRWRA dosages were needed for these hybrids to obtain a slump value compared to the FFC. The F6N3 had a lower slump value than the F6E3 because of the nature of nylon microfibers which is hydrophilic, nylon absorbs free water in fresh concrete which in turn reduces the workability and demands a higher dosage in HRWRA.

3. High HRWRA application—the highest tier comprises the F6U3 and F6S3 hybrids—both have polypropylene microfibers achieving a workable state at a 0.7% HRWRA dosage rate, among the highest in this parametric study. The major differences between the hybrids in this category

from the moderate dosage hybrids are the microfiber specifications, which have more robust form for fiber anchorage inside concrete, a higher interfacial fiber-concrete bonding power, and a longer fiber-length. These parameters affect the workability of fresh concrete—the microfibers clump wet concrete firmly together due to the dominant fiber characteristics and result in poor workability performance.

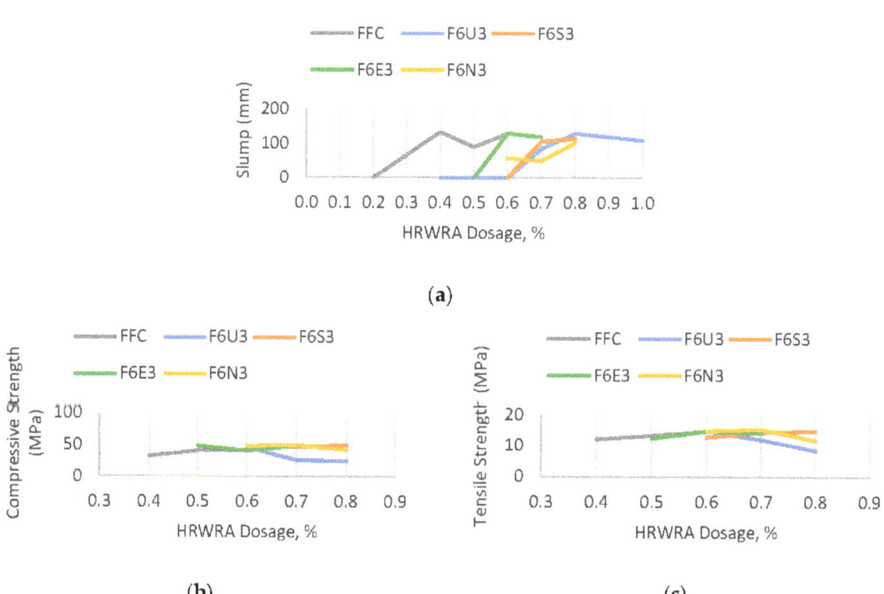


**Figure 4.** The (**a**) slump comparison and the (**b**) compressive and (**c**) tensile strength at 28-days.

The corresponding 28-day compressive strength of the developed HyFRC were tested and recorded as shown in Table 5. To obtain a good workability behavior is paramount in this study without losing significant loss in compressive strength. However, it was observed that the compressive strength for all the HyFRC deteriorated from the control plain concrete when a High Range Water Reducing Admixture (HRWRA) was introduced. The reduction in compressive strength was significant for the FFC at 0.4% dosage rate by as much as 40.92%, but the disparity was steadily diminishing as indicated in the 0.6% dosage rate where the strength-loss was halved to 23.19%.

Similar behavior was observed for the F6S3 hybrid, the decrease in compressive performance for this type of fiber combination was 13.77% at the starting dosage rate, but as more HRWRA were used, the cutback in performance was reduced to only 8.40%. Contrarily, the compressive strength of the F6U3 did not improve as dosage was increased—a 15.35% strength-loss was recorded at the starting dosage which further continued to a 53.91% reduction from plain concrete.

**Table 5.** Average 28-day compressive strength ($\sigma$ = standard deviation).

| HRWRA (%) | Average Compressive Strength (MPa) | | | | | |
|---|---|---|---|---|---|---|
| | FFC | F6U3 | F6S3 | F6E3 | F6N3 | C |
| 0.2 | 43.97 ($\sigma$: 2.18) | - | - | - | - | |
| 0.3 | - | - | - | - | - | |
| 0.4 | 31.87 ($\sigma$: 1.42) | - | - | - | - | 53.94 * ($\sigma$: 0.23) |
| 0.5 | 40.67 ($\sigma$: 1.44) | - | - | 48.52 ($\sigma$: 0.57) | - | |
| 0.6 | 41.43 ($\sigma$: 3.37) | 45.69 ($\sigma$: 2.38) | 46.51 ($\sigma$: 2.21) | 40.44 ($\sigma$: 0.18) | 48.21 ($\sigma$: 1.44) | |
| 0.7 | - | 25.93 ($\sigma$: 5.63) | 47.01 ($\sigma$: 1.20) | 49.15 ($\sigma$: 3.07) | 49.66 ($\sigma$: 0.90) | |
| 0.8 | - | 24.86 ($\sigma$: 0.38) | 49.41 ($\sigma$: 1.99) | - | 41.89 ($\sigma$: 2.00) | |
| 0.9 | - | - | - | - | - | |
| 1.0 | - | 26.47 ($\sigma$: 0.99) | - | - | - | |

* Compressive strength without HRWRAs.

The compressive behavior of the F6E3 and F6N3 hybrid showed a comparatively irregular pattern compared to the FFC, F6S3, and F6U3. The compressive strength for the F6E3 decreased by 10.05% at the initial dosage and continued to deteriorate by 25.03% as HRWRA dosage were increased. However, the steep strength-loss was dampened when the maximum dosage of 0.7% was used, only exhibiting an 8.89% decrease from control concrete. The opposite was observed for the F6N3, the compressive strength declined by 10.62% at the starting dosage but the loss was improved to 7.93% at 0.6% dosage rate. The hybrid recorded an abrupt fall in compressive strength from plain concrete by 22.34% when the maximum dosage rate of 0.8% was applied. The comparative results in compressive strength and the trend pattern can be observed in Figure 4b.

The reduction in compressive strength can be attributed to the aggregate segregation and possible concrete bleeding during concrete mixing, which resulted in a high amount of entrapped air inside concrete [35]. Consequently, the HyFRC would have a relatively lower unit weight—affecting the compressive strength directly when high amounts of fiber volumes were used.

The tensile strength of the HyFRC was tested at 28-days and the outcome showed reasonable improvement with the addition of the HRWRA as recorded in Table 6. The FFC design indicates a marginal 7.57% decrease in tensile strength at the starting dosage of 0.4% but regained the loss in strengths as the dosage was increased. At the maximum dosage of 0.6% the tensile strength was enhanced by 12.09% relative to plain concrete. A similar pattern was observed for the F6S3 fiber combination with a 1.76% decline in tensile strength at the initial dosage of 0.6% but gradually improved the strength by as much as 13% at the maximum dosage tier of 0.8%.

**Table 6.** Average 28-days tensile strength ($\sigma$ = standard deviation).

| HRWRA (%) | Average Tensile Strength (MPa) | | | | | |
|---|---|---|---|---|---|---|
| | **FFC** | **F6U3** | **F6S3** | **F6E3** | **F6N3** | **C** |
| 0.2 | 15.29 ($\sigma$: 0.69) | - | - | - | - | |
| 0.3 | - | - | - | - | - | |
| 0.4 | 12.08 ($\sigma$: 0.56) | - | - | - | - | 13.07 * ($\sigma$: 0.27) |
| 0.5 | 13.23 ($\sigma$: 2.26) | - | - | 12.39 ($\sigma$: 0.31) | - | |
| 0.6 | 14.65 ($\sigma$: 0.20) | 14.74 ($\sigma$: 1.33) | 12.84 ($\sigma$: 0.44) | 14.73 ($\sigma$: 1.18) | 14.76 ($\sigma$: 1.39) | |
| 0.7 | - | 11.98 ($\sigma$: 0.55) | 14.34 ($\sigma$: 0.84) | 14.13 ($\sigma$: 2.70) | 15.33 ($\sigma$: 0.60) | |
| 0.8 | - | 8.41 ($\sigma$: 0.69) | 14.77 ($\sigma$: 0.19) | - | 11.71 ($\sigma$: 0.62) | |
| 0.9 | - | - | - | - | - | |
| 1.0 | - | 10.66 ($\sigma$: 0.04) | - | - | - | |

* Tensile strength without HRWRAs.

It can be deduced that the FFC and F6U3 behave similarly in their response towards the inclusion of HRWRA—reducing the tensile strength when used in low dosage and improving it in high dosage. In contrast, the tensile behavior of the F6U3 hybrid is opposite to that of the FFC and F6S3 fiber-combinations. At the initial starting dosage of 0.6% the tensile strength was improved by 12.78%, however, the performance deteriorated with every increase in HRWRA dosage which eventually led to a 35.65% decline in performance at the maximum dosage limit of 0.8%.

For the case of the F6E3 and F6N3 hybrids, the tensile strength pattern is inconsistent as can be observed from Figure 4c. The F6E3 was initially weakened by 5.20% at the starting dosage but with further increase in HRWRA, the tensile strength was gradually improved by 12.70%. In addition, the F6N3 with HRWRA enhanced the tensile capability by 12.93% at 0.6% dosage rate. However, the results declined as the dosage was increased—the final dosage rate of 0.8% indicated a 10.41% decrease of tensile strength from the control specimen.

The enhancement in tensile strength was due to the presence of fibers inside cement matrix. These microfibers are useful in controlling micro-level cracks and prevent the nucleation of cracks which often propagates into single, larger macro-cracks [36]. A multi-crack hardening phenomenon increases the ultimate tensile strength of concrete as can be observed from the results of the developed HyFRC.

*4.2. Constitutive Modeling*

The FFC is a combination of the 54 and 38 mm Ferro macro synthetic fiber. The developed HyFRC produced a 40.0 MPa compressive strength at 28-days with a corresponding peak strain of 1992 $\mu\varepsilon$. As indicated in Figure 5a, a sixth order polynomial curve with the equation $y = 2E + 13x^6 - 1E + 12x^5 + 3E + 10x^4 - 3E + 08x^3 + 253{,}821x^2 + 10{,}110x$ was adopted to obtain an optimal trend line from the experimental compressive stress–strain curve. The correlation factor, $R^2$ for the polynomial trend line curve is in the range of 99.60%

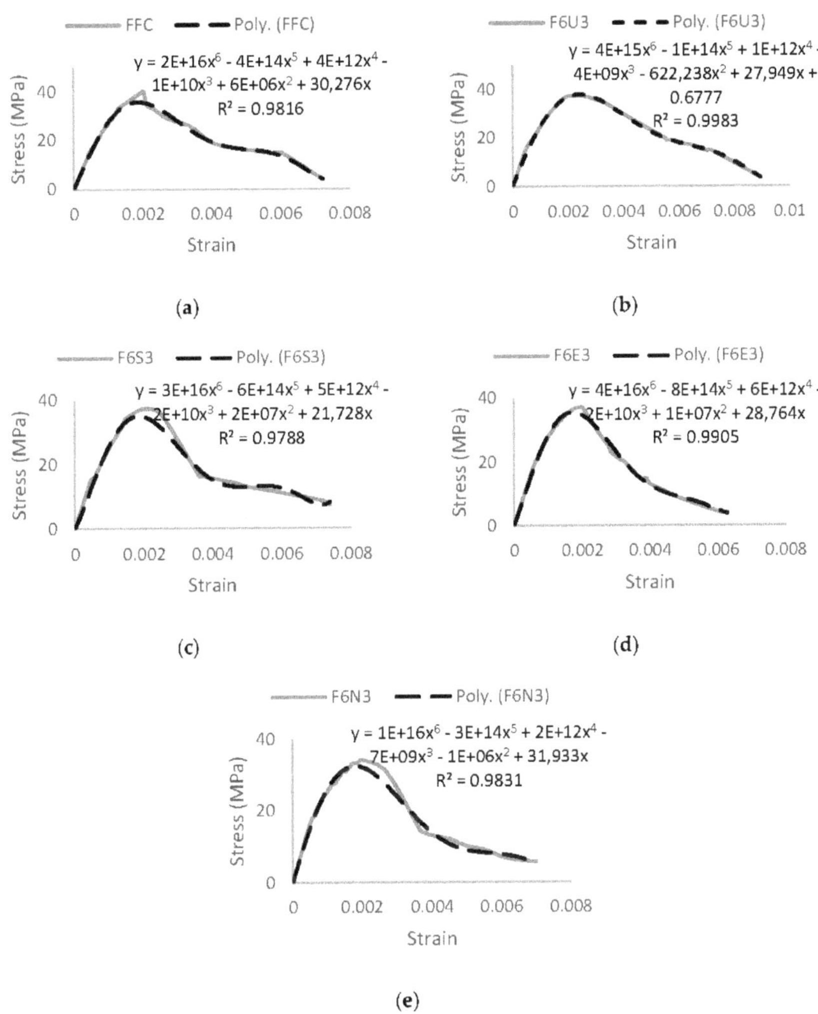

**Figure 5.** Uniaxial compression stress–strain curve: (**a**) FFC, (**b**) F6U3, (**c**) F6S3, (**d**) F6E3, (**e**) F6N3.

The tensile strength of the HyFRC at 28-days was 1.96 MPa with a corresponding peak strain of 59 $\mu\varepsilon$. In the elastic stage, a linear trend line with the equation $y = 34{,}057x$ and a correlation factor, $R^2$ of 99.23% was used to model the pre post-crack behavior of the HyFRC. In the plastic stage, a sixth order polynomial curve with the equation $y = -1E + 11x^6 + 1E + 10x^5 - 4E + 08x^4 + 7E + 06x^3 - 66{,}421x^2 + 212.77x + 0.375$ was adopted to obtain the best polynomial curve in modelling the strain-softening mode of failure. The correlation factor, $R^2$ for this polynomial trend line curve is in the range of 99.24% as shown in Figure 6.

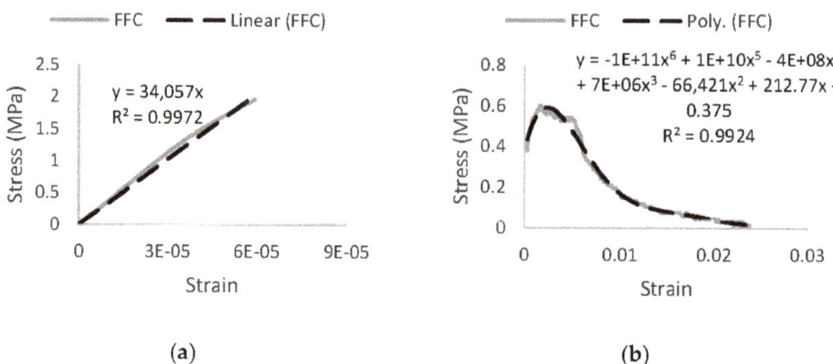

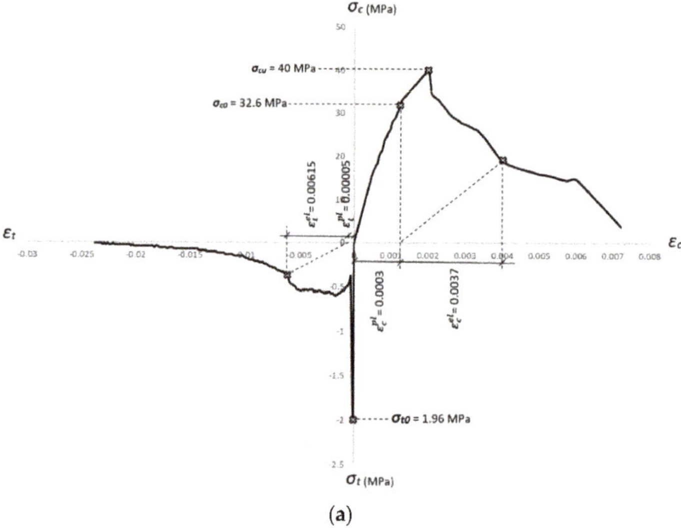



(a)

**Figure 7.** *Cont.*

**Figure 6.** FFC uniaxial tension stress–strain curve: (**a**) elastic stage and (**b**) plastic stage.

The compressive and tensile stress–strain curves of the developed FFC HyFRC were combined to form a constitutive model as shown in Figure 7a. In addition, the Concrete Damaged Plasticity (CDP) data collected from this model were used for verification in numerical analysis as presented in Table 7.

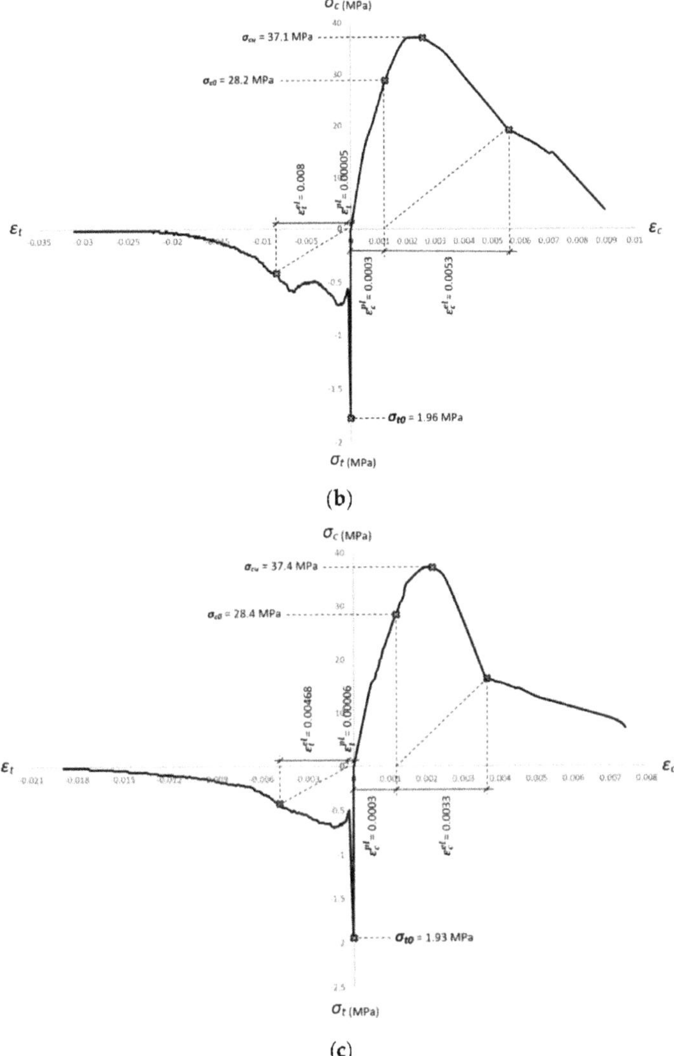

**(b)**

**(c)**

**Figure 7.** *Cont.*

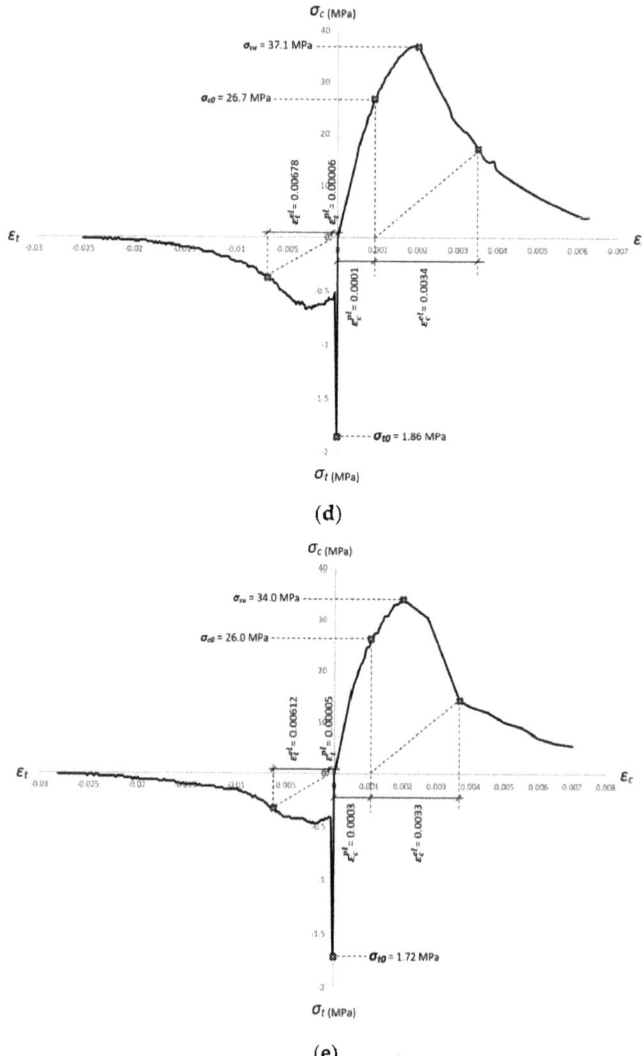



**(d)**



**(e)**

**Figure 7.** Constitutive models for the developed HyFRC in compression and tension: (**a**) FFC, (**b**) F6U3, (**c**) F6S3, (**d**) F6E3, (**e**) F6N3.

**Table 7.** Material properties of FFC HyFRC.

| Material Parameters | FFC | Plasticity Parameters | |
|---|---|---|---|
| Concrete Elasticity | | Dilation Angle | 31 |
| E (GPa) | 33 | Eccentricity | 0.1 |
| | | fb0/fc0 | 1.16 |
| N | 0.2 | K | 0.67 |
| | | Viscosity Parameter | 0 |
| Compressive Behavior | | Compression Damage | |
| Yield Stress (MPa) | Inelastic Strain | Damage Parameter C | Inelastic Strain |
| 15.2 | 0 | 0 | 0 |
| 20.7 | 0.0000698947 | 0 | 0.0000698947 |
| 26.3 | 0.0001791 | 0 | 0.0001791 |
| 30.7 | 0.000267441 | 0 | 0.000267441 |
| 32.6 | 0.000275877 | 0 | 0.000275877 |
| 33.1 | 0.000289388 | 0 | 0.000289388 |
| 40.0 | 0.000781074 | 0 | 0.000781074 |
| 38.0 | 0.000899545 | 0.05 | 0.000899545 |
| 35.4 | 0.001006598 | 0.11 | 0.001006598 |
| 34.7 | 0.001038464 | 0.13 | 0.001038464 |
| 33.8 | 0.001155271 | 0.16 | 0.001155271 |
| 30.0 | 0.001664856 | 0.25 | 0.001664856 |
| 25.3 | 0.002653893 | 0.37 | 0.002653893 |
| 20.1 | 0.003271853 | 0.5 | 0.003271853 |
| 15.2 | 0.004990053 | 0.62 | 0.004990053 |
| 3.7 | 0.007110412 | 0.91 | 0.007110412 |
| Tensile Behavior | | Tension Damage | |
| Yield Stress (MPa) | Cracking Strain | Damage Parameter T | Cracking Strain |
| 1.96 | 0 | 0.00 | 0 |
| 0.38 | 0.000289032 | 0.81 | 0.000289032 |
| 0.60 | 0.001634374 | 0.69 | 0.001634374 |
| 0.59 | 0.001803815 | 0.70 | 0.001803815 |
| 0.57 | 0.00188883 | 0.71 | 0.00188883 |
| 0.54 | 0.004903946 | 0.72 | 0.004903946 |
| 0.36 | 0.006195847 | 0.82 | 0.006195847 |
| 0.12 | 0.01254142 | 0.94 | 0.01254142 |
| 0.03 | 0.021962284 | 0.98 | 0.021962284 |

The Ultra-Net and Ferro fibers were combined to form the F6U4 hybrids. The Ultra-Net is a polypropylene fiber in a fibrillated twisted-bundle fashion while the Ferro is a fibrillated twisted bundle made of polypropylene and polyethylene fibers. The hybrid resulted in a 37.1 MPa compressive strength at 28-days with a corresponding peak strain of 2520 $\mu\varepsilon$. From Figure 5b, a sixth order polynomial curve was selected with an optimal trend line equation of $y = 4E + 15x^6 - 1E + 14x^5 + 1E + 12x^4 - 4E + 09x^3 - 622{,}238x^2 + 27{,}949x + 0.6777$. The correlation factor, $R^2$ for the trend line curve is in the range of 99.83%.

The peak stress of the hybrid in tension is 1.76 MPa at 28-days with a corresponding peak strain of 53 µε. The tensile behavior is divided into elastic and plastic stages to obtain more accurate trend line curves as depicted in Figure 8. A linear trend line with the equation y = 34,123x and a fifth order polynomial curve with the equation y = − 2E + 08x$^5$ + 6E + 06x$^4$ + 55,521x$^3$ − 3677.8x$^2$ − 5.8403x + 0.6496 was adopted for the elastic and plastic stage, respectively. The correlation factor, R$^2$ for both stages are in the range of 99.39% and 97.87%.

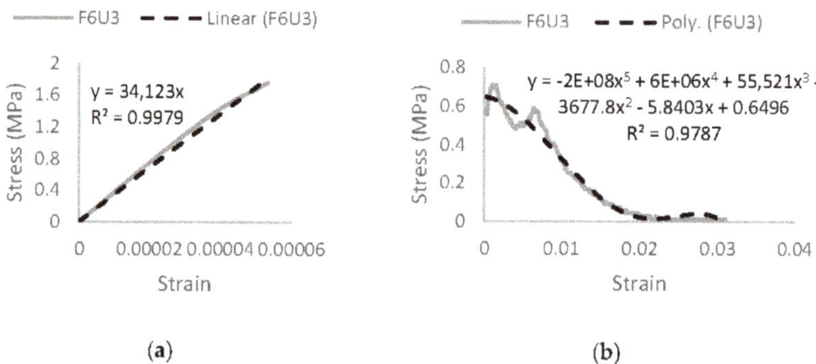


**Figure 8.** F6U3 uniaxial tension stress–strain curve: (a) elastic stage and (b) plastic stage.

Figure 7b shows the combined compressive and tensile constitutive data for the F6U3 hybrids. From these data, the F6U3 CDP material properties were derived, to be used in creating FE analytical models as tabulated in Table 8.

The Super-Net is a 38 mm polypropylene fiber in a fibrillated form. It is hybridized with the Ferro macrofibers to produce the F6S3 hybrid in an attempt to improve the mechanical properties of conventional concrete. The 28-day peak compressive strength produced is 37.4 MPa with a corresponding peak strain of 2000 µε. The equation y = 3E + 16x$^6$ − 6E + 14x$^5$ + 5E + 12x$^4$ − 2E + 10x$^3$ + 2E + 07x$^2$ + 21,728x with a correlation factor, R$^2$ of 97.88% was obtained from the sixth order polynomial curves as shown in Figure 5c.

In tension, the HyFRC results in 1.93 MPa tensile strength at 28-days with a corresponding peak strain of 59 µε. The tensile elastic stage behavior results in a trend line equation of y = 34,636x with a correlation factor, R$^2$ of 98.55% while the plastic stage behavior yields a y = − 7E + 11x$^6$ + 5E + 10x$^5$ − 1E + 09x$^4$ + 2E + 07x$^3$ − 106,957x$^2$ + 223.26x + 0.5141 equation with a correlation factor, R$^2$ in the range of 99.57%. A sixth order polynomial curve in the plastic stage curve was adopted as shown in Figure 9. The obtained compressive and tensile stress–strain curves for the F6S3 hybrid were combined to form a constitutive model as shown in Figure 7c. As tabulated in Table 9, the CDP data were derived from this constitutive curve to create F6S3 FE analytical models.

Table 8. Material properties of F6U3 HyFRC.

| Material Parameters | F6U3 | Plasticity Parameters | |
|---|---|---|---|
| Concrete Elasticity | | Dilation Angle | 31 |
| E (GPa) | 33 | Eccentricity | 0.1 |
| | | fb0/fc0 | 1.16 |
| N | 0.2 | K | 0.67 |
| | | Viscosity Parameter | 0 |
| Compressive Behavior | | Compression Damage | |
| Yield Stress (MPa) | Inelastic Strain | Damage Parameter C | Inelastic Strain |
| 15.2 | 0 | 0 | 0 |
| 22.5 | 0.000182435 | 0 | 0.000182435 |
| 27.8 | 0.000310719 | 0 | 0.000310719 |
| 33.7 | 0.000578202 | 0 | 0.000578202 |
| 34.8 | 0.000640472 | 0 | 0.000640472 |
| 35.7 | 0.000708915 | 0 | 0.000708915 |
| 36.9 | 0.000832921 | 0 | 0.000832921 |
| 37.1 | 0.001395297 | 0 | 0.001395297 |
| 35.9 | 0.001852203 | 0.03 | 0.001852203 |
| 34.2 | 0.002311373 | 0.08 | 0.002311373 |
| 28.2 | 0.003360657 | 0.24 | 0.003360657 |
| 18.0 | 0.005457189 | 0.52 | 0.005457189 |
| 16.9 | 0.005881804 | 0.54 | 0.005881804 |
| 14.9 | 0.006485556 | 0.60 | 0.006485556 |
| 3.7 | 0.008833224 | 0.90 | 0.008833224 |
| Tensile Behavior | | Tension Damage | |
| Yield Stress (MPa) | Cracking Strain | Damage Parameter T | Cracking Strain |
| 1.76 | 0 | 0.00 | 0 |
| 0.56 | 0.000347401 | 0.68 | 0.000347401 |
| 0.59 | 0.000425675 | 0.67 | 0.000425675 |
| 0.71 | 0.001530855 | 0.60 | 0.001530855 |
| 0.50 | 0.004055813 | 0.72 | 0.004055813 |
| 0.59 | 0.006460715 | 0.67 | 0.006460715 |
| 0.42 | 0.008319045 | 0.76 | 0.008319045 |
| 0.17 | 0.013007509 | 0.91 | 0.013007509 |
| 0.03 | 0.021398914 | 0.98 | 0.021398914 |

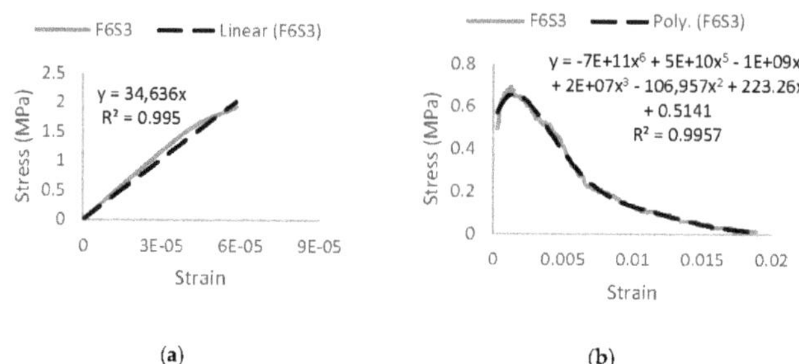



**Figure 9.** F6S3 uniaxial tension stress–strain curve: (**a**) elastic stage and (**b**) plastic stage.

**Table 9.** Material properties of F6S3 HyFRC.

| Material Parameters | F6S3 | Plasticity Parameters | |
|---|---|---|---|
| **Concrete Elasticity** | | Dilation Angle | 31 |
| E (GPa) | 33 | Eccentricity | 0.1 |
| | | fb0/fc0 | 1.16 |
| N | 0.2 | K | 0.67 |
| | | Viscosity Parameter | 0 |
| **Compressive Behavior** | | **Compression Damage** | |
| **Yield Stress (MPa)** | **Inelastic Strain** | **Damage Parameter C** | **Inelastic Strain** |
| 15.2 | 0 | 0 | 0 |
| 21.0 | 0.000139493 | 0 | 0.000139493 |
| 28.9 | 0.000317673 | 0 | 0.000317673 |
| 30.7 | 0.000347490 | 0 | 0.000347490 |
| 32.2 | 0.000401502 | 0 | 0.000401502 |
| 33.6 | 0.000411813 | 0 | 0.000411813 |
| 34.4 | 0.000427585 | 0 | 0.000427585 |
| 37.4 | 0.000868000 | 0 | 0.000868000 |
| 35.3 | 0.001442195 | 0.05 | 0.001442195 |
| 16.3 | 0.003112060 | 0.56 | 0.003112060 |
| 16.0 | 0.003346712 | 0.57 | 0.003346712 |
| 15.8 | 0.003372239 | 0.58 | 0.003372239 |
| 15.3 | 0.003784918 | 0.59 | 0.003784918 |
| 14.6 | 0.004080677 | 0.61 | 0.004080677 |
| 13.0 | 0.007189107 | 0.81 | 0.007189107 |
| **Tensile Behavior** | | **Tension Damage** | |
| **Yield Stress (MPa)** | **Cracking Strain** | **Damage Parameter T** | **Cracking Strain** |
| 1.93 | 0 | 0 | 0 |
| 0.50 | 0.000285633 | 0.74 | 0.000285633 |
| 0.54 | 0.000381286 | 0.72 | 0.000381286 |
| 0.60 | 0.000457034 | 0.69 | 0.000457034 |
| 0.69 | 0.001249656 | 0.64 | 0.001249656 |
| 0.54 | 0.003262186 | 0.72 | 0.003262186 |
| 0.38 | 0.005216852 | 0.80 | 0.005216852 |
| 0.17 | 0.008967449 | 0.91 | 0.008967449 |
| 0.03 | 0.017129274 | 0.98 | 0.017129274 |

The F6E3 hybrid is a blend between the Econo-Net microfiber and the Ferro macro synthetic fibers. The Econo-Net is classified as a medium-duty fibrillated polypropylene microfiber at 38 mm length. The combination with the Ferro macro synthetic fiber yields a 28-day compressive strength of 37.1 MPa with a corresponding peak strain of 2002 $\mu\varepsilon$. From Figure 5d, a sixth order polynomial curve with the equation $y = 4E + 16x^6 - 8E + 14x^5 + 6E + 12x^4 - 2E + 10x^3 + 1E + 07x^2 + 28{,}764x$ was adopted from the trend line of stress–strain experimental curves. The trend line curve has a correlation factor, $R^2$ in the range of 99.05%.

The 28-days tensile strength of the F6E3 hybrid was at 1.86 MPa with a corresponding peak strain of 56 $\mu\varepsilon$. The behavior in the elastic and plastic stage was divided to obtain a more accurate trend line. From Figure 10, it can be observed that a linear trend line with the equation $y = 35{,}614x$ and a correlation factor, $R^2$ of 96.89% was adopted to describe HyFRC behavior in the pre-cracking stage while the equation $y = -1E + 11x^6 + 1E + 10x^5 - 4E + 08x^4 + 8E + 06x^3 - 68{,}070x^2 + 205.71x + 0.4531$ with a correlation factor, $R^2$ of 99.75% was adopted to describe the post-cracking behavior, from a sixth order polynomial curve's trend line.

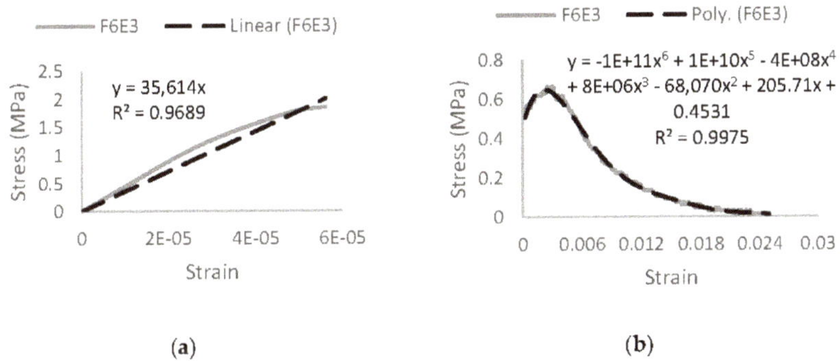


**Figure 10.** F6E3 uniaxial tension stress–strain curve: (**a**) elastic stage and (**b**) plastic stage.

As with the previous HyFRC designs, the stress–strain combined data obtained from uniaxial compressive and tensile tests are shown in Figure 7d. The data derived from the stress–strain curve determined the CDP material properties, which was essential in describing the behavior of the HyFRC in the proposed FE analytical models. The properties are tabulated below in Table 10.

The final design is the F6N3 combination—it is a mixture of the Nylo-Mono microfiber with the Ferro macro synthetic fibers. This microfiber is also the only nylon fiber used in this research; level 1 FORTA fibers are classified as a light-duty monofilament fiber that has a length of 19 mm. The combination with the Ferro produced a HyFRC with a compressive strength of 34.0 MPa at 28-days with a corresponding peak strain of 2008 $\mu\varepsilon$. As indicated in Figure 5e, the trend line equation obtained from the sixth order polynomial curve is $y = 1E + 16x^6 - 3E + 14x^5 + 2E + 12x^4 - 7E + 09x^3 - 1E + 06x^2 + 31{,}933x$ and has a correlation factor, $R^2$ of 98.31%.

The F6N3 design results in a concrete hybrid that has a 1.72 MPa tensile strength at 28-days with a corresponding peak strain of 52 $\mu\varepsilon$. The elastic behavior of the HyFRC yielded a linear trend line curve of $y = 34{,}066x$ with a correlation factor, $R^2$ of 98.94%. Additionally, a trend line curve of equation $y = -5E + 10x^6 + 5E + 09x^5 - 2E + 08x^4 + 4E + 06x^3 - 36{,}599x^2 + 109.69x + 0.3614$ was adopted from a sixth order polynomial curve to describe the best behavior in stress–strain of the HyFRC. The correlation factor, $R^2$ for the curve was is the range of 99.53% as shown in Figure 11.

The compressive and tensile stress–strain data of the F6N3 HyFRC design mix were combined to form a constitutive model as illustrated in Figure 7e. The data were then derived to obtain CDP material properties to be used in creating analytical FE models. The CDP material properties are tabulated in Table 11.

**Table 10.** Material properties of F6E3 HyFRC.

| Material Parameters | F6E3 | Plasticity Parameters | |
|---|---|---|---|
| Concrete Elasticity | | Dilation Angle | 31 |
| E (GPa) | 33 | Eccentricity | 0.1 |
| | | fb0/fc0 | 1.16 |
| N | 0.2 | K | 0.67 |
| | | Viscosity Parameter | 0 |
| Compressive Behavior | | Compression Damage | |
| Yield Stress (MPa) | Inelastic Strain | Damage Parameter C | Inelastic Strain |
| 15.2 | 0 | 0 | 0 |
| 20.6 | 0.000026910 | 0 | 0.000026910 |
| 30.8 | 0.000236497 | 0 | 0.000236497 |
| 35.0 | 0.000472281 | 0 | 0.000472281 |
| 36.9 | 0.000752355 | 0 | 0.000752355 |
| 37.0 | 0.000827543 | 0 | 0.000827543 |
| 37.1 | 0.000878977 | 0 | 0.000878977 |
| 29.7 | 0.001546411 | 0.20 | 0.001546411 |
| 25.0 | 0.002008789 | 0.32 | 0.002008789 |
| 23.3 | 0.002129371 | 0.37 | 0.002129371 |
| 21.8 | 0.002322170 | 0.41 | 0.002322170 |
| 20.1 | 0.002626371 | 0.46 | 0.002626371 |
| 17.4 | 0.002957642 | 0.53 | 0.002957642 |
| 15.1 | 0.003233765 | 0.59 | 0.003233765 |
| 14.6 | 0.003359927 | 0.61 | 0.003359927 |
| Tensile Behavior | | Tension Damage | |
| Yield Stress (MPa) | Cracking Strain | Damage Parameter T | Cracking Strain |
| 1.86 | 0 | 0 | 0 |
| 0.51 | 0.000284728 | 0.72 | 0.000284728 |
| 0.56 | 0.000383481 | 0.70 | 0.000383481 |
| 0.66 | 0.003113782 | 0.64 | 0.003113782 |
| 0.63 | 0.003191708 | 0.66 | 0.003191708 |
| 0.62 | 0.003269129 | 0.67 | 0.003269129 |
| 0.36 | 0.006826657 | 0.81 | 0.006826657 |
| 0.18 | 0.010705229 | 0.90 | 0.010705229 |
| 0.03 | 0.018748574 | 0.98 | 0.018748574 |

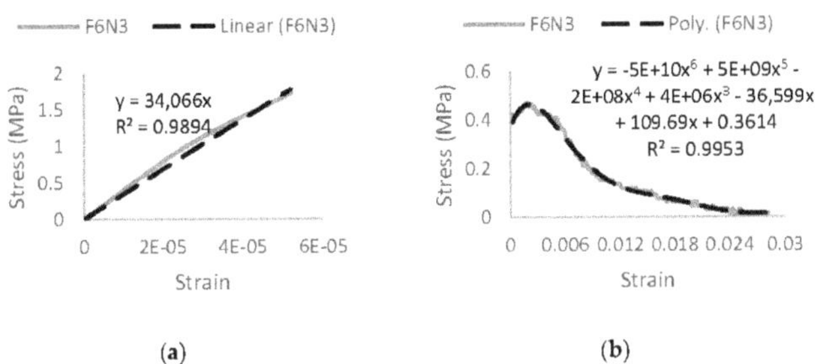



**Figure 11.** F6N3 uniaxial tension stress–strain curve: (**a**) elastic stage and (**b**) plastic stage.

**Table 11.** Material properties of F6N3 HyFRC.

| Material Parameters | F6N3 | Plasticity Parameters | |
|---|---|---|---|
| **Concrete Elasticity** | | Dilation Angle | 31 |
| E (GPa) | 33 | Eccentricity | 0.1 |
| | | fb0/fc0 | 1.16 |
| N | 0.2 | K | 0.67 |
| | | Viscosity Parameter | 0 |
| **Compressive Behavior** | | **Compression Damage** | |
| **Yield Stress (MPa)** | **Inelastic Strain** | **Damage Parameter C** | **Inelastic Strain** |
| 15.2 | 0 | 0 | 0 |
| 18.8 | 0.000051892 | 0 | 0.000051892 |
| 20.2 | 0.000064848 | 0 | 0.000064848 |
| 26.0 | 0.000255586 | 0 | 0.000255586 |
| 30.8 | 0.000616365 | 0 | 0.000616365 |
| 31.5 | 0.000630811 | 0 | 0.000630811 |
| 32.5 | 0.000699442 | 0 | 0.000699442 |
| 33.0 | 0.000717266 | 0 | 0.000717266 |
| 33.2 | 0.000848395 | 0 | 0.000848395 |
| 33.5 | 0.000873826 | 0 | 0.000873826 |
| 34.0 | 0.000976716 | 0 | 0.000976716 |
| 30.5 | 0.001801382 | 0.10 | 0.001801382 |
| 14.8 | 0.003185480 | 0.56 | 0.003185480 |
| 14.2 | 0.003237492 | 0.58 | 0.003237492 |
| 5.6 | 0.006830173 | 0.84 | 0.006830173 |
| **Tensile Behavior** | | **Tension Damage** | |
| **Yield Stress (MPa)** | **Cracking Strain** | **Damage Parameter T** | **Cracking Strain** |
| 1.72 | 0 | 0 | 0 |
| 0.42 | 0.000288805 | 0.75 | 0.000288805 |
| 0.41 | 0.000497226 | 0.76 | 0.000497226 |
| 0.47 | 0.002253059 | 0.73 | 0.002253059 |
| 0.41 | 0.005114826 | 0.76 | 0.005114826 |
| 0.32 | 0.006157604 | 0.82 | 0.006157604 |
| 0.20 | 0.008740424 | 0.89 | 0.008740424 |
| 0.11 | 0.015076801 | 0.94 | 0.015076801 |
| 0.03 | 0.023856674 | 0.98 | 0.023856674 |

## 5. Assessment of Results

### 5.1. Evaluation of HRWRA Effect

Each of the hybrids displayed different behavior in workability and also in its corresponding strength in both compression and tension. The main objective is to obtain ample workability behavior measured through the slump value without significant loss in compressive strength, while also procuring the best enhancement in tensile strength. These three criteria are superimposed against one another in a column and line graph to identify the effective applicational dosage.

The FFC mix-design exhibited an incline trend in both compressive and tensile strength with an increase in HRWRA dosage. As can be observed in Figure 12a, the best dosage to be applied for this HyFRC is 0.6%—the percentage difference with the highest slump in the 0.4% tier is only at a minimal 3.45%, providing the best workability with the highest corresponding compressive and tensile behavior.

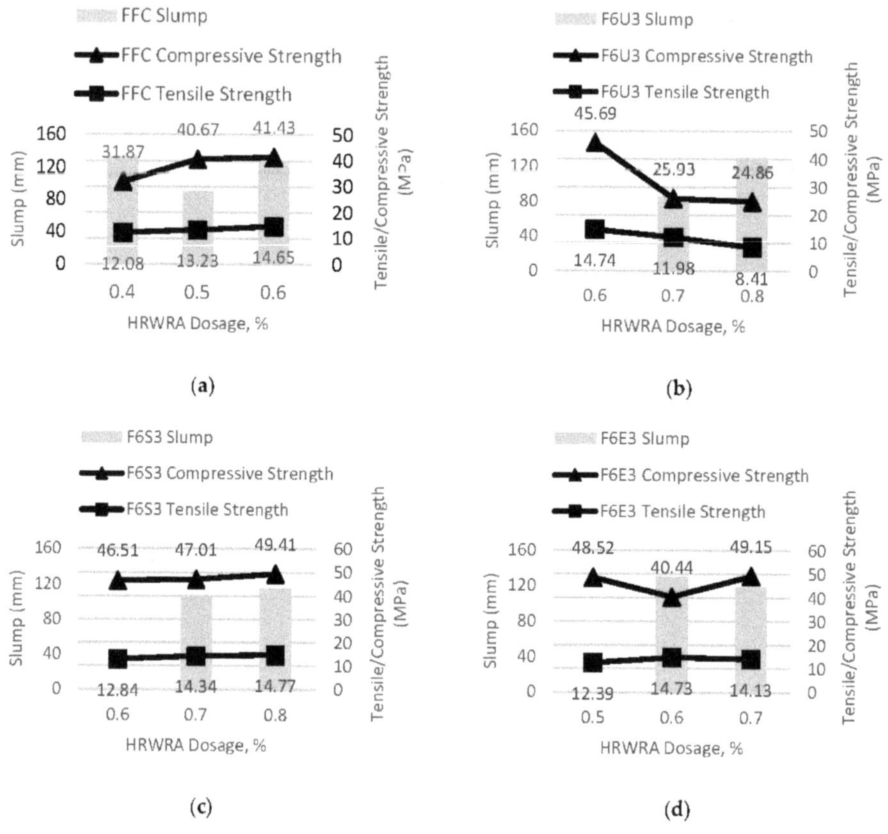


**Figure 12.** *Cont.*

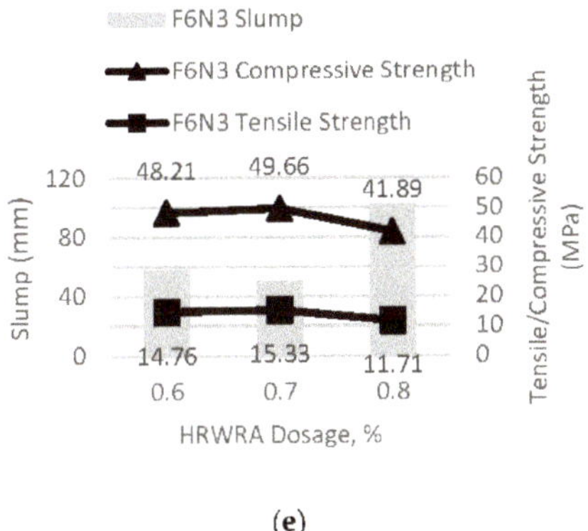



(**e**)

**Figure 12.** HyFRC dosage requirement (bar chart: slump, line graph: compressive/tensile strength): (**a**) FFC, (**b**) F6U3, (**c**) F6S3, (**d**) F6E3, (**e**) F6N3.

The F6U3 hybrids showed a decline in strength with the increase in HRWRA dosage as shown in Figure 12b. The most significant deterioration was from the 0.6–0.7% dosage rate by 43.25% for the compressive strength and 18.72% for the tensile strength which continued to decrease at the 0.8% dosage tier. In order to procure adequate workability without substantial loss in strength, the 0.7% HRWRA dosage was chosen.

In Figure 12c, the F6S3 fiber combination displayed an increased workability behavior with every increment in HRWRA dosage. The compressive and tensile strength is directly proportional to the increase in slump, herewith making 0.8% dosage the most suitable HRWRA prescription for this HyFRC mix design.

The irregular pattern in compressive strength and slump workability for the F6E3 hybrids are shown in Figure 12d. The best dosage rate to be applied into this HyFRC mix-design is the 0.7% dosage, which exhibited the best compressive strength along with a reasonable loss in tensile strength and slump. The decrease in tensile strength and slump from the 0.6% tier was only at a minimal 4.07% and 8.46% compared to the 17.72% decline in compressive strength from the 0.7% dosage rate to the 0.6% tier.

The F6N3 fiber-combination exhibited a decreasing compressive and tensile strength with incremental dosage of HRWRA. The most significant decline was from 0.7% to 0.8% dosage rate by 15.65% for compressive strength and 23.61% for tensile strength as indicated in Figure 12e. Therefore, the 0.7% dosage was considered for this mix design due to the best retention-loss of compressive strength while achieving the highest tensile strength. Although the slump is less than the 0.6% tier, the difference was minimal.

*5.2. Uniaxial Behavior Comparative Analysis*

The stress–strain behavior of all the developed HyFRC was compiled in Table 12 for comparison. It is indicated in Figure 13a that at 2000 με, the FFC showed the best compressive behavior in the elastic stage with a 40 MPa peak compressive strength and a corresponding strain of 1992 με; increasing the performance of plain concrete by 6.67%. The F6U3, F6S3, and F6E3 performed almost similarly to the controlled plain concrete with a slight decrease in performance by 1.07%, 0.27%, and 1.07%, respectively. However, the F6N3 combination weakens conventional concrete in compression by 9.33% with a corresponding strain of 2008 με. The FFC has no microfibers and only consists of macro-sized fibers. This proved beneficial because the addition of microfibers, as shown in the other designs have made concrete more brittle, due to the fiber-bridging of micro-level cracks [37]. This concrete hardening scenario in the elastic stage prevented a pull-out mode of failure for the Ferro fiber—breaking it during the rapid crack propagation without having sufficient opportunity to bridge the widening macro-sized gaps progressively.

**Table 12.** Comparison between the developed HyFRC.

| Mix Design | C | FFC | F6U3 | F6S3 | F6E3 | F6N3 |
|---|---|---|---|---|---|---|
| Compressive strength, MPa | 37.5 | 40.0 | 37.1 | 37.4 | 37.1 | 34.0 |
| Corresponding strain, με | 2000 | 1992 | 2520 | 2000 | 2002 | 2008 |
| Tensile strength, MPa | 1.43 | 1.96 | 1.76 | 1.93 | 1.86 | 1.72 |
| Corresponding strain, με | 43 | 59 | 53 | 59 | 56 | 52 |
| Max. strain deflection in compression, με | 3627 | 7224 | 8944 | 7406 | 6279 | 6999 |
| Peak tensile strain-hardening, MPa | 0.08 | 0.60 | 0.71 | 0.69 | 0.66 | 0.47 |
| Corresponding strain, με | 302 | 1653 | 1552 | 1271 | 3134 | 2267 |

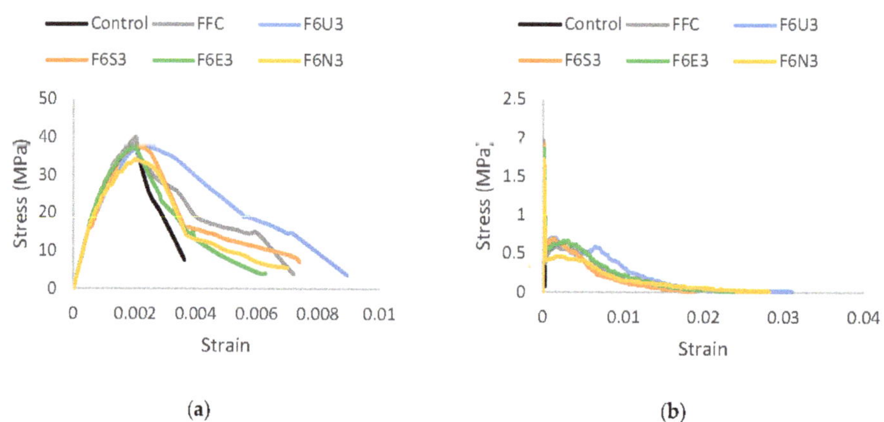


**Figure 13.** HyFRC tensile stress–strain curve comparison: (**a**) uniaxial compression, (**b**) uniaxial tension.

However, for the range greater than 2000 με, the F6U3 demonstrated the best behavior in the post-cracking region. Its maximum deflection reached to 8944 με, a 147% increase from plain concrete and a 26.51% increase from FFC. Although the micro-macro fiber relationship results in fiber-breakage in the elastic stage, the combination of the FORTA fibers proved efficient at a much higher deformation. Even though the Ultra-Net is a micro-class fiber, it has been shown in previous residual strength tests that it supplements the Ferro in fiber-bridging cracks even in the macro-level. This dampens the rate of crack propagation in concrete thus allowing sufficient opportunity for the Ferro fiber to achieve a gradual pull-out mode of failure. Henceforth, the added residual strength improved the compressive post-crack behavior—significantly better than just using macrofiber-only FFC design. The F6S3, F6E3, and F6N3 delivered a relatively poor performance in those regions but only at a minimal difference of 0.41%, 0.50%, and 0.8% from the FFC. Nevertheless, all of the developed HyFRC have improved the plastic behavior of plain concrete in compression substantially.

The compressive damage in the HyFRC cylinders is shown in Figure 14. It was observed that the control specimens absorbed the highest damage followed by the FFC. The control cylinder specimens were completely fractured and although the FFC-design was stronger in the elastic stage, it consumed a lot of damage in the post-cracking region, as can be seen with the numerous large crack-sizes on the specimens. The F6U3, F6S3, F6E3, and F6N3 specimens exhibited minimal differences in compressive damage with each other, but all displayed less cracks than the FFC and control specimens. This is because of the more effective crack-bridging effect of the macro-micro fiber combinations as shown in Figure 15.

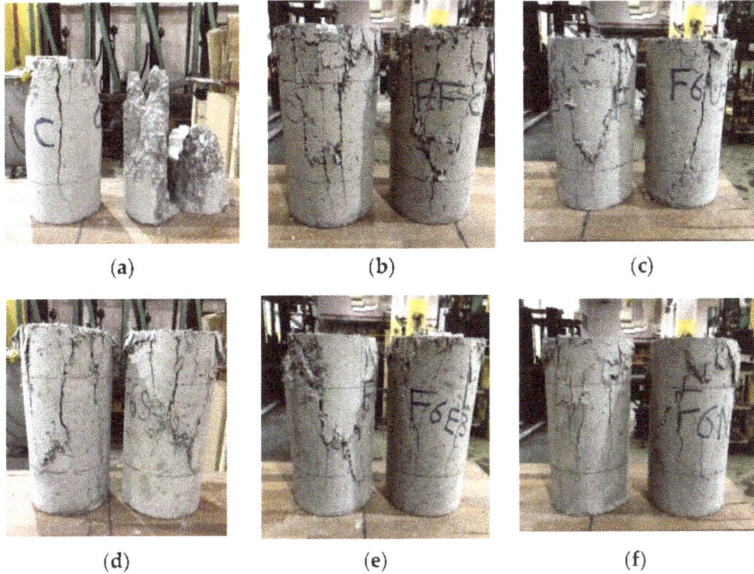

**Figure 14.** HyFRC damage in compression: (**a**) control, (**b**) FFC, (**c**) F6U3, (**d**) F6S3, (**e**) F6E3, (**f**) F6N3.

**Figure 15.** The fiber-bridging effect as shown in tension: (**a**) Uniaxial testing under tension for dog-bone specimens (**b**) Fiber-bridging effect during uniaxial tensile test.

In the case of tensile behavior, Figure 13b illustrates the different post-cracking performance for the developed HyFRC. It can be observed that plain concrete is brittle in nature, instantly failing to 0.08 MPa with a corresponding strain of 302 µε after achieving peak stress. However, the addition of fibers inside the cementitious composite changed the mode of failure from brittle to strain-softening with a surge of strain hardening in-between. At the range below 60 µε, the FFC displayed the best tensile behavior in the elastic stage with a 1.96 MPa strength and a corresponding strain of 59 µε. The advantage of lacking in microfibers for the FFC proved beneficial in tension as it was in compression—increasing the tensile strength of plain concrete by 37.06%. The subsequent best is the F6S3 with a 34.97% increase in performance followed by the F6E3, F6U3, and F6N3 with a 30.07%, 23.08%, and 20.28% increase, respectively.

However, the performance of the FFC deteriorates when the deformation is greater than 60 $\mu\varepsilon$ in the plastic region. The F6U3 produced the best post-cracking tensile behavior with a peak strain hardening at 0.71 MPa with a corresponding strain of 1552 $\mu\varepsilon$. This increased the control specimen plastic strength by an additional 0.63 MPa. The other developed HyFRC also provided additional residual strength to plain concrete by 0.52, 0.61, 0.58, and 0.39 MPa for the FFC, F6S3, F6E3, and F6N3 mix-design accordingly. In addition, the F6U3, F6S3, and F6E3 achieved a higher strain-hardening tensile strength than the FFC macrofiber-only design with a difference of 16.79%, 13.95%, and 9.52% in post-cracking performance.

The F6N3 is weaker than the FFC because the Nylo-Mono microfiber in the hybrid has the shortest fiber length in this study as well as having the least bonding power and anchorage capacity compared to the Ultra-Net, Super-Net, and Econo-Net microfibers. Although the Nylo-Mono has the highest tensile strength, nylons depend more on the fiber/cement interface rather than the elongation limit when in pull-out failure [38]. The monofilament form made the fiber inept to strongly anchor itself inside cementitious composites while the low-duty bonding power between the interfacial fiber surface did not provide adequate friction during pull-out to contribute in tensile strength. Furthermore, the 19 mm length was too short to fiber-bridge the widening crack gaps in the post-cracking region; and without the dampening of crack propagation in the elastic stage, the rapid crack localization would occur rapidly and transforms the mode of failure of the Ferro macrofiber from fiber pull-out to fiber-breakage, thus weakening the HyFRC in tension.

## 6. Concluding Remarks

In this study, the Hybrid Fiber Reinforced Concrete (HyFRC) materials were developed by hybridizing synthetic fibers from FORTA Corporation with the additional use of ADVA Cast 512 polymer-based High Range Water-Reducing Admixture (HRWRA) from GCP Applied Technologies during the concrete mixing process.

- The addition of HRWRA in the hybrid-mixes was studied, observed, and assessed to determine the best dosage requirement for improved workability behavior, compressive and tensile strengths. The optimal HRWRA dosage for the Ferro-Ferro hybrids are 0.6%, 0.7% for the Ferro-Ultra, Ferro-Econo, Ferro-Nylo hybrids, and 0.8% for the Ferro-Super hybrids.
- The developed HyFRC improved the compressive and tensile mechanical properties of cementitious composites reasonably. The Ferro-Ferro hybrids exhibited the best performance in the elastic stage in both compression and tension while in the plastics stage, the Ferro-Super hybrids displayed the best compressive strain-hardening while the Ferro-Super hybrids excelled the most in the tensile post-cracking stage.
- Constitutive models were developed for all five HyFRC materials for future works. Predictive works for structural application can be conducted from the constitutive laws while FE analyses can be accomplished by modeling RC structures using the Concrete Damaged Plasticity (CDP) data in the materials properties in this study.

**Author Contributions:** Conceptualization, S.M.I.S.Z. and F.H.; Data creation, S.M.I.S.Z. and F.H.; Formal analysis, S.M.I.S.Z.; Methodology, S.M.I.S.Z. and F.H.; Project administration, F.H.; Resources, F.H.; Software, S.M.I.S.Z. and F.H.; Supervision, F.H.; Writing—original draft, S.M.I.S.Z.; Writing—review and editing, F.H., F.N.A.A.A. and M.S.J. All authors have read and agreed to the published version of the manuscript.

**Funding:** This research was funded by Universiti Putra Malaysia (UPM) under the Putra Grant Research Project, No. 9531200.

**Acknowledgments:** The authors would like to thank Innofloor Sdn. Bhd. as the Malaysian FORTA Corporation supplier for providing the fibers to conduct this research. Their support is gratefully acknowledged.

**Conflicts of Interest:** The authors declare no conflict of interest.

## References

1.  ACI Committee 544. State-of-the-Art Report on Fiber Reinforced Concrete. *J. Proc.* **2002**, *70*, 729–744.
2.  Walton, P.L.; Majumdar, A.J. Cement Based Composites with Mixtures of Different Types of Fibres. *Composites* **1975**, *6*, 209–216. [CrossRef]
3.  Banthia, N.; Sheng, J. Micro-reinforced Cementitious Materials. *MRS Proc.* **1991**, *211*, 25–32. [CrossRef]
4.  Mobasher, B.; Li, C. Mechanical Properties of Hybrid Cement Based Composites. *ACI Mater. J.* **1996**, *93*, 284–292.
5.  Ramanalingam, N.; Paramasivam, P.; Mansur, M.A.; Maalej, M. Flexural Behaviour of Hybrid Fiber-Reinforced Cement Composites Containing High-Volume Fly Ash. *Symp. Pap.* **2001**, *199*, 147–162.
6.  Sun, W.; Chen, H.; Luo, X.; Qian, H. The effect of hybrid fibers and expansive agent on the shrinkage and permeability of high performance concrete. *Cem. Concr. Res.* **2001**, *31*, 595–601. [CrossRef]
7.  Banthia, N.; Nandakumar, N. Crack Growth Resistance of Hybrid Fiber Reinforced Cement Composites. *Cem. Concr. Compos.* **2003**, *25*, 3–9. [CrossRef]
8.  Banthia, N.; Soleimani, S.M. Flexural response of hybrid fiber-reinforced cementitious composites. *ACI Mater. J.* **2005**, *102*, 382–389.
9.  Banthia, N.; Gupta, R. Hybrid Fiber Reinforced Concrete: Fiber Synergy in High Strength Matrices. *RILEM Mater. Struct.* **2004**, *37*, 707–716. [CrossRef]
10. Yu, J.; Yao, J.; Lin, X.; Li, H.; Lam, J.Y.K.; Leung, C.K.Y.; Sham, I.M.L.; Shih, K. Tensile performance of sustainable Strain-Hardening Cementitious Composites with hybrid PVA and recycled PET fibers. *Cem. Concr. Res.* **2018**, *107*, 110–123. [CrossRef]
11. Roesler, J.R.; Altoubat, S.A.; Lange, D.A.; Rieder, K.A.; Ulreich, G.R. Effect of synthetic fibers on structural behavior of concrete slabs-on-ground. *ACI Mater. J.* **2006**, *103*, 3–10.
12. Greenough, T.; Nehdi, M. Shear Behaviour of Self-Consolidating Fibre-Reinforced Concrete Slender Beams. *ACI Mater. J.* **2008**, *105*, 468–477.
13. Altoubat, S.; Yazdanbakhsh, A.; Rieder, K.A. Shear behavior of macro-synthetic fiber-reinforced concrete beams without stirrups. *ACI Mater. J.* **2009**, *106*, 381–389.
14. Kawashima, K.; Zafra, R.G.; Sasaki, T.; Kajiwara, K.; Nakayama, M. Effect of Polypropylene Fiber Reinforced Cement Composite and Steel Fiber Reinforced Concrete for Enhancing the Seismic Performance of Bridge Columns. *J. Earthq. Eng.* **2011**, *15*, 1194–1211. [CrossRef]
15. Yu, K.; Wang, Y.; Yu, J.; Xu, S. A strain-hardening cementitious composites with the tensile capacity up to 8%. *Constr. Build. Mater.* **2017**, *137*, 410–419. [CrossRef]
16. Kawashima, K.; Zafra, R.G.; Sasaki, T.; Kajiwara, K.; Nakayama, M.; Unjoh, S.; Sakai, J.; Kosa, K.; Takahashi, Y.; Yabe, M. Seismic Performance of a Full-Size Polypropylene Fiber-Reinforced Cement Composite Bridge Column Based on E-Defense Shake Table Experiments. *J. Earthq. Eng.* **2012**, *16*, 463–495. [CrossRef]
17. Spadea, S.; Farina, I.; Berardi, V.P.; Dentale, F.; Fraternali, F. Energy dissipation capacity of concretes reinforced with recycled PET fibers. *Ing. Sismica* **2014**, *31*, 61–70.
18. Ganesan, N.; Indira, P.V.; Sabeena, M.V. Bond stress slip response of bars embedded in hybrid fibre reinforced high performance concrete. *Constr. Build. Mater.* **2014**, *50*, 108–115. [CrossRef]
19. Pakravan, H.R.; Ozbakkaloglu, T. Synthetic fibers for cementitious composites: A critical and in-depth review of recent advances. *Constr. Build. Mater.* **2019**, *207*, 491–518. [CrossRef]
20. Curosu, I.; Liebscher, M.; Mechtcherine, V.; Bellmann, C.; Michel, S. Tensile behavior of high-strength strain-hardening cement-based composites (HS-SHCC) made with high-performance polyethylene, aramid and PBO fibers. *Cem. Concr. Res.* **2017**, *98*, 71–81. [CrossRef]
21. Yin, S.; Tuladhar, R.; Collister, T.; Combe, M.; Sivakugan, N. Mechanical Properties and Post-crack Behaviours of Recycled PP Fibre Reinforced Concrete. In Proceedings of the Concrete 2015, 27th Biennial National Conference of the Concrete Institute of Australia in Conjunction with the 69th RILEM Week "Construction Innovations, Research into Practice, Melbourne, Australia, 30 August–2 September 2015.
22. Shen, L.; Worrell, E.; Patel, M.K. Open-loop recycling: A LCA case study of PET bottle-to-fibre recycling. *Resour. Conserv. Recycl.* **2010**, *55*, 34–52. [CrossRef]
23. Kheni, D.; Scott, R.H.; Deb, S.K.; Dutta, A. Ductility enhancement in beam-column connections using hybrid fiber-reinforced concrete. *ACI Struct. J.* **2015**, *112*, 167–178. [CrossRef]

24. ASTM C494/C494M-17. *Standard Specification for Chemical Admixtures for Concrete*; ASTM International: West Conshohocken, PA, USA, 2017.

25. ACI Committee 212. *Guide for the Use of High-Range Water-Reducing Admixtures (Superplasticizers) in Concrete*; ACI: Farmington Hills, MI, USA, 2017; p. 13.

26. ASTM C143/C143M-15a. *Standard Test Method for Slump of Hydraulic-Cement Concrete*; ASTM International: West Conshohocken, PA, USA, 2015.

27. BS EN 1992-2:2005. *Concrete. Complementary British Standard to BS EN 206. Specification for Constituent Materials and Concrete*; British Standards Institution: London, UK, 2019.

28. ASTM C496/C496M-11. *Standard Test Method for Splitting Tensile Strength of Cylindrical Concrete Specimens*; ASTM International: West Conshohocken, PA, USA, 2011.

29. ASTMC469/C469M-14. *Standard Test Method for Static Modulus of Elasticity and Poisson's Ratio of Concrete*; ASTM International: West Conshohocken, PA, USA, 2014.

30. Hassan, A.M.T.; Jones, S.W.; Mahmud, G.H. Experimental test methods to determine the uniaxial tensile and compressive behaviour of Ultra High Performance Fibre Reinforced Concrete (UHPFRC). *Constr. Build. Mater.* **2012**, *37*, 874–882. [CrossRef]

31. Kent, D.C.; Park, R. Flexural Members with Confined Concrete. *J. Struct. Eng. Div.* **1971**, *97*, 1969–1990.

32. Park, R. *Reinforced Concrete Structures*; John Wiley & Sons: New York, NY, USA, 1975.

33. Ananiev, S.; Ozbolt, J. A plastic-damage model for concrete. *Int. J. Solids Struct.* **1989**, *25*, 299–326.

34. JSCE Concrete Committee. *Recommendations for Design and Construction of High Performance Fiber Reinforced Cement Composites with Multiple Fine Cracks (HPFRCC)*; Japan Society of Civil Engineers, Concrete Committee: Tokyo, Japan, 2008.

35. Ramakrishnan, V.; Wu, G.Y.; Hosalli, G. Flexural Fatigue Strength, Endurance Limit, and Impact Strength of Fiber Reinforced Concretes. In *Proceedings of the International Symposium on Recent Developments in Concrete Fiber Composites*; Transportation Research Board: Washington, DC, USA, 1989; pp. 17–24.

36. Lawler, J.; Zampini, D.; Shah, S. Microfiber and Macrofiber Hybrid Fiber-Reinforced Concrete. *J. Mater. Civ. Eng.* **2005**, *17*, 595–604. [CrossRef]

37. Betterman, L.R.; Ouyang, C.; Shah, S.P. Fiber-matrix interaction in microfiber-reinforced mortar. *Adv. Cem. Based Mater.* **1995**, *2*, 53–61. [CrossRef]

38. Pakravan, H.R.; Jamshidi, M.; Latifi, M. Polymeric fibers pull out behavior and microstructure as cementitious composites reinforcement.pdf. *J. Text. Inst.* **2013**, *104*, 1056–1065. [CrossRef]

*Article*

# Capillary Water Absorption and Micro Pore Connectivity of Concrete with Fractal Analysis

**Xiangqun Ding [1], Xinyu Liang [1], Yichao Zhang [2,*], Yanfeng Fang [1], Jinghai Zhou [2] and Tianbei Kang [2]**

[1] School of Materials Science and Engineering, Shenyang Jianzhu University, Shenyang 110168, China; dingxiangqun@sjzu.edu.cn (X.D.); lunwenww@sina.com (X.L.); fangyf@sjzu.edu.cn (Y.F.)

[2] School of Civil Engineering, Shenyang Jianzhu University, Shenyang 110168, China; zhoujinghai@sjzu.edu.cn (J.Z.); kangtianbei@sjzu.edu.cn (T.K.)

* Correspondence: zhangyichao@sjzu.edu.cn

Received: 28 July 2020; Accepted: 28 September 2020; Published: 1 October 2020

**Abstract:** This study focuses on the relationship between the complexity of pore structure and capillary water absorption of concrete, as well as the connection behavior of concrete in specific directions. In this paper, the water absorption of concrete with different binders was tested during the curing process, and the pore structure of concrete was investigated by mercury intrusion porosimetry (MIP). The results show that the water absorption of concrete with mineral admixtures is lower, mainly due to the existence of reasonable pore structure. The effect of slag on concrete modification is more remarkable comparing with fly ash. In addition, the analysis shows that the pore with different diameters has different fractal characteristics. The connectivity probability and water absorption of unidirectional chaotic pore are linearly correlated with the pore diameter of 50–550 nm, and the correlation coefficient reaches a very significant level, and detailed analysis was undertaken to interpret these results based on fractal theory.

**Keywords:** concrete; pore structure; water absorption; MIP; fractal dimension; pore connectivity

## 1. Introduction

As one of the most commonly used building materials, concrete is inevitably subjected to various destructive factors, which stem from the production process or use environment of concrete. In addition, water is the most common substance accompanied by concrete, which always affects the durability of concrete [1,2]. Harmful ions dissolved in water (e.g., chloride ions, sulfate ions, or magnesium ions) are transported among the complex and disordered pores [3]. Both theoretical and experimental results show that the capillary suction of unsaturated concrete absorbs chloride ions in water and accelerates the corrosion of reinforced concrete under the condition of wetting and drying alternation.

Additionally, sulfate attack also depends on the moisture entering concrete during the process of water transport [4–6]. Water penetrating in and out actually exists in all the processes of deterioration resulting from the freeze-thaw cycle, and the damage extent of concrete shows an ultra-superposition effect under the alternation of wetting and drying [7]. Obviously, water absorption behavior plays an important role in durability of concrete.

The rate at which water is absorbed into concrete by capillary suction can provide useful information relating to the pore structure, the permeation characteristics, and the durability of concrete surface zone [8,9]. Meanwhile, the pore structure of concrete is of great significance to the durability of the material, and the complexity of pore structure has a significant impact on the permeability. The pore size of concrete varies from nanometer scales to micrometer scales, and the pore morphology shown by scanning electron microscopy (SEM) is rather complex and disordered [10].

Fractal theory, as a rising nonlinear science, may solve the problem of evaluating the complexity or roughness of self-similar or approximate self-similar objects effectively. The mathematical beauty of fractal is that it forms the infinite complexity with relatively simple equations and different information captured by test equipment and different definitions of fractal dimension (including the Hausdorff dimension, the box-counting dimension, etc.) [11,12]. Fractal theory has been applied to the study of concrete, and some valuable results have been obtained, especially in pore research, pore fractal dimension. It has a close relationship with the transport performance (chloride diffusion coefficient, depth of carbonation, etc.) as a parameter of the complex transport channel [13,14], which can be used as an index of pore structure damage [15,16]. If it is assumed that the pore can move by itself, the connecting behavior between the pore is a revolution from complexity to simplicity, so it can be inferred that the fractal dimension of the pore and the possibility of the connection correspond to each other. Microstructure parameters measured by MIP and X-ray CT images with different voxel sizes were compared [17].

Fractal theory can better characterize the connectivity of pores. At present, there are few studies on quantitative analysis of pore connectivity by fractal theory. In this study, concrete specimens with different pore structures were prepared by adding different mineral admixtures. In order to obtain the relationship between the complexity of pore structure and the absorption capillary water in specific directions, a series of tests were carried out on the properties of concrete specimens, and the experimental results were compared and discussed by processing data obtained from the test of water absorption and MIP. Finally, fractal theory is used to analyze the relationship of the pore connectivity and water absorption.

## 2. Materials and Mixture Proportions

The 32.5. R ordinary Portland cement was used which is produced by Shenyang Jidong Cement Company, China. The physical properties of cement are shown in Table 1. Additionally, the chemical compositions of fly ash and slag are shown in Table 2. There are three specimens in each group. If the deviation of the test results of three specimens in each group is within 5%, the test results are considered to be valid, and the average value of the test results of the three specimens is taken as the final test results. All experimental results in this paper are based on this method.

**Table 1.** Physical properties of cement.

| 80 um Sieve Reside/% | Setting Time/Min | | Soundness | Tensile Strength/MPa | | Compressive Strength/MPa | |
|---|---|---|---|---|---|---|---|
| | Initial | Final | | 3d | 28d | 3d | 28d |
| 1.4 | 160 | 220 | Qualified | 3.3 | 7.0 | 16.5 | 38.1 |

**Table 2.** Chemical composition of fly ash and slag (%).

| Chemical Composition | $SiO_2$ | $Al_2O_3$ | CaO | $MgO$ | $Fe_2O_3$ | $SO_3$ |
|---|---|---|---|---|---|---|
| Slag | 31.73 | 13.84 | 40.76 | 7.87 | 2.01 | 1.52 |
| Fly ash | 55.01 | 28.5 | 2.39 | 2.19 | 8.05 | – |

Low air-induced water-reducing agent was used which is produced by Shenyang Dongling concrete admixture company, China. It is a polycarboxylate superplasticizer with solid content of 20%. The mix proportions are shown in Table 3. The compressive strength test specimens were made of 100 mm × 100 mm × 100 mm. After curing for 24 h, the specimens were demolded and transported to the curing room for 7 and 28 days for testing. The workability and compressive strength of concrete are shown in Table 4.

**Table 3.** Mix proportions of concrete (kg/m$^3$).

| Binder Types | Water | Cement | Sand | Slag | Fly Ash | Coarse Aggregate | Water Reducing Agent |
|---|---|---|---|---|---|---|---|
| OPC | 215 | 420 | 660 | - | - | 1264 | 4.5 |
| OPC with slag | 180 | 340 | 798 | 100 | - | 1264 | 4.5 |
| OPC with fly ash | 180 | 380 | 798 | - | 70 | 1264 | 4.5 |
| OPC with slag and fly ash | 180 | 340 | 798 | 65 | 35 | 1264 | 4.5 |

**Table 4.** Workability and compressive strength of concrete.

| Binder Types | Slump/mm | Compressive Strength/MPa |
|---|---|---|
| OPC | 225 | 35.9 |
| OPC with slag | 200 | 41.5 |
| OPC with fly ash | 250 | 39.1 |
| OPC with compound admixture | 230 | 44.7 |

## 3. Water Absorption Test of Concrete

Duplicate specimens (40 mm × 40 mm × 160 mm) of mortar were used to determine the water absorption values of concrete at 7 and 28 days. The specimens were dried at 105 °C to constant, and the initial weight of all specimens was recorded. Then the surface of the specimens was coated with epoxy resin, except for an exposed surface. The cured specimens were immersed in water at 20 ± 2 °C, and the water head was maintained at 2.3 cm. Then the weight of the text specimens was recorded at a specified time interval, which can be set longer as the program continues. The schematic diagram of the water absorption test is shown in Figure 1.

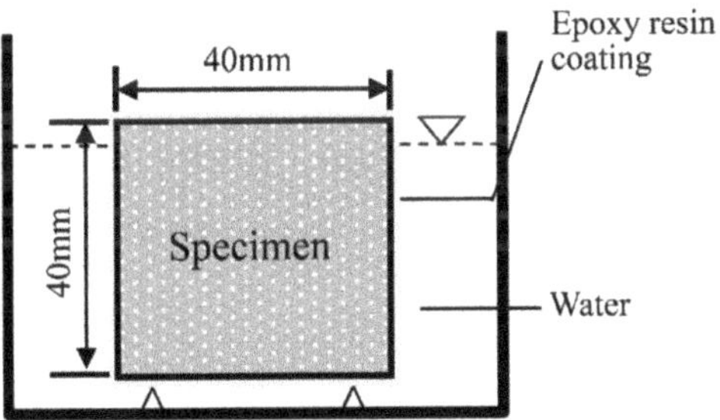


**Figure 1.** Schematic diagram of water absorption test.

Due to the strong barrier effect of epoxy resin, only one bottom surface can directly contact with water, which ensures the one-way transmission path of water. We all realize that only when the pores in concrete are connected with each other, the liquid will transfer freely and rapidly between the pores. Therefore, the volume of the permeable pore equals to the volume of the cumulative water absorption and the connecting porosity in one direction, which can be obtained by Equation (1).

$$\theta = m_f / (\rho_f \cdot V) \tag{1}$$

where $\theta$ is connecting porosity, $m_f$ is mass of fluid, $\rho_f$ is density of water, and $V$ is total volume of the tested specimen.

In addition, concrete is a kind of porous material, assuming that the capillary of concrete is multidimensional random parallel distribution, and the pore structure is cylindrical. The flow increment of concrete could be expressed as Equation (2) [18].

$$W(t) = n\pi r^2 \rho \sqrt{t\frac{r\gamma\cos\theta}{2\eta}} = s\sqrt{t} \tag{2}$$

where $W(t)$ is flow increment, $n$ is mass of pores, $r$ is pore size, $\rho$ is density of water, $t$ is time, $\gamma$ is surface tension of water, $\theta$ is contact angle of water, $\eta$ is viscosity of water, $s$ is coefficient of capillary absorption.

According to the characteristics of data collection and Equation (2), generalized least squares method could be used to calculate the capillary absorption rate of concrete. Figures 2 and 3 show the experimental imbibition data (expressed as cumulative absorbed volume/unit inflow area vs. $t_{1/2}$).

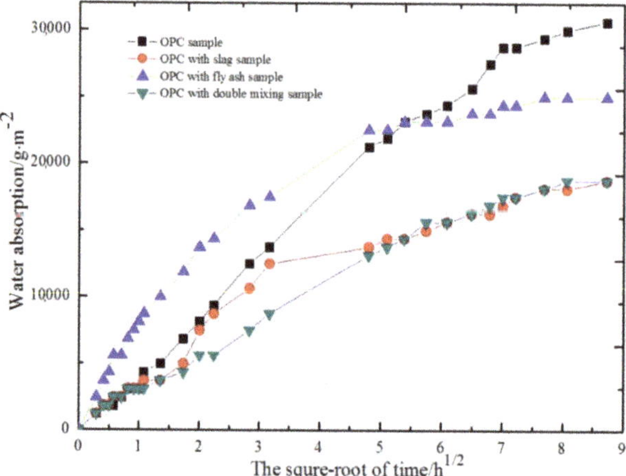


**Figure 2.** Relationship between water absorption and square root of time (7 days cured concrete).

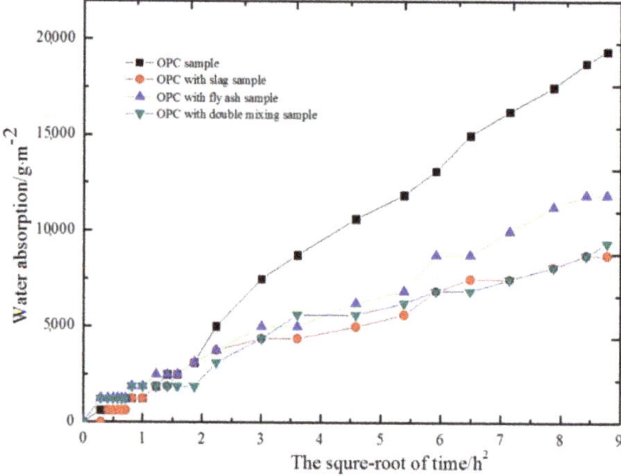


**Figure 3.** Relationship between water absorption and square root of time (28 days cured concrete).

As the curing time prolongs, the water absorption of all samples decreases (as shown in Figure 4). Comparison sample has the highest water absorption, followed by fly ash. Slag sample and double mixing sample have the best water absorption resistance.

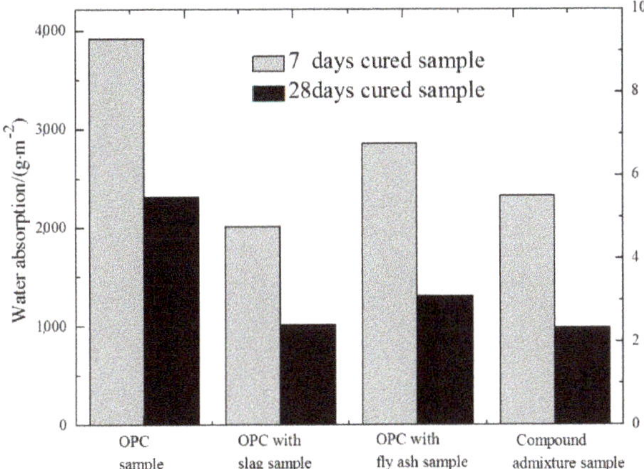


**Figure 4.** Water absorption of concrete.

The results show that the water absorption of concrete can be effectively reduced by adding two mineral additives (fly ash and slag), and the water absorption reduction effect of slag powder is more remarkable than that of fly ash.

## 4. Pore Fractal Dimension

### 4.1. MIP Test

After 28 days of standard curing, the concrete particles with the particle size of 3–5 mm were screened out by small hammer. The hydration reaction in the concrete particles was terminated by anhydrous ethanol. The treated samples were put into the oven and dried continuously at 105 ± 2 °C for 6 h to remove the excess moisture. The samples were put into the dilatometer and tested by mercury intrusion method.

In addition, samples were placed in an oven at 105 °C for 6 h to remove residual water or alcohol. Finally, the micropore structure was investigated using mercury intrusion porosimetry (MIP). The mercury injection test results of the four kinds of concrete after curing 28 d are shown in Figures 5 and 6.

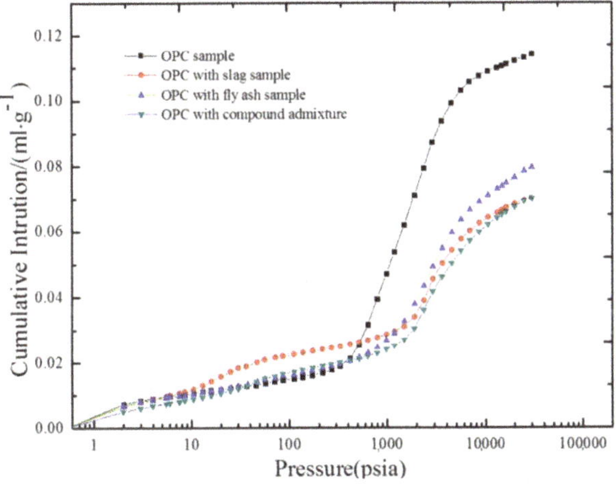


**Figure 5.** MIP curves of relationship between pressure and cumulative intrusion.

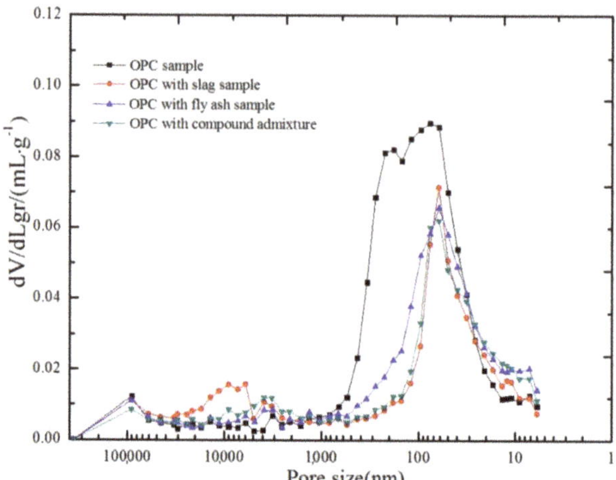

**Figure 6.** Pore size distribution of cement mortar.

Besides, connecting porosity as a one-dimension value of concrete, the proportion of the connecting porosity ($\theta$) in total porosity ($\theta_T$) of concrete can be seen as the connectivity probability among pores on a specific rotation direction. The schematic drawing of the pores distributing in concrete is shown in Figure 7.

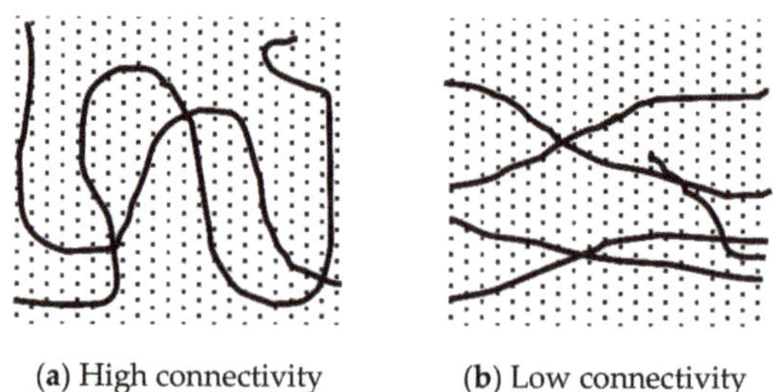

(a) High connectivity      (b) Low connectivity

**Figure 7.** Schematic drawing of the different connectivity probability.

It can be seen that the pore is intricately distributed in concrete, and their connection behavior occurs naturally and randomly. The permeability of porous concrete in different states under the same pore conditions is obviously different. Porosity has a direct physical meaning (void volume ratio), tortuosity is either an indirect or inferred measure of how the flow rate is affected (by twists and turns) within the porous medium. Highly interconnected connectivity as mentioned in the paper, is low tortuosity.

Therefore, the ratio ($\theta/\theta_T$) can be used to describe the connectivity trend in pore formation process. According to the data collected by MIP, the pore size distributions and connectivity probability ($\theta/\theta_T$) are shown in Table 5.

**Table 5.** Pore size distribution and connectivity probability of concrete.

| Binder Types | Porosity/% | | | | Total Porosity/% | Connectivity Probability/% |
|---|---|---|---|---|---|---|
| | >$10^3$ nm | $10^3$–$10^2$ nm | $10^2$–10 nm | <10 nm | | |
| OPC | 14.34 | 46.40 | 37.14 | 2.12 | 21.54 | 56.22 |
| OPC with slag | 33.82 | 13.82 | 49.20 | 3.17 | 14.78 | 37.01 |
| OPC with fly ash | 22.20 | 24.20 | 48.69 | 4.91 | 16.13 | 46.00 |
| OPC with compound admixture | 26.42 | 16.07 | 52.79 | 4.72 | 13.48 | 43.47 |

### 4.2. Calculation of Pore Fractal Dimension

Based on the fractal theory and the related micropore model, the deep information in concrete can be revealed. The Menger sponge model (shown in Figure 8) is an optimal fractal structure, which can be used to simulate the complex state of pore distribution in concrete [19]. The fractal model is established to simulate the fractal pores of materials. The cube with side length $R$ is divided into equal-sized small cubes. We selected a rule, removed some of these small cubes, and the remaining small cubes were $N(m)$. With this operation, the size of the remaining cubes decreases and the number increases. The remaining infinitely small cubes form the matrix of the material, while the small cube spaces with different orders are removed to form the pores of different orders in the material. After $k$ operations, the remaining cube size is $r_k = R/m_k$. The following equations can be obtained:

$$N_k = (r_k/R)^{-D} \tag{3}$$

where $D$ is dimension of pore volume, and the structure volume can be deduced as follows:

$$V_k \propto r_k^{3-D} \tag{4}$$

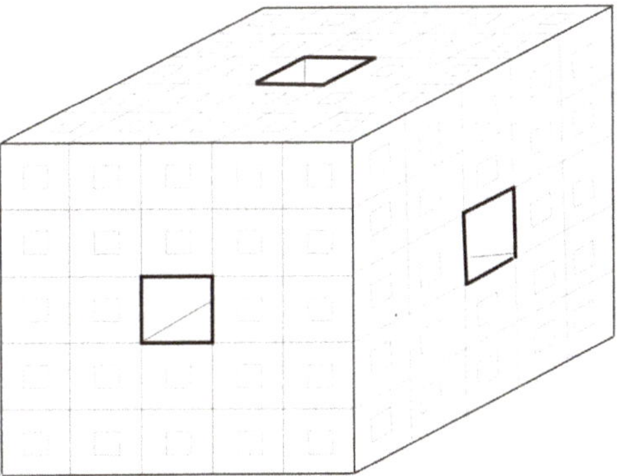

**Figure 8.** Menger sponge model.

According to Equation (4) and the relation between pressure and pore size, the d$V$ and d$r$ are taken as logarithmic, respectively, to draw the curve, the fractal dimension of porous volume can be determined by the slope of the curve. Lg$r$ and Lg$V$ are the logarithm to $r$ and the logarithm to $V$, respectively. The relationship between Lg$r$ and Lg$V$ are shown in Figures 9–12.

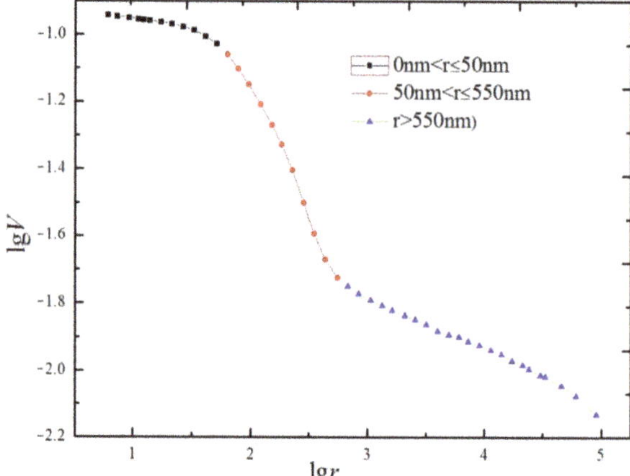

**Figure 9.** Relationship between Lg*r* and Lg*V* of comparing concrete.

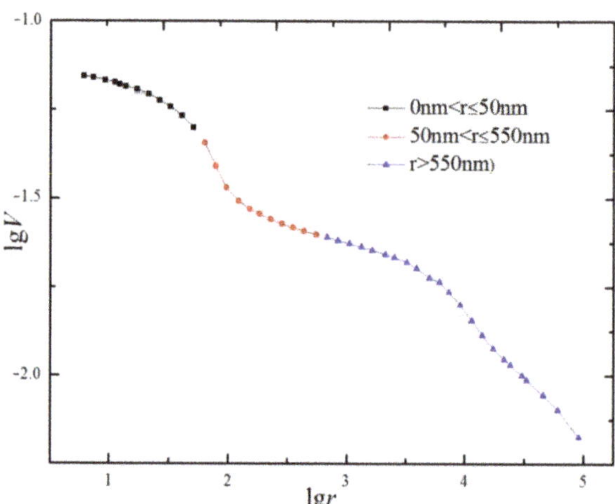

**Figure 10.** Relationship between Lg*r* and Lg*V* of slag concrete.

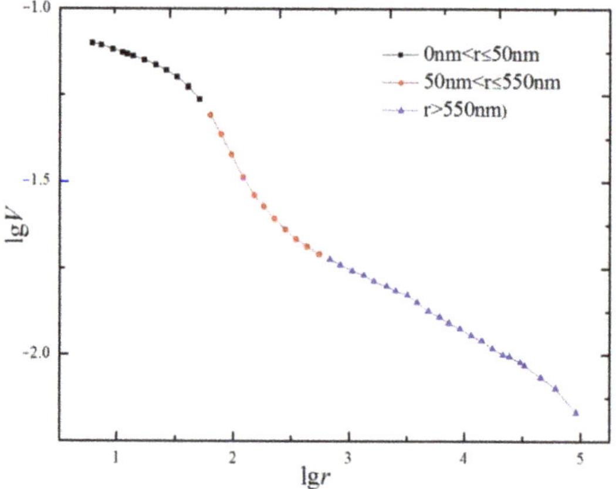

**Figure 11.** Relationship between Lg*r* and Lg*V* of fly ash concrete.

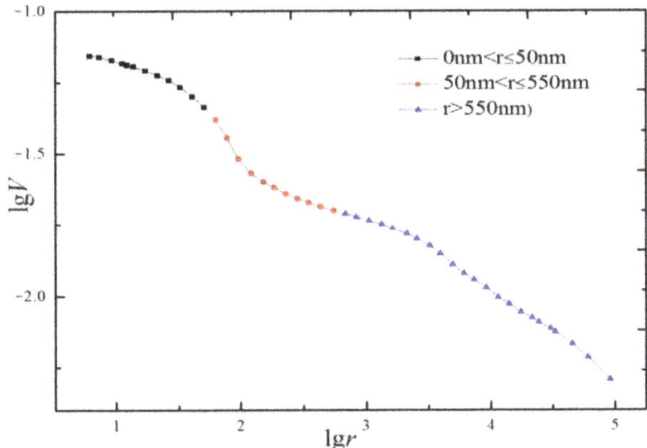

**Figure 12.** Relationship between L*gr* and L*gV* of OPC with compound admixture.

The above figure shows that there are inflection points near the aperture of 50 and 550 nm, so the relationship between L*gr* and L*gV* cannot be described as a straight line, but two straight lines. In fact, when trying to establish a relationship between fractal values and macroscopic properties, the polytrope of fractal dimension should be taken into consideration among different size scope, because it turns out to be different fractal properties with different size scope of an object, then the fractal dimension of pores in concrete should be achieved segmentally by the relationship according to Equation (4), and the pore fractal volume dimension obtained through the calculation of mercury intrusion data are shown in Table 6.

**Table 6.** Fractal dimension of pores and correlation coefficients.

|  |  | $d_1$ (<50 nm) | $d_2$ (50~550 nm) | $d_3$ (>550 nm) | $d$ |
|---|---|---|---|---|---|
| OPC | $D$ | 2.9188 | 2.2584 | 2.8394 | 2.6651 |
|  | $R$ | 0.9432 | 0.9933 | 0.9947 | 0.9677 |
| OPC with slag | $D$ | 2.8525 | 2.7553 | 2.7269 | 2.7693 |
|  | $R$ | 0.9632 | 0.8289 | 0.9762 | 0.9856 |
| OPC with fly ash | $D$ | 2.8367 | 2.5721 | 2.8042 | 2.7327 |
|  | $R$ | 0.9722 | 0.9822 | 0.9954 | 0.9843 |
| OPC with compound admixture | $D$ | 2.8118 | 2.6849 | 2.7244 | 2.7308 |
|  | $R$ | 0.9730 | 0.9473 | 0.9919 | 0.9939 |

## 5. Results and Discussion

The investigations performed in this study showed that alterations of mineral admixture can seriously affect the water absorption of concrete as is shown in Figure 13. Alterations of mineral admixture are not limited to macroscopic effect but act in the microstructure evolution of concrete. Mineral admixtures could refine the pore structures indeed, which is quantified as fractal dimension in this paper. Additionally, thinking about pore volume distribution from the perspective of the ideal fractal model, higher pore volume dimension means more complex pores distributed in concrete, under the same porosity conditions. The more complex the pore is, the longer the transmission path of water is, and also the transportation time which makes it a lower water absorption value of concrete [20]. Due to these assumptions, total pore fractal dimension should have a good agreement with water absorption value. The relation between water absorption values and total pore dimension is as shown below:

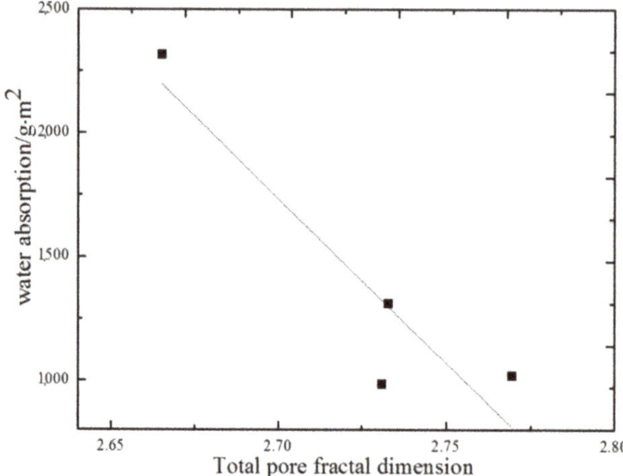

**Figure 13.** Relationship between water absorption values and total pore dimension (28 days cured concrete).

However, this is not an ideal case. As shown in Figure 13, there is no good correlation between the total pore volume size and the water absorption value (correlation coefficient is 0.78). Especially when the fractal dimension is 2.7308 and 2.7327, the water absorption rate is higher with the increase of the fractal dimension. It is concluded that the fractal dimension of total pore volume does contain all the information of the complex state of pores. In Figures 9–12, there are inflection points in the relationship curve between Lg$r$ and Lg$V$. Therefore, the total pore fractal dimension cannot accurately reflect the complex state of the pore. Therefore, the key factor to establish the objective relationship between pore complexity and water absorption value is to select the appropriate section of pore size, accurately characterize its complexity by pore fractal dimension, and explain the most possible pore size of concrete. The relationship of water absorption value and fractal dimension of 50~550 nm is shown in Figures 14 and 15.

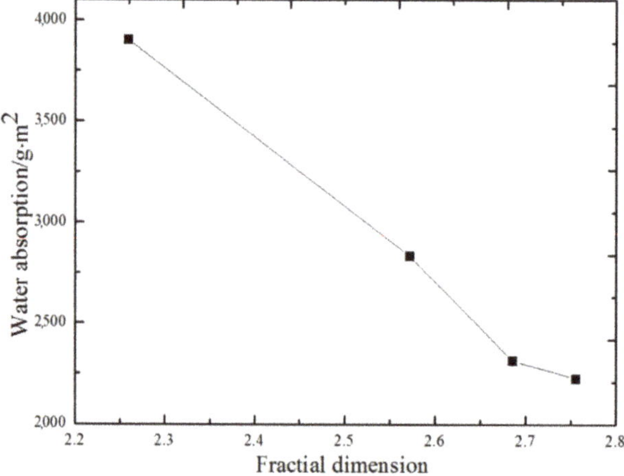

**Figure 14.** Relationship of water absorption value and fractal dimension of 50~550 nm (7 days cured concrete).

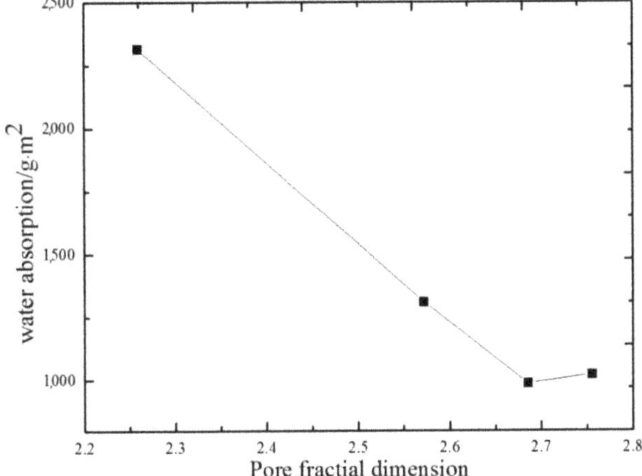

**Figure 15.** Relationship of water absorption value and fractal dimension of 50~550 nm (28 days cured concrete).

The fractal dimension of the comparing sample is obviously higher than that of the other samples with a most probable aperture near 671 nm. From another point of view, it is concluded that the fractal dimension of pore size of the comparing sample is the lowest between 50 and 550 nm, so there is another most possible pore size between 50 and 550 nm. From the mercury injection data, it can be seen the results of this analysis. Therefore, due to the inherent defects of MIP (high pressure, solid particle breakage, etc.), all samples have the maximum possible pore size between 50 and 550 nm. Compared with other sections, the middle section of pore diameter can always accurately describe the complex state of the pore, so that the fractal dimension of the pore between 50 and 550 nm has a good linear relationship with the water absorption value. The correlation coefficients of concrete cured on 7 and 28 d are greater than 0.95. Although the samples were tested at 28 days, there was a significant correlation between the absorption value of 7 days and the fractal volume dimension. The results show that the pore volume fractal dimension of 50–550 nm can be used to evaluate the water absorption of concrete. Furthermore, the pores with small fractal dimension correspond to low complexity of pores, which indicates that the pore tends to be simplified. Therefore, with the increase of pore fractal dimension, the probability of pore connectivity decreases. The relationship between connectivity probability of pores and pore volume dimension of 50~550 nm is shown in Figure 16.

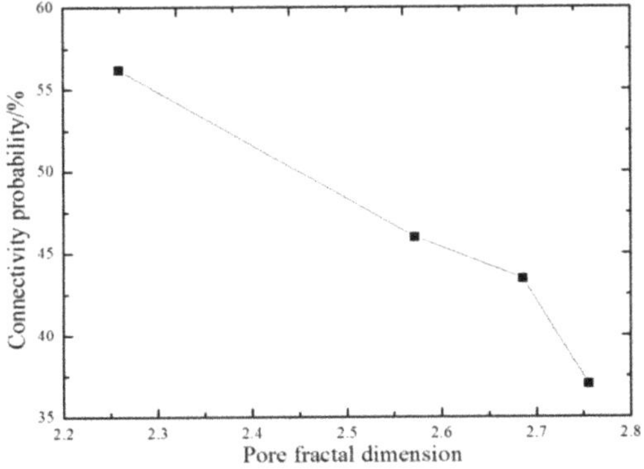

**Figure 16.** Relationship between connectivity probability of pores and pore volume dimension of 50~550 nm.

As shown in the figure above, the fractal dimension of pores in concrete is closely related to the probability of pore connectivity, and the correlation coefficient can reach 0.935. The above correlation can be explained by the quantification of pore fractal dimension between 50 and 550 nm, which quantifies the complexity of internal pores in concrete. The discovery of the relationship between fractal dimension and probability of pores distribution plays an important role in studying the internal possibility of concrete with the theory of chaos and fractal.

## 6. Conclusions

(1)   The pore structure of different regions shows different fractal characteristics. Fractal theory can analyze and evaluate the pore structure characteristics of concrete, especially when evaluating the permeability of concrete, the complexity of pore structure can be described in detail and quantitatively.

(2)   With the extension of curing time, the water absorption value of all samples will decrease. Adding two mineral additives (fly ash and slag) can effectively reduce the water absorption of concrete, and the water absorption reduction effect of slag powder is more significant than that of fly ash.

(3)   Compared with the total pore fractal dimension, the pore fractal dimension of 50–550 nm can accurately describe the complex state of the pore, and the pore fractal dimension of 50–550 nm has a good linear relationship with the water absorption value.

(4)   Pore fractal dimension between 50 and 550 nm has a close correlation with connectivity probability of pores inside concrete. The fractal theory could be applied for researching the probability of pore distribution inside concrete. The pore fractal dimension of 50–550 nm is closely related to the probability of pore connectivity in concrete. Fractal theory can be used to study the internal probability of concrete.

**Author Contributions:** Conceptualization, X.D. and J.Z.; methodology, X.D.; software, T.K.; validation, Y.Z., Y.F. and T.K.; formal analysis, Y.Z.; investigation, Y.Z.; resources, Y.Z.; data curation, T.K.; writing—original draft preparation, Y.Z. and X.L.; writing—review and editing, Y.F.; visualization, Y.F.; supervision, J.Z.; project administration, J.Z.; funding acquisition, J.Z. All authors have read and agreed to the published version of the manuscript.

**Funding:** This work was financially supported by the National Natural Science Foundation of China (51678374), the Liaoning Province Natural Science Foundation of China (20180550092), the Innovation Team Project of Liaoning (LT2019011) State Key Laboratory of Silicate Materials for Architectures (SYSJJ2020-15).

**Acknowledgments:** Special thanks are extended anonymous reviewers for their value comments.

**Conflicts of Interest:** The authors declare no conflict of interest.

## References

1.   Neville, A. Consideration of durability of concrete structures: Past, present, and future. *Mater. Struct.* **2001**, *34*, 114–118. [CrossRef]

2.   Basheer, L.; Cleland, D.J. Durability and water absorption properties of surface treated concretes. *Mater. Struct.* **2011**, *44*, 957–967. [CrossRef]

3.   Juenger, M.C.G.; Siddique, R. Recent advances in understanding the role of supplementary cementitious materials in concrete. *Cem. Concr. Res.* **2015**, *78*, 71–80. [CrossRef]

4.   Bremner, T.; Hover, K.; Poston, R.; Broomfield, J.; Joseph, T.; Price, R.; Clear, K.; Khan, M.; Reddy, D.; Clifton, J. *ACI 222R-01 Protection of Metals in Concrete Against Corrosion*; American Concrete Institute: Farmington Hills, MI, USA, 2001.

5.   Li, K.; Li, C. Modeling hydroionic transport in cement-based porous materials under drying-wetting actions. *J. Appl. Mech.* **2013**, *80*, 20904. [CrossRef]

6.   Bassuoni, M.T.; Nehdi, M.L. Durability of self-consolidating concrete to different exposure regimes of sodium sulfate attack. *Mater. Struct.* **2009**, *42*, 1039–1057. [CrossRef]

7.   Zhao, X.-l.; Wei, J.; Liu, Z.-k. Durability of concrete under multi-damage action. *J. Wuhan Univ. Technol. Mater. Sci. Ed.* **2004**, *19*, 73–75. [CrossRef]

8.   Safiuddin, M.; Mahmud, H.B.; Jumaat, M.Z. Efficacy of ASTM saturation techniques for measuring the water absorption of concrete. *Arab. J. Sci. Eng.* **2011**, *36*, 761. [CrossRef]

9.   Khan, M.I.; Mourad, S.M.; Charif, A. Utilization of supplementary cementitious materials in HPC: From rheology to pore structure. *KSCE J. Civ. Eng.* **2017**, *21*, 889–899. [CrossRef]

10.   Tang, M.; Li, X. Current situation and development of concrete fractal characteristic. *Concrete* **2004**, *12*, 8–11. [CrossRef]

11.   Dubuc, B.; Quiniou, J.F.; Roques, C.C.; Tricot, C.; Zucker, S.W. Evaluating the fractal dimension of profiles. *Phys. Rev. A* **1989**, *39*, 1500–1512. [CrossRef] [PubMed]

12.   Guo, W.; Qin, H.; Chen, H.; Sun, W. Fractal theory and its applications in the study of concrete materials. *J. Chin. Ceram. Soc.* **2010**, *38*, 1362–1368.

13.   Tang, M.; Chen, Z.; Yang, F. Research on characteristics of pore fractal and chloride diffusion in C50 pumped concrete. *Concrete* **2010**, 92–95. [CrossRef]

14.   Yin, H.; Lv, H.; Zhao, Y. Fractal characteristics of cement paste by carbonation. *Concrete* **2009**, 97–99. [CrossRef]

15.   Jin, S.; Zhang, J.; Huang, B. The relationship between freeze-thaw resistance and pore structure of concrete. *Pavement Geotech. Eng. Transp.* **2013**, 60–67. [CrossRef]

16.   Chen, X.; Zhou, J.; Ding, N. Fractal characterization of pore system evolution in cementitious materials. *KSCE J. Civ. Eng.* **2015**, *19*, 719–724. [CrossRef]

17.   Xue, S.; Zhang, P.; Bao, j.; He, L.; Hu, Y.; Yang, S. Comparison of Mercury Intrusion Porosimetry and multi-scale X-ray CT on characterizing the microstructure of heat-treated cement mortar. *Mater. Charact.* **2020**, *160*, 110085. [CrossRef]

18.   Gummerson, R.J.; Hall, C.; Hoff, W.D. Water movement in porous building materials—II. Hydraulic suction and sorptivity of brick and other masonry materials. *Build. Environ.* **1980**, *15*, 101–108. [CrossRef]

19.   Tang, M.; Wang, J.; Li, L. Research on fractal characteristics of concrete materials pore with MIP. *J. Shenyang Archit. Civ. Eng. Inst.* **2001**, *17*, 272–275. [CrossRef]

20.   Xie, C.; Wang, Q.; Li, S.; Hui, B. Relations of pore fractral dimension to pore structure and compressive strength of concrete under different water to binder ratio and curing condition. *Bull. Chin. Ceram. Soc.* **2015**, *34*, 3695–3702.

*Article*

# Numerical Investigation on Dynamic Response of RC T-Beams Strengthened with CFRP under Impact Loading

**Huiling Zhao [1], Xiangqing Kong [1,*], Ying Fu [2,*], Yihan Gu [1] and Xuezhi Wang [1]**

[1]  School of Civil Engineering, Liaoning University of Technology, Jinzhou 121001, China; zgbxzhl@163.com (H.Z.); emailgyh@126.com (Y.G.); myemailwxz@163.com (X.W.)
[2]  College of Engineering, Bohai University, Jinzhou 121001, China
*  Correspondence: xqkong@lnut.edu.cn (X.K.); 5117832@163.com (Y.F.)

Received: 20 August 2020; Accepted: 26 September 2020; Published: 1 October 2020

**Abstract:** To precisely evaluate the retrofitting effectiveness of Carbon Fiber Reinforced Plastic (CFRP) sheets on the impact response of reinforced concrete (RC) T-beams, a non-linear finite element model was developed to simulate the structural response of T-beams with CFRP under impact loads. The numerical model was firstly verified by comparing the numerical simulation results with the experimental data, i.e., impact force, reaction force, and mid-span displacement. The strengthening effect of CFRP was analyzed from the section damage evaluation. Then the impact force, mid-span displacement, and failure mode of CFRP-strengthened RC T-beams were studied in comparison with those of un-strengthened T-beams. In addition, the influence of the impact resistance of T-beams strengthened with FRP was investigated in terms of CFRP strengthening mode, CFRP strengthening sizes, CFRP layers and FRP material types. The numerical simulation results indicate that the overall stiffness of the T-beams was improved significantly due to external CFRP strips. Compared with the un-strengthened beam, the maximum mid-span displacement of the CFRP-strengthened beam was reduced by 7.9%. Additionally, the sectional damage factors of the whole span of the CFRP-strengthened beam were reduced to less than 0.3, indicating that the impact resistance of the T-beams was effectively enhanced.

**Keywords:** impact resistance; T-shaped reinforced concrete beams; CFRP; numerical analysis

## 1. Introduction

Reinforced concrete (RC) beams are important load-bearing components in building structures that might be subjected to unexpected impact loads. Compared with the high cost of rebuilding, reinforcing and transforming old structures can bring more beneficial social benefits, the safety and durability of the structure, reduced costs and impact on the surrounding environment, etc. Fiber-reinforced polymer (FRP) materials are an essential and vibrant structural composite material, owing to their high strength, light weight, corrosion resistance, and magnetic resistance properties [1]. Compared with traditional structural reinforcement methods, e.g., stick steel strengthening, increasing the sectional law, concrete replacement, additional fulcrum, and the external prestressing strengthening method, the external FRP reinforcement method is an ideal choice due to its advantages of easy construction and barely increased structural dead weight [2].

Many researchers have carried out studies on the flexural and shearing properties of FRP-strengthened concrete beams under static load [3–7]. In recent years, numerous experimental and numerical studies have been conducted in order to evaluate the effectiveness of FRP reinforcement on the impact resistance performance of RC beams. Tang and Saadatmanesh [8,9] conducted a set of drop hammer tests on external bonding FRP of RC beams and found that the FRP strips could constrain the development of shear cracks and improve their impact resistance. They also found that impact resistance was related to fiber type, thickness, quality and strength. Soleimani et al. [10] investigated a low-speed impact test on 12 RC beams strengthened with jet Glass Fiber Reinforced Plastics (GFRP). The results showed that the beams' impact resistance could not be improved by increasing the thickness of FRP if they adopted a two-sided bonding reinforcement method. However, increasing the GFRP thickness could increase the beam's impact resistance if they adopted a U-shaped reinforcement method. Pham and Hao [11,12] successfully studied the influence of the bonding mode of FRP strips and section modification on RC beams' impact resistance and found that a 45° inclined bonding beam had a higher bearing capacity and deformation resistance than a U-shaped bonding beam. In another study, Liu et al. [13] reported using carbon fiber reinforced plastics (CFRP) to strengthen RC beams without stirrups under static load and impact loading, respectively. FRP could not prevent the shear failure of beams under static load, but could prevent the shear failure of beams under impact load to reach four times of the energy under static load. In terms of numerical simulation, Bhatti and Kishi [14,15] performed an analysis using LS-DYNA to investigate the shear failure behavior of ordinary RC beams and rock shed beams with sand cushion under different impact velocities. Tu et al. [16] compared the concrete damage model (MAT72) in ANSYS/LS-DYNA and the RHT model in AUTODYN, and carried out unit tests on the two models under different pressure conditions. The results showed that MAT72 better conformed to the requirements of simulation accuracy. Meng et al. [17] used the CSCM model in ANSYS/LS-DYNA to simulate the process of axial impact on concrete cylindrical specimens. Then they [18,19] further optimized the model parameters, carried out numerical analysis on reinforced concrete beams under static load and impact load, and the axial impact resistance behavior of cylinder. The default value of software parameters overestimated the plasticity of concrete. Pham and Hao [20] used the 72R3 model of LS-DYNA to simulate RC beam under impact load, and found that boundary conditions had little effect on impact force, but significantly affected its mid-span displacement and failure mode. Zhao et al. [21] proposed the section damage factor to evaluate the damage of RC beams, and used LS-DYNA to study the dynamic response and damage degree of RC beams under impact load with different stirrup spacing, boundary conditions, shape of hammer head, and impact action position.

Although previous studies have made great progress in the dynamic response of FRP strengthened RC members, most of them have mainly been centered on common RC beams. Limited work has been conducted on the impact resistance of T-shaped reinforced concrete beams, especially the strengthening effect of FRP on RC T-beams under impact loads. RC T-beams are widely used in bridges, civil air defense works, and other building structures because of their light weight and fine bending strength. Therefore, in order to precisely evaluate the retrofitting effectiveness of the FRP on the impact response of RC T-beams, in this paper, a series of numerical studies are conducted aiming to study the dynamic behaviors of T-beams strengthened with CFRP patches under the impact loads. Firstly, three-dimensional models of RC T-beams models with CFRP are developed in ANSYS/LS-DYNA and validated with early experimental data. Then the numerical results in terms of the impact force, mid-span displacement and crack development process are compared for the un-strengthened and strengthened T-beams to quantitatively evaluate the strengthening effect of the CFRP sheets. Finally, the effects of different influence factors (e.g., CFRP strengthening mode, CFRP strengthening sizes, CFRP layers and FRP material types) on the impact resistance of FRP-strengthened RC T-beams are comprehensively analyzed.

## 2. Review of the Available Experimental Procedure

This paper was based on the drop hammer impact test conducted by Liu [22]. The dimensions of the CFRP-strengthened T-beam are presented in Figure 1. The CFRP strips were uniformly arranged along the beam span with a width of 90 mm and a spacing of 90 mm. The strength grade of concrete in the specimen was C30, with cylindrical compressive strength was 24.3 MPa, and the concrete cover was 25 mm. The longitudinal reinforcement and auxiliary steel bar were 3D16 and 4D8, with a yield strength of 477.3 MPa and 443 MPa. The drop hammer impact test setup is shown in Figure 2. The hammer head diameter was 200 mm, and the impact position was in the T-beam's mid-span. The mass of the drop-weight was 296 kg, and the velocity when it impacted was 3780 mm·s$^{-1}$.

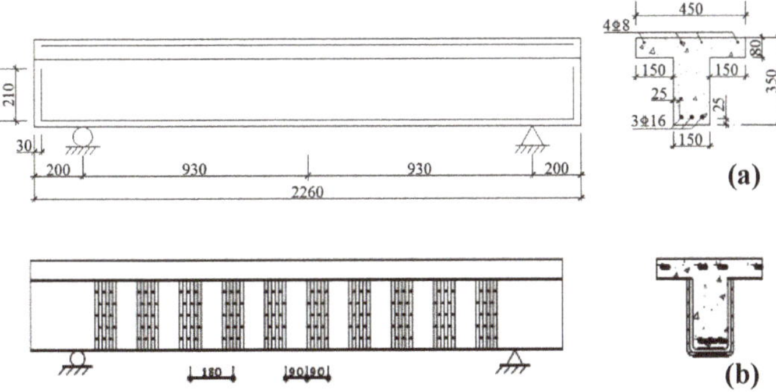

**Figure 1.** Details of the tested beam (obtained from Liu [22]). (**a**) Cross-section size and rebar arrangement. (**b**) Strengthening details for the beam.

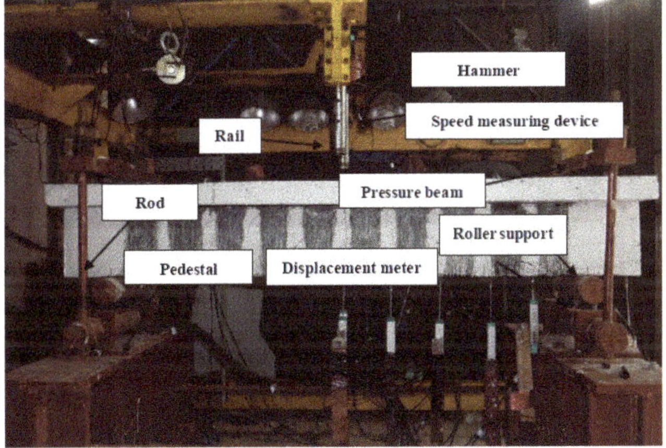


**Figure 2.** Drop hammer impact test setup (obtained from Liu [22]).

## 3. Numerical Model of RC T-Beams Strengthened with CFRP

### 3.1. Finite Element Model

A three-dimensional (3D) elastic plastic finite element (FE) analysis was performed in ANSYS/LS-DYNA. The 3D FE model for the T-beam with CFRP is shown in Figure 3. In the finite element model, the concrete, drop hammer and supports were modeled using 3D solid element (Solid164). The 3D truss element (Link160) was used for longitudinal steel bars and auxiliary steel bars. The spring damping element (Combi165) was used to simulate the rod between the upper and lower supports. CFRP sheet was modeled using thin shell element (Shell163). After convergence verification, there was a total number of 81,958 elements, with 54,360 concrete elements and 1113 steel elements. The structural model mesh sizes were 15 mm. Automatic face-to-face contact was adopted in this paper, in order to model the contact behavior between the beam and the drop hammer, and the beam and supports; a description of the parameters can be found using the Keyword *CONTACT_AUTOMATIC_SURFACE _TO_SURFACE of LS-DYNA. To simplify the calculation, the bond was assumed between concrete and steel bars or CFRP strips by sharing the common nodes, which have been widely used for simulation of the impact resistance of RC structures. Roller support was modeled by constraining the vertical and axial translational freedom of the center axis of lower support, other nodes only constrain in the axial direction, and constrain the axial translational freedom of all upper support nodes in the pressure beam. The impact force can be modeled using the keyword *INITIAL _VELOCITY, where the initial velocity was applied to the hammer.

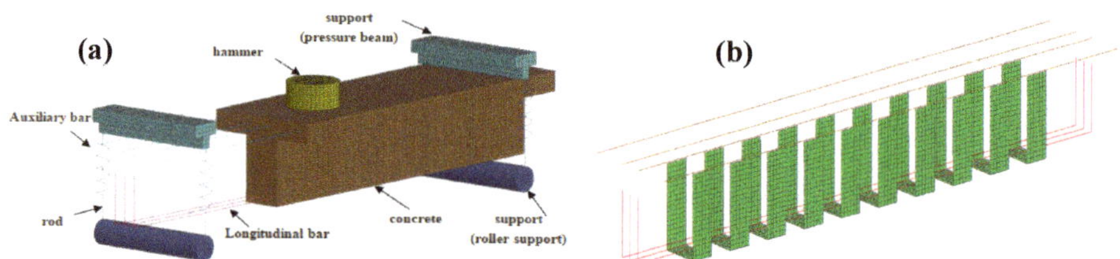


**Figure 3.** 3D finite element model of CFRP-strengthened T-beam. (**a**) Details of RC T-beam. (**b**) CFRP U-wraps.

### 3.2. Constitutive Model and Material Parameters

The Continuous Surface Cap Model (CSCM) is compared with the other concrete damage model to verify better to simulate the low-speed impact behavior under low confining pressure and low hydrostatic pressure. Therefore, CSCM (MAT_CSCM_CONCRETE) is used to model the dynamic properties of concrete; the theoretical formula of this model is presented in [23]. The specific material parameters of concrete are provided in Table 1. The Nonlinear Plasticity Kinematic Model (MAT_PLASTIC_KINEMATIC) is used to model the steel material. The steel strain rate effect is expressed by the Cowper–Symonds constitutive equation (Equation (1)):

$$\frac{\sigma'_0}{\sigma_0} = 1 + \left(\frac{\dot{\varepsilon}}{C}\right)^{\frac{1}{p}} \tag{1}$$

where $\sigma_0$ represents the static flow stress, $\sigma'_0$ represents the dynamic flow stress when the plastic strain rate is $\dot{\varepsilon}$, $C$ and $p$ represent the strain rate parameters, which are 40.4 and 5, respectively, for mild steel [24]. The specific material parameters of steel are provided in Table 2. To save run-time costs during the transit analysis, the hammer is defined as *MAT_RIGID. Linear elastic material (MAT_ELASTIC) is used to model the support, with the elastic modulus and Poisson's ratio are $2 \times 10^5$ MPa and 0.3. The spring-damper element is given by the elastic spring material (MAT_SPRING_ELASTIC), the elastic stiffness is $5 \times 10^4$ N/mm. The *MAT_ENHANCED_COMPOSITE_DAMAGE material constitutive model is used to model CFRP sheets. This model can describe arbitrary orthotropic materials and based on Chang–Chang laminate failure criteria [25]. The specific material model parameters of FRP are shown in Table 3 [26–28].

**Table 1.** The material parameters of concrete.

| Density RO (ton·mm$^{-3}$) | Calculated Control Parameters NPLOT | Maximum Strain Increment INCRE | Strain Rate Switch IRATE | Unit of Erosion ERODE | Pressure Recovery Parameter RECOV | Cap Options IRETRC | Pre-Damage PRED | Uniaxial Compressive Strength $f'_c$ (MPa) | Largest Aggregate Diameter DAGG (mm) |
|---|---|---|---|---|---|---|---|---|---|
| $2.4 \times 10^{-9}$ | 1 | 0 | 1 | 1.4 | 0 | 0 | 0 | 25 | 15 |

**Table 2.** The material parameters of steel.

| Density RO (ton·mm$^{-3}$) | Elastic Modulus E (MPa) | Poisson Ratio PR (MPa) | Yield Strength SIGY (MPa) | Tangent Modulus ETAN (MPa) | Hardening Parameter BETA | Strain Rate Parameter SRC | Strain Rate Parameter SRP | Failure Strain FS |
|---|---|---|---|---|---|---|---|---|
| $7.8 \times 10^{-7}$ | $2 \times 10^5$ | 0.3 | 477 | $2 \times 10^3$ | 0 | 40.4 | 5 | 0.12 |

**Table 3.** The material parameters of FRP.

| Material Parameters | CFRP | AFRP | GFRP |
|---|---|---|---|
| Density $\rho$ (ton·mm$^{-3}$) | $1.53 \times 10^{-9}$ | $1.44 \times 10^{-9}$ | $1.80 \times 10^{-9}$ |
| Longitudinal modulus of elasticity $E_a$ (MP) | $1.28 \times 10^5$ | $6.7 \times 10^4$ | $3.09 \times 10^4$ |
| Transverse modulus for composite $E_b$ (MPa) | $8.4 \times 10^3$ | $4.7 \times 10^3$ | $8.3 \times 10^3$ |
| Poisson's ration $v_{ba}$ | 0.0218 | 0.0280 | 0.0866 |
| In-plane shear modulus $G_{ab}$ (MPa) | $4.0 \times 10^3$ | $2.0 \times 10^3$ | $2.8 \times 10^3$ |
| Traverse shear modulus $G_{bc}$ (MPa) | $4.0 \times 10^3$ | $1.586 \times 10^3$ | $2.8 \times 10^3$ |
| Longitudinal shear modulus $G_{ca}$ (MPa) | $4.0 \times 10^3$ | $1.586 \times 10^3$ | $2.8 \times 10^3$ |
| Longitudinal compressive strength $X_c$ (MPa) | 1060 | 312 | 480 |
| Longitudinal tensile strength $X_t$ (MPa) | 2093 | 1420 | 983 |
| Transverse compressive strength $Y_c$ (MPa) | 198 | 145 | 140 |
| Transverse tensile strength $Y_t$ (MPa) | 50 | 36 | 40 |
| In-plane shear strength $S_c$ (MPa) | 104 | 53 | 70 |

## 4. Validation and Analysis

### 4.1. Impact Force and Reaction Force

The impact force time history curve of the specimen beams is shown in Figure 4. The impact force peak and the impact duration are consistent between the finite element simulation and the experimental test results. The impact force curve can be divided into three stages: pulse stage, storming stage and recovery stage. In the pulse stage, the impact force rises to a peak steeply and falls rapidly, forming the main peak of impact force, lasting about 1.2 ms. The simulation and experimental results of the maximum impact force are 664 KN and 623 KN, respectively, with a calculation error of 6.6%. Subsequently, several secondary impact force peaks can be obviously found in the shock stage. The maximum mid-span displacement of the specimen can be obtained, and the shock duration is about 14 ms. The impact force drops to zero during the recovery phase. The comparison of numerical simulation and test results of the reaction force for RC T-beams under impact load is shown in Figure 5. The maximum reaction force simulation result of the CFRP-strengthened beam is 316 kN, the test value is 345 kN, and the discrepancy is 8.4%. The numerical results are highly consistent with the experimental results.

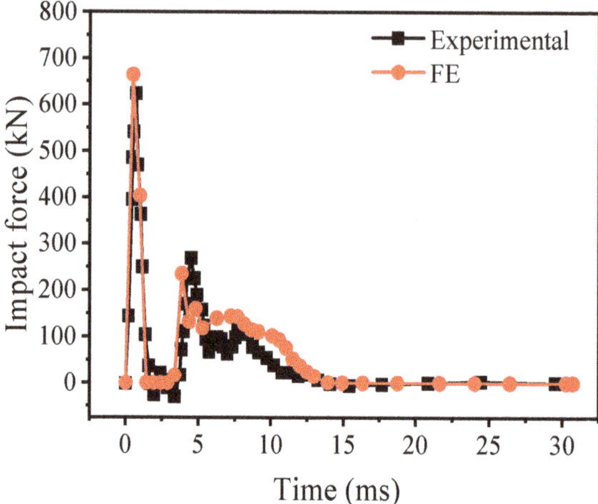

**Figure 4.** Comparison of impact force time histories obtained from the FE simulation and experimental test.

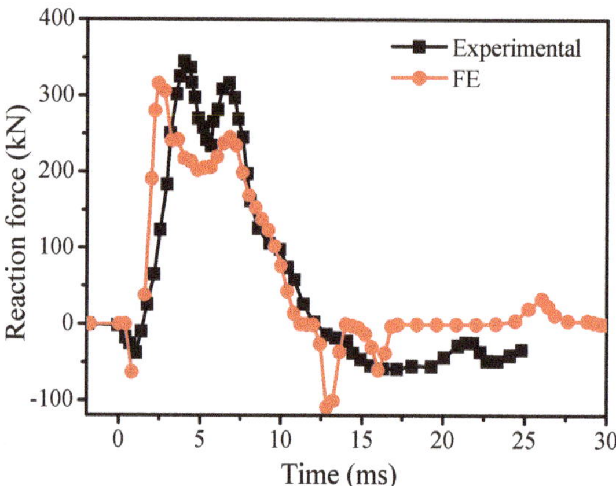

**Figure 5.** Comparison of reaction force time histories obtained from the FE simulation and experimental test.

## 4.2. Displacement

The comparison of mid-span displacement obtained from the FE simulation and the experimental test are shown in Figure 6. It can be noted that mid-span displacement time history presents a half-sine wave. This indicates the partial elastic recovery when the displacement achieves maximum, with the dissipated energy that the structural plastic deformation and damage caused residual displacement. The maximum mid-span displacement of the numerical model is 10.54 mm and the maximum mid-span displacement of the test is 10.88 mm. The difference between them is only 3.1%. In summary, the numerical simulation results are in good agreement with the experimental results.

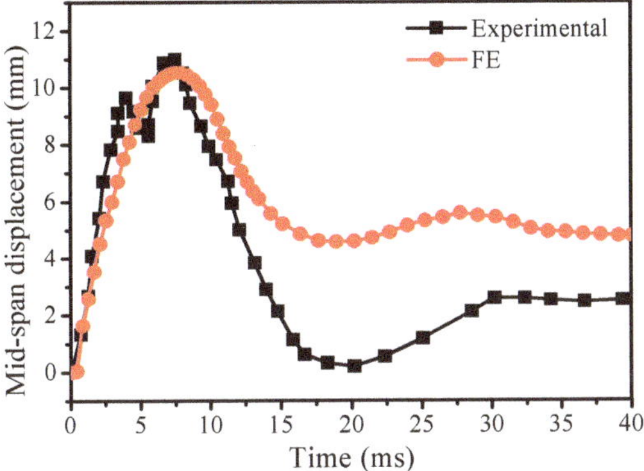

**Figure 6.** Comparison of mid-span displacement obtained from the FE simulation and the experimental test.

### 4.3. Fracture Development Process and Failure Mode

The comparison of numerical simulation and test results of failure modes for RC T-beam with CFRP under impact load is shown in Figure 7. The failure mode of the CFRP-strengthened beam is mainly subject to oblique cracks caused by shear force, which shows a typical shear failure mode. When the impact force reaches its peak (0.5 ms), two 45° diagonal cracks appear symmetrically on the beam web. At the end of the pulse phase (1.2 ms), the main oblique crack develops downward to the horizontal height of longitudinal reinforcement. When the impact force reaches its second peak (3 ms), short shear cracks appear at the CFRP strips' spacing. When the displacement reaches the maximum (7.5 ms), the main change is that the number of small shear cracks between the CFRP strips in the mid-span area increases. At the end of impact (40 ms), some small cracks in the mid-span are closed due to the beam's rebound. Overall, the simulation model shows a favorable agreement with the test results for the failure mode.

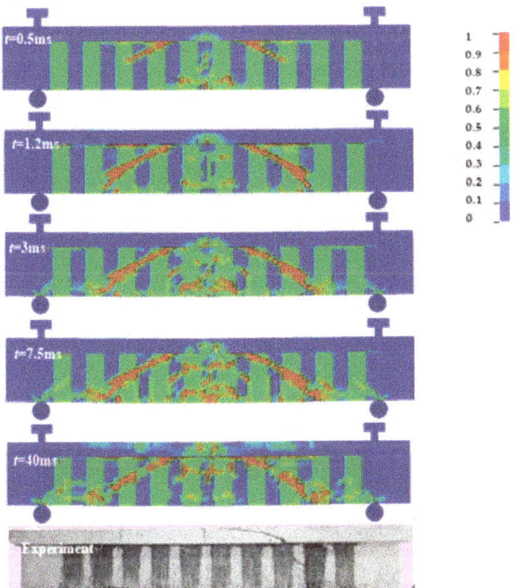

**Figure 7.** Failure pattern of RC T-beam with CFRP from test and simulation.

The strain–time history of CFRP is shown in Figure 8. The initial rapid increase of strain of CFRP strips divided by the corresponding time interval was taken as the CFRP average strain rate. In this study, the average strain rate of CFRP is 9 s$^{-1}$. When the strain rate is less than 10 s$^{-1}$, the increase of dynamic tensile strength and elastic modulus of CFRP materials is only about 3% [29,30]. Therefore, the influence of strain rate on CFRP is in dynamic debonding or fracture strain [13].

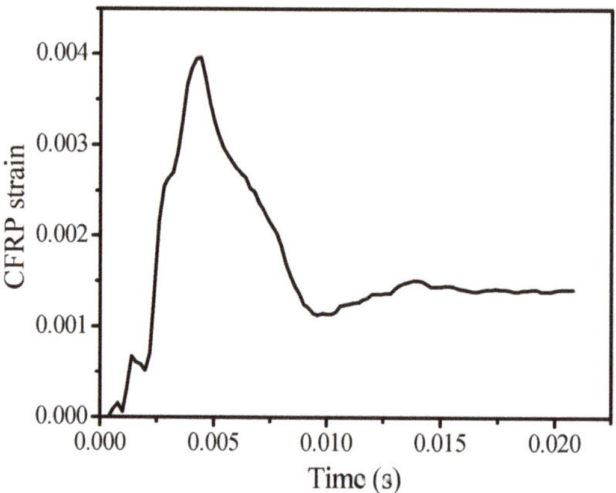

**Figure 8.** Strain–time history of CFRP.

*4.4. Section Damage Assessment*

The concrete CSCM constitutive model reflects the damage of concrete under impact loads. The damage stress $\sigma_{ij}^{d}$ is defined as follows:

$$\sigma_{ij}^{d} = (1-d)\sigma_{ij}^{vp} \tag{2}$$

where $d$ is damage factor; $d = 0$ is no damage; $d = 1$ is complete damage; and $\sigma_{ij}^{vp}$ is the no damage stress tensor. Equation (2) expresses the isotropic and proportional reduction of the volume modulus and shear modulus of the material. The failure mode generally shows a limitation of the RC beam structure under impact loads, which has a short acting time and limited area of action. In this study, the damage assessment method [21] is used to determine the most serious damage section in the T-beam, while damage factor $d_s$ is the damage factors average value in T-beam cross-section:

$$d_s = \frac{1}{n}\sum_{1}^{n} d \tag{3}$$

where $d$ is the unit damage factor; $n$ is the sum of section elements. To conveniently and intuitively measure the degree of damage in each section, the section damage factor $d_s$ is specified as follows [21], where $d_s = 0{\sim}0.3$ is slight damage; $d_s = 0.3{\sim}0.6$ is moderate damage; $d_s = 0.6{\sim}0.9$ is severe damage; $d_s = 0.9{\sim}1$ is component failure.

Figure 9 shows the distribution of section damage factor ($d_s$) along the beam axis of the beam with CFRP, it can reflect that each section impact damage degree and CFRP impact resistance effect. $d_s$ is calculated by Equation (3). Overall, the T-beam with CFRP has exhibited a lightly damaged stage and the $d_s$ of all sections are below 0.3. In general, the T-beam with CFRP U-wraps can obviously reduce the section damage degree, this confirms an effective mode for improving the structure impact resistance.

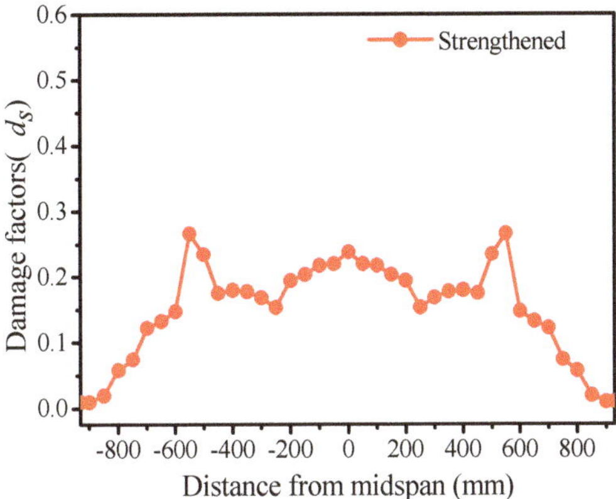

**Figure 9.** Sectional damage factors of the CFRP-strengthened T-beam.

## 5. Parametric Study

From the current literature, it is clear that when external FRP is used to strengthen the structure, e.g., thickness, strength and strengthening mode of FRP play central roles [8,13]. In this paper, the dynamic responses of CFRP-strengthened beams and un-strengthened beams under impact load are analyzed based on impact force, reaction force and crack patterns. To have a deeper understanding of the impact resistance of T-beams strengthened with FRP, different CFRP strengthening modes, CFRP strengthening dimensions, CFRP strengthening layers and FRP material types are used to study the dynamic response of T-beam without stirrup.

### 5.1. Strengthening Effect of FRP

The displacement of the T-beam with and without CFRP at the mid-span is given in Figure 10a. According to the numerical simulation result, the maximum mid-span displacement and residual displacement of the un-strengthened beam are 11.44 mm and 6.23 mm. Simultaneously, the maximum mid-span displacement and residual displacement of beams with CFRP are reduced by 7.9% and 23.3%, respectively. The reason for this is that the CFRP U-wraps can improve the shear capacity and restrain the shear cracks generated by the T-beam under impact load, thus improving the structural stiffness and reducing the overall deformation [8,22]. Figure 10b shows the reaction force time history for the T-beams with and without CFRP. It can be seen that the maximum reaction force of the CFRP-strengthened beam increases by 3.6% with respect to that of the un-strengthened beam. In addition, the semi-sinusoidal fluctuations of un-strengthened beams are violent and massive compared with the CFRP-strengthened beams. This is consistent with the fluctuation of the reaction force observed in the impact test of CFRP-strengthened and un-strengthened beams [22].

Figure 11 illustrates the crack patterns of the RC T-beam without CFRP. The RC T-beam cracks very quickly on account of there being no stirrup. It can be seen that, similar to the RC T-beam strengthened with CFRP, the damage of the un-strengthened beam was also mainly caused by two 45-degree inclined cracks on the web. However, compared with the beam strengthened with CFRP, the un-strengthened beam's damage under impact load is more severe. In addition, only one shear crack appears in the span of the un-strengthened beam, but several thin and dense cracks appear in the web of the strengthened beam. This indicates that the CFRP can not only restrain oblique fracture development, but also dissipates the impact of kinetic energy by developing many small cracks, thus improving the impact resistance [22].

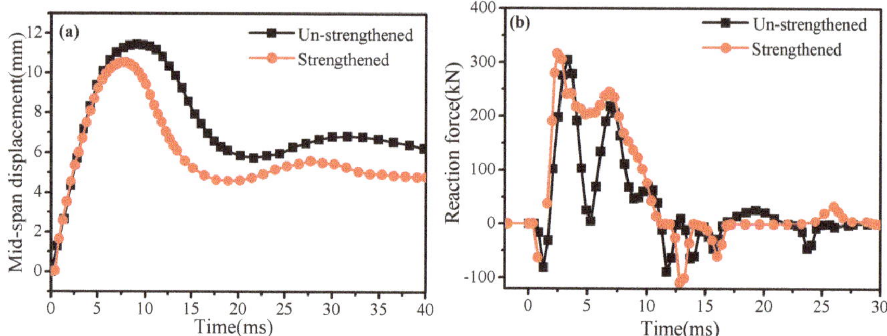

**Figure 10.** Time histories of T-beam with and without CFRP. (**a**) Mid-span displacement. (**b**) Reaction force.

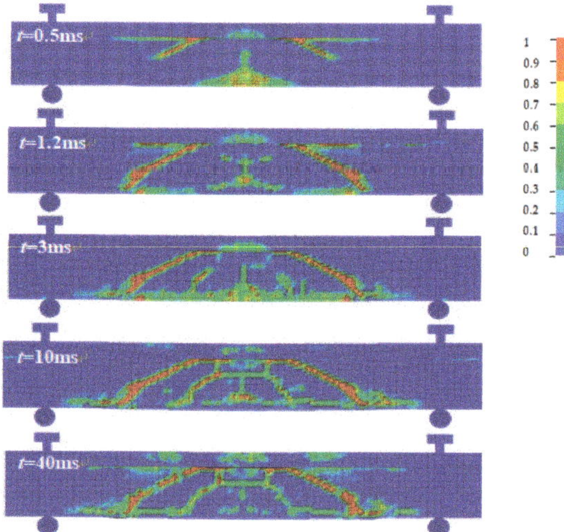

**Figure 11.** Crack patterns of RC T-beam without CFRP.

## 5.2. Effect of Strengthening Modes

In this section, the effect of different U-shaped strengthening modes on the impact response of the T-beam is analyzed. The details of four different FRP strengthening modes (i.e., vertical U-wraps, 45° U-wraps, U-wraps in the mid-span and U-wraps in the shear-span, which are named A, B, C and D, respectively), which are common strengthening techniques, are illustrated in Figure 12. For the four models, the other conditions are all the same.

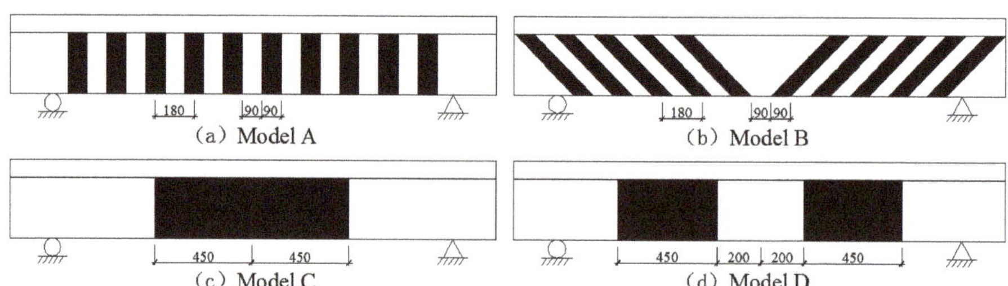


**Figure 12.** Four kinds of CFRP strengthening modes.

The mid-span displacement time histories under four CFRP strengthening modes are given in Figure 13. The maximum displacement and residual displacement under different strengthening modes are shown in Table 4. It can be seen that the maximum mid-span displacement for the RC T-beams in the four modes A, B, C and D are 10.54 mm, 8.31 mm, 9.07 mm and 9.49 mm, respectively. Compared with the un-strengthened beam, the maximum mid-span displacement decreased by 7.9%, 27.4%, 20.7% and 17.0%, respectively. Among the four different strengthening types, mode A has the lowest reinforcement effect, while mode B is the highest. The longitudinal tensile strength is about 40 times the transverse tensile strength due to CFRP is an anisotropic material. Therefore, the 45° U-wraps of the fiber strip can reflect the advantages of the high longitudinal tensile strength of CFRP, and this can constrain the development of oblique cracks. The CFRP is perpendicular to the beam axis in the longitudinal direction for reinforcement modes A, C and D, equivalent to adding external stirrups in the reinforcement area [31]; the difference is the additional stirrup position. Therefore, it is observed that the CFRP should be placed in the severe damage section, which can restrain fracture development and reduce impact deformation.

**Table 4.** Numerical results of strengthened beams with various parameters.

| Strengthening Modes | FRP Strengthening Sizes (mm²) | FRP Strengthening Layers | FRP Material Types | Maximum Mid-Pan Displacement (mm) | Residual Displacement (mm) |
|---|---|---|---|---|---|
| Un-strengthened | - | 0 | - | 11.44 | 6.23 |
| Model A | 621,000 | 1 | CFRP | 10.54 | 4.78 |
| Model B | 621,000 | 1 | CFRP | 8.31 | 3.17 |
| Model C | 621,000 | 1 | CFRP | 9.07 | 4.55 |
| Model D | 621,000 | 1 | CFRP | 9.49 | 4.82 |
| Model A | 372,600 | 1 | CFRP | 10.73 | 5.89 |
| Model A | 869,400 | 1 | CFRP | 9.38 | 4.56 |
| Model A | 1,242,000 | 1 | CFRP | 8.96 | 3.57 |
| Model A | 621,000 | 2 | CFRP | 10.20 | 4.53 |
| Model A | 621,000 | 3 | CFRP | 10.05 | 4.45 |
| Model A | 621,000 | 4 | CFRP | 9.98 | 4.35 |
| Model A | 621,000 | 5 | CFRP | 9.93 | 4.30 |
| Model A | 621,000 | 1 | AFRP | 10.88 | 4.98 |
| Model A | 621,000 | 1 | GFRP | 11.01 | 5.13 |

Note 1: Model A is the vertical U-shaped reinforcement, Model B is the 45° U-shaped reinforcement, Model C is the U-shaped reinforcement in the mid-span, and Model D is the U-shaped reinforcement in the shear-span. Note 2: Reinforcement size 621,000 mm², FRP strip 90 mm, spacing 90 mm; Reinforcement size 372,600 mm², FRP strip 90 mm, spacing 180 mm; Reinforcement size 869,400 mm², FRP strip 180 mm, spacing 90 mm; Reinforcement size 1,242,000 mm², no spacing reinforcement.

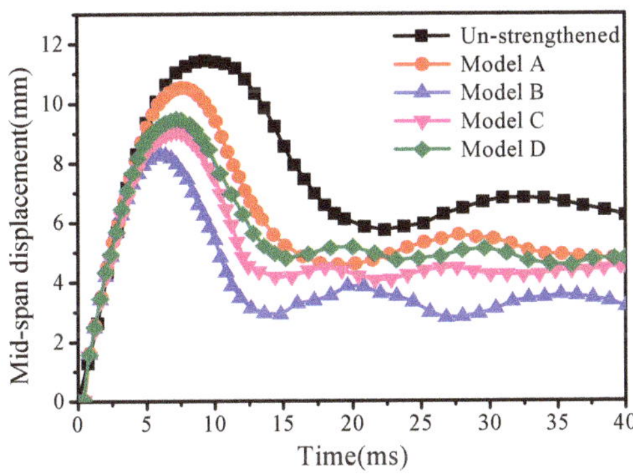


**Figure 13.** Mid-span displacement-time histories of T-beams strengthened with different CFRP strengthening modes.

### 5.3. Effect of Strengthening Sizes

An investigation is performed on the CFRP sizes, which can affect reinforced beams subjected to impact loads. As can be seen in Figure 14, four beams are reinforced with U-shaped CFRP strips, the dimensions of CFRP are 372,600 mm$^2$, 621,000 mm$^2$, 869,400 mm$^2$ and 1,242,000 mm$^2$.

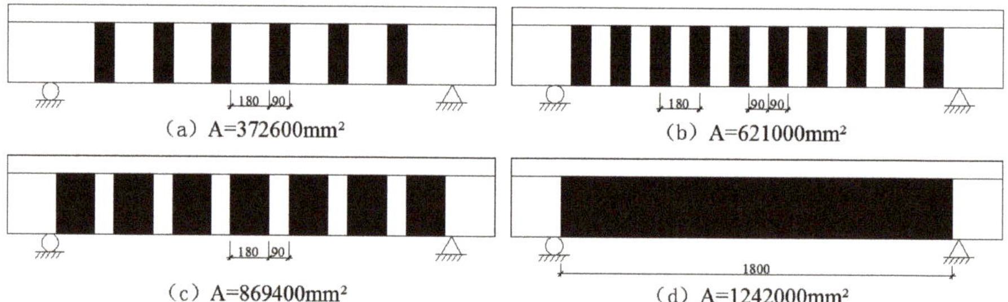


**Figure 14.** Four kinds of CFRP strengthening sizes.

Figure 15 and Table 4 present the dynamic response of different CFRP strengthening sizes of T-beams under impact loads. The mid-span displacement of the strengthened beam decreases with the CFRP size and amount increase. This indicates the CFRP reinforcement has significant advantages in impact resistance. The maximum mid-span displacement and residual displacement of the un-strengthened beam are 11.44 mm and 6.23 mm, respectively. The strengthening sizes of the T-beams are 372,600 mm$^2$, 621,000 mm$^2$, 869,400 mm$^2$ and 1,242,000 mm$^2$; the maximum mid-span displacement decreases by 6.2%, 7.9%, 18.0% and 21.7%, respectively; and the residual displacement decreases by 5.5%, 23.3%, 26.8% and 42.7%, respectively. With the increase of CFRP strengthening size, the stiffness, flexural and shear capacities of the beams increase, the structural deformation reduces.

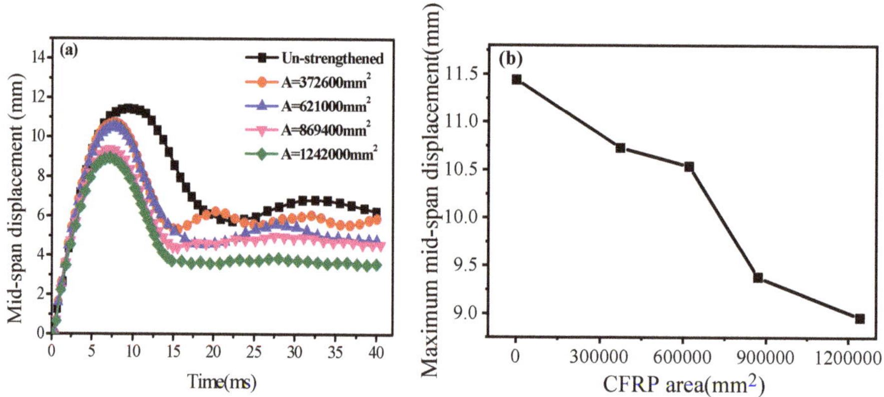


**Figure 15.** (**a**) Mid-span displacement-time histories and (**b**) maximum displacements of T-beams strengthened with different CFRP strengthening sizes.

### 5.4. Effect of Strengthening Layers

The effect of the different CFRP strengthening layers on the impact response of the T-beams is studied in this part. CFRP-strengthened beams with different numbers of layers (1–5-layer), all of which have the same strengthening mode A (size of CFRP is 621,000 mm$^2$), are analyzed.

Figure 16a and Table 4 show the results of the mid-span displacement for beams with different layers. It is found that the maximum mid-span displacement and residual displacement decrease with the increase in CFRP layers. This is attributed to the fact that impact resistance improves with the increase in CFRP layers. However, the increase in CFRP layers is limited in improving the impact resistance. Figure 16b presents the mid-span maximum displacement of T-beams with different CFRP-strengthened layers. It is clear that the mid-span displacement decreases by 4.7% as the number of CFRP layers increases from 1 to 3. Nevertheless, the mid-span displacement only decreases by 0.7% and 0.5% with the increase in layers from 3 to 4 and 5. The main reason for this is that it is difficult for the many layers to work collaboratively. Therefore, by considering the strengthening effectiveness as well as economic considerations, for the strengthened RC beams in this case, the T-beams with 2–3-layer CFRP are adequate for improving the impact resistance.

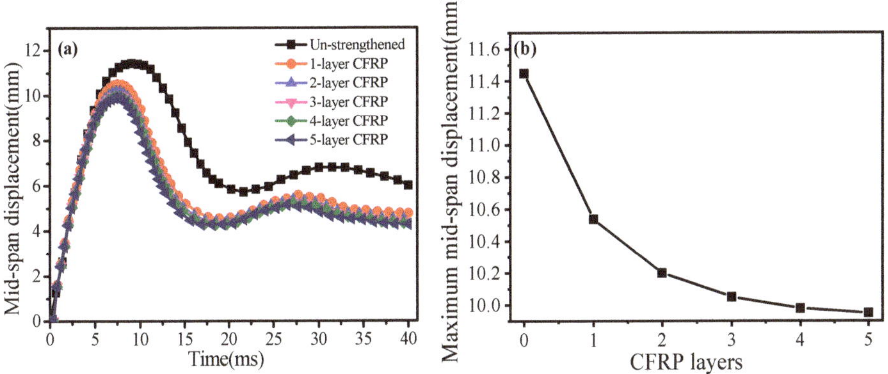


**Figure 16.** (**a**) Mid-span displacement-time histories and (**b**) maximum displacements of T-beams strengthened with different CFRP strengthening layers.

## 5.5. Effect of Strengthening Material Types

Considering the high cost of CFRP, other FRP materials can be adopted to strengthen the structure. This study compared the influence of CFRP, Aramid Fiber Reinforced Polymer (AFRP), GFRP on the impact resistance of T-beam. The FRP main mechanical properties parameters are shown in Table 3.

The mid-span displacement of three numerical models with different FRP materials is summarized in Figure 17 and Table 4. As can be seen, the tendencies of the displacement–time histories of the three T-beams strengthened by different FRP are similar. The maximum mid-span displacement and residual displacement of AFRP strengthened beams are 10.88 mm and 4.98 mm, which are 3.2% and 4.2% more than CFRP-strengthened beams. The maximum mid-span displacement and residual displacement of GFRP strengthened beams are 11.01 mm and 5.13 mm, which are increased by 4.5% and 7.3% compare with CFRP-strengthened beams. This indicates that the CFRP is superior to AFRP and GFRP at improving the impact resistance of the RC beam. However, it should be emphasized that the displacement change is within 8%; there is little difference in the reinforcement effect of different FRP materials. In consequence, it is more economical to choose AFRP and GFRP.

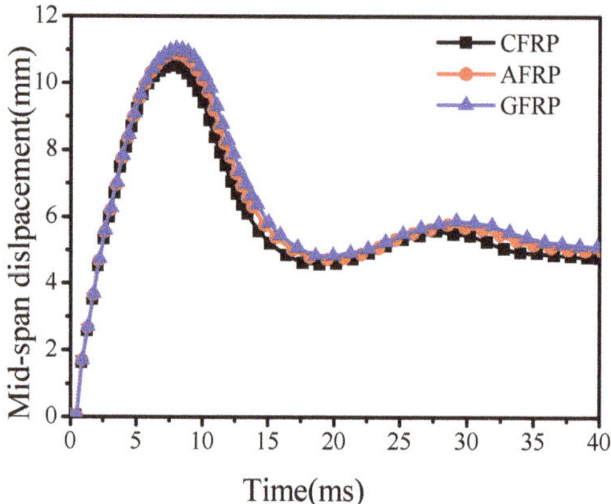

**Figure 17.** Mid-span displacement-time histories of T-beams strengthened with different FRP material.

## 6. Conclusions

In this study, ANSYS/LS-DYNA is used to study numerical simulation on the dynamic response of CFRP-strengthened T-beam under impact loads. This paper compared the beams with and without CFRP in terms of mid-span displacement, reaction force and failure mode. Moreover, the effects of CFRP strengthening modes, CFRP strengthening sizes, CFRP strengthening layers and different FRP material types on the impact behavior of RC T-beams were explored. The following can be concluded.

Within the range of 700 mm at the mid-span, it can be observed that all sections $d_s$ within the CFRP-strengthened beams are below 0.3 and are slight damage. Therefore, CFRP reinforcement can significantly reduce the beam damage degree under impact loads.

Compared with the un-strengthened beams, the maximum displacement and residual displacement of the CFRP-strengthened beams are reduced by 7.9% and 23.3%, respectively. In addition, the beams with CFRP can dissipate the impact energy by developing small cracks. Thus, CFRP reinforcement enhances the beam impact resistance.

Furthermore, comprehensive parametric studies on FRP-strengthened RC T-beams are carried out by considering various effect parameters under impact loads. It is found for the four different strengthening modes, the strengthened T-beams with 45° U-wraps has the best reinforcement effect. Otherwise, the mid-span displacement of beams decreased with the CFRP sizes increased. Through a set of simulations, the CFRP strengthening layers can improve impact resistance of the beam. However, as the number of CFRP layer increased there was not a proportional decrease in mid-span displacement. Therefore, it is recommended the T-beams with 2-3-layer CFRP are adequate in enhancing the impact resistance. Additionally, compared the influence of CFRP, AFRP and GFRP for T-beam impact resistance, the best reinforcement material is CFRP, AFRP and GFRP are inferior.

**Author Contributions:** Conceptualization, X.K.; methodology, X.K.; software, H.Z. and Y.G.; validation, X.W.; formal analysis, Y.F.; investigation, H.Z. and Y.G.; resources, Y.F.; data curation, H.Z.; writing—original draft preparation, H.Z.; writing—review and editing, Y.F., Y.G. and X.W.; visualization, Y.G.; supervision, X.K.; project administration, X.K.; funding acquisition, Y.F. All authors have read and agreed to the published version of the manuscript.

**Funding:** This research was funded by the National Nature Science Foundation of China (51704029) and LiaoNing Revitalization Talents Program (XLYC1807044, XLYC1807050).



## References

1. Teng, J.G.; Ye, L.P. *New Composite Materials and Structures*; Science Press: Beijing, China, 2006; Volume 7, pp. 177–192.

2. Zhu, K.J. Experimental Analysis on the Impact Resistance Performance of RC Beams Strengthened with Externally Bonded AFRP Fabric. Master's Thesis, Henan Polytechnic University, Jiaozuo, China, 2014.

3. Chen, J.F.; Teng, J.G. Shear capacity of FRP-strengthened RC beams: FRP debonding. *Constr. Build. Mater.* **2003**, *17*, 27–41. [CrossRef]

4. Deng, Z.G.; Zhang, P.F.; Li, J.H. Fatigue and Static Behaviors of RC Beams Strengthened with Prestressed AFRP. *China J. Highw. Transp.* **2007**, *20*, 49–55.

5. Pan, J.L.; Chen, Z.F.; Liang, Z.N. Experimental investigation on flexural and flexural/shear crack induced debonding failure in FRP strengthened concrete beams. *J. Southeast Univ. (Nat. Sci. Ed.)* **2007**, *37*, 229–234.

6. Chen, G.M.; Teng, J.G.; Chen, J.F. Process of debonding in RC beams shear-strengthened with FRP U-strips or side strips. *Int. J. Solids Struct.* **2012**, *49*, 1266–1282. [CrossRef]

7. Dong, J.; Wang, Q.; Guan, Z. Structural behaviour of RC beams with external flexural and flexural–shear strengthening by FRP sheets. *Compos. Part B Eng.* **2013**, *44*, 604–612. [CrossRef]

8. Tang, T.; Saadatmanesh, H. Behavior of concrete beams strengthened with fiber-reinforced polymer laminates under impact loading. *J. Compos. Constr.* **2003**, 209–218. [CrossRef]

9. Saadatmanesh, H.; Tang, T. Analytical and Experimental Studies of Fiber-Reinforced Polymer-Strengthened Concrete Beams Under Impact Loading. *ACI Struct. J.* **2005**, *102*, 139–149.

10. Soleimani, S.M.; Banthia, N.; Mindess, S. Sprayed GFRP shear-strengthened reinforced concrete Beams under Impact Loading. *Adv. Constr. Mater.* **2007**, 279–286. [CrossRef]

11. Pham, T.M.; Hao, H. Impact behavior of FRP-strengthened RC beams without stirrups. *J. Compos. Constr.* **2016**, *20*, 1–13. [CrossRef]

12. Pham, T.M.; Hao, H. Behavior of fiber-reinforced polymer-strengthened reinforced concrete beams under static and impact loads. *Int. J. Prot. Struct.* **2017**, *8*, 3–24. [CrossRef]

13. Liu, T.; Xiao, Y. Impact Behavior of CFRP-Strip-Wrapped RC Beams without Stirrups. *J. Compos. Constr.* **2017**, *21*, 37–43. [CrossRef]

14. Bhatti, A.Q.; Kishi, N.; Mikami, H.; Anto, T. Elasto-plastic impact response analysis of shear-failure-type RC beams with shear rebars. *Mater. Des.* **2009**, *30*, 502–510. [CrossRef]

15. Bhattia, A.Q.; Kishi, N. Impact response of RC rock-shed girder with sand cushion under falling load. *Nucl. Eng. Des.* **2010**, *240*, 2626–2632. [CrossRef]

16. Tu, Z.; Lu, Y. Evaluation of typical concrete material models used in hydrocodes for high dynamic response simulations. *Int. J. Impact Eng.* **2009**, *36*, 132–146. [CrossRef]

17. Meng, Y.; Yi, W.J. Dynamic behavior of concrete cylinder specimens under low velocity impact. *J. Vib. Shock* **2011**, *30*, 205–210.

18. Meng, Y. Experiment and Numerical Simulation Study on Reinforced Concrete Beam under Impact Loading. Ph.D. Thesis, Hunan University, Changsha, China, 2012.

19. Meng, Y.; Hang, F.L. Numerical simulation study on dynamic behavior of concrete cylinder under drop hammer impact loading. *J. Railw. Sci. Eng.* **2017**, *14*, 2427–2434.

20. Pham, T.M.; Hao, H. Plastic hinges and inertia forces in RC beams under impact loads. *Int. J. Impact Eng.* **2017**, *103*, 1–11. [CrossRef]

21. Zhao, W.C.; Qian, J.; Zhang, W.N. Performance and damage assessment of reinforced concrete beams under impact loading. *Explos. Shock Waves* **2019**, *39*, 111–122.

22. Liu, J.T. Experimental Study on Shear Behavior of RC T Beams without Stirrups Strengthened with CFRP Sheets under Impact Loading. Master's Thesis, Hunan University, Changsha, China, 2016.

23. Murray, Y.D. *Users Manual for LS-DYNA Concrete Material Model 159. Report No. FHWA-HRT-05-062*; Federal Highway Administration: McLean, VA, USA, 2007.

24. Zeng, X. *Experimental and Numerical Study of Behaviors of RC Beams and Columns under Impact Loadings and Rapid Loadings*; Hunan University: Changsha, China, 2014.

25. Zhao, S.Y.; Xue, P. New two-dimensional polynomial failure criteria for composite materials. *Adv. Mater. Sci. Eng.* **2014**, *2014*, 1–7. [CrossRef]

26. Yang, L.; Yan, Y.; Kuang, N. Experimental and numerical investigation of aramid fibre reinforced laminates subjected to low velocity impact. *Polym. Test.* **2013**, *32*, 1163–1173. [CrossRef]

27. Mou, H.L.; Zhou, T.C.; Feng, Y.F.; Feng, Z.Y. Experiment and Simulation of Composite Corrugated Plate under Quasi-static Crushing and Analysis of Material Model Parameters. *Mech. Sci. Technol. Aerosp. Eng.* **2015**, *34*, 618–622.

28. Han, H.; Taheri, F.; Pegg, N.; Lu, Y. A numerical study on the axial crushing response of hybrid pultruded and ±45° braided tubes. *Compos. Struct.* **2007**, *80*, 253–264. [CrossRef]
29. Al-Zubaidy, H.; Zhao, X.L.; Al-Mahaidi, R. Mechanical Characterisation of the Dynamic Tensile Properties of CFRP Sheet and Adhesive at Medium Strain Rates. *Compos. Struct.* **2013**, *96*, 153–163. [CrossRef]
30. Zhang, X.H.; Hao, H.; Shi, Y.C.; Cui, J.; Zhang, X. Static and Dynamic Material Properties of CFR/Epoxy Laminates. *Constr. Build. Mater.* **2016**, *114*, 638–649. [CrossRef]
31. Rodriguez-Nikl, T.; Lee, C.S.; Hegemier, G.A.; Seible, F. Experimental performance of concrete columns with composite jackets under blast loading. *J. Struct. Eng.* **2012**, *138*, 81–89. [CrossRef]

*Article*

# Compressive Strength Forecasting of Air-Entrained Rubberized Concrete during the Hardening Process Utilizing Elastic Wave Method

**Zhi Heng Lim [1], Foo Wei Lee [1,*], Kim Hung Mo [2], Jee Hock Lim [1], Ming Kun Yew [1] and Kok Zee Kwong [1]**

[1]   Department of Civil Engineering, Universiti Tunku Abdul Rahman, Kampar 31900, Negeri Perak, Malaysia; Leon2810-lzh@hotmail.com (Z.H.L.); limjh@utar.edu.my (J.H.L.); yewmk@utar.edu.my (M.K.Y.); kwongkz@utar.edu.my (K.Z.K.)

[2]   Department of Civil Engineering, University of Malaya, Kuala Lumpur 50603, Wilayah Persekutuan Kuala Lumpur, Malaysia; khmo@um.edu.my

*   Correspondence: leefw@utar.edu.my; Tel.: +60-149-643-833

Received: 17 June 2020; Accepted: 16 July 2020; Published: 5 August 2020

**Abstract:** Conventional compressive strength test of concrete involves the destruction of concrete samples or existing structures. Thus, the focus of this research is to ascertain a more effective method to assess the compressive strength of concrete, especially during the hardening process. One of the prevalent non-destructive test (NDT) methods that involves the employment of elastic wave has been proposed to forecast the compressive strength development of air-entrained rubberized concrete. The change of the properties, such as wave amplitude, velocity and dominant frequency of the wave that propagates within the concrete is investigated. These wave parameters are then correlated with the compressive strength data, obtained using the conventional compressive strength test. It has been certified that both correlation between wave amplitude and concrete compressive strength, as well as the correlation between velocity and concrete compressive strength, have high regression degrees, which are 0.9404 and 0.8788, respectively. On the contrary, dominant wave frequency has been proved imprecise to be used to correlate with the concrete compressive strength development, as a low correlation coefficient of 0.2677 is reported. In a nutshell, the correlation data of wave amplitude and velocity could be used to forecast the compressive strength development of an air-entrained rubberized concrete in the future.

**Keywords:** compressive strength; non-destructive test (NDT); elastic wave; air-entrained rubberized concrete

## 1. Introduction

Concrete serves as the essential essence in almost every variety of typical construction. Concrete can be found in almost every building structure, typical examples of which are pavement, bridge, house, tunnel or dam [1]. This material is widely applied in construction, due to its excellent ability to resist compression loadings. Following the concrete development to suit to different construction purposes, rubberized concrete has been given a spotlight in the civil industry. The ground crumb rubber particles are usually mixed in cement concrete for various civil applications, such as geotechnical works, in road construction, in agriculture to seal silos, in onshore and offshore breakwaters and in retaining walls [2]. Numerous established studies have shown that the incorporation of crumb rubber alters the physical and mechanical properties of the concrete. The inclusion of crumb rubber particles as concrete aggregates present concrete with lower ultimate strength, modulus of elasticity. Nevertheless, they also enhance the ductility, energy and force absorbing potential, as well as the lightweight nature of the concrete [3–5]. The promising lightweight nature of rubberized concrete is capable of satisfying various non-load bearing civil applications. Apart from that, the employment of air-entraining agents in rubberized concrete can further lengthen the life span of rubberized concrete, particularly when the concrete is continuously exposed to the external environment. Recent research has shown that air-entraining agents modify the rheology of concrete and enhances its durability in freezing and thawing cycles [6]. However, the compressive strength of air-entrained rubberized concrete is somewhat inconsistent, compared to that of conventional concrete. Therefore, the strength analysis and interpretation of air-entraining rubberized concrete are somewhat tedious and time-consuming. Thus, a more effective method of determining the concrete strength needs to be invented.

Generally, concrete compressive strength can be classified into two main categories, which are destructive test (DT) and non-destructive test (NDT). While DT is being carried out, the concrete is generally destroyed to evaluate its strength properties [7]. It is always essential to perform this test, because of the importance of ensuring the quality of the manufactured concrete for long-lasting performance. The process is rather simple and straightforward, and results can be obtained without consuming much time. In general cases, this test is compulsory in every construction process, especially just before the concrete is being cast. On the contrary, NDT, as an alternative measure, gives a rather remarkable and effective way of assessing the concrete compressive strength. It has the main advantage of obtaining the properties of concrete rapidly without destroying the specimen at a moderate cost. This method puts less concern on the powered performance of the concrete, while it focuses heavily on the evaluation of physical characteristics [8]. This unique method also provides a more suitable way of assessing the concrete strength of the existing structures. Nowadays, the use of NDT is encouraged, due to its effectiveness in achieving the purposes of examining both the external and internal states, as well as the current condition of the concrete structures, particularly in completed buildings. The process of acquiring the characteristics of the specimen, as well as existing structures utilizes the application of ultrasonic and sonic as the facilitators of the test. Additionally, some methods of NDT adopt the radar, as well as infrared technology, to achieve the investigation of properties that are not visible to the naked eye [9].

Over the past few years, there have been abundant researches carried out relating to the application of NDTs in the evaluation of the mechanical properties of the concrete, particularly compressive strength. Common examples of NDTs include the impact echo and ultrasonic pulse velocity tests. The wave frequency obtained from hammering the concrete specimen, along with the ultrasonic wave velocity when it travels through the concrete specimen, could be used in concrete compressive strength forecasting. One finding worth mentioning is that the variations of frequency and ultrasonic pulse velocity under constant applied load can be adopted to predict the concrete compressive strength under damaged and undamaged states [10]. The relationship between the ultrasonic pulse velocity and concrete compressive strength has been investigated based on the propagation of mechanical wave through the concrete specimens in another study. Although there is no direct physical relationship between the compressive strength and the wave properties, the relationship curves can be established beforehand to correlate the wave parameters and the concrete compressive strength [11]. Furthermore, the effects of the water to cement ratio, maximum aggregate size, aggregate type and fly ash addition on the modulus of elasticity of low-quality concrete was investigated using ultrasonic pulse velocity method, by Yildirim et al. [12]. According to their results, a strong relationship was established between the modulus of elasticity and ultrasonic pulse velocity. However, the ultrasonic pulse velocity test has been generally applied to detect the voids and discontinuities in hardened concrete, and is more sensitive to the internal structure, including the density of concrete [13]. The concrete density always displayed a close relationship with the concrete compressive strength, and this is the reason why the ultrasonic pulse velocity test is eligible in the evaluation of the concrete compressive strength. Moreover, Amini et al. [14] developed the concrete compressive strength predictive models that are independent of concrete's past history and mixed proportions. Ultrasonic pulse velocity and rebound hammer tests were adopted in their studies. The compressive strength of the concrete specimens obtained using conventional compression tests were used in the correlation with the mechanical wave parameters. The constructed predictive models present a significant accuracy in estimating the concrete strength, regardless of its past history and mix proportion. Sreenivasulu et al. [15] studied the evaluation of compressive strength of geopolymer concrete using some common NDT techniques, including Schmidt rebound hammer ultrasonic pulse velocity and combined methods. It was found that rebound number obtained using rebound hammer test depicts a better correlation degree with the compressive strength of geopolymer concrete.

In this study, the aim is to investigate the compressive strength development of different mixed proportions of air-entrained rubberized concrete through two approaches, namely DT and NDT. The correlation between the concrete compressive strength data obtained on day-1, day-7, and day-28 by DT and NDT will be investigated by using the informal parameters from the elastic wave data. Furthermore, this study also aims to substantiate the eligibility and feasibility of NDT to be employed in the compressive strength forecasting of air-entrained rubberized concrete. This is to divert the use of conventional concrete compression tests, to the application of NDT, which is less time consuming and cost-saving in the future. Moreover, NDT is a more practical method of assessing the compressive strength development of air-entrained rubberized concrete, as wave analysis can be repeatedly conducted on the same specimen on different maturity days of concrete.

## 2. Materials and Methods

### 2.1. Mix Proportion

The main purpose of this study is to ascertain the development of compressive strength of air-entrained rubberized concrete by employing the two approaches mentioned in the earlier part. Thus, five different mix proportions are produced by varying the replacement level of fine aggregate by powdered crumb rubber. The granulometric curves of the sand and powdered crumb rubber used in the production of air-entrained rubberized concrete are depicted in Figure 1.

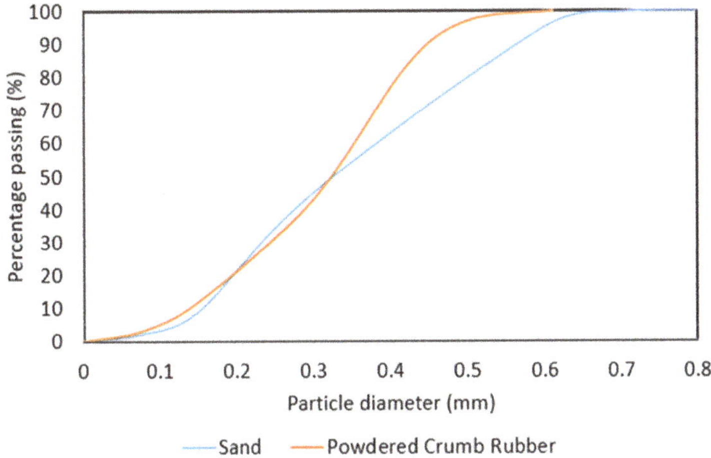


**Figure 1.** Granulometric curves of sand and powdered crumb rubber.

The incorporation of crumb rubber in concrete usually aims to produce the concrete with good lightweight nature, whose density falls below 2000 kg/m$^3$. For each concrete mix proportion, few concrete cube specimens with a dimension of 100 mm × 100 mm × 100 mm are produced for compressive strength assessment at different stages of concrete development. The mix proportion details of the air-entrained rubberized concrete specimens are summarised in Table 1. S0, S20, S40,

S60 and S80 denote air-entrained rubberized concrete, with 0%, 20%, 40%, 60% and 80% volume replacement level of fine aggregate by powdered crumb rubber, respectively.

**Table 1.** Material details of different mix proportions.

| Designation | S0 | S20 | S40 | S60 | S80 |
|---|---|---|---|---|---|
| Cement (kg/m$^3$) | 836.95 | 836.95 | 836.95 | 836.95 | 836.95 |
| Fine Aggregate (kg/m$^3$) | 836.95 | 669.56 | 502.17 | 334.78 | 167.39 |
| Water (kg/m$^3$) | 418.47 | 418.47 | 418.47 | 418.47 | 418.47 |
| Powdered Crumb Rubber (kg/m$^3$) | 0.00 | 69.48 | 138.96 | 208.45 | 277.93 |
| Air-entraining Agent (%) | 0.50 | 0.50 | 0.50 | 0.50 | 0.50 |
| Expected bulk density (kg/m$^3$) | 1918.17 | 1828.60 | 1739.02 | 1649.44 | 1559.87 |

### 2.2. Preparation of Concrete Cube Specimen

The concrete construction steps and procedures are conducted consistently for all cube specimens. The steps generally include three stages, namely concrete mixing, demolding, and curing. At first, the cementitious and aggregate materials, such as cement, sand and powdered crumb rubber, are mixed evenly for 1 min in a ball-mixer, and this stage is also known as a dry mix. After that, water is poured into the dry concrete mix, and the mixing continues until the homogeneous state is attained. Superplasticizer is then added into the mix to improve the workability of the concrete. Lastly, the air-entraining agent is incorporated into the concrete mix to exert air voids in the concrete, to lower the fresh density of the concrete. After the unstable air bubbles are eliminated, and the homogeneity state of concrete is confirmed, the fresh concrete is then cast into the cube mold with the dimension of 100 mm × 100 mm × 100 mm.

Demolding of the concrete cube specimens is performed on the next day. The samples are then placed in a water storage tank for concrete curing to maintain the presence of free water on the exterior surface of the samples. The water temperature is maintained within the range of 25 °C to 28 °C. During concrete curing, the elements of cement experience a chemical reaction with water, and the process of binding of aggregates starts, causing heat to be emitted from the concrete. By the time the concrete has been cured for 28 days, the specimens are removed from the water tank, as they have achieved optimum maturity

The concrete cube specimens of each mix proportion are assessed on day-1, day-7 and day-28. Therefore, a certain amount of cube specimen is required for each mix proportion. The number of cube specimen needs to be constructed throughout the whole study is summarized in Table 2.

**Table 2.** Testing methods and the total number of concrete cube specimen required.

| Testing Methods | | No. of Samples Required | | | | |
|---|---|---|---|---|---|---|
| | | S0 | S20 | S40 | S60 | S80 |
| Compressive Strength Test (control) | Day-1 | 3 | 3 | 3 | 3 | 3 |
| | Day-7 | 3 | 3 | 3 | 3 | 3 |
| | Day-28 | 3 | 3 | 3 | 3 | 3 |
| NDT elastic wave method | Day-1 | | | | | |
| | Day-7 | 3 | 3 | 3 | 3 | 3 |
| | Day-28 | | | | | |
| Total | | 12 | 12 | 12 | 12 | 12 |

### 2.3. Testing of the Concrete Cube Specimens

In this study, two main approaches are implemented to assess the concrete strength development of these five different mix proportions on different days of concrete maturity. In the elastic wave approach, the wave result is analyzed in velocity, frequency and amplitude forms. Then, these parameters are then correlated to the concrete compressive strength and compared to that obtained by utilizing the conventional approach, which is compressive strength test using AD 300/EL Digital Readout 3000 kN concrete compression testing machine (Selangor, Malaysia). This is to guarantee the suitability and precision of NDT approaches in assessing the compressive strength of the concrete.

### 2.3.1. Compressive Strength Test

Compressive strength test utilizing a concrete compression machine is a conventional approach adopted to evaluate the concrete compressive or characteristic strength, and it is conducted in accordance with the standard ASTM C109. The concrete cube with a dimension of 100 mm × 100 mm × 100 mm is fabricated as the compressive strength test specimen. The axial compressive load is applied on the sample at the loading rate of 0.02 mm/s. The peak load attained prior to the ultimate failure of the specimen is deemed as the compressive strength of the specimen.

### 2.3.2. Elastic Wave Method

This testing method is classified as NDT, and it involves the utilization of electronic instruments such as data logger, sensors and piezoelectric transducers. The data logger is connected to a portable computer containing Labview Signal Express software (United States). Upon the completion of equipment set up, the piezoelectric transducers are attached to the concrete cube specimen surface with coupling agent wax. After that, a steel ball with a diameter of 15 mm is used to impact the specimen, causing the production of the elastic wave moving through the sample. The complete set up is illustrated in Figure 2.

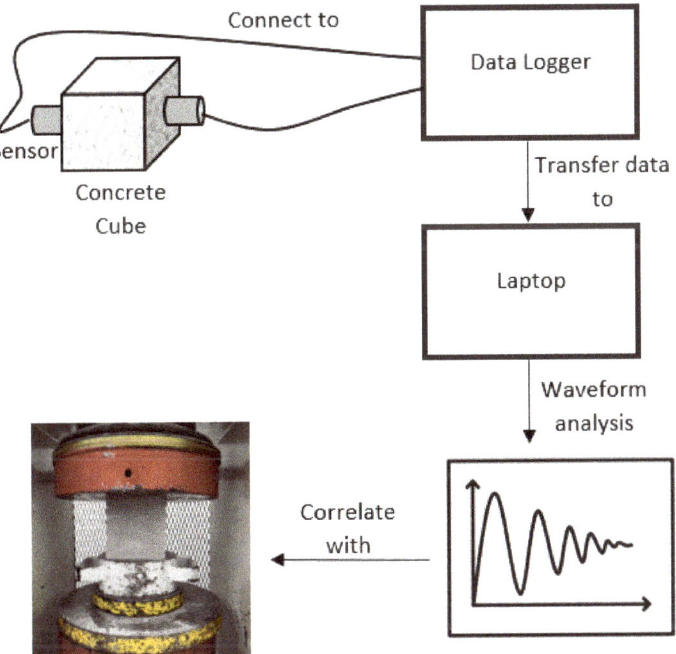


**Figure 2.** Sketch of the experimental set up.

## 3. Experimental Results and Discussion

### 3.1. Compressive Strength Test

The results yielded from the compressive strength test on air-entrained rubberized concrete cube specimens are summarized in Table 3.

**Table 3.** Average compressive strength of air-entrained rubberized concrete with different mix proportions on day-1, day-7 and day-28.

| Sample | Average Oven-Dry Density (kg/m$^3$) | | | Average Compressive Strength (MPa) | | |
|---|---|---|---|---|---|---|
| | Day 1 | Day 7 | Day 28 | Day 1 | Day 7 | Day 28 |
| S0 | 1784 | 1769 | 1788 | 4.28 | 15.16 | 16.24 |
| S20 | 1715 | 1702 | 1704 | 4.09 | 12.84 | 15.11 |
| S40 | 1588 | 1583 | 1568 | 3.26 | 11.89 | 13.28 |
| S60 | 1529 | 1521 | 1532 | 2.79 | 9.15 | 10.94 |
| S80 | 1461 | 1451 | 1454 | 2.51 | 7.57 | 8.39 |

Based on the results tabulated in Table 3, it can be clearly seen that the concrete compressive strength decreases as the crumb rubber proportion increases from 0% to 80%. The same trend is observed in day-1, day-7 and day-28 concrete compressive strength. It is known that crumb rubber is a soft and elastic material which is lighter than sand. The strength of the concrete is highly influenced by the individual strength of each mixing material. Structure wise, powdered crumb rubber is not as strong as sand, and, therefore, the concrete strength tends to drop as the crumb rubber replacement level increases. The identical behavior of rubber particles in concrete is also reported in a study, which states that the rubber particles possess low strength, and this characteristic will be inherited by concrete incorporated with rubber particles [3].

Furthermore, the decline in concrete compressive strength is also due to the lack of adhesion between the smooth crumb rubber particles and cement paste [4]. During the loading phase, cracks will be formed and developed rapidly around the rubber particles, which will lead to concrete rupturing at an accelerated rate. Moreover, the employment of crumb rubber particles in concrete will also contribute to the generation of voids in the hardened concrete. The packing of crumb rubber particles can be complicated and inefficient at high substitution level of sand due to the generation of voids [16].

All the explanations above, on the behavior of powdered crumb rubber in the concrete lead to a conclusion, which is the decline in concrete compressive strength as the crumb rubber proportion increases. According to Table 3, the concrete compressive strength on day-28 drops from 16.24 MPa to 8.39 MPa when the crumb rubber proportion increases from 0% to 80%.

### 3.2. Non-Destructive Impact Echo Test

In this particular test, data of the three main parameters are extracted from the raw waveform and analyzed. These parameters include wave amplitude, velocity and frequency. Each of these wave parameters is then computed individually and explicitly studied to predict the concrete compressive strength of each concrete mix proportion.

#### 3.2.1. Wave Amplitude

As the elastic wave is formed and propagated throughout a medium, its energy level decreases gradually, due to the dissipation of energy that occurs due to travelling in the medium. As a result, the longer the distance of travel of the wave, the higher is the energy loss experienced by that particular wave. The energy level is indicated as the amplitude of the wave in the analysis. Typical examples graphs of amplitude versus time are illustrated in Figure 3, from which the amplitude values can be extracted directly from the raw waveforms. Each of the cube specimens is hammered for ten times, and thus, ten waveforms are acquired and were stacked together to calculate the average values.

Taking concrete cube specimen S0 as an example, the amplitude of P-wave undergoes a noticeable increase from day-1 to day-7, and eventually to day-28. It is relatively certain that P-wave amplitude grows over the concrete cube maturity. The P-wave behavioral changes from day-1 to day-7 and day-28 are depicted as a relationship of acceleration against time in Figure 3a–c.

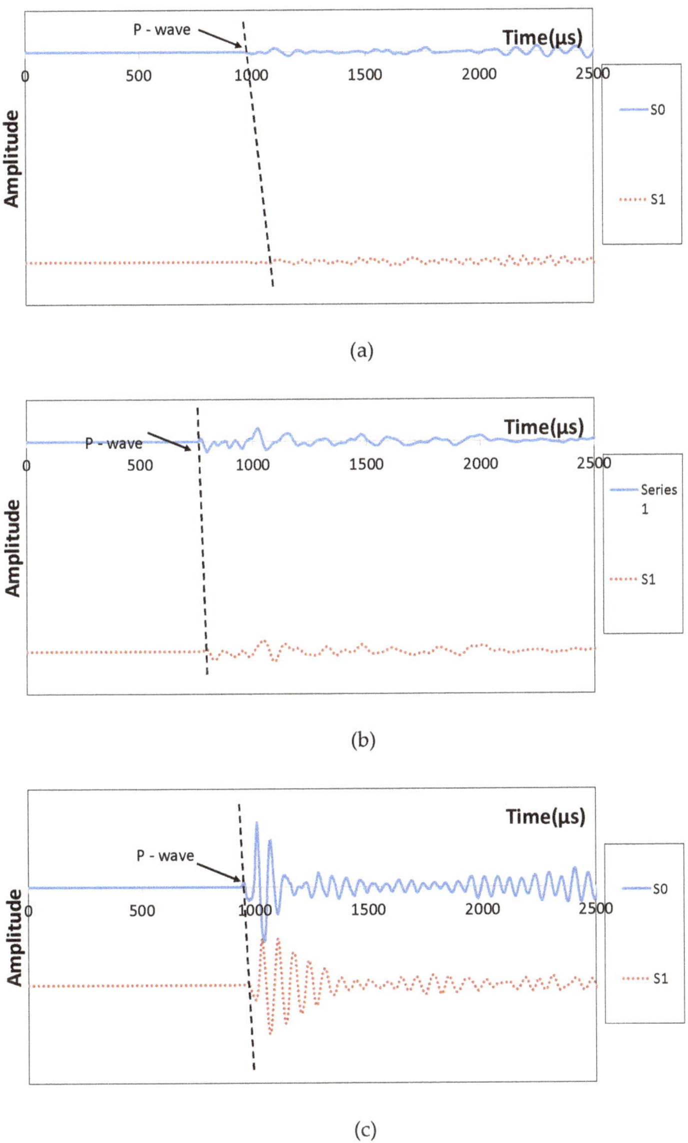


**Figure 3.** Amplitude versus time graph obtained from S0 concrete cube specimen on (**a**) day-1, (**b**) day-7 and (**c**) day-28.

### 3.2.2. Wave Velocity

The velocity of the wave is a parameter that indicates how fast the wave travels within a medium. P-wave is the fastest wave, but it has the lowest amount of energy. To compute the P-wave velocity, the arrival of the P-wave component is determined from the raw waveform captured from the experiment. Once the P-wave component is identified for both sensor 0 and sensor 1, the travel time will be known. Then, the distance travelled by the wave is acquired by averaging the width of the concrete cube specimen (assumed to be 100 mm). By diving the distance travelled by the time taken for the wave to travel to its destination point, the P-wave velocity is then computed.

One observation worth mentioning is that the P-wave velocity generally increases with the maturity time of air-entrained rubberized concrete cube specimen of all mix proportions. The average velocity of P-wave when it travels in each concrete cube specimen is tabulated in Table 4.

**Table 4.** P-wave velocity in each concrete cube specimen on different maturity dates

| Maturity | Average Velocity (m/s) | | | | |
|---|---|---|---|---|---|
| | **S0** | **S20** | **S40** | **S60** | **S80** |
| **Day 1** | 2876 | 3204 | 2981 | 2804 | 3008 |
| **Day 7** | 4296 | 3521 | 3682 | 3310 | 3198 |
| **Day 28** | 4304 | 3821 | 3764 | 3512 | 3589 |

### 3.2.3. P-Wave Dominant Frequency

Frequency of a wave can be defined as the number of the wave that propagates through a fixed point in a particular period. The obtained raw data were transformed to power spectrum using fast Fourier transform (FFT). The fundamental unit of this parameter is Hertz (Hz), which is one wave per unit second. This graph is in the form of amplitude versus frequency, which depicts the distribution of energy levels according to different wave periods. The frequency at which the amplitude is at its peak level is known as the dominant frequency. The power spectrums from the P-wave analysis on concrete cube specimen S0 are displayed in Figure 4. It is discovered that the dominant frequency range for air-entrained rubberized concrete with this particular mix proportion lies within the range of 10,000 to 20,000 Hz.

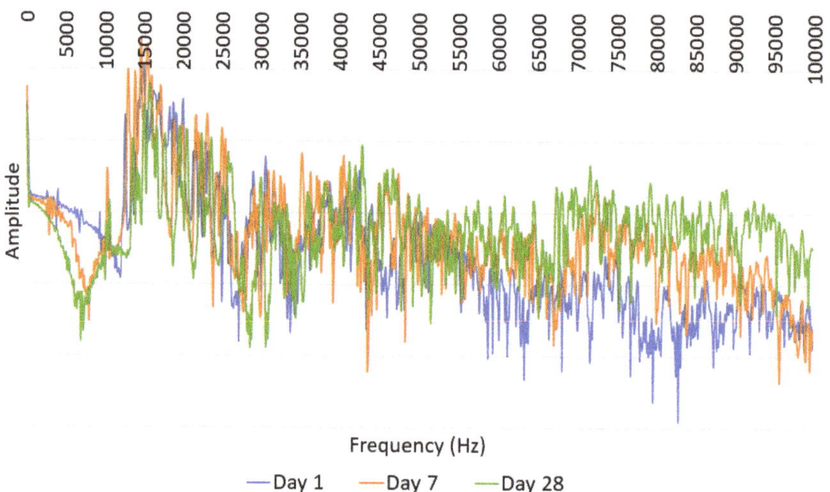


**Figure 4.** Power spectrums generated from P-wave analysis on S0 concrete cube specimen.

### 3.3. Correlation between P-Wave Parameters and Concrete Compressive Strength

The relationships between the concrete compressive strength and these P-wave parameters are certain to be inevitable. The characteristics of these parameters, which are obtained from the testing on concrete cube specimens on different days of maturity (1-day, 7-day and 28-day), are then correlated to that obtained from conventional concrete compressive strength tests. The higher regression values obtained from the correlations could be promising to be used to predict compressive strength, once the P-wave parameters were calculated, such as amplitude and velocity.

### 3.3.1. Correlation with P-Wave Amplitude

The average values of amplitude are inserted into a single chart according to their respective compressive strengths, without scattering them into different charts. This is an effective approach to observe whether all these amplitudes values explicit a similar pattern. The graph that illustrates the correlation with P-wave amplitude is shown in Figure 5.

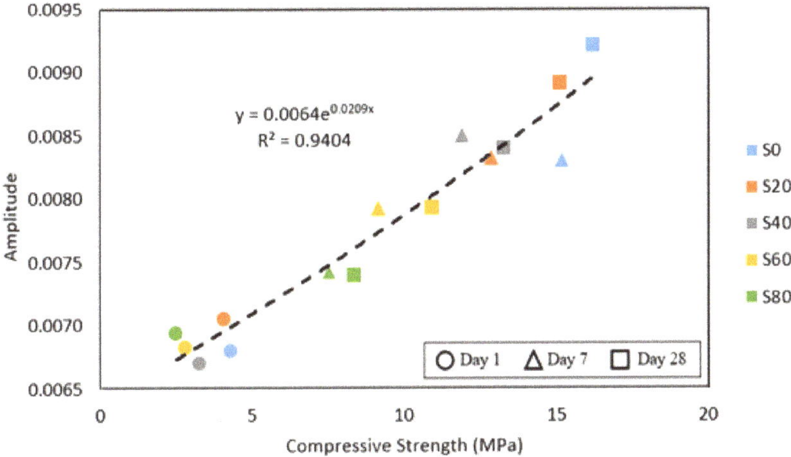


**Figure 5.** Correlation between P-wave amplitude and the concrete compressive strength of all mix proportions on day-1, day-7 and day-28 maturity ages.

Based on the graph in Figure 5, the P-wave amplitude increases corresponding to the increase in compressive strength. The trend line depicts a smooth rising curve from day-1 to day-7, and eventually to day-28 for all air-entrained rubberized concrete mix proportions. This also implies that the energy level of P-wave rises as the compressive strength of the concrete cube specimens increases, regardless of other environmental factors. To rephrase this statement in terms of P-wave analysis, the energy level of P-wave increases due to the lower rate of energy dissipation as the wave travels in a specimen with higher compressive strength. The trend line can be described by an exponential equation $y = 0.0064e^{0.0209x}$, with a high degree of correlation, which is 0.9404. This high R2 coefficient substantiates wave amplitude as a suitable parameter to be adopted in predicting the compressive strength of air-entrained rubberized concrete.

The hardening period has a significant influence on the wave amplitude as the wave passes through the specimen. This is because concrete undergoes hardening, and it triggers a chemical reaction that binds water with cement, sand and powdered crumb rubber particles. As time passes, the hydration process leads to the formation of calcium-silicate-hydrate (C-S-H), which fills the voids within the concrete. The void percentage will be significantly reduced upon the completion of the hardening process, typically on day-28. As the P-wave travels through the matured concrete, less energy is dissipated to the air in void spaces, and the output wave has a more significant portion of remaining energy. This explains why the amplitude of P-wave increases from day-1 to day-7, and eventually day-28 produces a concrete cube specimen with the same mix proportion.

Furthermore, the wave amplitude tends to decline as well, when the crumb rubber proportion increases from 0% (S0) to 80% (S80). This is because the employment of powdered crumb rubber particles in concrete will lead to the generation of more air voids within the concrete. The P-wave disperses more energy to the air voids when it passes through the concrete medium during the P-wave analysis test. Thus, the output P-wave has a smaller portion of the remaining energy.

### 3.3.2. Correlation with P-Wave Velocity

The graph of the correlation between P-wave velocity and compressive strength of the concrete cube specimen is displayed in Figure 6.

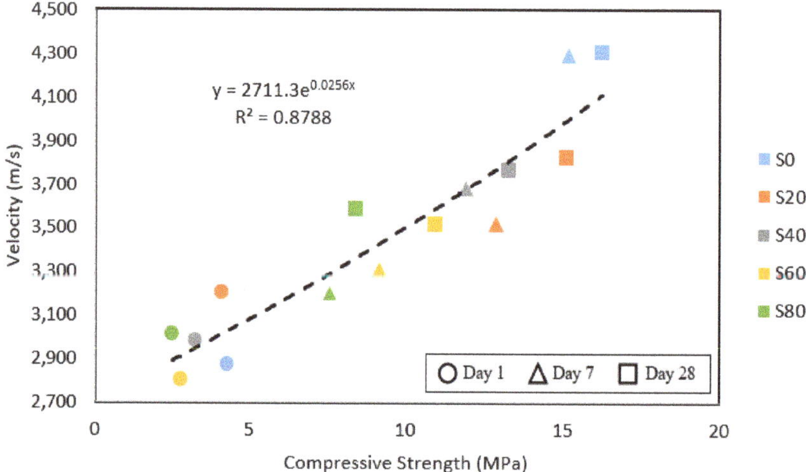


**Figure 6.** Correlation between P-wave velocity and the concrete compressive strength of all mix proportions on day-1, day-7 and day-28 maturity ages.

As can be seen from Figure 6, the P-wave velocity gradually increases responding to the increase in concrete compressive strength. In addition, it is also evident that the P-wave velocity rises as the maturity age of concrete cube specimen changes from day-1 to day-7 and to day-28. For day-1 concrete cube specimens, the velocities range from 2876 to 3204 m/s. For day-7 concrete cube specimens, it has a range with higher velocities, which is from 3198 to 4296 m/s, while day 28 concrete cubes demonstrate a further increase in average velocity, which ranges between 3512 and 4304 m/s. The trend line that expresses the relationship between the P-wave velocity, and the concrete compressive strength follows an exponential function, which is $y = 2711.3e^{0.0256x}$. The regression degree of the trend line curve is excellent as well, which is 0.8788, above the minimum requirement of 0.8 [17].

Concrete porosity and air void percentage are the concrete properties that have a tremendous influence on the P-wave velocity when it passes through a concrete medium. It is noted that since concrete is considered a porous compound, the porosity has a greater tendency of affecting its material strength as it has a strong connection with the presence of air voids too [18]. As concrete ages, it hydrates even further, and the matrix skeleton that links the aggregates together is formed, which, in turn, declines the porosity degree of the concrete medium. P-waves are known to travel faster in a solid medium, and slower in air medium. The pore filling by the concrete material as the result of hydration will cause the P-wave to travel faster in the matured concrete medium.

Another notable finding is that the P-wave tends to move slower in concrete with higher crumb rubber proportion. The P-wave velocity declines from 4303 to 3589 m/s as the crumb rubber proportion increases from 0% to 80% in matured concrete (day-28). This is because the adoption of powdered crumb rubber will contribute to the formation of air voids during the concrete hardening process. Thus, the P-wave consumes more time to reach its destination when it is travelling in a concrete medium, which is composed of a high percentage of powdered crumb rubber.

### 3.3.3. Correlation with P-Wave Dominant Frequency

To perform the correlation analysis, the dominant frequency is extracted from the power spectrum and used in correlation with the concrete compressive strength. The graph that demonstrates the relationship between the P-wave dominant frequency and the concrete compressive strength of all mix proportions on different maturity ages is displayed in Figure 7.

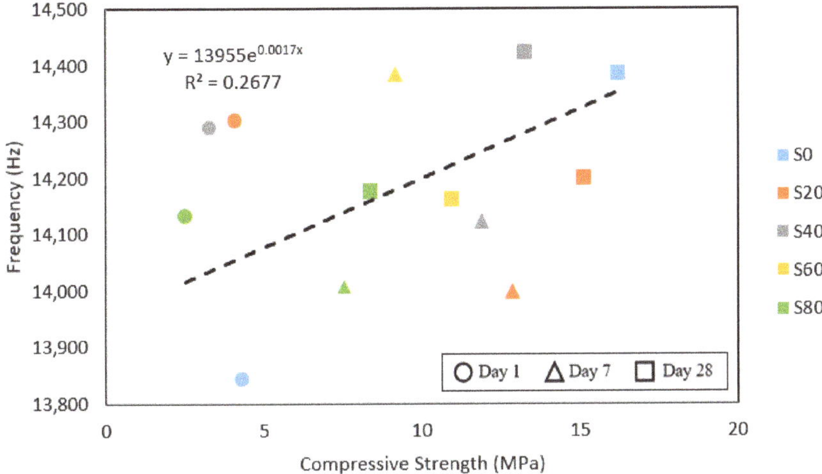


**Figure 7.** Correlation between P-wave dominant frequency and the concrete compressive strength of all mix proportions on day-1, day-7 and day-28 maturity ages.

According to the graph illustrated in Figure 7, the trend line does not seem to represent the scattered points, and it is incurred with many uncertainties. The trial function that is adopted to represent the parametric relationship is $y = 13955e^{0.0017x}$. The coefficient of correlation between the dominant frequency of the P-wave and concrete compressive strength is very low—only 0.2677. The dominant frequency of the P-wave does not establish a solid relationship with the concrete compressive strength during the concrete hardening stage. This is because the dominant frequency is defined as the frequency in which the P-wave amplitude is at its peak level. The distribution of amplitude over the range of frequency when the P-wave travels through the concrete cube specimen is highly dependent on the nature of P-wave itself. This particular P-wave property will not be influenced by the rheology of air-entrained rubberized concrete. Therefore, the dominant frequency is concluded to be inaccurate in estimating the compressive strength of air-entrained rubberized concrete.

## 4. Conclusions

In this study, elastic P-wave amplitude, velocity, and dominant frequency are used in correlation with the compressive strength of air-entrained rubberized concrete that is obtained by using the conventional approach, which is the concrete compressive strength test. Both the P-wave amplitude and the velocity display close linkage to the compressive strength of air-entrained rubberized concrete. This is because both correlation between wave amplitude and concrete compressive strength, as well as the correlation between velocity and concrete compressive strength, have high regression degrees, which are 0.9404 and 0.8788, respectively. Therefore, these two parameters could be adopted in an NDT approach to assess the concrete strength of air-entrained rubberized concrete in the future. However, correlation with the P-wave amplitude seems to be the most precise one. The exponential equation obtained from the correlations with P-wave amplitude (y = $0.0064e^{0.0209x}$) can be utilized to forecast the compressive strength of the air-entrained rubberized concrete with most of the mix proportions. Another point worth mentioning is that only one specimen is required. The P-wave analysis can be repeated on the same specimen on different days of maturity, thus, yielding different amplitude values. By inserting these amplitude values into the exponential equation, the compressive strength development of that particular concrete specimen during the hardening phase can be known. On the contrary, dominant frequency fails to establish a stable and consistent relationship with concrete compressive strength. This is because the relationship between the dominant frequency of P-wave and concrete strength of air-entrained rubberized concrete has a weak regression coefficient, which is only 0.2677, and this figure is far from satisfying the minimum requirement. Thus, the dominant frequency is not suitable to be employed to estimate the concrete strength of air-entrained rubberized concrete. In summary, NDT is an efficient and less time-consuming approach to assess the concrete strength, as it does not involve the demolition of the specimen upon testing and, therefore, a lower quantity of the specimen is required.

**Author Contributions:** Conceptualization, Z.H.L., F.W.L. and K.Z.K.; methodology, K.H.M. and M.K.Y.; software, F.W.L.; validation, J.H.L., Z.H.L. and K.H.M.; formal analysis, M.K.Y., F.W.L. and K.Z.K.; investigation, Z.H.L.; resources, F.W.L.; data curation, K.H.M. and J.H.L.; writing—original draft preparation, Z.H.L. and K.Z.K.; writing—review and editing, F.W.L. and K.H.M.; visualization, J.H.L.; supervision, M.K.Y. and F.W.L.; project administration, Z.H.L. and F.W.L.; funding acquisition, F.W.L. and K.H.M. All authors have read and agreed to the published version of the manuscript.

**Funding:** This research received no external funding.

**Acknowledgments:** The authors would like to express their gratitude to Universiti Tunku Abdul Rahman (UTAR) for providing facilities and resources for the research to progress smoothly.



# References

1. Benghida, D. Concrete as a sustainable construction material. *Key Eng. Mater.* **2017**, *744*, 196–200. [CrossRef]
2. Thomas, B.S.; Gupta, R.C. A comprehensive review on the applications of waste tire rubber in cement concrete. *Renew. Sustain. Energy Rev.* **2016**, *54*, 1323–1333. [CrossRef]
3. Atahan, A.O.; Yücel, A.Ö. Crumb rubber in concrete: Static and dynamic evaluation. *Constr. Build. Mater.* **2012**, *36*, 617–622. [CrossRef]
4. Bisht, K.; Ramana, P.V. Evaluation of mechanical and durability properties of crumb rubber concrete. *Constr. Build. Mater.* **2017**, *155*, 811–817. [CrossRef]
5. Khaloo, A.R.; Dehestani, M.; Rahmatabadi, P. Mechanical properties of concrete containing a high volume of tire–rubber particles. *Waste Manag.* **2008**, *28*, 2472–2482. [CrossRef] [PubMed]
6. Gagné, R. Air entraining agents. In *Science and Technology of Concrete Admixtures*; Pierre-Claude Aïtcin, Robert J Flatt Woodhead Publisher: Amsterdam, The Netherlands, 2016; pp. 379–391.
7. Shankar, S.; Joshi, H.R. Comparison of concrete properties determined by destructive and non-destructive tests. *J. Inst. Eng.* **2014**, *10*, 130–139. [CrossRef]
8. Breysse, D. Nondestructive evaluation of concrete strength: An historical review and a new perspective by combining NDT methods. *Constr. Build. Mater.* **2012**, *33*, 139–163. [CrossRef]
9. Sack, D.A.; Olson, L.D. Advanced NDT methods for evaluating concrete bridges and other structures. *NDT Int.* **1995**, *28*, 349–357. [CrossRef]
10. Arundas, P.H.; Dewangan, U.K. Compressive strength of concrete based on ultrasonic and impact echo test. *Indian J. Sci. Technol.* **2016**, *9*, 1–7. [CrossRef]
11. Hannachi, S.; Guetteche, M.N. Review of the ultrasonic pulse velocity evaluating concrete compressive strength on site. Proceedings of Scientific Cooperation International Workshops on Engineering Branches, Istanbul, Turkey, 8–9 August 2014; pp. 103–112.

12.  Yıldırım, H.; Sengul, O. Modulus of elasticity of substandard and normal concretes. *Constr. Build. Mater.* **2011**, *25*, 1645–1652. [CrossRef]
13.  Rojas-Henao, L.; Fernández-Gómez, J.; López-Agüí, J.C. Rebound Hammer, Pulse Velocity, and Core Tests in Self-Consolidating Concrete. *ACI Mater. J.* **2012**, *109*, 235–243.
14.  Amini, K.; Jalalpour, M.; Delatte, N. Advancing concrete strength prediction using non-destructive testing: Development and verification of a generalizable model. *Constr. Build. Mater.* **2016**, *102*, 762–768. [CrossRef]
15.  Sreenivasulu, C.; Jawahar, J.G.; Sashidhar, C. Predicting compressive strength of geopolymer concrete using NDT techniques. *Asian J. Civil. Eng.* **2018**, *19*, 513–525. [CrossRef]
16.  Benazzouk, A.; Douzane, O.; Langlet, T.; Mezreb, K.; Roucoult, J.M.; Quéneudec, M. Physico-mechanical properties and water absorption of cement composite containing shredded rubber wastes. *Cem. Concr. Compos.* **2007**, *29*, 732–740. [CrossRef]
17.  Frost, J. *Introduction to Statistics: An Intuitive Guide*; **2019**; p. 104. Available online: https://statisticsbyjim.com/basics/introduction-statistics-intuitive-guide/ (accessed on 2 March 2020).
18.  Lian, C.; Zhuge, Y.; Beecham, S. The relationship between porosity and strength for porous concrete. *Constr. Build. Mater.* **2011**, *25*, 4294–4298. [CrossRef]

*Article*

# Analysis of Concrete Failure on the Descending Branch of the Load-Displacement Curve

**Gennadiy Kolesnikov**

Institute of Forestry, Mining and Construction Sciences, Petrozavodsk State University, Lenin pr., 33, 185910 Petrozavodsk, Russia; kolesnikovgn@ya.ru

Received: 11 August 2020; Accepted: 9 October 2020; Published: 12 October 2020

**Abstract:** In this paper, load-displacement and stress-strain diagrams are considered for the uniaxial compression of concrete and under three-point bending. It is known that the destruction of such materials occurs on the descending branch of the load-displacement diagram. The attention of the presented research is focused on the explanation of this phenomenon. Fracture mechanics approaches are used as a research tool. The method for determining effective stresses and modulus of elasticity of concrete based on the results of uniaxial compression tests has been substantiated. The ratios necessary for the calculation were obtained without any assumptions about the reinforcement of concrete and the mechanical properties of its components. However, the effect of these properties is considered indirectly, using the stress and strain peaks determined by standard concrete compression tests. It was found that the effective stresses increase both on the ascending branch and on the descending branch of the load-displacement diagram. This explains the destruction of concrete on the descending branch of the load-displacement diagram. The results of determining the stresses and modulus of elasticity under uniaxial compression are comparable with the results obtained in experiments known in the literature.

**Keywords:** concrete; stress-strain curve; concrete failure; damage of material; effective modulus of elasticity; effective stress

## 1. Introduction

Numerous studies are focused on improving the technical, economic and environmental characteristics of concrete, the continuous flow of which confirms the relevance and complexity of the problems of increasing the competitiveness of this material [1–28]. As a result of research, there are more economical components of concrete, new materials for reinforcement, and improved technologies for the manufacture of concrete as a multicomponent composite material [5–7]. Experimental studies are required to ensure sufficient structural reliability, but the cost remains high. Modern methods of mathematical modeling can reduce the cost of studying concrete and other materials [8,9]. The most complete information about concrete behavior is obtained using stress-strain models, reviews of which can be found in the articles [5,10]. For a sufficiently accurate prediction of the behavior of concrete under various influences, a significant number of studies have been carried out to improve the corresponding models, which, however, remain the subject of discussion. When constructing stress-strain models, methods of analytical mechanics, numerical methods and methods of fracture mechanics can be used [11,12], as well as methods of curve fitting to the experimentally obtained data [13–16]. A typical load–displacement curve is shown in Figure 1.

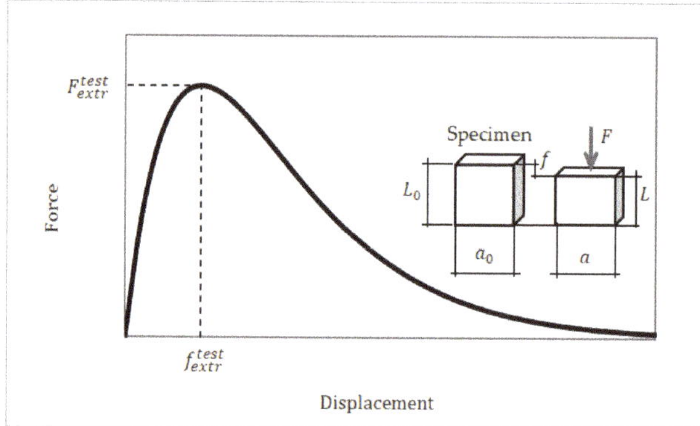


**Figure 1.** Diagram load (*F*)–displacement (*f*) for concrete under uniaxial compression.

In this study, the application of the basic concepts of fracture mechanics to the substantiation of the total stress-strain relations for concrete under uniaxial compression is considered. Such relationships have been studied in numerous works, reviews of which can be found in the articles [12–14]. However, an analysis of the literature showed that the following questions that determine the purpose of this work remained insufficiently studied: How does one explain that concrete samples under uniaxial compression fail on the descending branch of the stress-strain diagram? Is it necessary to take into account the Poisson effect in the proposed approach?

Purpose of the work: development of a methodology for theoretical analysis of the interdependencies of load and displacement, stresses and strains, deformations and material damage, material damage and effective stresses.

In this work, an analytical model is theoretically substantiated, which coincides with the experimentally substantiated Furamura model [17] known from the literature (an analysis of the Furamura model is also available in [14,18]). In addition to the data known from the literature, the model was analyzed using definitions known in fracture mechanics, such as damage function, effective area, effective stresses, and effective elastic modulus. For the listed definitions, in accordance with their physical meaning, analytical relationships are obtained. Using the obtained relations, the coefficient for determining the effective modulus of elasticity was refined: instead of the experimentally found 2.18 [18], a theoretically substantiated coefficient of 2.72 is proposed. In addition, it is substantiated that the effective stresses, in accordance with Hooke's law, continuously increase, both on the ascending and descending branches of the stress-strain diagram under the uniaxial compression. This theoretically explains the destruction of concrete samples on the descending branch of the diagram.

## 2. Materials and Methods

### 2.1. Mechanical Model: A Brief Description

1. Concrete is viewed as a structure composed of interacting mesoscale elements.
2. The material of each element obeys Hooke's law.
3. The modulus of elasticity, strength and other physical and mechanical properties of the material of each element do not depend on its size and do not change over time.
4. With an increase in the external load, and hence displacement, individual mesoscale elements are destroyed, as a result of which the effective area decreases, and the load is redistributed to the elements that remain intact. As a result, the average statistical value of the effective stresses in the material of the remaining intact mesoscale elements increases.
5. The destruction of mesoscale elements and their conglomerates leads to a decrease in the effective area and a decrease in the resistance of the macrostructure to external force, which corresponds

to the descending branch of the "load-displacement" diagram. However, effective stresses (i.e., stresses in the material of mesoscale elements) increase. The growth of effective stresses is limited by the ultimate strength of the material of mesoscale elements.

6.  Stresses determined without taking damage into account can be called apparent stresses [12].

7.  The Poisson effect can cause some growth in the transverse dimensions and a corresponding change in the cross-sectional area of the sample under uniaxial compression. Thus, two trends should be analyzed: first, a decrease in cross-sectional area due to destruction of mesoscale elements and, second, an increase in area due to the Poisson effect.

8.  The primary source of information for the mathematical description of the model and obtaining numerical results is the load-displacement diagram (Figure 1).

### 2.2. Mathematical Description of the Mechanical Model

Let the initial length and cross-sectional area of the sample be equal to $L_0$ and $A_0$, respectively. Effective cross-sectional area is $0 < \widetilde{A} \le A_0$.

It follows from the above that a certain displacement value $f$ corresponds to $\widetilde{A} = A_0 - \widetilde{A}_D + \widetilde{A}_\mu$. Here, $\widetilde{A}_D$ and $\widetilde{A}_\mu$ are partial cross-sectional areas depending on the destruction of mesoscale elements and the Poisson effect, respectively.

The value $f + \Delta f$ corresponds to $\widetilde{A} = A_0 - (\widetilde{A}_D + \Delta\widetilde{A}_D) + (\widetilde{A}_\mu + \Delta A_\mu)$.

### 2.2.1. Determination of $\Delta\widetilde{A}_D$

If $\Delta f$ is a small enough value, then we can assume that

$$\Delta\widetilde{A}_D = -\frac{\Delta f}{f_{extr}^{test}}\widetilde{A} \tag{1}$$

With an increase in axial strain, mesoscale elements are destroyed, and the effective area $\widetilde{A}$ decreases, i.e., the area increment is negative, which is taken into account in (1) by the minus sign. The ratio $\Delta f / f_{extr}^{test}$ is the normalized displacement increment. Axial strain $\varepsilon = f/L_0$ and $\varepsilon_{extr}^{test} = f_{extr}^{test}/L_0$. Then, instead of (1), we write (2):

$$\Delta\widetilde{A}_D = -\frac{\Delta\varepsilon}{\varepsilon_{extr}^{test}}\widetilde{A}. \tag{2}$$

### 2.2.2. Determination of $\Delta A_\mu$

Let $\widetilde{a}$ be the characteristic size of the cross section, such that $\widetilde{A} = \widetilde{a}^2$. If displacement increases from $f$ to $f + \Delta f$, then the effective area under uniaxial compression will be equal to $\widetilde{A} + \Delta\widetilde{A} = (\widetilde{a} + \mu\varepsilon\widetilde{a})^2 \approx \widetilde{A}(1 + 2\mu\varepsilon)$, where $\mu$ is the Poisson's ratio, $\varepsilon$ is axial strain. Thus, $\widetilde{A}_\mu = 2\mu\varepsilon\widetilde{A}$ and

$$\Delta\widetilde{A}_\mu = 2\mu\Delta\varepsilon\widetilde{A}. \tag{3}$$

### 2.2.3. Determination of $\Delta\widetilde{A}$

Using relations (2) and (3), and taking into account that for concrete $\varepsilon_{extr}^{test} \ll 1$, let's define $\Delta\widetilde{A} = \Delta\widetilde{A}_D + \Delta\widetilde{A}_\mu$:

$$\Delta\widetilde{A} = -\widetilde{A}\frac{\Delta\varepsilon}{\varepsilon_{extr}^{test}}(1 - 2\mu\varepsilon_{extr}^{test}) \approx -\widetilde{A}\frac{\Delta\varepsilon}{\varepsilon_{extr}^{test}}. \tag{4}$$

### 2.2.4. Effective Area, Damage Function and Effective Modulus of Elasticity

We transform relation (4) to dimensionless form, dividing both parts by $A_0$. We denote $\widetilde{A}/A_0 = \Theta$ and $\Delta\widetilde{A}/A_0 = \Delta\Theta$. When $\Delta\varepsilon \to 0$, instead of Equation (4), we obtain:

$$\frac{d\Theta}{\Theta} = -\frac{d\varepsilon}{\varepsilon_{extr}^{test}}. \tag{5}$$

Taking into account that if $\varepsilon = 0$, then $\widetilde{A} = A_0$ and $\Theta = 1$, from Equation (5), we find $\Theta = e^{-\frac{\varepsilon}{\varepsilon_{extr}^{test}}}$. Then, the effective area

$$\widetilde{A} = A_0 e^{-\frac{\varepsilon}{\varepsilon_{extr}^{test}}}. \tag{6}$$

It follows from relation (6) that the function $\Theta(\varepsilon)$ is a function of damage: $\widetilde{A} = A_0\Theta$. If $\varepsilon = 0$, then there is no damage, $\widetilde{A} = A_0$, $\Theta = 1$. If $\varepsilon \gg \varepsilon_{extr}^{test}$, then $\widetilde{A} \approx 0$, $\Theta = 0$.

Let's substitute relation (6) into the Equation $F = \varepsilon \widetilde{E}\widetilde{A}$, where $\widetilde{E}$—effective modulus of elasticity:

$$F = \varepsilon \widetilde{E} A_0 e^{-\frac{\varepsilon}{\varepsilon_{extr}^{test}}}. \tag{7}$$

If, then $\varepsilon = \varepsilon_{extr}^{test}$ then $F_{extr}^{test}/A_0 = \sigma_{extr}^{test}$ and from Equation (7), we find

$$\widetilde{E} = \frac{\sigma_{extr}^{test}}{\varepsilon_{extr}^{test}} e. \tag{8}$$

Note that Formula (8) in form and physical meaning coincides with the formula for determining the modulus of elasticity from the paper [18] p. 3; only the coefficients differ, namely, $e \approx 2.72$ in Formula (8), and 2.18 in the cited paper [18].

### 2.2.5. Curve Equation $\sigma(\varepsilon)$

Substituting (8) into (7), we obtain the equation of the curve $\sigma = \sigma(\varepsilon)$:

$$\sigma = \sigma_{extr}^{test} \frac{\varepsilon}{\varepsilon_{extr}^{test}} e^{1-\frac{\varepsilon}{\varepsilon_{extr}^{test}}}. \tag{9}$$

We rewrite (9) in the normalized form (10):

$$\frac{\sigma}{\sigma_{extr}^{test}} = \frac{\varepsilon}{\varepsilon_{extr}^{test}} e^{1-\frac{\varepsilon}{\varepsilon_{extr}^{test}}}. \tag{10}$$

Model in the form (10) coincides with the above-mentioned Furamura model, the analysis of which can be found in the paper [18].

The values of $\sigma_{extr}^{test}$ and $\varepsilon_{extr}^{test}$ are determined from the results of standard compression tests.

## 3. Results and Discussion

### 3.1. Some Features of the Model

If the test is carried out at a constant speed of movement of the crosshead of the testing machine $v$ during time $t$, then $\varepsilon = f/L_0 = vt/L_0$. We denote $S = \sigma/\sigma_{extr}^{test}$, $T = t/t_{extr}^{test}$. Then, using (10), we write

$$S = Te^{1-T}. \tag{11}$$

The speed and acceleration of process (11) are determined by the corresponding derivatives $(n = 1, 2, 3)$:

$$\frac{d^n S}{dT^n} = (-1)^{n-1}(n-T)^{1-T}.$$

(12)

If, in Equation (12), $T = 1, 2, 3$, then $\frac{d^n S}{dT^n} = 0$ (Figure 2).

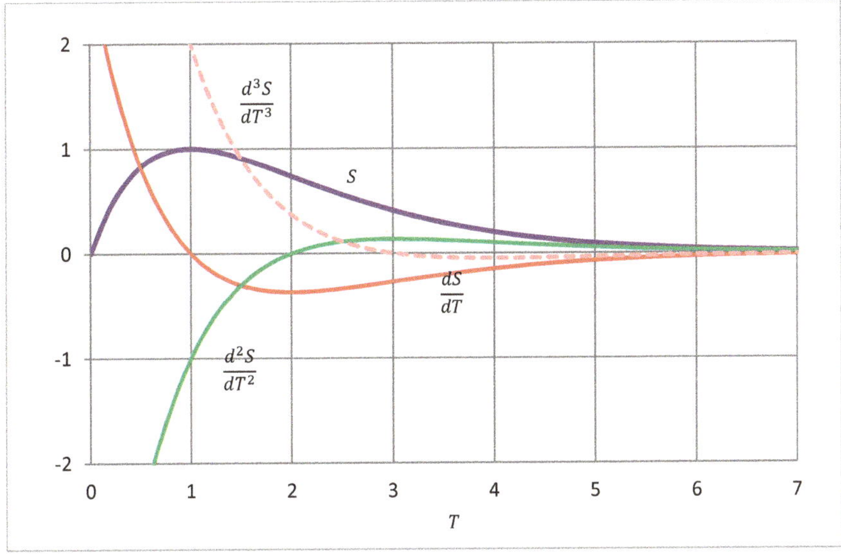


**Figure 2.** Dimensionless characteristics of the process (12).

On the ascending branch of the diagram, the process speed and the absolute value of the acceleration decrease. On the descending branch of the diagram, the speed $dS/dT$ and acceleration $d^2S/dT^2$ of the process are extreme if $S = 2$ and $S = 3$, respectively. It can be assumed that destruction is most likely in the vicinity of these points, which is shown, for example, by diagrams known from the literature [16].

It should be noted that relations (9), (10), and (11) determine changes in the apparent stress $\sigma$ (in the terminology of [12]), i.e., excluding damage (6). Therefore, contradictions are possible when analyzing the physical meaning of the curves in Figure 2.

The actual question is: why does the sample collapse on the descending branch of the diagram? To find the answer, it is necessary to consider the effect of damage (6) and the effective characteristics of the material. Known results in this area are presented, for example, in articles [9,11,12,19–21]. Furthermore, in this work, the above relations (1)–(10) are used.

### 3.2. Effective Stress

Force and displacement are determined experimentally, taking into account (directly or indirectly) all factors that influence the behavior of the sample. Therefore, force and displacement can be attributed to effective characteristics. However, we are interested in predicted characteristics that include effective area $\widetilde{A}$ and damage function $\Theta = \exp(-\varepsilon / \varepsilon_{extr}^{test})$ (6), effective modulus $\widetilde{E}$ (8) and effective stress.

Using (8), we calculate the effective stress $\widetilde{\sigma} = \varepsilon \widetilde{E}$:

$$\widetilde{\sigma} = \varepsilon \frac{\sigma_{extr}^{test}}{\varepsilon_{extr}^{test}} e.$$

(13)

How are apparent and effective stresses interrelated? It follows from relations (9) and (13) that

$$\sigma = \widetilde{\sigma} e^{-\frac{\varepsilon}{\varepsilon_{extr}^{test}}}.$$ (14)

$$\widetilde{\sigma} = \sigma e^{\frac{\varepsilon}{\varepsilon_{extr}^{test}}}.$$ (15)

Let's denote $\widetilde{s} = \widetilde{\sigma}/\sigma_{extr}^{test}$, $\widetilde{e} = \varepsilon/\varepsilon_{extr}^{test}$. Then, using (13), we write $\widetilde{e} = \varepsilon e$. The behavior of the effective and apparent stresses is modeled by the graphs of the functions $\widetilde{s}(t)$ and $s(t)$, respectively (Figure 3).

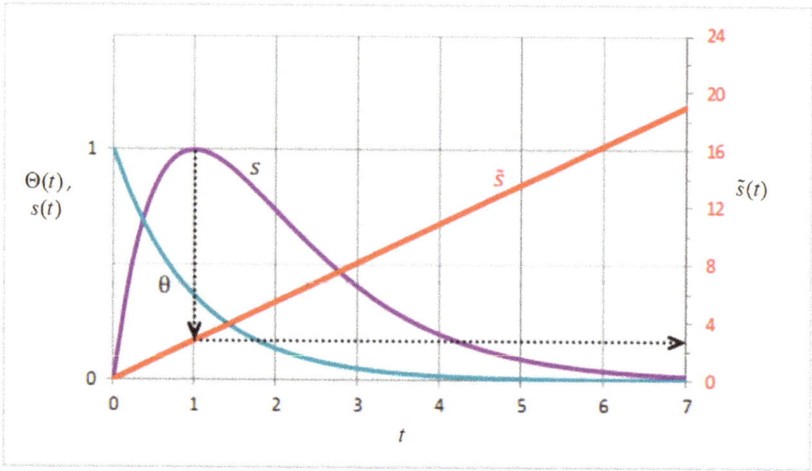


**Figure 3.** Functions $\Theta(t)$, $\widetilde{s}(t)$ and $s(t)$.

If the magnitude of the external force during compression tests changes at a constant rate, then the dependences $\Theta(t)$, $\widetilde{s}(t)$ and $s(t)$ are similar to the graphs in Figure 3.

Figure 3, as well as Formula (13), shows that, with an increase in deformation, the effective stresses only increase. This explains the phenomenon of the destruction of concrete on the descending branch of the load-displacement diagram. The growth of effective stresses is limited by the strength of the material.

It is necessary to pay attention to the fact that the above relations (1)–(15) were obtained without direct assumptions about concrete reinforcement, granulometric composition of the aggregate and about other material properties. Obviously, however, these properties are taken into account indirectly by means of $F_{extr}^{test}$ and $f_{extr}^{test}$ (Figure 1) (either $\sigma_{extr}^{test}$ and $\varepsilon_{extr}^{test}$). Consequently, the presented dependences can be used to simulate other brittle materials, the compression curves of which are similar to the curve in Figure 1. To test this assumption, let us compare the simulation results with experimental data known from the literature [22].

*3.3. Comparison with Experiments Known in the Literature*

In the article [22] (Table 6), stress-strain characteristics for low and high strength concrete (30 and 70 MPa) are presented. In this study, in particular, stress-strain curves are obtained, and peak values of stresses and strains during uniaxial compression of steel-fiber-reinforced concrete are given. The fiber volume fraction varied from 0.00 to 1.25%.

The initial data for calculating the effective modulus of elasticity $\widetilde{E}$ and effective stress $\widetilde{\sigma}$ are given in Table 1.

**Table 1.** Comparison with literature data [18,22].

| Number of Samples: | 1 | 2 | 3 | 4 | 5 | 6 | 7 | 8 | 9 |
|---|---|---|---|---|---|---|---|---|---|
| Fiber volume, % | 0.00 | 0.50 | 0.75 | 1.00 | 0.00 | 0.50 | 0.75 | 1.00 | 1.25 |
| $\sigma_{extr}^{test}$, MPa [22] | 28.19 | 29.34 | 29.94 | 30.87 | 54.65 | 54.86 | 57.94 | 59.82 | 56.91 |
| $\varepsilon_{extr}^{test}$, m$\varepsilon$; [22] (1 m$\varepsilon$ = 0.001) | 1.950 | 2.657 | 2.931 | 2.954 | 2.050 | 3.08 | 3.000 | 3.080 | 3.080 |
| $\widetilde{\sigma}$, MPa, (13) (if $\varepsilon = \varepsilon_{extr}^{test}$) | 76.63 | 79.75 | 81.39 | 83.91 | 148.55 | 149.12 | 157.50 | 162.61 | 154.70 |
| $\widetilde{E}$, MPa; (8) | 39,297 | 30,017 | 27,767 | 28,407 | 72,465 | 48,417 | 52,499 | 52,795 | 50,226 |
| $E$, MPa; [22] | 25,260 | 25,090 | 25,900 | 25,990 | 45,210 | 46,570 | 47,160 | 47,400 | 46,540 |
| $^1E$, MPa; [18] | 31,515 | 24,073 | 22,269 | 22,782 | 58,116 | 38,829 | 42,103 | 42,340 | 40,280 |

$^1$ Calculated by the formula $E = 2.18\frac{\sigma_{extr}^{test}}{\varepsilon_{extr}^{test}}$ according to [18].

The dependences $\sigma(\varepsilon)$ and $\widetilde{\sigma}(\varepsilon)$ for the initial data presented in Table 1 are shown in Figures 4 and 5, respectively.

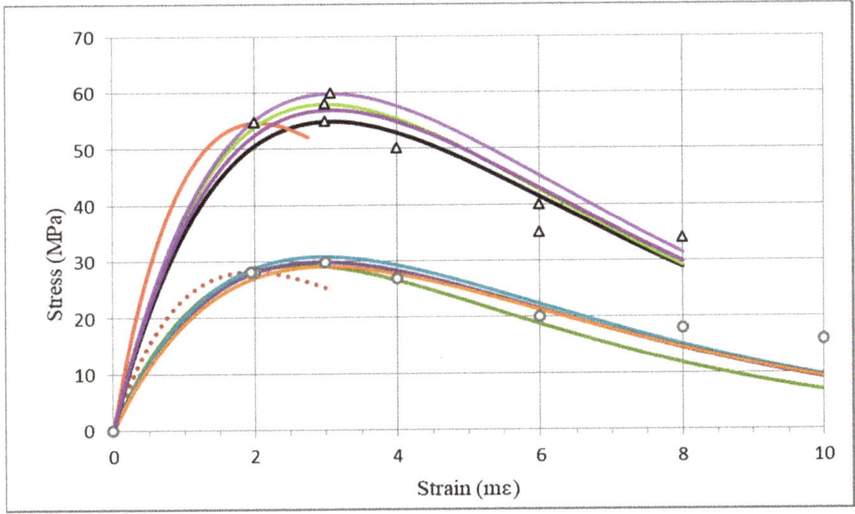


**Figure 4.** Stress: experiments [22] (markers) and calculation (9) (solid lines).

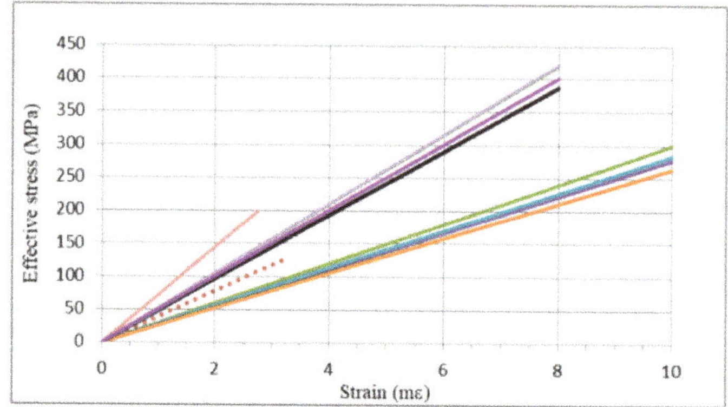

**Figure 5.** Effective stress $\widetilde{\sigma}(\varepsilon)$. Lines 1–10 indicate the numbers of samples according to Table 1.

Figure 5 clearly shows that reinforcement reduces the effective stresses in concrete and, as a result, hinders cracking.

The results presented in Table 1 and Figures 4 and 5 are comparable with the results obtained in experiments [18,22]. With an increase in the reinforcement from 0.5% to 1.25%, the differences between the experimental and calculated values of the elastic modulus (Table 1) decrease from 16% (sample No. 1) to 4% (sample No. 7). These deviations are permissible for practical use. In this case, the determination of effective stresses and elastic modulus is easy to implement using Formulas (8) and (13). Peak values of stresses and strains are used as the initial data for calculations, which are determined experimentally using standard methods.

### 3.4. Relationship between Load and Displacement and Bending Stress-Strain

Using the above approach (Section 2.2.4), we investigate the load-displacement and stress-strain relationships during the bending of the beam. The purpose of this part of the work is to substantiate the statement: the extrema of the load-displacement and stress-strain curves do not coincide. Thus, new (but not exhaustive) data will be obtained on the causes of the destruction of concrete and other brittle materials when the load decreases after passing the extremum on the load-displacement curve.

The current section contains a small theoretical framework for the analysis methodology and an example of analysis of a beam from frozen sandy soil. Note that frozen soil can be viewed as an analogue of concrete, in which ice acts as a binder. The material in this section is a development of an earlier study [25]. This section discusses a new version of the model for analyzing effective tensile stresses in a beam during three-point bending.

Let $B$ and $H_0$—respectively, the width and height of the cross-section of the beam, $f$—the vertical displacement of the point of application of the force $F$ (Figure 6).

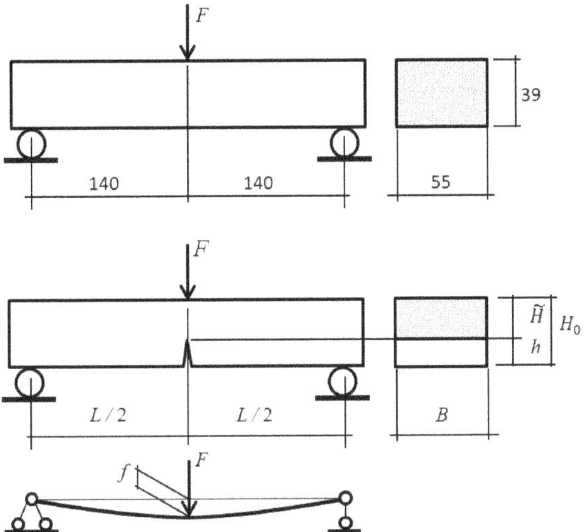

**Figure 6.** Beam before and after cracking (dimensions in mm).

With increasing load $F$, crack of length h appear and grow, as a result of which, the effective section height $\widetilde{H}$ decreases. If the displacement $f$ changes by some value $\Delta f$, then the change in effective height is equal to $\Delta\widetilde{H}$. For sufficiently small increments, the dependence of $\Delta\widetilde{H}$ on $\Delta f$ can be written as a linear function with a constant proportionality coefficient $K_1$:

$$\Delta\widetilde{H} = \frac{\Delta f}{f_{extr}} K_1 \widetilde{H}. \tag{16}$$

We divide both sides of Equality (16) by $H_0$ and pass to the dimensionless parameters $\Theta$ and $\Delta\Theta$:

$$\Theta = \frac{\widetilde{H}}{H_0}. \; \Delta\Theta = \frac{\Delta\widetilde{H}}{H_0}. \tag{17}$$

The parameter $\Theta$ can be considered as a dimensionless characteristic of the effective cross-sectional area of the beam. The values range from 0 to 1; the value $\Theta = 0$ corresponds to complete destruction; the value $\Theta = 1$ corresponds to a condition without damage (no cracks).

If $\Delta\theta \to 0$, then instead of Equality (16), we write:

$$\frac{d\Theta}{\Theta} = \frac{df}{f_{extr}} K_1. \tag{18}$$

Integrating both sides of Equality (18), we determine the integration constant from the conditions: if $f = 0$, then $\widetilde{H} = H_0$, i.e., $\Theta = 1$. We obtain, after transformations:

$$\widetilde{H} = H_0 e^{\frac{f}{f_{extr}} K_1}. \tag{19}$$

Using (19), we determine $\widetilde{I}$—the effective moment of inertia of the cross section with an evolving crack and $\widetilde{W}$—the effective moment of resistance of the same cross section:

$$\widetilde{I} = \frac{B\widetilde{H}^3}{12}. \tag{20}$$

$$\widetilde{W} = \frac{B\widetilde{H}^2}{6}. \tag{21}$$

Experiments have shown that the ratio load $F$—displacement $f$ can be adequately represented as (22):

$$F = \frac{48\widetilde{E}If}{L^3}.$$

(22)

Using (19) and (20), we write (22) as a function $F = F(f)$:

$$F = \frac{48\widetilde{E}I_0 f}{L^3} e^{\frac{3fK_1}{f_{extr}}}.$$

(23)

Here $I = BH_0^3/12$.

Let us consider such a case when function (23) has an extremum (Figure 7). We also assume that the $f_{extr}$ and $F_{extr}$ values are determined from the three-point bend test. Then, from the condition $dF/df = 0$, we find: $K_1 = -1/3$. Knowing $K_1$, using (23), from the condition $F = F_{extr}$ at $f = f_{extr}$, we determine the effective modulus of elasticity $\widetilde{E}$:

$$\widetilde{E} = \frac{F_{extr}L^3}{48I_0 f_{extr}} e = Ee.$$

(24)

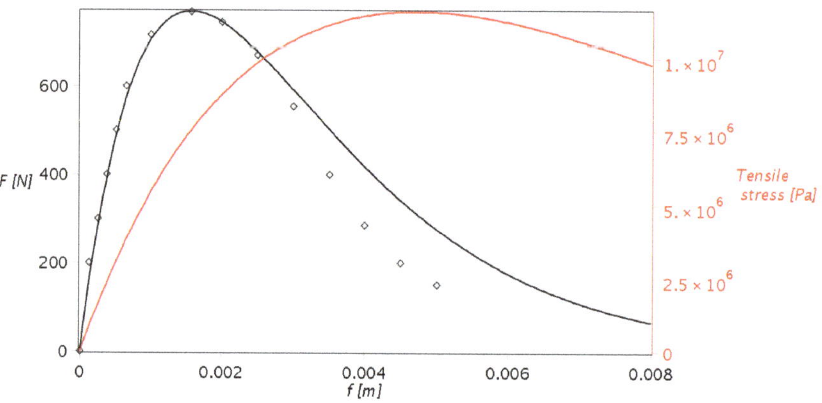


**Figure 7.** The results of tests (markers), load-displacement curve (black line), and stress-strain curve (red line).

Here, $E$ is the modulus of elasticity, which is usually used in the formula for calculating the deflection of a beam in the middle of the span: $f = (FL^3)/(48EI_0)$. Thus, Equality (24) establishes the relationship between the effective modulus of elasticity $\widetilde{E}$ and the ordinary secant modulus of elasticity $E$.

Using the found values $K_1$, $\widetilde{E}$ and taking into account relation (23), after transformations, we explicitly define the dependence $F = F(f)$:

$$F = F_{extr} \frac{f}{f_{extr}} e^{1 - \frac{f}{f_{extr}}}.$$

(25)

For the beam according to Figure 6 in experiments, it was found: $F_{extr} = 769$ N and $f_{extr} = 1.574$ mm. The test results are shown in Figure 7 with markers. Load-displacement curve (25) in Figure 7 with black line. Tests [26] were performed on a SHIMADZU AGS-300kNX STD.

The effective tensile stresses in the section with a crack (Figure 6), presented in Figure 7 (red line), were calculated using the Formula (26):

$$\widetilde{\sigma} = \frac{FL}{4\widetilde{W}}.$$

(26)

Here, $\widetilde{W}$ is determined by (21), taking into account relation (19).

## 4. Conclusions

1. The method for determining the effective stresses and modulus of elasticity of concrete based on the results of uniaxial compression tests, taking into account changes in the cross-sectional area, has been substantiated. The ratios necessary for the calculation were obtained without direct assumptions about the reinforcement of concrete, the granulometric composition of the aggregate and about other properties of the material. However, these properties are taken into account indirectly, using stress and strain peaks. Therefore, the presented dependencies can be used to model a certain class of brittle materials.

2. It was found that the effective stresses increase both on the ascending branch and on the descending branch of the load-displacement diagram. This explains the physical meaning of the phenomenon that the destruction of concrete (and other brittle materials) occurs on the descending branch of the load-displacement diagram.

3. The results of determining the stresses and modulus of elasticity under uniaxial compression are comparable with the results obtained in experiments known in the literature. It has been confirmed that, with an increase in reinforcement, the differences between the experimental and calculated values of the elastic modulus decrease from 16% to 4%. These deviations are permissible for practical use. The effective stress plots clearly show that the reinforcement reduces the effective stresses in the concrete and, as a consequence, hinders the formation of cracks.

**Funding:** This research received no external funding.

**Conflicts of Interest:** The author declares no conflict of interest.

## References


1. Makul, N. Modern sustainable cement and concrete composites: Review of current status, challenges and guidelines. *Sustain. Mater. Technol.* **2020**, *25*, e00155. [CrossRef]
2. Wangler, T.; Roussel, N.; Bos, F.P.; Salet, T.A.; Flatt, R.J. Digital concrete: A review. *Cem. Concr. Res.* **2019**, *123*, 105780. [CrossRef]
3. Miller, S.A. The role of cement service-life on the efficient use of resources. *Environ. Res. Lett.* **2020**, *15*, 024004. [CrossRef]
4. Tam, V.W.; Soomro, M.; Evangelista, A.C.J. A review of recycled aggregated in concrete applications (2000–2017). *Constr. Build. Mater.* **2018**, *172*, 272–292. [CrossRef]
5. Matos, L.M.; Barros, J.A.; Ventura-Gouveia, A.; Calçada, R.A. Constitutive model for fibre reinforced concrete by coupling the fibre and aggregate interlock resisting mechanisms. *Cem. Concr. Compos.* **2020**, *111*, 103618. [CrossRef]
6. Inácio, M.M.; Lapi, M.; Ramos, A.P. Punching of reinforced concrete flat slabs–Rational use of high strength concrete. *Eng. Struct.* **2020**, *206*, 110194. [CrossRef]
7. Xu, Y.; Chen, H.; Wang, P. Effect of Polypropylene Fiber on Properties of Alkali-Activated Slag Mortar. *Adv. Civ. Eng.* **2020**, *2020*, 4752841. [CrossRef]
8. Shafieifar, M.; Farzad, M.; Azizinamini, A. Experimental and numerical study on mechanical properties of Ultra High Performance Concrete (UHPC). *Constr. Build. Mater.* **2017**, *156*, 402–411. [CrossRef]
9. Wang, X.; Zhang, M.; Jivkov, A.P. Computational technology for analysis of 3D meso-structure effects on damage and failure of concrete. *Int. J. Solids Struct.* **2016**, *80*, 310–333. [CrossRef]
10. Aslani, F.; Jowkarmeimandi, R. Stress–strain model for concrete under cyclic loading. *Mag. Concr. Res.* **2012**, *64*, 673–685. [CrossRef]
11. Kurumatani, M.; Terada, K.; Kato, J.; Kyoya, T.; Kashiyama, K. An isotropic damage model based on fracture mechanics for concrete. *Eng. Fract. Mech.* **2016**, *155*, 49–66. [CrossRef]
12. Liu, D.; He, M.; Cai, M. A damage model for modeling the complete stress-strain relations of brittle rocks under uniaxial compression. *Int. J. Damage Mech.* **2018**, *27*, 1000–1019. [CrossRef]

13. Gao, Y.; Ren, X.; Zhang, J.; Zhong, L.; Yu, S.; Yang, X.; Zhang, W. Proposed Constitutive Law of Uniaxial Compression for Concrete under Deterioration Effects. *Materials* **2020**, *13*, 2048. [CrossRef]

14. Stojković, N.; Perić, D.; Stojić, D.; Marković, N. New stress-strain model for concrete at high temperatures. *Teh. Vjesn.* **2017**, *24*, 863–868. [CrossRef]

15. Benin, A.V.; Semenov, A.S.; Semenov, S.G.; Beliaev, M.O.; Modestov, V.S. Methods of identification of concrete elastic-plastic-damage models. *Mag. Civ. Eng.* **2017**, *76*, 279–297. [CrossRef]

16. Liu, X.; Wu, T.; Liu, Y. Stress-strain relationship for plain and fibre-reinforced lightweight aggregate concrete. *Constr. Build. Mater.* **2019**, *225*, 256–272. [CrossRef]

17. Furamura, F. Stress-strain curve of concrete at high temperatures. *Trans. Archit. Inst. Jpn.* **1966**, *7004*, 686.

18. Baldwin, R.; North, M.A. Stress-strain curves of concrete at high temperature—A review. *Fire Saf. Sci.* **1969**, *785*, 1. Available online: http://www.iafss.org/publications/frn/785/-1/view/frn_785.pdf (accessed on 10 October 2020).

19. Sevostianov, I.; Levin, V.; Radi, E. Effective properties of linear viscoelastic microcracked materials: Application of Maxwell homogenization scheme. *Mech. Mater.* **2015**, *84*, 28–43. [CrossRef]

20. Li, Y.; Chen, Y.F.; Zhou, C.B. Effective stress principle for partially saturated rock fractures. *Rock Mech. Rock Eng.* **2016**, *49*, 1091–1096. [CrossRef]

21. Mao, X.; Liu, Y.; Guan, W.; Liu, S.; Li, J. A new effective stress constitutive equation for cemented porous media based on experiment and derivation. *Arab. J. Geosci.* **2018**, *11*, 337. [CrossRef]

22. Koniki, S.; Prasad, D.R. Mechanical properties and constitutive stress-strain behaviour of steel fiber reinforced concrete under uni-axial stresses. *J. Build. Pathol. Rehabil.* **2019**, *4*, 6. [CrossRef]

23. Zhou, Y.; Liu, J.; Yang, H.; Ji, H. Failure Patterns and Energy Analysis of Shaft Lining Concrete in Simulated Deep Underground Environments. *J. Wuhan Univ. Technol. Mater. Sci. Ed.* **2020**, *35*, 418–430. [CrossRef]

24. Lemery, J.; Ftima, M.B.; Leclerc, M.; Wang, C. The disturbed fracture process zone theory for the assessment of the asymptotic fracture energy of concrete. *Eng. Fract. Mech.* **2020**, *231*, 107022. [CrossRef]

25. Gavrilov, T.A.; Kolesnikov, G.N. Evolving crack influence on the strength of frozen sand soils. *Mag. Civ. Eng.* **2020**, *94*, 54–64. [CrossRef]

26. Ahmed, M.; Liang, Q.Q.; Patel, V.I.; Hadi, M.N. Local-global interaction buckling of square high strength concrete-filled double steel tubular slender beam-columns. *Thin-Walled Struct.* **2019**, *143*, 106244. [CrossRef]

27. Alatshan, F.; Osman, S.A.; Hamid, R.; Mashiri, F. Stiffened concrete-filled steel tubes: A systematic review. *Thin-Walled Struct.* **2020**, *148*, 106590. [CrossRef]

28. Micelli, F.; Cascardi, A.; Aiello, M.A. A Study on FRP-Confined Concrete in Presence of Different Preload Levels. In Proceedings of the 9th International Conference on Fibre-Reinforced Polymer (FRP) Composites in Civil Engineering—CICE, Paris, France, 17–19 July 2018; p. 493.

*Article*

# Numerical Analysis of a Novel Shaft Lining Structure in Coal Mines Consisting of Hybrid-Fiber-Reinforced Concrete

Xuesong Wang, Hua Cheng, Taoli Wu, Zhishu Yao * and Xianwen Huang

School of Civil Engineering and Architecture, Anhui University of Science and Technology, Huainan 232001, China; xuesongaust@163.com (X.W.); hcheng@aust.edu.cn (H.C.); taoliwu816@163.com (T.W.); huangxianwen194@163.com (X.H.)
* Correspondence: zsyao@aust.edu.cn; Tel.: +86-139-5541-8561

Received: 17 August 2020; Accepted: 11 October 2020; Published: 12 October 2020

**Abstract:** To address the temperature cracking of concrete in frozen shaft linings in extra-thick alluvial layers in coal mines, a novel shaft lining structure of coal mines consisting of hybrid-fiber-reinforced concrete (HFRC) was developed. Using the Finite Element Method (FEM), a numerical simulation test of the HFRC shaft lining structure with four factors and three levels was carried out, and the mechanical characteristics of the shaft lining structure were obtained. The results show that under a uniform surface load, the maximum hoop stress position of the HFRC shaft lining presents a transition trend from the inside surface to the outside surface; the hoop strain of shaft lining concrete is always a compressive strain, and the inside surface is greater than the outside surface. The empirical formula for the ultimate capacity of this new type of shaft lining structure was obtained by fitting. Compared with the model test results, the maximum relative error of the calculated value is only 6.69%, which provides a certain reference value for designing this kind of shaft lining structure.

**Keywords:** hybrid-fiber-reinforced concrete; shaft lining; numerical simulation; orthogonal test; ultimate capacity

---

## 1. Introduction

The shafts of coal mines are very significant to the safety of production in coal mines. Once the shaft lining breaks, it will not only cause great economic losses, but also pose a serious threat to the safety of underground miners. The artificial freeze method is usually used to construct vertical shafts passing through extra-thick alluvial layers, which is also called freezing shaft sinking. However, water gushing, or leakage, often occurs after the frozen soil thaws. For mass concrete pouring projects, such as freezing shafts, the internal temperature of the concrete can reach 60 to 80 °C [1,2], and at this time, the temperature inside the freezing wall is about −10 to −5 °C, and the air temperature inside the shaft is about 0 °C. The huge temperature difference will cause the shaft lining concrete to undergo additional temperature shrinkage, and the shrinkage of the shaft lining concrete is bound to be constrained by the freezing wall or the outside shaft lining, which will produce large temperature stress in the shaft lining concrete [3,4]. However, the early tensile strength of high-strength concrete is low. When the tensile stress caused by temperature stress is greater than its tensile strength, the shaft lining will produce temperature cracks. When the frozen wall is completely thawed, under the action of high-pressure water in the extra-thick alluvium, the temperature cracks in the shaft lining concrete will continue to expand and even penetrate, and then water leakage will occur [5,6], which will seriously threaten the mine safety production. Because of the freezing shaft lining of freezing shaft sinking in extra-thick alluvium, it is urgent to apply high-performance shaft lining concrete with good crack resistance and strong toughness.

Certain amounts of polypropylene plastic steel fiber (PPSF) and polyvinyl alcohol fiber (PVAF) can be added to concrete, which will greatly enhance its crack resistance and toughness [7–9]. At the same time, the synergistic effect and superposition effect resulting from multi-scale and multi-element mixing of fibers can delay the formation of cracks in the hardening stage of concrete, enhance the impermeability and crack resistance of concrete [10–13] and improve the mechanical properties of hybrid-fiber-reinforced concrete (HFRC) compared with those of concrete mixed with single fibers [14,15]. Because of its good anti-cracking and impermeability properties, many scholars have applied fiber concrete to shaft lining structure model tests and studied the force characteristics and failure mechanism of the fiber concrete shaft lining structure model. Yao et al. [16] carried out a model failure test of HFRC shaft lining based on similarity theory and verified that, as a shaft lining construction material, HFRC is superior to ordinary concrete. Qin et al. [17] studied the compressive strength and failure characteristics of ordinary high-strength concrete and HFRC shaft lining models through indoor model tests. The mixed-use of steel fiber and PPSF restricts the expansion of cracks, improves the stress performance of the shaft lining, and mitigates the problems of hoop cracking, water permeability, and uneven deformation during use. However, when mixing shaft lining concrete, the dispersion of steel fibers is difficult to control, and steel fibers will greatly reduce the fluidity of concrete, which is not convenient for engineering applications. Taking into account the workability of concrete during the construction of the project, Yao et al. [18] studied the mechanical properties of concrete with hybrid PPSF and PVAF and then carried out a similar model test on the shaft lining structure based on similarity theory. By comparing it with the ordinary concrete group, it was found that hybrid-fiber had little effect on improving the uniaxial compressive strength of concrete. However, it can greatly reduce the cracking of concrete and improve the ultimate capacity of the shaft lining.

With the rapid development of computer technology, many scholars have carried out numerical simulations of fiber-reinforced concrete, thereby enriching the research theory and research methods of fiber-reinforced concrete. Qureshi et al. [19] calibrated the mechanical parameters of concrete and steel bars by basic mechanical property tests and carried out finite element modeling. It was found that the FEM was in good agreement with the experimental results. This research provided a reference for the numerical simulation of concrete structures. The interaction between the fiber and concrete interface is also a major difficulty in the process of modeling. Carozzi et al. [20] used the tangential stress of the non-linear interface to characterize the interaction between the fibers and the surrounding mortar. The meshes were divided by non-linear truss elements. This research provided a reference for defining the interaction between the fiber and concrete interface. Radtke et al. [21] modeled fibers by applying discrete force to the meshes. The background meshes represent the matrix, while the discrete force represents the interaction between the fibers and the matrix. This is a novel calculation method for describing fiber-reinforced concrete.

The above research is mainly focused on mechanical property tests of fiber-reinforced concrete specimens or shaft lining structure models, but numerical simulation research on the HFRC shaft lining structure is less involved. Therefore, using the ANSYS finite element analysis software, this research first studies the influence of the thickness-diameter ratio, concrete design strength, PVAF content, and PPSF content on the HFRC shaft lining structure and explores the mechanical characteristics of the shaft lining structure. Next, according to the simulation results, the empirical calculation formula for the ultimate capacity of this new type of shaft lining structure is obtained by fitting. Then, the rationality of the empirical calculation formula is verified through a shaft lining structure model test. Finally, through a range analysis of the ultimate capacity of the shaft lining structure, the order of influence on the ultimate capacity of the shaft lining structure is analyzed. The research results are expected to provide a certain reference for designing this kind of shaft lining structure.

## 2. Establishment of the Numerical Model of the HFRC Shaft Lining Structure

In the analysis of reinforced concrete structures, ANSYS can not only provide data on the basic mechanical characteristics analysis, including the displacement, strain, and stress caused by the structure under load, but also record and analyze the concrete compression yield, plastic creep of steel bars and bond-slip between steel bars and concrete. Therefore, it is a feasible method to simulate the freezing shaft lining structures by ANSYS [20–24].

### 2.1. Element Type

For modeling the shaft lining structure, separate models were adopted. In this process, HFRC was simulated by SOLID65 element, which can be used to simulate reinforced composite materials (such as steel bars and fibers), concrete cracking (three orthogonal directions), crushing, plastic deformation, etc. The steel bars were simulated by LINK180 element, which is a spatial rod element with functions, such as plasticity, creep, large deformation, large strain, etc. It has a wide range of engineering applications and can be used to simulate steel bars, trusses, springs, etc. It was assumed that there was no relative slip between the two types of elements, and the coordination of the displacement of the concrete elements and the steel bar elements was realized by sharing the nodes [25,26].

### 2.2. Material Constitutive Model and Parameters

In this numerical simulation, the uniaxial compression constitutive relation of HFRC was selected according to Formula (1), which can better reflect the rising and falling parts of the stress–strain relationship curve [27].

$$y = \alpha_a x + (3 - 2\alpha_a)x^2 + (\alpha_a - 2)x^3, \text{ when } x \leq 1$$
$$y = \frac{x}{\alpha_d(x-1)^2 + x}, \text{ when } x \geq 1 \tag{1}$$

In the above equations,

$$x = \frac{\varepsilon}{\varepsilon_c}, y = \frac{\sigma}{f_c} \tag{2}$$

In these formulas, $\sigma$ and $\varepsilon$ stand for the stress and strain of concrete; $\alpha_a$ and $\alpha_d$ stand for the parameter values of the rising and falling parts of the concrete stress–strain curve and were calculated by Formulas (3) and (4); $f_c$ stands for the axial compressive strength of concrete; $\varepsilon_c$ stands for the peak strain of concrete under compression.

$$\alpha_a = 2.4 - 0.01f_{cu} \tag{3}$$

$$\alpha_d = 0.132f_{cu}^{0.785} - 0.905 \tag{4}$$

In these formulas, $f_{cu}$ stands for the cubic compressive strength.

On the falling part of the concrete uniaxial compressive stress–strain curve, when the stress was reduced to $0.5f_c$, the corresponding compressive strain was $\varepsilon_u$. When calculating and analyzing the concrete structures, the uniaxial compressive strain should not exceed $\varepsilon_u$, which is given by Formula (5):

$$\frac{\varepsilon_u}{\varepsilon_c} = \frac{1}{2\alpha_d}\left(1 + 2\alpha_d + \sqrt{1 + 4\alpha_d}\right) \tag{5}$$

Before the numerical simulation, a uniaxial compression test and an axial compression test of HFRC were carried out. The obtained mechanical parameters of concrete are shown in Table 1. In this table, C-1 to C-9 represents the number of concretes selected for the nine shaft linings in this numerical simulation. The elastic strain modulus is the ratio between a load of 40% of the axial compressive strength and the corresponding strain. In addition, Poisson's ratio was uniformly taken as 0.2.

**Table 1.** Mechanical parameters of concretes for shaft lining structures.

| NO. | Compressive Strength/MPa | Axial Compressive Strength/MPa | Peak Compressive Strain/με | Elastic Modulus/MPa |
|---|---|---|---|---|
| C-1 | 80.6 | 62.16 | $2.124 \times 10^3$ | $3.787 \times 10^4$ |
| C-2 | 86.8 | 68.83 | $2.203 \times 10^3$ | $3.838 \times 10^4$ |
| C-3 | 89.7 | 73.11 | $2.246 \times 10^3$ | $3.860 \times 10^4$ |
| C-4 | 81.5 | 62.57 | $2.141 \times 10^3$ | $3.784 \times 10^4$ |
| C-5 | 84.1 | 68.12 | $2.182 \times 10^3$ | $3.832 \times 10^4$ |
| C-6 | 92.4 | 74.54 | $2.258 \times 10^3$ | $3.875 \times 10^4$ |
| C-7 | 78.2 | 61.45 | $2.137 \times 10^3$ | $3.798 \times 10^4$ |
| C-8 | 82.9 | 67.69 | $2.195 \times 10^3$ | $3.825 \times 10^4$ |
| C-9 | 90.7 | 73.68 | $2.269 \times 10^3$ | $3.873 \times 10^4$ |

The axial compressive strength of C-1 concrete was 62.16 MPa, the cubic compressive strength was 80.6 MPa, and the peak compressive strain of concrete was $2.124 \times 10^3$ $\mu\varepsilon$. The uniaxial compressive strength of concrete can be obtained from Formula (1) to Formula (5). The pressure constitutive relationship curve is shown in Figure 1. Similarly, the uniaxial compression constitutive relationship curves from C-2 to C-9 can be obtained.

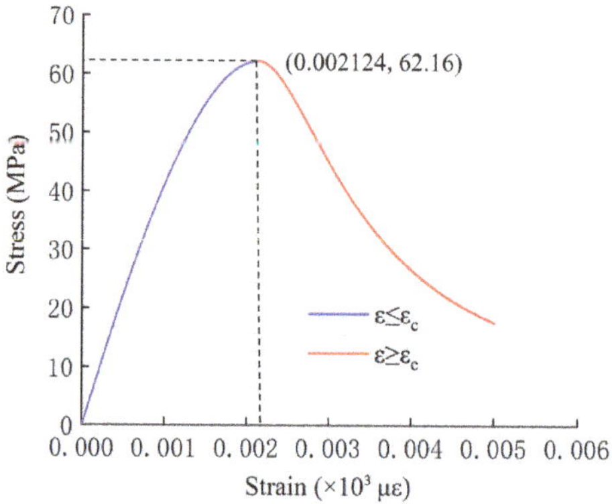


**Figure 1.** Constitutive relationship of C-1 concrete under axial compression.

The selected failure criterion of concrete was Willam and Warnke's five-parameter strength criterion, as shown in Formula (6), which takes into account the multiaxial stress state of concrete.

$$\frac{F}{f_c} - S = 0 \qquad (6)$$

In the above formula, $F = F(\sigma_1, \sigma_2, \sigma_3)$ is the principal stress state function; $S = S(f_t, f_c, f_{cb}, f_1, f_2)$ is the failure surface function; $f_t$, $f_c$, $f_{cb}$, $f_1$ and $f_2$ are the uniaxial tensile strength, uniaxial compressive strength, biaxial compressive strength under hydrostatic pressure, and multiaxial compressive strength under hydrostatic pressure of concrete, respectively, and $f_{cb} = 1.2f_c$, $f_1 = 1.45f_c$, and $f_2 = 1.725f_c$.

The ideal elastic–plastic model was adopted for the steel bars, and the yield condition obeys the Mises criterion. In the numerical calculation, the elastic modulus was $2.1 \times 10^5$ MPa, Poisson's ratio was 0.3, and the yield strength was 240 MPa.

### 2.3. Shaft Lining Simulation Scheme and Boundary Conditions

In the numerical simulation of this research, the only four factors considered were the thickness–diameter ratio, shaft concrete strength, PVAF content, and PPSF content, and it mainly analyzed the stress–strain characteristics of the HFRC shaft lining structure model under loads. The value of the thickness–diameter ratio was selected according to the thickness–diameter ratio of the shaft lining at the engineering site. The three levels of the thickness–diameter ratio were 0.2675,

0.2908, and 0.3140. The concrete of the shaft lining was selected from C-1 to C-9 in Table 1. The PVAF content was 0.728 kg/m$^3$, 1.092 kg/m$^3$ and 1.456 kg/m$^3$. The PPSF content was 4 kg/m$^3$, 5 kg/m$^3$ and 6 kg/m$^3$. According to the four-factor three-level orthogonal test method, it was necessary to carry out the numerical simulation on the shaft lining structure with nine different parameters. The specific design parameters are shown in Table 2.

**Table 2.** Orthogonal design parameters for numerical simulation of shaft lining structures. PVAF, polyvinyl alcohol fiber; PPSF, polypropylene plastic steel fiber.

| NO. | Thickness–Diameter Ratio/λ | Design Strength of Concrete/MPa | PVAF Content/kg·m$^{-3}$ | PPSF Content/kg·m$^{-3}$ |
|---|---|---|---|---|
| D-1 | 0.2675 | C70 | 0.728 | 4 |
| D-2 | 0.2675 | C75 | 1.092 | 5 |
| D-3 | 0.2675 | C80 | 1.456 | 6 |
| D-4 | 0.2908 | C70 | 1.092 | 6 |
| D-5 | 0.2908 | C75 | 1.456 | 4 |
| D-6 | 0.2908 | C80 | 0.728 | 5 |
| D-7 | 0.3140 | C70 | 1.456 | 5 |
| D-8 | 0.3140 | C75 | 0.728 | 6 |
| D-9 | 0.3140 | C80 | 1.092 | 4 |

In this simulation of the shaft lining, the hoop reinforcement ratio was 0.6%, the vertical reinforcement ratio was 0.3%, and the reinforcement diameter was 5 mm. According to the size of the test equipment, the outside diameter of the model was 925 mm, the height was 562.5 mm, and the thicknesses of the shaft lining corresponding to thickness–diameter ratios of 0.2675, 0.2908, and 0.3140 were 97.6 mm, 104.2 mm and 110.5 mm, respectively.

Considering the axial symmetry of the shaft lining structure in this numerical simulation, the $\frac{1}{4}$ 3D finite element calculation model was established according to the design parameters of the shaft lining [28]. In the process of mesh generation, it is necessary to pay attention to the node sharing of steel bars and concrete. The surface load was simulated by applying a larger horizontal load. The boundary condition of the shaft lining model was that the upper and lower end faces were constrained by longitudinal displacement, two $\frac{1}{4}$ cross-sections were constrained by hoop displacement, and then a uniform surface load was applied on the outside surface of the model. The mesh division diagram and boundary conditions of the shaft lining model are shown in Figure 2.

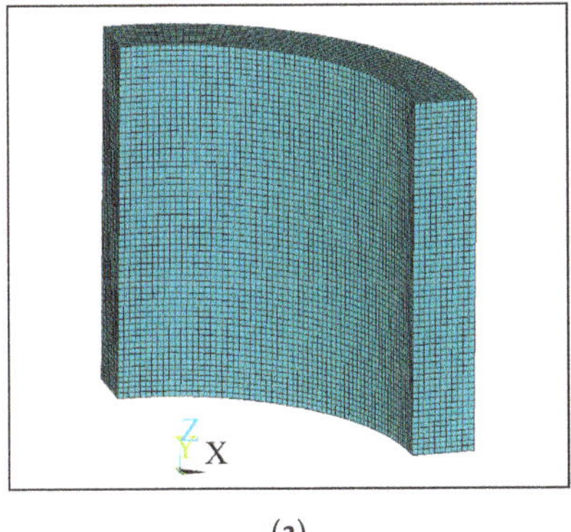


(a)

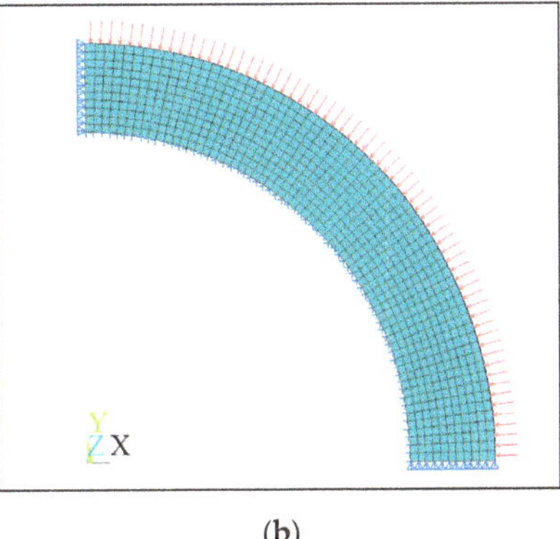


(b)

**Figure 2.** (a) Mesh division of the shaft lining model; (b) boundary conditions of the shaft lining model.

## 3. Analysis of Numerical Simulation Results

### 3.1. Relationship between Hoop Stress and Surface Load

In the general postprocessor, the data on hoop stress and the surface load of the shaft lining under each sub-step were obtained to analyze and study the mechanical characteristics and failure mechanism of the shaft lining structure. Based on the control variable method and the idea of taking the median, the representative D-2, D-5, and D-8 were selected to analyze the stress characteristics of shaft lining concrete under a surface load. The hoop stress cloud diagrams of shaft lining concrete obtained by ANSYS simulation calculations are shown in Figures 3–5.

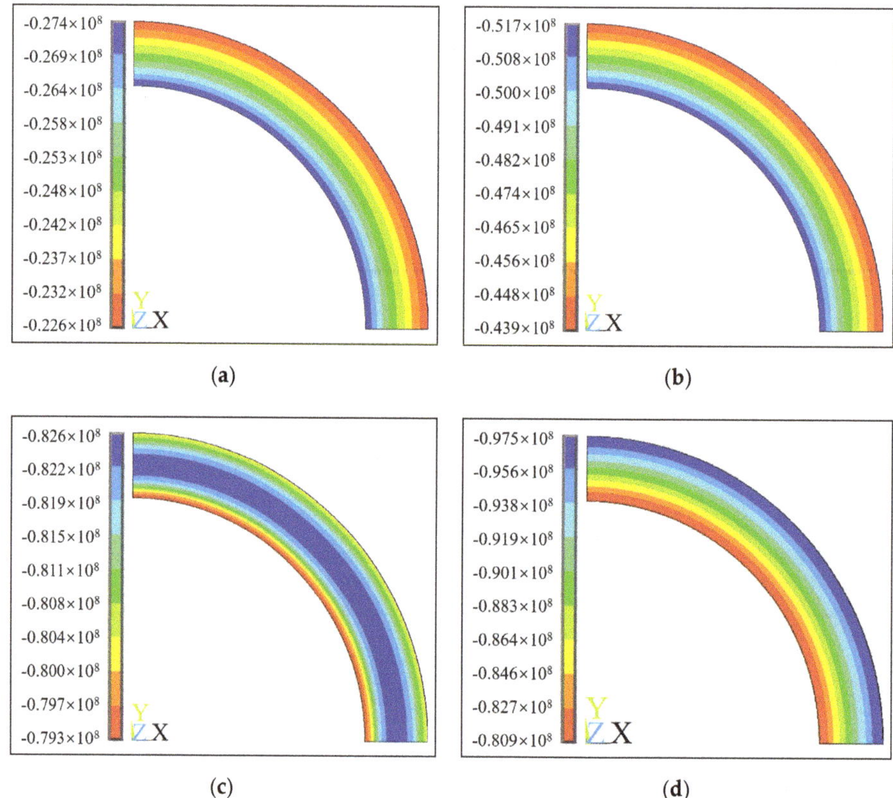

Figure 3. Hoop stress cloud diagrams of D-2 model shaft lining concrete under a surface load: (a) 5 MPa; (b) 10 MPa; (c) 17 MPa; (d) 20.9 MPa.

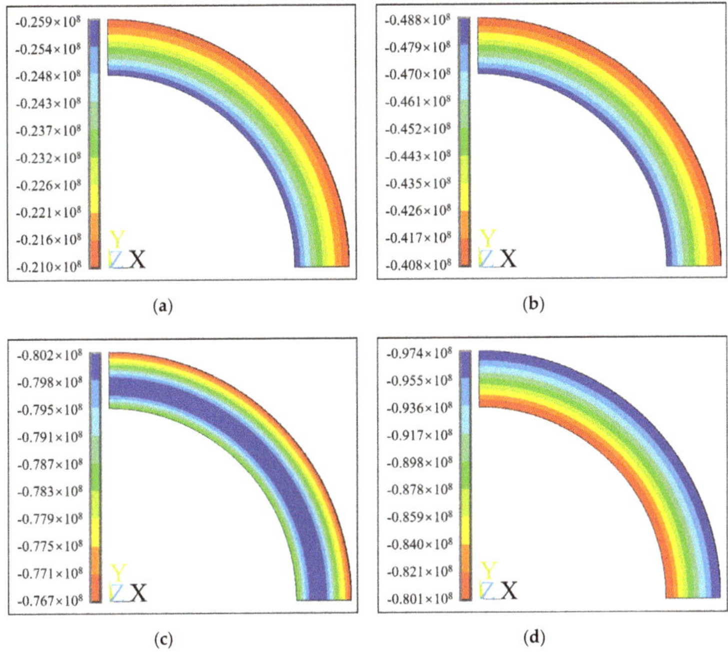

**Figure 4.** Hoop stress cloud diagrams of D-5 model shaft lining concrete under a surface load: (**a**) 5 MPa; (**b**) 10 MPa; (**c**) 18 MPa; (**d**) 23.3 MPa.

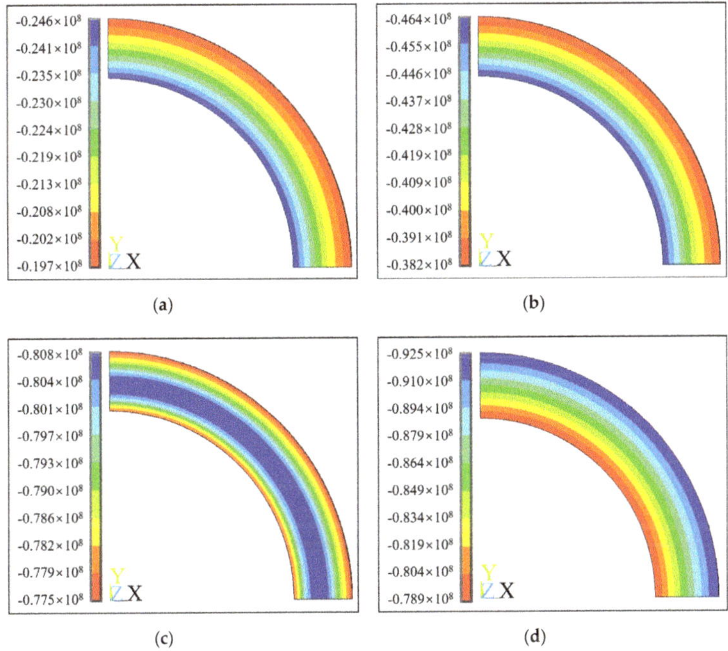

**Figure 5.** Hoop stress cloud diagrams of D-8 model shaft lining concrete under a surface load: (**a**) 5 MPa; (**b**) 10 MPa; (**c**) 19 MPa; (**d**) 25.3 MPa.

It can be seen from Figures 3–5 that under the action of a uniform surface load, the hoop stress of shaft lining concrete is not always unchanged. From Figures 3c, 4c and 5c, it can be seen that when the corresponding surface load reaches 17 MPa, 18 MPa, and 19 MPa, the maximum hoop stress of shaft lining concrete appears in the middle of the shaft lining structure, and the hoop stress of the inside surface reaches the minimum. From Figures 3d, 4d and 5d, it can be seen that the maximum hoop stress presents a transition trend from the inside surface to the outside surface with the increasing surface load.

In order to clearly and intuitively visualize the change in the hoop stress of the shaft lining concrete during the whole loading process, the relationship between the hoop stress and surface load of the inside and outside surfaces of the shaft lining with different shaft thicknesses is drawn, as shown in Figure 6.

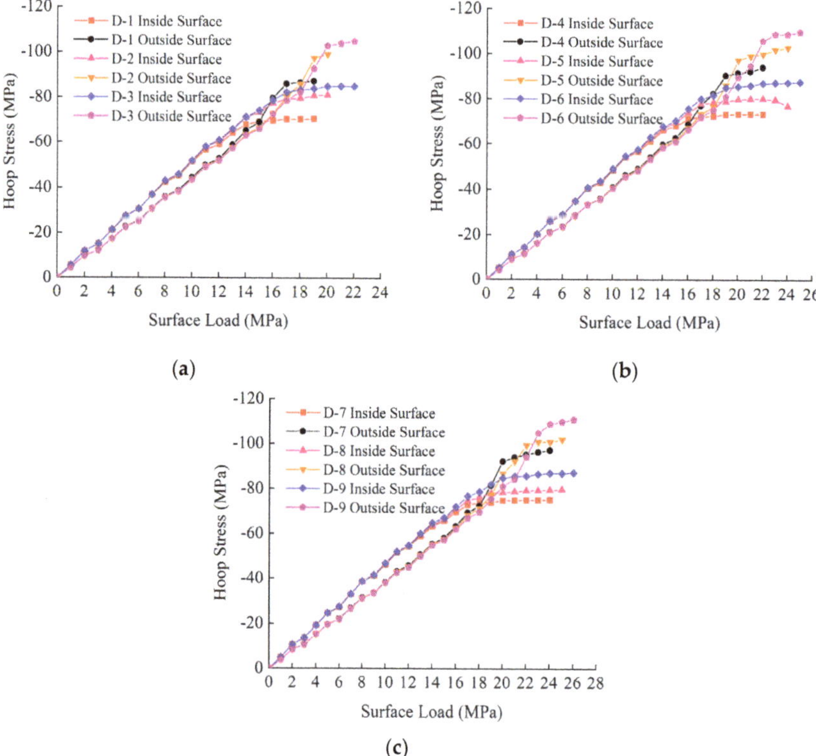


**Figure 6.** Relationship between hoop stresses and surface loads of shaft lining: (**a**) D-1~3 (shaft thickness 97.6 mm); (**b**) D-4~6 (shaft thickness 104.2 mm); (**c**) D-7~9 (shaft thickness 110.5 mm).

It can be seen from Figure 6 that the relationship curve between the hoop stress and surface load of the inside and outside surfaces of concrete can be roughly divided into two sections. In the first stage, the hoop stresses of the inside and outside surfaces of the concrete present highly similar linear growth, and the hoop stress of the inside surface of the concrete is greater than that of the outside surface. In the second stage, with the further increase in the surface load, the hoop stress at the outside surface of the concrete continues to increase, but the growth rate is significantly reduced, while the hoop stress at the inside surface of the concrete tends to stabilize. The outside surface is greater than the inside surface. When the surface load of D-1–D-9 exceeds 15 MPa, 17 MPa, 18 MPa, 17 MPa, 18 MPa, 20 MPa, 18 MPa, 19 MPa, and 21 MPa, respectively, the hoop stress of the outside surface begins to exceed that of the inside surface. In addition, when the shaft lining is damaged, the maximum hoop stress values of each model concrete reach 87.8 MPa, 97.5 MPa, 105 MPa, 92.5 MPa, 103 MPa, 111 MPa, 97.3 MPa, 103 MPa, and 111 MPa, respectively, which all exceed the uniaxial compressive strength of the concrete. According to the analysis, when the surface load acts on the shaft lining structure model, the concretes of the inside and outside surfaces of the shaft lining model are in a bidirectional compression state and three-dimensional compression state, respectively. According to the "Code for Design of Concrete Structures" (GB 50010-2010) [29], the compressive strength of concrete under multiaxial stress conditions would increase relative to that under uniaxial stress conditions.

### 3.2. Relationship between Hoop Strain and Surface Load

D-2, D-5, and D-8 are selected to study the strain characteristics of shaft lining concrete under a surface load. The strain cloud diagrams of the shaft lining concrete obtained by ANSYS simulation calculations are shown in Figures 7–9.

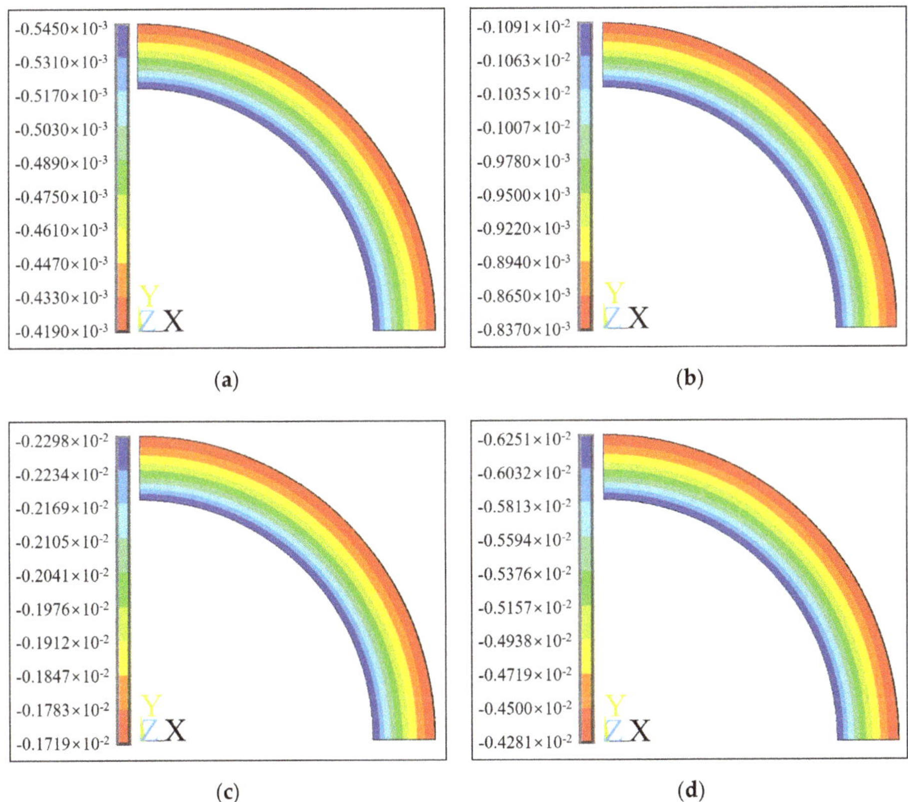

**Figure 7.** Hoop strain cloud diagrams of D-2 model shaft lining concrete under a surface load: (**a**) 5 MPa; (**b**) 10 MPa; (**c**) 17 MPa; (**d**) 20.9 MPa.

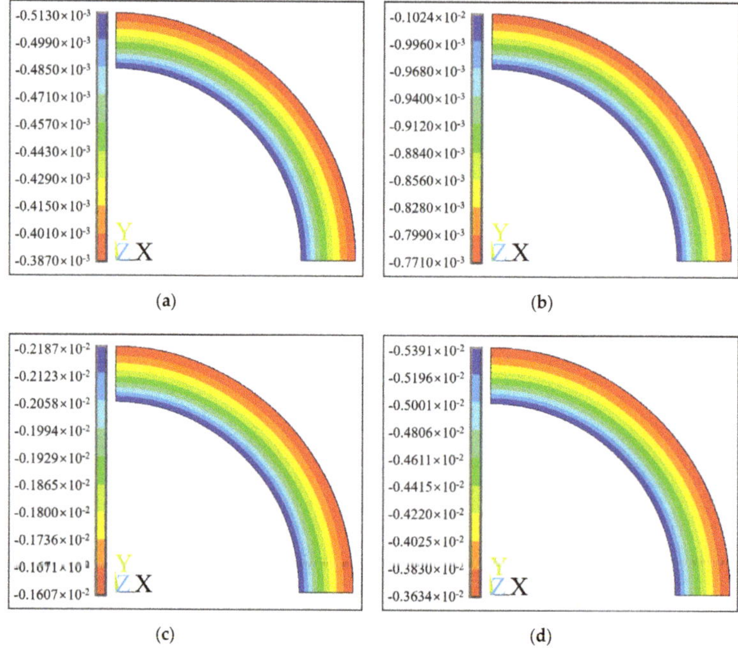

**Figure 8.** Hoop strain cloud diagrams of D-5 model shaft lining concrete under a surface load: (**a**) 5 MPa; (**b**) 10 MPa; (**c**) 18 MPa; (**d**) 23.3 MPa.

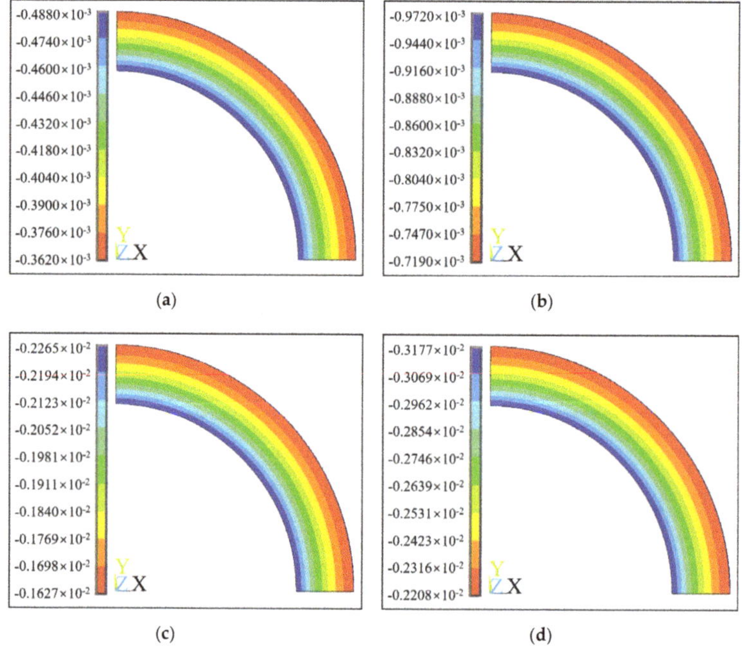

**Figure 9.** Hoop strain cloud diagrams of D-8 model shaft lining concrete under a surface load: (**a**) 5 MPa; (**b**) 10 MPa; (**c**) 19 MPa; (**d**) 25.3 MPa.

It can be seen from Figures 7–9 that, firstly, under the action of the surface load, the hoop strain produced by the concrete of the shaft lining structure model is always a compressive strain. Secondly, with the hoop strain value of the inside and outside surfaces of the concrete during the entire process from loading to failure, the inside surface is always larger than the outside surface, which is different from the regular hoop stresses on the inside and outside surfaces of concrete.

In order to more clearly and intuitively visualize the changes in the hoop strain of the shaft lining concrete during the entire loading process, the relationship between the hoop strain and the surface load of the inside and outside surfaces of the shaft lining concrete with different shaft thicknesses is drawn, as shown in Figure 10.

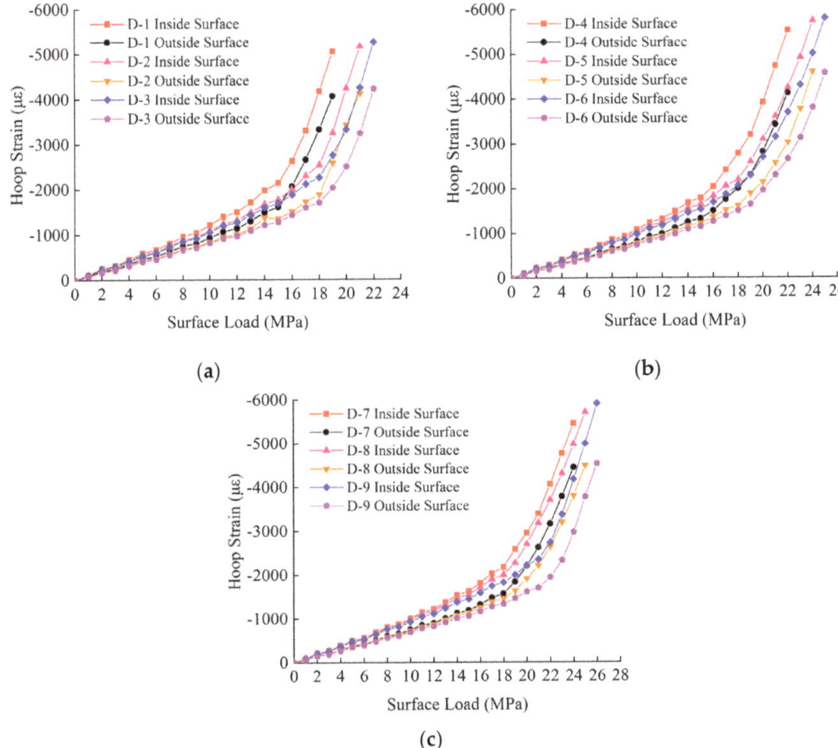


**Figure 10.** Relationship between hoop strain and surface loads of the shaft lining: (**a**) D-1~3 (shaft thickness 97.6 mm); (**b**) D-4~6 (shaft thickness 104.2 mm); (**c**) D-7~9 (shaft thickness 110.5 mm).

It can be seen from Figure 10 that the hoop strain can be roughly divided into two stages for the entire loading process. In the first stage, when the surface load values of the different groups do not exceed 15 MPa, 18 MPa, 19 MPa, 18 MPa, 19 MPa, 21 MPa, 19 MPa, 20 MPa, and 22 MPa, respectively, the hoop strain and surface load show a linear relationship. In the second stage, as the surface load increases until the shaft lining breaks, the hoop strain and the surface load show a non-linear relationship, and the hoop strain value rises rapidly. The analysis shows that the first stage can be regarded as the elastic deformation stage, which corresponds to the rising part of the stress–strain curve in the concrete constitutive model, so the hoop strain of the concrete in each relationship curve is highly similar. In the second stage, due to the action of the hybrid-fiber, the trend of the stress–strain curve of the concrete is continuous and gentle. Therefore, when the failure is approaching, the concrete undergoes substantial plastic deformation, resulting in a rapid increase in the hoop strain of the concrete in the figures.

### 3.3. Analysis of the Ultimate Capacity of the Shaft Lining Structure

Table 3 summarizes the ultimate capacity of the shaft lining structure models calculated by ANSYS numerical simulation.

**Table 3.** The numerical ultimate capacity of shaft lining structure.

| NO. | Thickness–Diameter Ratio | Concrete Strength Grade/MPa | PVAF Volume Fraction/% | PPSF Volume Fraction/% | Axial Compressive Strength of Concrete/MPa | Ultimate Capacity/MPa |
|---|---|---|---|---|---|---|
| D-1 | 0.2675 | C70 | 0.0564 | 0.4396 | 62.16 | 19.5 |
| D-2 | 0.2675 | C75 | 0.0847 | 0.5495 | 68.83 | 20.9 |
| D-3 | 0.2675 | C80 | 0.1129 | 0.6593 | 73.11 | 21.7 |
| D-4 | 0.2908 | C70 | 0.0847 | 0.6593 | 62.57 | 22.2 |
| D-5 | 0.2908 | C75 | 0.1129 | 0.4396 | 68.12 | 23.3 |
| D-6 | 0.2908 | C80 | 0.0564 | 0.5495 | 74.54 | 24.6 |
| D-7 | 0.3140 | C70 | 0.1129 | 0.5495 | 61.45 | 24.5 |
| D-8 | 0.3140 | C75 | 0.0264 | 0.6593 | 67.69 | 25.3 |
| D-9 | 0.3140 | C80 | 0.0847 | 0.4396 | 73.68 | 26.8 |

The next part of this study examines the relationship between the ultimate bearing capacity of the shaft lining and the thickness–diameter ratio, the axial compressive strength of the HFRC, and the volume fraction of the hybrid-fibers. According to the dimensional analysis theory, combined with the results of the numerical simulation of the shaft lining model, the empirical formula for calculating the ultimate capacity of the HFRC shaft lining can be derived as follows:

$$P_b = a\rho_1^b \rho_2^c \lambda^d f_c^e \tag{7}$$

In this formula, $P_b$ stands for the ultimate capacity of the shaft lining, MPa; $\rho_1$ stands for the volume ratio of PVAF, %; $\rho_2$ stands for the volume ratio of PPSF, %; $\lambda$ stands for the thickness–diameter ratio, which is the ratio of the thickness of the shaft lining to the inner radius of the shaft lining; $f_c$ stands for the axial compressive strength of concrete, MPa; $a$, $b$, $c$, $d$ and $e$ stand for the constant to be solved.

From the data in Table 3, the empirical calculation formula for the ultimate capacity of the HFRC shaft lining was obtained by using Origin fitting:

$$P_b = 12.09677\rho_1^{0.01652} \rho_2^{0.0049} \lambda^{1.32888} f_c^{0.57726} \tag{8}$$

### 3.4. Model Test Verification of the Shaft Lining Structure

#### 3.4.1. Model Test of the Shaft Lining Structure

In order to verify the correctness of Formula (8), a high-strength shaft lining structure high-pressure loading device that was independently developed by Anhui University of Science and Technology was used to conduct a model test of the shaft lining structure. The outer diameter and height of the shaft lining model are 925 mm and 562.5 mm, respectively. The parameters of the shaft lining structure model are shown in Table 4.

**Table 4.** Design parameters of the hybrid-fiber-reinforced concrete (HFRC) shaft lining model.

| NO. | Inner Radius/mm | Thickness/mm | Thickness–Diameter Ratio | Design Strength of Concrete/MPa | Reinforcement Ratio/% | PVAF Content/kg·m$^{-3}$ | PPSF Content/kg·m$^{-3}$ |
|---|---|---|---|---|---|---|---|
| D-I | 729.8 | 97.6 | 0.2675 | C70 | 0.6 | 1.092 | 5 |
| D-II | 716.6 | 104.2 | 0.2908 | C75 | 0.6 | 1.092 | 5 |
| D-III | 704.0 | 110.5 | 0.3140 | C80 | 0.6 | 1.092 | 5 |

Note: The reinforcement ratio in the table is the hoop reinforcement ratio; the vertical reinforcement ratio is 0.3%; the reinforcement diameter is 5 mm.

The loading device and the damaged shaft lining model are shown in Figures 11 and 12.

**Figure 11.** Shaft lining structure high-pressure loading device.

**Figure 12.** The failure mode of the HFRC shaft lining model.

It can be seen from Figure 12 that the HFRC shaft lining has no obvious fracture surface when it is destroyed, and it is only partially damaged. The outside surface of the shaft lining structure is relatively complete, and there is a little damage to the concrete where it is cracking and falling off from the inside surface. As a result of the addition of PVAF and PPSF to the concrete, the brittleness of the shaft lining concrete is improved, and the deformability of the shaft lining structure is significantly improved. Therefore, the use of the HFRC shaft lining will greatly improve the deformation characteristics of the shaft lining structure, and at the same time, improve the anti-cracking and seepage resistance performance of the shaft lining structure, which has a significant effect on preventing the shaft lining from flooding.

### 3.4.2. The Relationship between Shaft Lining Hoop Strain and Surface Load

According to the test results collected by the strain gauge, the relationship curve between the hoop strain of the inside and outside surfaces of the HFRC shaft lining model and the surface load is drawn, as shown in Figure 13.

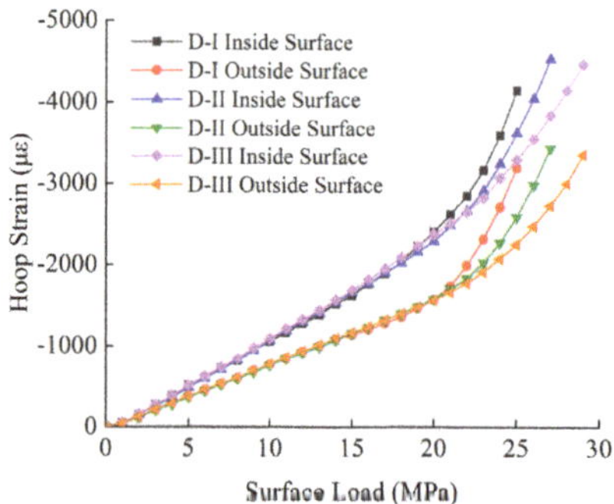


**Figure 13.** Relationship between hoop strain and surface loads of the shaft lining.

It can be seen from Figure 13 that the relationship curve between the hoop strain and surface load of the hybrid-fiber shaft lining concrete can be roughly divided into two stages. In the first stage, when the surface load is 0~20 MPa, the hoop strain values of the inside and outside surfaces increase linearly with the gradual increase in the surface load. In the second stage, the surface load exceeds 20 MPa until the shaft lining model breaks. The growth rate of the hoop strain value is significantly increased. The results are basically consistent with the relationship curve between the hoop strain and surface load of the inside and outside surfaces of the shaft lining obtained by the numerical simulation in Figure 10, which verifies that the parameter selection was reasonable when the shaft lining model was simulated in this research. When the shaft lining model specimen is damaged, the maximum hoop strain of the inside surface concrete reaches $-4539$ $\mu\varepsilon$. The HFRC shaft lining structure shows obvious plastic characteristics, which is beneficial for improving the brittleness of the ordinary high-strength concrete shaft lining structure and plays a positive role in improving the deformation resistance of the frozen shaft lining and the safety and reliability of the shaft lining structure.

### 3.4.3. Verification of the Empirical Formula for the Ultimate Capacity of the Shaft Lining

The relevant parameters of shaft lining models were substituted into Formula (8), and the calculated values of the shaft lining ultimate capacity were obtained, as shown in Table 5.

**Table 5.** The ultimate capacity of the shaft lining structure in the model test.

| NO. | Thickness-Diameter Ratio/λ | Design Strength of Concrete/MPa | PVAF Volume Rate/% | PPSF Volume Rate/% | Axial Compressive Strength of Concrete/MPa | Ultimate Capacity/MPa | | Relative Error/% |
|---|---|---|---|---|---|---|---|---|
| | | | | | | Test Value/MPa | Calculated Value/MPa | |
| D-I | 0.2675 | C70 | 0.0847 | 0.5495 | 63.35 | 23.6 | 22.02 | 6.69 |
| D-II | 0.2908 | C75 | 0.0847 | 0.5495 | 69.23 | 26.2 | 25.89 | 1.18 |
| D-III | 0.3140 | C80 | 0.0847 | 0.5495 | 74.59 | 28.7 | 29.94 | 4.32 |

It can be seen from Table 5 that the maximum relative error between the calculated value of the ultimate capacity of the shaft lining model and the test value is only 6.69%, indicating that the fitted empirical formula for the ultimate capacity of the shaft lining can better estimate this type of shaft lining structure. This provides a certain reference value for this kind of shaft lining structure design.

## 4. Range Analysis of the Ultimate Capacity of the Shaft Lining Structure

In order to study the influence of various factors on the ultimate capacity of the shaft lining, a range analysis of the ultimate bearing capacity was carried out. The analysis results are shown in Table 6. The thickness–diameter ratio, shaft concrete design strength, PVAF content, and PPSF content are reported as factors A, B, C, and D, respectively.

**Table 6.** Factor optimization analysis.

| Level of Factors | | A | B | C | D | Optimal Combination | Order of Factors |
|---|---|---|---|---|---|---|---|
| | $k_1$ | 20.70 | 22.07 | 23.13 | 23.20 | | |
| Ultimate | $k_2$ | 23.37 | 23.17 | 23.30 | 23.33 | $A_3B_3C_2D_2$ | ABDC |
| capacity | $k_3$ | 25.53 | 24.37 | 23.17 | 23.07 | | |
| | $R$ | 4.83 | 2.30 | 0.17 | 0.26 | | |

From the comparison of the R value and *k* value in Table 6, it can be seen that the order of influence on the ultimate capacity of the shaft lining structure is the thickness–diameter ratio, the design strength of the concrete, the content of PPSF and the content of PVAF. The greater the thickness–diameter ratio and the design strength of the shaft lining concrete, the greater the *k* value, that is, the greater the ultimate capacity of the shaft lining structure. In addition, the *k* value of the hybrid-fiber reaches the maximum under the second-level combination.

In order to more intuitively observe the influence of each factor level on the ultimate capacity of the shaft lining structure model, the relationship curve between each factor and the ultimate capacity of the shaft lining structure is drawn, as shown in Figure 14.

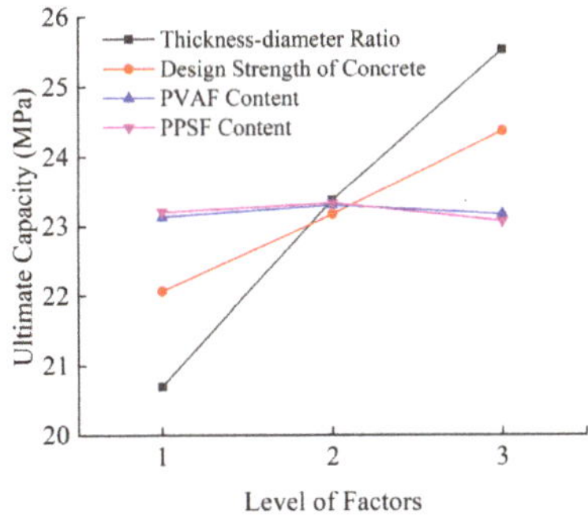


**Figure 14.** Relation curve of each factor and ultimate capacity of the shaft lining structure.

## 5. Conclusions

Temperature cracking of concrete is easy to occur in frozen shaft linings in extra-thick alluvial layers in coal mines. This research proposed to use the novel shaft lining structure of coal mines consisting of HFRC to solve this problem. Blending PPSF and PVAF into concrete, the synergistic effect and superimposition effect resulting from multi-scale and multi-element mixing of fibers greatly enhance the crack resistance and toughness of the shaft lining concrete, and delay the formation of cracks in the hardening stage of concrete. It can effectively solve the temperature cracking problem in frozen shaft linings in extra-thick alluvial layers in coal mines. With the use of the FEM, a numerical simulation test of the HFRC shaft lining structure was carried out. The following conclusions can be drawn:

1. From the hoop stress surface load curve and hoop strain surface load curve obtained from the numerical simulation, it can be seen that the maximum hoop stress gradually increases with the surface load, showing a trend of gradual transition from the inside surface to the outside surface. When the shaft lining is broken, the hoop stress at the outside surface is greater than that at the inside surface. The hoop strain at the inside and outside surfaces of the shaft lining is always a compressive strain, and the hoop strain at the inside surface is always greater than that at the outside surface.

2. The results of numerical simulation show that the compressive strength of concrete is improved to a certain extent because the concrete in the inside and outside surfaces of the shaft lining model is in a bidirectional compression state and three-dimensional compression state, respectively. When the shaft lining is damaged, the maximum hoop stress values of concrete on the inside and outside surfaces of each shaft lining reach 87.8 MPa, 97.5 MPa, 105 MPa, 92.5 MPa, 103 MPa, 111 MPa, 97.3 MPa, and 103 MPa, respectively. The maximum hoop stress exceeds the uniaxial compressive strength of this kind of concrete.

3. According to the results of numerical simulation, an empirical formula for the ultimate capacity of the HFRC shaft lining was obtained by fitting, and the rationality of the empirical formula was verified by the shaft lining model test. The maximum relative error between the calculated value of the shaft lining ultimate capacity and the test value is only 6.69%, and the relative error is small, which provides a reference for designing this type of shaft lining structure.

4. The order of influence on the ultimate capacity of this kind of HFRC shaft lining structure is the thickness–diameter ratio, design strength of concrete, PPSF content and PVAF content. The ultimate capacity of the shaft lining structure markedly increases with the increase in the thickness–diameter ratio and the design strength of the concrete. However, the increase in the amount of hybrid-fiber cannot significantly enhance the ultimate capacity of the shaft lining structure, and the effect of adding more fibers is slightly worse. Therefore, the optimal blending amount of hybrid-fiber is 1.092 kg/m$^3$ of PVAF and 5 kg/m$^3$ of PPSF.

**Author Contributions:** Design and manuscript writing, X.W.; methodology and validation, H.C.; data analysis and collection, T.W.; guidance and revision of manuscript, Z.Y.; drawing and tabulating, X.H. All authors have read and agreed to the published version of the manuscript.

**Funding:** This research was supported by the National Natural Science Foundation of China (No. 51674006) and the Anhui University discipline professional talented person (No. gxbjZD09), Anhui Provincial Natural Science Foundation Youth Project (1908085QE185), Anhui Provincial College of Natural Science Research Key Project (KJ2018A0098), Project funded by China Postdoctoral Science Foundation (2018M642502), and the Science Research Foundation for Young Teachers in Anhui University of Science and Technology (QN2017211).

**Conflicts of Interest:** The authors declare no conflict of interest.

## References

1. Zhang, T.; Yang, W.H.; Chen, G.H.; Huang, J.H.; Han, T.; Zhang, C. Monitoring and analysis of hydration heat temperature field for high performance mass concrete freezing shaft lining. *J. Min. Saf. Eng.* **2016**, *33*, 290–296.

2. Yao, Z.S.; Cheng, H.; Zhang, G.Y.; Wang, X.Q. Research on site measurement of outer shaft wall stressing in freezing sinking shaft in special thick alluvium. *Coal Sci. Technol.* **2004**, *32*, 49–52.

3. Song, L.; Yang, W.H.; Li, H.P. Monitoring of freezing shaft sinking in ultra-deep alluvium of guotun coal mine. *J. Min. Saf. Eng.* **2010**, *27*, 19–23.

4. Xue, W.P.; Zhu, Q.; Yao, Z.S.; Song, H.Q.; Yang, L. Influence of high permeability pore water pressure on physical and mechanical properties of shaft lining concrete. *J. Mater. Eng.* **2019**, *37*, 262–265, 270.

5. Xue, W.P.; Yao, Z.S.; Kong, G.; Xie, D.Z.; Wu, H. Experimental study on mechanical properties of shaft lining concrete under stress-seepage coupling. *Bull. Chin. Ceram. Soc.* **2018**, *37*, 985–989.

6. Yao, Z.S.; Gao, Y.; Song, H.Q. Experimental study on crack control and anti-permeability of high performance concrete in freezing shaft lining. *Bull. Chin. Ceram. Soc.* **2014**, *33*, 918–922.

7. Barr, B.; Newman, P.D. Toughness of polypropylene fibre-reinforced concrete. *Composites* **1985**, *16*, 48–53. [CrossRef]

8. Noushini, A.; Samali, B.; Vessalas, K. Effect of poly vinyl alcohol (PVA) fibre on dynamic and material properties of fibre reinforced concrete. *Constr. Build. Mater.* **2013**, *49*, 374–383. [CrossRef]

9. Zhang, R.; Jin, L.; Tian, Y.; Dou, G.; Du, X. Static and dynamic mechanical properties of eco-friendly polyvinyl alcohol fiber-reinforced ultra-high-strength concrete. *Struct. Concr.* **2019**, *20*, 1051–1063. [CrossRef]

10. Liang, N.H.; Liu, X.R.; Sun, J. Experimental study of crack resistance of multi-scale polypropylene fiber reinforced concrete. *J. China Coal Soc.* **2012**, *37*, 1304–1309.

11. Banthia, N.; Gupta, R. Hybrid fiber reinforced concrete (HyFRC): Fiber synergy in high strength matrices. *Mater. Struct.* **2004**, *37*, 707–716. [CrossRef]

12. Kang, S.T.; Jeong, C.; Kyung-Taek, K.; Seok, L.K.; Yeon, L.B. Hybrid effects of steel fiber and microfiber on the tensile behavior of ultra-high performance concrete. *Compos. Struct.* **2016**, *145*, 37–42. [CrossRef]

13. Sanchayan, S.; Foster, S.J. High temperature behaviour of hybrid steel–PVA fibre reinforced reactive powder concrete. *Mater. Struct.* **2016**, *49*, 769–782. [CrossRef]

14. Yao, Z.S.; Li, X.; Fu, C.; Xue, W.P. Mechanical Properties of Polypropylene Macrofiber-Reinforced Concrete. *Adv. Mater. Sci. Eng.* **2019**, *2019*, 7590214. [CrossRef]

15. Ramesh, B.; Gokulnath, V.; Krishnan, S.V. Influence of M-Sand in self compacting concrete with addition of glass powder in M-25 grade. *Mater. Today Process.* **2020**, *22*, 535–540.

16. Yao, Z.S. An experimental study on steel fiber reinforced high strength concrete shaft lining in deep alluvium. *Chin. J. Rock Mech. Eng.* **2005**, *24*, 1253–1258.

17. Qin, B.D.; Liu, X.L.; Liu, S.F. Experimental study on ultimate strength and failure features of shaft wall with hybrid fiber reinforced high strength concrete. *Rail. Eng.* **2017**, 156–159.

18. Yao, Z.S.; Li, X.; Wu, T.L.; Yang, L.; Liu, X.H. Hybrid-Fiber-Reinforced Concrete Used in Frozen Shaft Lining Structure in Coal Mines. *Materials* **2019**, *12*, 3988. [CrossRef]

19. Qureshi, L.A.; Muhammad, U. Effects of Incorporating Steel and Glass Fibers on Shear Behavior of Concrete Column-Beam Joints. *KSCE J. Civ. Eng.* **2018**, *22*, 2970–2981. [CrossRef]

20. Carozzi, F.G.; Milani, G.; Poggi, C. Mechanical properties and numerical modeling of Fabric Reinforced Cementitious Matrix (FRCM) systems for strengthening of masonry structures. *Compos. Struct.* **2014**, *107*, 711–725. [CrossRef]

21. Radtke, F.K.F.; Simone, A.; Sluys, L.J. A computational model for failure analysis of fibre reinforced concrete with discrete treatment of fibres. *Eng. Fract. Mech.* **2010**, *77*, 597–620. [CrossRef]

22. Jiao, L.; Tao, F. Bonding property and seismic behavior of reinforced concrete and finite element analysis. *Arab. J. Geosci.* **2020**, *13*, 45. [CrossRef]

23. Palanivelu, S. Flexural Behaviour of a Cold-Formed Steel-Concrete Composite Beam with Channel Type Shear Connector—An Experimental and Analytical Study. *Civ. Environ. Eng. Rep.* **2019**, *29*, 228–240. [CrossRef]

24. Xue, J. Crack Analysis of Reinforced Concrete Structure based on ANSYS. *Int. J. Civ. Eng. Mach. Manu.* **2019**, *4*, 27–33.

25. Yao, Z.S.; Gui, J.G.; Cheng, H.; Rong, C.X. Numerical simulation of composite shift lining of inner steel cylinder and high strength reinforced concrete. *J. Guangxi Univ. Nat. Sci. Ed.* **2010**, *35*, 35–38.

26. Wu, J. Study on Mechanical Characteristics of inner Shaft Lining at Frozen Shaft in West Area. Master's Thesis, Anhui University Science Technology, Huainan, China, 2013.

27. Lu, H.L.; Cui, G.X. Mechanical mechanism of shaft lining structure in thick alluvium. *J. China Univ. Min. Technol.* **1999**, *28*, 539–543.

28. Cai, H.B.; Li, X.F.; Cheng, H.; Yao, Z.S.; Wang, F.; Qian, W.W. Research on the structure of shaft lining with asphalt block and steel-reinforced concrete at deep shaft ingate and its application. *J. Min. Saf. Eng.* **2018**, *35*, 27–33.

29. Chinese Standard GB50010-2010. China Ministry of Housing and Urban-Rural Development. In *Code for Design of Concrete Structures*; China Architecture and Building Press: Beijing, China, 2013. (In Chinese)

*Editorial*

# Numerical Study of Concrete

**Vipulkumar Ishvarbhai Patel**

School of Engineering and Mathmatical Sciences, La Trobe University, Bendigo, Victoria 3552, Australia; V.Patel@latrobe.edu.au

**Citation:** Patel, V.I. Numerical Study of Concrete. *Crystals* **2021**, *11*, 74. https://doi.org/10.3390/cryst11010074

Received: 15 January 2021
Accepted: 15 January 2021
Published: 18 January 2021

**Publisher's Note:** MDPI stays neutral with regard to jurisdictional claims in published maps and institutional affiliations.

This Special Issue, "Numerical Study of Concrete", consists of 22 research articles.

Wang et al. [1] developed a finite element model for the numerical simulation of a hybrid-fiber-reinforced concrete (HFRC) shaft lining structure. The numerical results indicate that the maximum hoop stress position of the HFRC shaft lining presents a transition trend from the inside surface to the outside surface; the hoop strain of shaft lining concrete is always compressive, and the inside surface is greater than the outside surface. Kolesnikov [2] presents the load–displacement and stress–strain responses of concrete under uniaxial compression as well as three-point bending. The non-destructive test (NDT) method was proposed by Lim et al. [3] for the measurement of concrete's compressive strength. The water absorption of concrete with different binders was tested by Ding et al. [4]. The pore structure of concrete was investigated by mercury intrusion porosimetry. It was found that the water absorption of concrete with mineral admixtures is lower. This is due to the existence of a reasonable pore structure. Zhao et al. [5] used the finite element method for modeling the fundamental behavior of T-beams with carbon fiber-reinforced plastic under impact loads. The results show that the overall stiffness of the T-beams was significantly improved due to the carbon fiber-reinforced plastic strips.

Zainal et al. [6] conducted an experimental study for predicting the behavior of hybrid fiber-reinforced concrete materials with a high-range water-reducing admixture. It was concluded that the Ferro with a Ferro mix combination improved the performance of concrete in the elastic stage, while the Ferro with the ultra-net combination had the highest compressive strain surplus in the plastic stage. Ahmad et al. [7] utilized artificial neural networks for determining the properties of fiber-reinforced polymers-confined concrete. Lelovic et al. [8] presented a new method for the experimental determination of cohesion at pre-set angles of shear deformation. Alrshoudi et al. [9] developed the concept of a new pre-packed aggregate fiber-reinforced concrete which is reinforced with polypropylene (PP) waste carpet fibres, investigating its mechanical properties and impact resistance under drop weight impact loads. Mohammadyan-Yasouj et al. [10] investigated the thermal performance of alginate concrete reinforced with basalt fiber. The effects of the admixtures, erosion age, concentration of sulfate solution, and sulfate erosion on the mechanical properties of mortar were investigated by Liu et al. [11]. Benbow et al. [12] presented the development of a coupled modeling simulator for assessing the evolution of a geological repository in the near field for radioactive waste disposal where concrete is used as backfill.

Muhtar et al. [13] predicted the stiffness of bamboo-reinforced concrete beams from an experimental results database using artificial neural networks. Karam et al. [14] carried out an analytical investigation on the concrete damage progress of the Perfobond shear connector under the influence of various lateral pressures. Song et al. [15] simulated the adsorption characteristics of five types of common alkanol-amine inhibitors on C-S-H gel in the alkaline liquid environment using the molecular dynamics and grand canonical Monte Carlo methods. Javed et al. [16] developed a model for predicting the ultimate axial strength of concrete-filled steel tubular columns under axial compression. Javed et al. [17] utilized novel Gene Expression Programming and regression techniques for determining the compressive strength of sugarcane bagasse ash concrete. Phutthimethakul et al. [18]

used flue gas desulfurization gypsum, construction and demolition waste, and oil palm waste trunks to produce concrete bricks. Chen et al. [19] experimentally and numerically studied the blast-resistant performance of steel fiber-reinforced concrete and polyvinyl alcohol fiber-reinforced concrete panels with a contact detonation test.

The study presented by Alyousef et al. [20] aims to investigate the resistance of concrete composites reinforced with waste metalized plastic fibres to sulphate and acid attacks. Yehia et al. [21] studied the effect of aggregate type on concrete's compressive strength. The durability of polyvinyl alcohol fiber-reinforced cementitious composite containing nano-SiO$_2$ was evaluated by Liu et al. [22] using the adaptive neuro-fuzzy inference system.

**Funding:** This research received no external funding.

**Informed Consent Statement:** Informed consent was obtained from all subjects involved in the study.

**Acknowledgments:** The contribution of all the authors is gratefully acknowledged. The editor would like to express his thanks to the *Crystals* Editorial Office, and, on top of that, to Debbie Yang (a Technical Coordinator of the issue) for the excellent communication, support, and friendly and professional attitude.

**Conflicts of Interest:** The author declares no conflict of interest.

# References


1. Wang, X.; Cheng, H.; Wu, T.; Yao, Z.; Huang, X. Numerical analysis of a novel shaft lining structure in coal mines consisting of hybrid-fiber-reinforced concrete. *Crystals* **2020**, *10*, 928. [CrossRef]
2. Kolesnikov, G. Analysis of concrete failure on the descending branch of the load-displacement curve. *Crystals* **2020**, *10*, 921. [CrossRef]
3. Lim, Z.H.; Lee, F.W.; Mo, K.H.; Lim, J.H.; Yew, M.K.; Kwong, K.Z. Compressive strength forecasting of air-entrained rubberized concrete during the hardening process utilizing elastic wave method. *Crystals* **2020**, *10*, 912. [CrossRef]
4. Ding, X.; Liang, X.; Zhang, Y.; Fang, Y.; Zhou, J.; Kang, T. Capillary water absorption and micro pore connectivity of concrete with fractal analysis. *Crystals* **2020**, *10*, 892. [CrossRef]
5. Zhao, H.; Kong, X.; Fu, Y.; Gu, Y.; Wang, X. Numerical investigation on dynamic response of RC T-Beams strengthened with CFRP under impact loading. *Crystals* **2020**, *10*, 890. [CrossRef]
6. Zainal, S.M.I.S.; Hejazi, F.; Aziz, F.N.A.A.; Jaafar, M.S. Constitutive modeling of new synthetic hybrid fibers reinforced concrete from experimental testing in uniaxial compression and tension. *Crystals* **2020**, *10*, 885. [CrossRef]
7. Ahmad, A.; Plevris, V.; Khan, Q.Z. Prediction of properties of FRP-confined concrete cylinders based on artificial neural networks. *Crystals* **2020**, *10*, 811. [CrossRef]
8. Lelovic, S.; Vasovic, D. Determination of Mohr-coulomb parameters for modelling of concrete. *Crystals* **2020**, *10*, 808. [CrossRef]
9. Alrshoudi, F.; Mohammadhosseini, H.; Alyousef, R.; Tahir, M.; Alabduljabbar, H.; Mohamed, A.M. The impact resistance and deformation performance of novel pre-packed aggregate concrete reinforced with waste polypropylene fibres. *Crystals* **2020**, *10*, 788. [CrossRef]
10. Mohammadyan-Yasouj, S.E.; Ahangar, H.A.; Oskoei, N.A.; Shokravi, H.; Koloor, S.S.R.; Petrů, M. Thermal performance of alginate concrete reinforced with basalt fiber. *Crystals* **2020**, *10*, 779. [CrossRef]
11. Liu, P.; Chen, Y.; Yu, Z. Effects of erosion form and admixture on cement mortar performances exposed to sulfate environment. *Crystals* **2020**, *10*, 774. [CrossRef]
12. Benbow, S.J.; Kawama, D.; Takase, H.; Shimizu, H.; Oda, C.; Hirano, F.; Takayama, Y.; Mihara, M.; Honda, A. A coupled modeling simulator for near-field processes in cement engineered barrier systems for radioactive waste disposal. *Crystals* **2020**, *10*, 767. [CrossRef]
13. Muhtar Gunasti, A.; Suhardi Nursaid Irawati Dewi, I.C.; Dasuki, M.; Ariyani, S.; Fitriana Mahmudi, I.; Abadi, T.; Rahman, M.; Hidayatullah, S.; Nilogiri, A.; Galuh, S.D.; et al. The prediction of stiffness of bamboo-reinforced concrete beams using experiment data and Artificial Neural Networks (ANNs). *Crystals* **2020**, *10*, 157.
14. Karam, M.S.; Yamamoto, Y.; Nakamura, H.; Miura, T. Numerical evaluation of the Perfobond (PBL) shear connector subjected to lateral pressure using coupled Rigid Body Spring Model (RBSM) and Nonlinear Solid Finite Element Method (FEM). *Crystals* **2020**, *10*, 743. [CrossRef]
15. Song, Z.; Cai, H.; Liu, Q.; Liu, X.; Pu, Q.; Zang, Y.; Xu, N. Numerical simulation of adsorption of organic inhibitors on C-S-H Gel. *Crystals* **2020**, *10*, 742. [CrossRef]
16. Javed, M.F.; Farooq, F.; Memon, S.A.; Akbar, A.; Khan, M.A.; Aslam, F.; Alyousef, R.; Alabduljabbar, H.; Rehman, S.K. New prediction model for the ultimate axial capacity of concrete-filled steel tubes: An evolutionary approach. *Crystals* **2020**, *10*, 741. [CrossRef]

17. Javed, M.F.; Amin, M.N.; Shah, M.I.; Khan, K.; Iftikhar, B.; Farooq, F.; Aslam, F.; Alyousef, R.; Alabduljabbar, H. Applications of Gene Expression Programming and Regression Techniques for estimating compressive strength of Bagasse ash based concrete. *Crystals* **2020**, *10*, 737. [CrossRef]
18. Phutthimethakul, L.; Kumpueng, P.; Supakata, N. Use of flue gas desulfurization gypsum, construction and demolition waste, and oil palm waste trunks to produce concrete bricks. *Crystals* **2020**, *10*, 709. [CrossRef]
19. Chen, L.; Sun, W.; Chen, B.; Xu, S.; Liang, J.; Ding, C.; Feng, J. A comparative study on blast-resistant performance of steel and PVA fiber-reinforced concrete: Experimental and numerical analyses. *Crystals* **2020**, *10*, 707. [CrossRef]
20. Alyousef, R.; Mohammadhosseini, H.; Alrshoudi, F.; Tahir, M.; Alabduljabbar, H.; Mohamed, A.M. Enhanced performance of concrete composites comprising waste metalised polypropylene fibres exposed to aggressive environments. *Crystals* **2020**, *10*, 696. [CrossRef]
21. Yehia, S.; Abdelfatah, A.; Mansour, D. Effect of aggregate type and specimen configuration on concrete compressive strength. *Crystals* **2020**, *10*, 625. [CrossRef]
22. Liu, T.Y.; Zhang, P.; Li, Q.F.; Hu, S.W.; Ling, Y.F. Durability assessment of PVA fiber-reinforced cementitious composite containing nano-SiO$_2$ using adaptive neuro-fuzzy inference system. *Crystals* **2020**, *10*, 347. [CrossRef]

MDPI

St. Alban-Anlage 66

4052 Basel

Switzerland

Tel. +41 61 683 77 34

Fax +41 61 302 89 18

www.mdpi.com

*Crystals* Editorial Office

E-mail: crystals@mdpi.com

www.mdpi.com/journal/crystals